A. S. Wightman
Palmer Physical Lab
8 January 1966

Lectures on Differential Geometry

PRENTICE-HALL INTERNATIONAL, INC. London
PRENTICE-HALL OF AUSTRALIA, PTY., LTD. Sydney
PRENTICE-HALL OF CANADA, LTD. Toronto
PRENTICE-HALL OF INDIA (PRIVATE) LTD. New Delhi
PRENTICE-HALL OF JAPAN, INC. Tokyo
PRENTICE-HALL DE MEXICO, S.A. Mexico City

SHLOMO STERNBERG

Associate Professor of Mathematics, Harvard University

Lectures on Differential Geometry

PRENTICE-HALL, INC. ENGLEWOOD CLIFFS, N.J.

Prentice-Hall Mathematics Series

Library of Congress Catalog Card Number 64-7993
Printed in the United States of America. C52715

To my parents

Preface

This book is based on lectures delivered at Harvard during the academic year 1960–61. The book assumes a knowledge of the elements of modern algebra (groups, vector spaces, etc.) and point set topology, as well as some amount of elementary analysis. Beyond this it is self contained. It is by no means complete, however. I do not presume to give either a "foundations of . . . " giving all the basic information in the field, or a "survey" touching on every topic. Rather, I treat various selected topics in differential geometry, which interest me, from a viewpoint which I hope the reader will find consistent. The reader who wishes to learn differential geometry should also read a number of other books, both to get a different slant on the material covered here and for a treatment of subject matter that we don't discuss. I will give some suggestions for further reading in the chapter-by-chapter analysis that follows. For general reference I recommend four books — Lang's *Introduction to Differentiable Manifolds*, for a clean treatment of the foundations of differentiable manifolds, and Nomizu's *Lie Groups and Differential Geometry*, for the theory of connections. Most of the material in these two books is presented here in one form or another, but it would be instructive for the reader to see the development of these subjects in a manner independent of the rest of the subjects we deal with. For a treatment of deRham's and Hodge's theorems and the related facts in Riemann geometry see deRham's *Variétés Différentiable.* Finally, the reader should get a hold of a book on classical differential geometry. I have no preference here.

A few words are in order on the nature of my exposition. Some subjects are treated curtly and others at length. My general policy is to be concise about material that is "standard" (such as the theory of connections in Section 1 of Chapter VII, for example). Also, this book was written over a period of two years, so that my loquaciousness varied with the weather, moods, etc. There are exercises spread throughout the book. They are to be regarded as part of the text. Very few of the exercises are hard. They usually represent additional pieces of information or motivation that I found more convenient or instructive to present in exercise form. Frequently, the proof of a theorem will depend on a previous exercise.

I hope that there are no serious logical errors in the book. I do not have any similar aspirations on such matters as signs, factors of 2π, etc. I have a difficult time keeping such matters straight so although I've tried hard there are undoubtedly errors of this order cropping up in many places. My only advice is to check carefully. The more complicated a formula or equation is, the more susceptible it is to error. . . .

Now to a more detailed analysis. In Chapter I we round up algebraic machinery we will need throughout the book. All of the material is standard — Reference — Bourbaki's *Algebra Multilinaire.* Our notation is fairly close to Bourbaki's.

In Chapter II we present the basic facts about differentiable manifolds and maps. For another treatment of most of this material (except Sard's theorem and the approximation theorems) see Lang's *Introduction to Differentiable Manifolds.* My purpose in introducing Sard's theorem is two-fold. First of all, it has great intrinsic interest and usefulness (and is slightly difficult to dig out of the literature). Secondly, it is a highly non-trivial theorem which is elementary in the sense that it uses only the notion of differentiable map. I, for one, find it difficult to learn a whole batch of definitions one after the other without some reasonably hard theorems to tie them down. For readers with the same inclination I introduced Sard's theorem and the various approximation theorems even before defining the tangent space.

My treatment of "quantities," i.e., regarding the bundle of frames (or coframes) as the primary object, is probably not the neatest way of going about things (cf. Lang again). My purpose is to motivate the notion of principal bundles; also to point out that the tensors, etc. are "first order objects" and there are corresponding "higher order" objects — although we don't study them.

Chapter III contains the standard material on integration on manifolds. For a different treatment and for deRham's theorem and the theory of harmonic forms see deRham's *Variétés Différentiables.* For a further discussion of the material of Sections 4 and 5 with many beautiful applications see E. Cartan's *Leçons sur les Invariants Intégraux.* This book, like all of Cartan's work, is rich in geometrical ideas. It is also quite readable — once the reader gets accustomed to the language of infinitesimals. As to Section 6 — there are many good references on the transformation theory of mechanics. My favorite is to the last two chapters of Levi-Civita and Amaldi, *Lezioni di Mechanica Razionale.* For a modern reference (in English) cf. Goldstein's *Classical Mechanics.*

In Chapter IV we treat the calculus of variation as the study of a geometrical structure. The course that we steer is sort of midway between that suggested by Cartan in *Invariants Intégraux* and Caratheodory in his *Variationsrechung.* Carathéodory's influence can be seen most strongly in our treatment of the sufficient conditions. Our emphasis on staying with geometrical objects intrinsically associated to the tangent or cotangent bundle tends to obscure some of the ideas proper to the calculus of variations. In particular, we do not give fair treatment to the second variation and we do not present the Morse index theorem or Morse theory in general. For a good treatment of Morse theory see Bott's notes at the University of Bonn (1960) or Milnor's notes at Princeton (1961).

Chapter V gives the standard treatment of the elements of Lie groups. The classical reference is Chevalley's *Lie Groups.*

Chapter VI is the most disconnected chapter in the book. I try to give applications of the results of the previous chapters to the differential geometry of Euclidean space, in particular to show how the structure equations of the Euclidean group play a crucial role. I also want to motivate the notion of connection introduced in Chapter VII. The reader will get a fairly lopsided view of Euclidean differential geometry from this chapter. He should supplement it by reading a text on classical differential geometry. For a book with many references whose notation is close to ours perhaps Blashke's *Einfuhrung in die Differentialgeometrie* is a good reference.

In Chapter VII, I present the theory of G-structures. Section 1 gives the theory of connections. For this, Nomizu's book mentioned above is a good source. The rest of the chapter is a treatment of what Cartan calls the general equivalence problem. A large part of Cartan's collected works (still not fully appreciated) is devoted to this subject. I try to present some of his main ideas and tools. The treatment given here is based almost entirely on part of a joint paper (as yet unpublished) on the infinite Lie groups with I. M. Singer. Section 4 relies somewhat on the exposition given in the Ph.D. thesis of Guillemin (Harvard, 1962). For a different approach to G-structures see Bernard [3].

Not all the chapters are dependent on all the preceding material. Here is a rough diagram of the dependency relations:

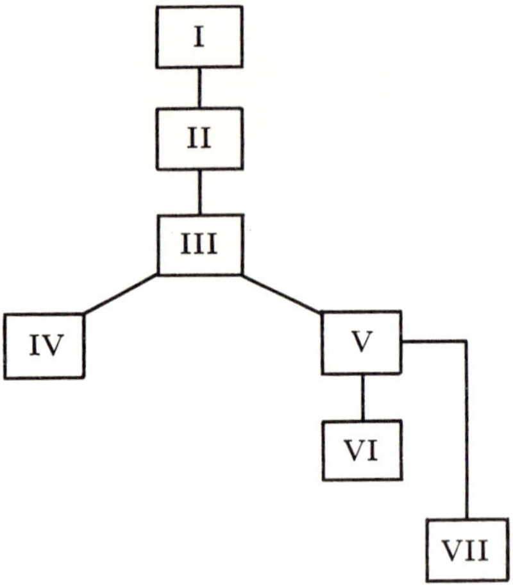

The writing of this book was a long and difficult task, and I received help from many people. My primary obligation is to my teachers: The influence of the late Professor Wintner can be seen in the strong tendency toward classical mechanics. My first introduction to the works of Élie Cartan, and indeed, to modern differential geometry, came at the hands of Professor Chern.

I have had many profitable conversations regarding various subjects treated here with Professors Bott, Gleason, Hermann, Lang, Mackey and Palais. I am indebted to Professors Nomizu and Smale for pointing out a number of errors that occurred in the first draft. As already mentioned, the heart of Chapter VII is based on joint work with Professor Singer. My special thanks are due to Dr. Han Sah for help throughout the preparation of this book — ranging from simplifications of proofs and exposition to the dirty work of correcting typographical errors, etc. I am also indebted to Mrs. Mary McQuillin, who not only did an excellent job of typing the whole manuscript, but also took complete charge of the hard editorial work, thus relieving me of a heavy burden. I also wish to express my thanks to the Sloan Foundation for financial support during the preparation of the manuscript.

Above all I must express my love and gratitude to my wife, Aviva. Without her patience, love, and support this book would not have been finished.

ויהי נועם וכו׳
תושלב״ע

Cambridge (Mass.); København; Jerusalem.
November, 1962

SHLOMO STERNBERG

Contents

Lectures on Differential Geometry

CHAPTER I

Algebraic Preliminaries

The main idea of differential geometry is to apply the tools of analysis to the solution of geometrical problems. An important device in analysis is to study objects by studying their "infinitesimal parts." Thus a curve is studied by examining its tangent vectors, a function by studying its differentials, and so on. The crucial advantage of the passing to the "infinitesimal" is that it linearizes everything. Thus every curve is "infinitesimally" a straight line (meaning we look at the tangent), every differentiable map is "infinitesimally" linear (we look at its Jacobian matrix), etc. Since we will be interested in studying such objects as curves, surfaces, and in general, higher dimensional "differentiable objects," we first study their "infinitesimal" analogues. This leads to a study of various algebraic problems which all fall under the heading of what is known as *multilinear algebra.*

In this chapter we will collect various algebraic facts (mainly about tensor algebra) that will be needed, and fix our terminology. Since much of this material is standard, we will be more concise than usual, referring to standard texts on linear algebra for a more detailed exposition.

We assume that the reader is familiar with the concept of vector space, dual space, linear transformation, group, and so on. (In most of this chapter the ground field of our vector spaces will not matter. In some cases characteristic 2 would cause its usual difficulties. In any event, we will be interested in only the real numbers as ground field.)

1. Tensor products of vector spaces. Let V and W be vector spaces of dimension m and n, respectively. A map f of $V \times W$ into a vector space Z is called bilinear if it is linear in each variable separately, i.e., if

$$f(a_1v_1 + a_2v_2, b_1w_1 + b_2w_2) = a_1b_1f(v_1, w_1) + a_1b_2f(v_1, w_2) + a_2b_1f(v_2, w_1) + a_2b_2f(v_2, w_2)$$

for all vectors v_1, v_2 in V, w_1, w_2 in W and any numbers a_1, a_2, b_1, and b_2. Notice that if $v_1, \ldots, v_m$ is a basis of V and $w_1, \ldots, w_n$ is a basis of W, then any bilinear map is uniquely determined by the values $f(v_i, w_j)$.

EXERCISE 1.1. Let f be a bilinear map from $V \times W$ to Z and let g be a linear map from the vector space Z to the vector space Y. Show that $g \circ f$ is a bilinear map of $V \times W$ to Y.

We denote by $\mathrm{Hom}(V, W)$ the space of linear maps from V to W. Recall that $\mathrm{Hom}(V, W)$ is a vector space of dimension mn. Denote by V^* the space of all linear functions on V, i.e., the dual space of V. Define the map, φ, of $V \times W \to \mathrm{Hom}(V^*, W)$ as follows: To any pair (v, w) we must assign a linear map $\varphi(v, w)$ from V^* to W. To give $\varphi(v, w)$ it is enough to say what $\varphi(v, w)$ applied to v^* is, for all $v^* \in V^*$. Put

$$\varphi(v, w)v^* = \langle v, v^* \rangle w$$

(where $\langle v, v^* \rangle$ denotes the value of the linear function v^* on the vector v. Thus $\langle v, v^* \rangle$ is a number).

EXERCISE 1.2. Show that φ is a bilinear map of $V \times W$ into $\mathrm{Hom}(V^*, W)$.

EXERCISE 1.3. Let $v_1, \ldots, v_m$ be a basis of V and $w_1, \ldots, w_n$ a basis of W. Show that

$$\{\varphi(v_i, w_j)\} \qquad i = 1, \ldots, m, \quad j = 1, \ldots, n$$

is a basis of $\mathrm{Hom}(V^*, W)$.

Let f be any bilinear map from $V \times W \to Z$. Define a *linear* map g_f from $\mathrm{Hom}(V^*, W) \to Z$ by setting

$$g_f(\varphi(v_i, w_j)) = f(v_i, w_j),$$

where $v_1, \ldots, v_m$ is a basis of V and $w_1, \ldots, w_n$ is a basis of W. This defines the map g_f on the basis $\{\varphi(v_i, w_j)\}$ of $\mathrm{Hom}(V^*, W)$. Extend g_f to all of $\mathrm{Hom}(V^*, W)$ so as to be linear.

EXERCISE 1.4. Show that for any $v \in V$ and $w \in W$, $g_f(\varphi(v, w)) = f(v, w)$. Conclude that the definition given above for g_f is independent of the choice of basis $v_1, \ldots, v_m$ and $w_1, \ldots, w_n$.

Thus we have associated to every bilinear map, f, of $V \times W$ into any Z a linear map, g_f, of $\mathrm{Hom}(V^*, W) \to Z$ such that

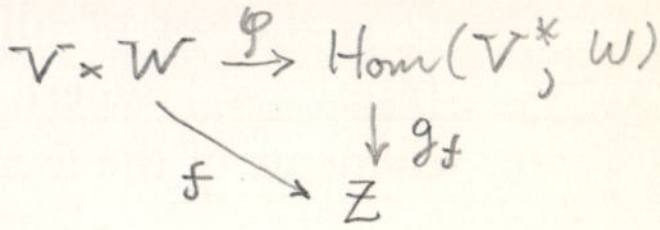

$$f = g_f \circ \varphi. \tag{1}$$

It is clear that f is defined once we know g_f.

EXERCISE 1.5. Show that g_f is uniquely determined by (1). That is, if $h \circ \varphi = h' \circ \varphi$ then $h = h'$.

More generally, let $V_1, V_2, \ldots, V_k$ be finite dimensional vector spaces. A map, f, of $V_1 \times V_2 \times \cdots \times V_k$ into a vector space W is called multilinear if it is linear in each of the variables separately. That is, if

$$f(u_1, \ldots, au_i + bv_i, u_{i+1}, \ldots, u_k) = af(u_1, \ldots, u_i, \ldots, u_k) + bf(u_1, \ldots, v_i, \ldots, u_k).$$

We wish to find a vector space which is universal with respect to multilinear maps of $V_1 \times \cdots \times V_k$. That is, we want to construct a vector space, U, and a multilinear map $\varphi: V_1 \times \cdots \times V_k \to U$ such that for any multilinear map $f: V_1 \times V_2 \times \cdots \times V_k \to W$ we can find a unique linear map of $g: U \to W$ with $f = g \circ \varphi$.

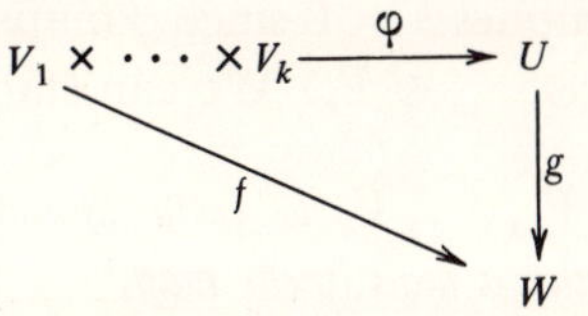

More succinctly, we wish to be able to factor any f through φ. First of all, it is clear that the vector space U and the map φ is determined up to isomorphism. Given two such spaces U_1 and U_2 with maps φ_1 and φ_2, we can factor φ_1 through φ_2

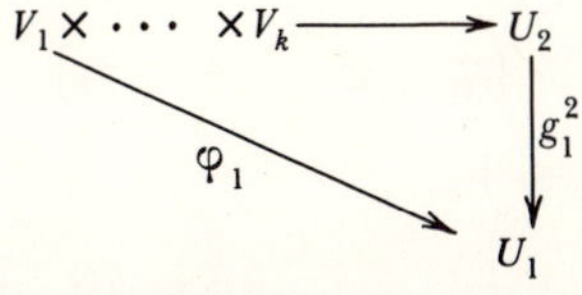

so $\varphi_1 = g_1^2 \circ \varphi_2$ where g_1^2 is uniquely determined. Similarly $\varphi_2 = g_2^1 \circ \varphi_1$ so $\varphi_1 = g_1^2 \circ g_2^1 \circ \varphi_1$. By the uniqueness requirement, $g_1^2 \circ g_2^1 = \text{id}$. Thus g_1^2 provides an isomorphism of U_2 onto U_1.

We now establish the existence of such a space U. Let $\bar{U}$ denote the free vector space whose generators are the elements of $V_1 \times \cdots \times V_k$. $\bar{U}$ is the set of all finite linear combinations of symbols of the form $(v_1, \ldots, v_k)$ where $v_j \in V_j$. Let $\bar{R}$ be the subspace of $\bar{U}$ generated by elements of the form

$$a(v_1, \ldots, v_k) - (v_1, \ldots, av_i, v_{i+1}, \ldots, v_k)$$

$$(v_1, \ldots, v_i + w_i, \ldots, v_k) - (v_1, \ldots, v_i, \ldots, v_k) - (v_1, \ldots, w_i, \ldots, v_k).$$

Then the space $U = \bar{U}/\bar{R}$ will be denoted by $V_1 \otimes \cdots \otimes V_k$ and called the *tensor product of the* V_i. We define the multilinear map, φ, of $V_1 \times \cdots \times V_k$ into $V_1 \otimes \cdots \otimes V_k$ by sending $(v_1, \ldots, v_k)$ viewed as an element of $\bar{U}$ into its coset mod $\bar{R}$. We shall write

$$\varphi(v_1, \ldots, v_k) = v_1 \otimes \cdots \otimes v_k.$$

Our claim is that the space $V_1 \otimes \cdots \otimes V_k$ together with the map φ is universal with respect to multilinear maps. In fact, let

$$f: V_1 \times \cdots \times V_k \to Z$$

be a multilinear map. Define the map $\bar{g}: \bar{U} \to Z$ by $\bar{g}(v_1, \ldots, v_k) = f(v_1, \ldots, v_k)$ and extend linearly. The assumption that f is multilinear implies that $\bar{g}$ vanishes on $\bar{R}$ and therefore defines a map g of $V_1 \otimes \cdots \otimes V_k$ into Z such that $f = g \circ \varphi$. That g is unique follows from the fact that $h \circ \varphi = 0$ implies $h = 0$ since φ maps $V_1 \times \cdots \times V_k$ onto a set of generators for $V_1 \otimes \cdots \otimes V_k$. We can state therefore,

THEOREM 1.1. *Let $V_1, \ldots, V_k$ be vector spaces. There exists a vector space $V_1 \otimes \cdots \otimes V_k$ and a multilinear map,*

$$\varphi: V_1 \times \cdots \times V_k \to V_1 \otimes \cdots \otimes V_k$$

which is universal with respect to multilinear maps. For any other such universal space U there is a canonical isomorphism of U onto $V_1 \otimes \cdots \otimes V_k$. If we write $\varphi(v_1, \ldots, v_k) = v_1 \otimes \cdots \otimes v_k$ and if $\{e_{i_r, r}\}$ is a basis of V_r $(1 \leqslant i_r \leqslant \dim V_r)$ then

$$\{e_{i_1, 1} \otimes \cdots \otimes e_{i_k, k}\}$$

is a basis for $V_1 \otimes \cdots \otimes V_k$.

The only fact we haven't yet established is the last statement of the theorem which we shall leave as an exercise to the reader.

EXERCISE 1.6. Show that there is an isomorphism of

$$(V_1 \otimes \cdots \otimes V_l) \otimes (V_{l+1} \otimes \cdots \otimes V_k)$$

onto $V_1 \otimes \cdots \otimes V_k$. We shall denote this isomorphism by As (for associative).

EXERCISE 1.7. Show that there is an isomorphism of $V_1 \otimes V_2$ onto $V_2 \otimes V_1$.

EXERCISE 1.8. Show that $K \otimes V = V$ if K is the ground field.

Let $T_i: V_i \to W_i (i = 1, \ldots, k)$ be linear maps where V_i and W_i are vector spaces. Then the mapping of $V_1 \times \cdots \times V_k$ into

$$W_1 \otimes \cdots \otimes W_k$$

sending $(v_1, \ldots, v_k) \to (Tv_1) \otimes \cdots \otimes (Tv_k)$ is clearly multilinear and therefore defines a map, $T_1 \otimes \cdots \otimes T_k$, of $V_1 \otimes \cdots \otimes V_k \to W_1 \otimes \cdots \otimes W_k$. If S_i map the vector spaces $U_i \to V_i$ then

$$(T_1 \otimes \cdots \otimes T_k) \circ (S_1 \otimes \cdots \otimes S_k) = (T_1 \circ S_1) \otimes \cdots \otimes T_k \circ S_k.$$

In particular, if ρ_i are representations[1] of the groups G_i on the vector spaces V_i, we can form the representation $\rho_1 \otimes \cdots \otimes \rho_k$ of

$$G_1 \times \cdots \times G_k$$

on $V_1 \otimes \cdots \otimes V_k$ by setting $\rho_1 \otimes \cdots \otimes \rho_k(g_1, \ldots, g_k) = \rho_1(g_1) \otimes \cdots \otimes \rho_k(g_k)$. If $G_1 = \cdots = G_k = G$ we can map $G \to G \times \cdots \times G$ by the diagonal map $g \to (g, \ldots, g)$ and compose this with $\rho_1 \otimes \cdots \otimes \rho_k$ to obtain a representation of G on $V_1 \otimes \cdots \otimes V_k$ which sends $g \to \rho_1(g) \otimes \cdots \otimes \rho_k(g)$. This representation is called the *tensor product of the representations* $\rho_1, \ldots, \rho_k$.

EXERCISE 1.9. If $S: V \to V$ and $T: W \to W$ where the matrix of S relative to the basis $\{e_1, \ldots, e_m\}$ of V is s_{ij} and the matrix of T relative to the basis $\{f_1, \ldots, f_n\}$ of W is t_{kl}, what is the matrix of $S \otimes T$ relative to the basis $\{e_i \otimes f_k\}$ of $V \otimes W$?

2. The tensor algebra of a vector space. Let V be a finite dimensional vector space and V^* the space of linear forms on V. If $v \in V$ and $w^* \in V^*$,

---

[1] Recall that a representation of a group G on a vector space V is a homomorphism, ρ, of G into the group of all linear transformations of V. Thus $\rho(g)$ is a linear transformation and $\rho(g_1 g_2) = \rho(g_1)\rho(g_2)$.

we will denote the value of w^* on v by $\langle v, w^*\rangle$. Of particular interest are the tensor products of copies of V and V^*. For example, the space $V \otimes V^*$ can be identified with the space of all linear transformations of V (cf. Exercise 1.4). In fact, we can define $(f \otimes g^*)(v) = \langle v, g^*\rangle f$ so that every element of $V \otimes V^*$ of the form $f \otimes g^*$ defines a linear transformation, and we extend by linearity to all of $V \otimes V^*$. On the other hand, let $\{e_1, \ldots, e_n\}$ be a basis of V and $\{e^{1^*}, \ldots, e^{n^*}\}$ the corresponding basis for V^* so that

$$\langle e_i, e^{j^*}\rangle = \delta^i_j = \begin{cases} 1 & \text{if } i = j \\ 0 & \text{if } i \neq j. \end{cases}$$

A linear transformation whose matrix is a^i_j with respect to the basis $(e_1, \ldots, e_n)$ is clearly equal to $\sum a^i_j e_j \otimes e^{i^*}$. Similarly it is easy to see that the space $V^* \otimes V^*$ can be viewed as the space of bilinear forms on V.

We wish to study those spaces of the form $V_1 \otimes \cdots \otimes V_k$ where each V_i is either V or V^*. If there are r copies of V and s copies of V^* then the space is called a *tensor space of type* (r, s); r is called the *contravariant degree* and s is called the *covariant degree.* Of course, the integers (r, s) do not determine the space. The order in which the spaces occur matters; we distinguish between $V \otimes V^*$ and $V^* \otimes V$. Given two tensor spaces, $U = V_1 \otimes \cdots \otimes V_k$ of type (r, s) and $U' = V_1' \otimes \cdots \otimes V_{k'}'$ of type (r', s') the associative law for tensor products defines a bilinear map of $U \times U' \to V_1 \otimes \cdots \otimes V_k \otimes V_1' \otimes \cdots \otimes V_{k'}'$, a tensor space of type $(r + r', s + s')$. This map defines a multiplicative structure on the (weak) direct sum of all the tensor products of V and V^*. We denote this space by $T(V)$, i.e.,

$$T(V) = K + V + V^* + V \otimes V + V \otimes V^* \\ + V^* \otimes V + V^* \otimes V^* + \cdots.$$

It (with its multiplicative structure) is called the *tensor algebra of the vector space.*

There is another useful operation, known as *contraction*, which maps a tensor space of type (r, s) into one of type $(r - 1, s - 1)$ (if $r, s \geqslant 1$). It is defined as follows: Let $U = V_1 \otimes \cdots \otimes V_k$ where $V_i = V$ and

$$V_j = V^* \qquad (1 \leqslant i \leqslant k, 1 \leqslant j \leqslant k).$$

Let U' denote the tensor product of all the terms of U in the same order *omitting* V_i and V_j. The map of $V_1 \times \cdots \times V_k$ into U' defined by

$$(v_1, \ldots, v_k) \to \langle v_i, v_j\rangle v_1 \otimes \cdots \otimes v_k$$

(where the tensor product on the right does not contain v_i and v_j) is easily seen to be bilinear. It therefore defines a map of $U \to U'$ called *contraction* with respect to i and j.

For any linear transformation, S, of V we will denote by S^*, the transformation of V^* defined by

$$\langle v, S^*w^* \rangle = \langle Sv, w^* \rangle \qquad \text{for all } v \in V.$$

Let $\{e_1, \ldots, e_n\}$ be a basis of V and $\{e^{1*}, \ldots, e^{n*}\}$ be the dual basis. If A is a non-singular linear transformation of V, it will take the basis $\{e_i\}$ into a basis $\{f_i\}$ where $f_i = Ae_i$. The dual basis f^{i*} is given by $f^{i*} = (A^{-1})^*e^{i*}$. Suppose that $\{e_i\}$ and $\{f_j\}$ are related by

$$e_i = \sum \alpha_i^j f_j, \qquad f_i = \sum \beta_i^j e_j,$$

so that

$$\sum \alpha_k^j \beta_i^k = \delta_i^j.$$

Then

$$f^{j*} = \sum \alpha_i^j e^{i*}, \qquad e^{j*} = \sum \beta_i^j f^{i*}.$$

Thus if $v \in V$ is given by $v = \sum v^i e_i = \sum w^i f_i$

$$w^i = \sum \alpha_j^i v^j$$

and if, for $v^* \in V^*$, $v^* = \sum x_i e^{i*} = \sum y_i f^{i*}$ then

$$y_i = \sum \beta_i^j x_j.$$

Corresponding to the basis $\{e_1, \ldots, e_n\}$ and its dual, Theorem 1.1 chooses for us a basis of all the tensor products of V and V^*. For example, the basis of $V \otimes V^* \otimes V \otimes V$ corresponding to $\{e_1, \ldots, e_n\}$ is

$$\{e_{i_1} \otimes e^{i_2^*} \otimes e_{i_3} \otimes e_{i_4}\} \qquad 1 \leqslant i_1, i_2, i_3, i_4 \leqslant n.$$

Any tensor $t \in V \otimes V^* \otimes V \otimes V$ can be written as

$$(2.1) \qquad t = \sum_{i_1, i_2, i_3, i_4} t^{i_1}{}_{i_2}{}^{i_3 i_4} e_{i_1} \otimes e^{i_2^*} \otimes e_{i_3} \otimes e_{i_4}.$$

If we express t in terms of the basis corresponding to $(f_1, \ldots, f_n)$

$$t = \sum s^{j_1}{}_{j_2}{}^{j_3 j_4} f_{j_1} \otimes f^{j_2^*} \otimes f_{j_3} \otimes f_{j_4},$$

then the change of coordinates is given by

(2.2) $$s^{j_1}{}_{j_2}{}^{j_3 j_4} = \sum \alpha^{j_1}_{i_1} \beta^{i_2}_{j_2} \alpha^{j_3}_{i_3} \alpha^{j_4}_{i_4} t^{i_1}{}_{i_2}{}^{i_3 i_4}.$$

The tensor product of t with the tensor

$$r = \sum r^{i_1}{}_{i_2 i_3} e_{i_1} \otimes e^*_{i_2} \otimes e^*_{i_1}$$

is given by

$$t \otimes r = \sum t^{i_1}{}_{i_2}{}^{i_3 i_4} r^{i_5}{}_{i_6 i_7} e_{i_1} \otimes e^*_{i_2} \otimes e_{i_3} \otimes e_{i_4} \otimes e_{i_5} \otimes e^*_{i_6} \otimes e^*_{i_7}.$$

Since the position of the indices indicates which tensor products of e's and e^*'s occur, we will sometimes write (2.1*)

(2.1*) $$t = (t^{i_1}{}_{i_2}{}^{i_3 i_4})$$

instead of the longer expression (2.1). Thus the equation above can be written as

$$(t^{i_1}{}_{i_2}{}^{i_3 i_4}) \otimes (r^{j_1}{}_{j_2 j_3}) = (t^{i_1}{}_{i_2}{}^{i_3 i_4} r^{i_5}{}_{i_6 i_7}).$$

Similarly, the contraction can be made explicit in terms of coordinates. In fact

EXERCISE 2.1. Show that contracting $(t^{i_1}{}_{i_2}{}^{i_3 i_4})$ with respect to the first two positions yields $(\sum_i t^i{}_i{}^{i_3 i_4})$.

The index symbolism involved in replacing (2.1) by (2.1*) can be viewed from a slightly different standpoint. Let $\mathscr{F}(V)$ denote the set of all frames (i.e., bases) of V. Let $GL(n)$ (general linear group in n-variables) denote the group of all non-singular $n \times n$ matrices, where the product of (α^i_j) and (α'^i_j) is $(\sum_k \alpha^i_k \alpha'^k_j)$.

We can define a mapping of $\mathscr{F}(V)$ onto itself corresponding to any matrix $A = (\alpha^i_j)$ as follows: If $\underline{e} = (e_1, \ldots, e_n) \in \mathscr{F}(V)$ then $\underline{e}A = \underline{f} = (f_1, \ldots, f_n)$, where

(2.3) $$f_j = \sum \alpha^i_j e_i.$$

If $A' = (\alpha'^i_j)$ and $\underline{f}A = \underline{g}$ then

(2.4) $$g_i = \sum \alpha'^k_j f_k = \sum_i \left(\sum_k \alpha'^k_j \alpha^i_k \right) e_i.$$

But $(\sum \alpha'^k_j \alpha^i_k) = AA'$. Thus we can write (2.4) as

(2.5) $$\underline{e}(AA') = (\underline{e}A)A'.$$

This, of course, is why we write the action of A on $\underline{e}$ as $\underline{e}A$ instead of the usual convention of writing the function to the left of the argument. We paraphrase Equation (2.5) by saying that $GL(n)$ acts on the right as a group of transformations of $\mathscr{F}(V)$. We should also note that this action is *free*: Two distinct matrices will take any frame into two distinct frames. It also is *transitive*: Any frame can be carried into any other frame by a suitable matrix. Now let Q be any space on which $GL(n)$ acts on the left, i.e.,

$$(AA')q = A(A'q). \tag{2.6}$$

On the space $\mathscr{F}(V) \times Q$ introduce the following equivalence relation

$$(e, p) \sim (f, q)$$

if there is an $A \in GL(n)$ such that

$$\underline{f} = \underline{e}A \quad \text{and} \quad p = Aq.$$

It is easy to check that this is indeed an equivalence relation. The cosets of $\mathscr{F}(V) \times Q$ under this equivalence relation are called *quantities* (over V) *of type* Q. Examples of this construction:

1. Let $Q_V = R^n =$ the space of all n-tuples $X = (x^1, \ldots, x^n)$. If $A = (\alpha^i_j)$, let $A(x^1, \ldots, x^n) = (y^1, \ldots, y^n)$ where $y^i = \sum \alpha^i_j x^j$. If we map $\mathscr{F}(V) \times Q_V \xrightarrow{\varphi} V$ by setting $\varphi(e, X) = \sum x^j e_j$. Then

$$\varphi(eA, A^{-1}X) = \sum_{i,j,k} (\beta^i_j x^j)(\alpha^k_i e_k) = \sum x^j e_j.$$

Thus φ is constant on the cosets of $\mathscr{F}(V) \times Q_V$. Therefore, it maps the quantities of type Q_V into V. On the other hand, the map which assigns to every $v \in V$ and $e \in \mathscr{F}(V)$ the coordinates of v relative to e and the rules for change of coordinates show that every $v \in V$ corresponds to a unique quantity of type Q_V.

2. Let $V_{V^*} = R^n =$ all n-tuples $(x_1, \ldots, x_n)$ and define the action of $GL(n)$ on Q_{V^*} by $A(x_1, \ldots, x_n) = (y_1, \ldots, y_n)$

$$y_i = \sum \beta^j_i x_j$$

where $A = (\alpha^i_j)$ and $\sum \alpha^i_j \beta^j_k = \delta^k_i$. We define a map $\varphi\colon \mathscr{F}(V) \times Q_{V^*} \to V^*$ by

$$\varphi(e_j(x_1, \ldots, x_n)) = \sum x_i e^{i*}.$$

As above, this map is constant on cosets and allows us to identify quantities of type Q_{V^*} with V^*.

3. If we take Q as a tensor product of various copies of the representations Q_V and Q_{V^*} we obtain a space which we can identify with the corresponding tensor product of V's and V^*. This is what is behind the symbolism of (2.1*). It is a useful point of view because frequently in practice one is given an n^k-tuplet for every frame and the law for change of coordinates, i.e., a quantity of a certain type. If the law for change of coordinates is a tensor product of the representations Q_V and Q_{V^*} then the quantity is (can be identified with) a tensor.

4. $Q_{\delta_p} = R$ and $Ax = |\alpha^i_j|^p x$. Quantities of this type are called *(scalar) densities of weight* p.

5. $Q_\rho = R^k$ and the action of $GL(n)$ on Q_ρ is some k-dimensional *(linear)* representation, ρ, of $GL(n)$. A quantity of this type is called a *linear quantity*. Each of the previous four examples are linear quantities. Their importance lies in the fact that *any* linear quantity can be obtained from their irreducible parts by elementary combinations (i.e., tensor products and direct sums).

Exercise 2.2. Suppose that for every frame $(e_1, \ldots, e_n)$ there is given a set of numbers ${}_et^{i_1, \ldots, i_k}$. Show that in order for these to represent a tensor, it is necessary and sufficient that for any tensor ${}_es_{i_1 i_2 i_3}$ the expression

$$\sum_{i_1, i_2, i_3} {}_et^{i_1, \ldots, i_k}\, {}_es_{i_1 i_2 i_3}$$

represent a tensor. (The number or position of the indices involved has nothing to do with the criterion.)

3. The contravariant and symmetric algebras. There are certain algebras associated with the tensor algebra which are of considerable importance. First of all, the set of all tensors of covariant degree zero, $C(V)$, is clearly a subalgebra of $T(V)$

$$\begin{aligned} C(V) &= K + V + V \otimes V + \cdots \\ &= C^0(V) + C^1(V) + C^2(V) + \cdots \end{aligned}$$

where $C^p(V) = V \otimes \cdots \otimes V$ (p-times). It is called the *algebra of contravariant tensors*.

Exercise 3.1. The algebra $C(V)$ has a "universal" property: Show that for any map, f, of V into an associative algebra A, there is a unique homomorphism, φ, of $C(V) \to A$ such that $f = \varphi \circ i$ where i is the injection of V into $C(V)$.

Any linear map $f: V \to W$ induces a linear map

$$f \otimes \cdots \otimes f: V \otimes \cdots \otimes V \to W \otimes \cdots \otimes W$$

which we shall call $C^p(f)$ and thus a homomorphism $C(f): C(V) \to C(W)$.[2] It is clear that $C(f \circ g) = C(f) \circ C(g)$.

We can define a pairing of $C(V)$ and $C(V^*)$ into R as follows:

$$\langle x, y^* \rangle = 0 \quad \text{if} \quad x \in C^p(V), \quad y \in C^q(V^*) \qquad q \neq p$$

if $x \in C^p(V)$, $y \in C^p(V^*)$ define $\langle x, y^* \rangle$ to be the complete contraction of $x \otimes y^*$ with respect to all indices in order. In terms of coordinates, if $x = (x^{i_1,\ldots,i_p})$ and $y^* = (y_{i_1,\ldots,i_p})$ set

$$\langle x, y^* \rangle = \sum_{i_1,\ldots,i_p} x^{i_1,\ldots,i_p} y_{i_1,\ldots,i_p}.$$

It is easy to see that with this pairing, $C^p(V^*)$ can be identified with $(C^p(V))^*$. Furthermore, this identification is natural in the following sense: For any map $f: V \to W$, $x \in C(V)$, $y^* \in C(W^*)$

$$\langle C(f)x, y^* \rangle = \langle x, C(f^*)y^* \rangle \tag{3.1}$$

where f^* is the adjoint map of $W^* \to V^*$. Therefore, we shall identify $C^p(V^*)$ with $(C^p(V))^*$.

Exercise 3.2. Show that

$$\langle x_1 \otimes \cdots \otimes x_p, y^1 \otimes \cdots \otimes y^p \rangle = \langle x_1, y^{1*} \rangle \cdots \langle x_p, y^{p*} \rangle.$$

The permutation group on p letters, π_p, acts on the space $C^p(V)$: Given any permutation $\sigma \in \pi_p$ and any tensor of the form $x_1 \otimes \cdots \otimes x_p$ we define

$$\sigma(x_1 \otimes \cdots \otimes x_p) = x_{\sigma(1)} \otimes \cdots \otimes x_{\sigma(p)}$$

and extend by linearity to all of $C^p(V)$. In terms of a basis

$$(t^{i_1,\ldots,i_p}) = (t^{i_{\sigma^{-1}(1)},\ldots,i_{\sigma^{-1}(p)}}).$$

---

[2] Thus the rule which assigns to each vector space V the algebra $C(V)$ is a covariant functor. It is an unfortunate historical accident that $C(V)$ is called contravariant.

Thus, we have a representation of π_p on $C^p(V)$. It is easy to see that this representation commutes with the tensor product representation of $GL(n)$; i.e., for any $t \in C^p(V)$, $\sigma \in \pi$ and $g \in GL(n)$

$$g\sigma t = \sigma g t \tag{3.2}$$

where gt is $g \otimes \cdots \otimes g$ (p times) acting on t.

It follows from (3.2) that any simultaneous eigenvector of π_p, that is, any tensor satisfying

$$\sigma t = \rho(\sigma) t \tag{3.3}$$

where ρ is some numerical function on π_p is taken by g into another tensor satisfying the same equation:

$$\sigma g t = g \sigma t = g \rho(\sigma) t = \rho(\sigma) g t.$$

The two important cases are

$$\rho(\sigma) = 1$$

and

$$\rho(\sigma) = \operatorname{sgn}(\sigma) = \begin{cases} +1 & \text{if} \quad \sigma \text{ is even} \\ -1 & \text{if} \quad \sigma \text{ is odd.} \end{cases}$$

The subspace of $C^p(V)$ consisting of those tensors satisfying $\sigma t = t$ will be denoted by $S^p(V)$ and will be called the *space of symmetric* (*contravariant*) *tensors* of degree p. There is a natural projection, $\mathscr{S}$, of $C^p(V)$ onto $S^p(V)$ given by

$$\mathscr{S}(t) = \frac{1}{p!} \sum_{\sigma \in \pi_p} \sigma t.$$

It is easy to check that $\mathscr{S}^2 = \mathscr{S}$ and that $\mathscr{S}(C^p(V)) = S^p(V)$.

Let

$$S(V) = K + S^1(V) + S^2(V) + \cdots.$$

We define a product on $S(V)$ by

$$a \odot b = \mathscr{S}(a \otimes b).$$

EXERCISE 3.3. Show that $\mathscr{S}(V)$ is a commutative associative algebra.

EXERCISE 3.4. Show that $S(V)$ is universal with respect to maps of V into commutative algebras. That is, if $f: V \to A$ then there is a unique homomorphism of $S(V) \to A$ such that $f = \varphi \circ i$ where i is the injection of V into $S(V)$.

EXERCISE 3.5. Show that $S(V^*)$ can be identified with the ring of polynomial functions over V.

EXERCISE 3.6. Let V operate on $S(V^*)$ by

$$\begin{aligned} v[x] &= 0 \text{ for } x \in S^0(V) \\ v[x] &= \text{the tensor } pv \otimes x \text{ contracted with respect} \\ &\quad\ \text{to any index if } x \in S^p(V^*). \end{aligned}$$

Show that $v[s \circ t] = v[s] \circ t + s \circ v[t]$ so that v is a derivation of $S(V^*)$. Show that in terms of the identification of Exercise 3.5 this action is just derivation in the v direction.

EXERCISE 3.7. Let $f: V \to W$. Show that $\mathscr{S}C^p(f) = C^p(f)\mathscr{S}$, so that f induces a map $S^p(f)$ of $S^p(V) \to S^p(W)$ and a homomorphism $S(f)$ of $S(V) \to S(W)$.

EXERCISE 3.8. Show that if $x \in C^p(V)$ and $y^* \in C^p(V^*)$ that $\langle \mathscr{S}x, y^* \rangle = \langle x, \mathscr{S}y^* \rangle$. Conclude that we can identify $S^p(V^*)$ with $(S^p(V))^*$ and that this identification is natural.

The space $S^2(V^*)$ is the space of all symmetric bilinear forms on V. An element, b of $S^2(V^*)$ is sometimes called an *indefinite metric on* V. It assigns to every $v \in V$ the "length" $b(v, v)$. Due to the isomorphism of $V^* \otimes V^*$ with the space $\mathscr{L}(V, V^*)$, b induces a linear map of $V \to V^*$. To say that this map is an isomorphism means that for every fixed $v \neq 0$, the linear function $b(v, \cdot)$ is not identically zero. A metric, b, with this property is called *non-singular*. A choice of non-singular b (which induces an isomorphism of V onto V^*, and therefore of V^* onto V) gives us a way of mapping tensors of contravariant type into covariant type and conversely. Let $b = (g_{ij})$ (relative to some basis) and the element of $V \otimes V$ giving the inverse isomorphism of $V^* \to V$ (to that induced by b) be given by (g^{ij}). Then, for example, we have an isomorphism of $V \otimes V^* \otimes V \to V \otimes V \otimes V^*$ given by:

$$(t^{i_1}{}_{i_2}{}^{i_3}) \to \left(\sum g^{i_2 j_2} g_{i_3 j_3} t^{i_1}{}_{j_2}{}^{i_3}\right).$$

These operations [multiplication by (g^{ij}) or (g_{ij}) followed by contraction] are sometimes called *raising* (*or lowering*) *the indices*.

By a well known theorem of algebra every metric has, in a suitable coordinate system, the form

$$b(v, v) = \sum_{i=1}^{k} (v^i)^2 - \sum_{i=k+1}^{r} (v^i)^2 \qquad \text{where } v = (v^i).$$

Here the integers k and r are invariants of b. If $r = n$ then b is non-singular, and the converse is true. If $k = n$, then $b(v, v) > 0$ for all $v \neq 0$ and the metric is called *positive definite* or *Euclidean.* If $k = n - 1$ the metric is called *Lorentzian.*

4. Exterior algebra. We now study the subspace of $C^p(V)$ corresponding to the second simultaneous eigenvalue in (3.3), i.e., the space of anti-symmetric tensors, $\bigwedge^p(V)$. As in the case of the symmetric tensors, there is an obvious projection of $C^p(V)$ onto $\bigwedge^p(V)$ given by

$$\mathscr{A}t = \frac{1}{p!} \sum (\operatorname{sgn} \sigma)\sigma(t)$$

where $\operatorname{sgn} \sigma = \pm 1$ according as σ is even or odd. As before, we introduce a multiplication on

$$\bigwedge(V) = \bigwedge^0(V) + \bigwedge^1(V) + \bigwedge^2(V) + \cdots$$

[where $\bigwedge^0(V) = K, \bigwedge^1(V) = V$] by

$$s \wedge t = \mathscr{A}(s \otimes t) \qquad s \in \bigwedge^p(V), t \in \bigwedge^b(V).$$

Thus $s \wedge t \in \bigwedge^{p+q}(V)$. The sign of the permutation on $p + q$ letters which moves the first p letters past the last q is $(-1)^{pq}$ (the motion is obtained by moving each of the p letters starting with the last of the p's through the q's thus by $p \cdot q$ interchanges of adjoining letters). Thus $\mathscr{A}(t \otimes s) = (-1)^{pq} \mathscr{A}(s \otimes t)$ so that

$$s \wedge t = (-1)^{pq} t \wedge s. \tag{4.1}$$

Exercise 4.1. Show that $\bigwedge(V)$ is an associative algebra.

Exercise 4.2. Characterize $\bigwedge^p(V)$ and $\bigwedge(V)$ by universal properties.

THEOREM 4.1. *The exterior algebra of an n-dimensional vector space is of dimension* 2^n. *If* $e_1, \ldots, e_n$ *is a basis of* V *then a base of* $\bigwedge(V)$ *is given by* 1 *and the elements*

$$e_{i_1} \wedge \cdots \wedge e_{i_r} \qquad 1 \leq i_1 < \cdots < i_r \leq n, \quad r = 1, \ldots, n. \tag{4.2}$$

Proof. The elements $\mathscr{A}(e_{j_1} \otimes \cdots \otimes e_{j_r}) = e_{j_1} \wedge \cdots \wedge e_{j_r}$ obviously span $\bigwedge^r(V)$. By the anticommutativity relation (4.1), it follows that the $e_{i_1} \wedge \cdots \wedge e_{i_r}$ with $i_1 < \cdots < i_r$ span $\bigwedge^r(V)$ and that $\bigwedge^r(V) = 0$ for $r > n$. All that we must show is that the elements (4.2) are linearly independent. We can restrict attention to a fixed r as terms corresponding to different values of r are obviously independent. Now for $r = n$, $e_1 \wedge \cdots \wedge e_n = \mathscr{A}(e_1 \otimes \cdots \otimes e_n) \neq 0$ as the different terms $e_{i_1} \otimes \cdots \otimes e_{i_n}$ are independent. For $r < n$ suppose there be a linear relation of the form

$$\sum a^{i_1,\ldots,i_r} e_{i_1} \wedge \cdots \wedge e_{i_r} = 0. \tag{4.3}$$

For a fixed set of indices $i_1, \ldots, i_r$ let $j_{r+1}, \ldots, j_n$ be the complementary set. Multiply (4.3) by $e_{j_{r+1}} \wedge \cdots \wedge e_{j_n}$. With the exception of $e_{i_1} \wedge \cdots \wedge e_{i_r}$ all terms in the product will have repeated factor and therefore vanish by anticommutativity. Thus $a^{i_1,\ldots,i_r} e_{i_1} \wedge \cdots \wedge e_{i_r} \wedge e_{j_{r+1}} \wedge \cdots \wedge e_{j_n} = \pm a^{i_1,\ldots,i_r} e_1 \wedge \cdots \wedge e_n = 0$, so $a^{i_1,\ldots,i_r} = 0$. This proves the theorem.

Depending on the situation, we will sometimes express an antisymmetric tensor in terms of the bases (4.2) or in terms of its representation as a tensor whichever is more convenient. Thus we may write

$$t = \sum_{i_1 < \cdots < i_r} t^{i_1,\ldots,i_r} e_{i_1} \wedge \cdots \wedge e_{i_r}$$

or

$$t = (t^{i_1,\ldots,i_r}).$$

We now study some geometrical applications of the exterior algebra.

THEOREM 4.2. *A necessary and sufficient condition that the vectors* $x_1, \ldots, x_r$ *be linearly dependent is that*

$$x_1 \wedge \cdots \wedge x_r = 0. \tag{4.4}$$

Proof. If the vectors are dependent, then we can express one in terms of the other, say

$$x_r = \sum_{k=1}^{r-1} a_k x_k.$$

Thus

$$\begin{aligned} x_1 \wedge \cdots \wedge x_r &= x_1 \wedge \cdots \wedge \left(\textstyle\sum a_k x_k\right) \\ &= \textstyle\sum a_k x_1 \wedge \cdots \wedge x_{r-1} \wedge x_k = 0 \end{aligned}$$

as every term in the last sum has a repeated factor.

On the other hand, if $x_1, \ldots, x_r$ are linearly independent, we can always find $x_{r+1}, \ldots, x_n$ so that $x_1, \ldots, x_n$ form a base of V. Then $x_1 \wedge \cdots \wedge x_n \neq 0$ so *a fortiori* $x_1 \wedge \cdots \wedge x_r \neq 0$.

The elements of $\bigwedge^r(V)$ are sometimes called *r-vectors*. Those r-vectors which can be written as the exterior product of elements of V (every r-vector is a linear combination of such) are called *decomposable* r-vectors. We can then state

THEOREM 4.3. *To each r-dimensional linear subspace W of V there corresponds a decomposable r-vector $w \neq 0$, determined up to a (non-zero) scalar factor, such that*

$$x \in W \quad \text{if and only if} \quad x \wedge w = 0.$$

Let W_1 and W_2 be subspaces of dimension r_1 and r_2 with w_1 and w_2 corresponding r_1 and r_2 vectors. In order that $W_1 \subset W_2$ it is necessary and sufficient that there exist an $(r_2 - r_1)$ vector v such that $w_2 = w_1 \wedge v$. In order that $W_1 \cap W_2 = \{0\}$ it is necessary and sufficient that $w_1 \wedge w_2 \neq 0$. If this condition is satisfied then $w_1 \wedge w_2$ is an $(r_1 + r_2)$-vector corresponding to $W_1 + W_2$.

Proof. If $x_1, \ldots, x_r$ is a basis for W then set $w = x_1 \wedge \cdots \wedge x_r$. If $y_1, \ldots, y_r$ is any other basis with

$$y_i = \sum a_{ij} x_j$$

then

$$y_1 \wedge \cdots \wedge y_r = \det (a_{ij}) x_1 \wedge \cdots \wedge x_r$$

so that w is determined up to a scalar factor. In view of the previous theorem, $x \in W$, i.e., is a linear combination of $x_1, \ldots, x_r$ if and only if $x_1 \wedge \cdots \wedge x_r \wedge x = w \wedge x = 0$. A necessary and sufficient condition that W_1 be contained in W_2 is that any basis $x_1, \ldots, x_{r_1}$ of W_1 can be extended to basis $x_1, \ldots, x_{r_1}, x_{r_1+1}, \ldots, x_{r_2}$ of W_2. Then $w_1 = x_1 \wedge \cdots \wedge x_{r_1}$, $w_2 = x_1 \wedge \cdots \wedge x_{r_2}$ and $v = x_{r_1+1} \wedge \cdots \wedge x_{r_2}$. A necessary and sufficient condition that $W_1 \cap W_2 = \{0\}$ is that $x_1, \ldots, x_{r_1}, y_1, \ldots, y_{r_2}$ be independent where $y_1, \ldots, y_{r_2}$ is a basis for W_2. But if $w_1 = x_1 \wedge \cdots \wedge x_{r_1}$ and $w_2 = y_1 \wedge \cdots \wedge y_{r_2}$ this condition is equivalent to $w_1 \wedge w_2 \neq 0$. If $w_1 \wedge w_2 \neq 0$, then $w_1 \wedge w_2 = x_1 \wedge \cdots \wedge x_{r_1} \wedge y_1 \wedge \cdots \wedge y_{r_2}$ is clearly an r-vector for $W_1 + W_2$. Q.E.D.

The results of Theorem 4.3 can be paraphrased as saying that p-dimensional subspaces correspond to equivalence classes of p-vectors; two p-vectors being regarded as equivalent if they differ by a non-zero factor. This, of course, is the higher dimensional version of the fact that a line through the origin determines any vector lying on it up to a *non-zero* factor. In elementary geometry,

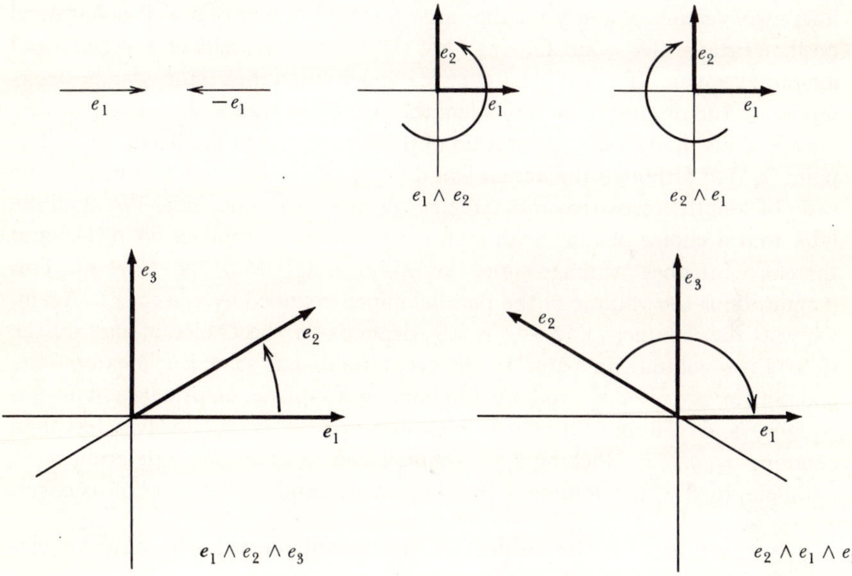

one also studies the notion of a "directed" line. Every line has two "directions." There is no way, of course, of choosing one direction in preference to the other but the two differ in "sign." A precise way of formulating this notion is to introduce a stronger equivalence relation between vectors; namely, regard two vectors as equivalent if they differ by a *positive* factor. A directed line is such an equivalence class. It is clear that two of the new equivalence classes correspond to each of the old ones and the source of the difference in sign becomes transparent. Similarly, in higher dimensions, we will speak of oriented p-spaces. They are equivalence classes of p-vectors with the stronger equivalence relation "two p-vectors are equivalent if they differ by a positive factor." Again each p-dimensional space has two orientations. (In two dimensions, for example, the orientations amount to giving a sense of rotation: Clockwise versus anti-clockwise. In three dimensions we distinguish right- and left-handedness, as shown in the figure on page 17.)

Choosing a p-vector involves choosing more than the oriented p-plane. Two different p-vectors determining the same oriented p-plane differ by a constant factor. This factor is an invariant of the full linear group. By this we mean that if $x_1 \wedge \cdots \wedge x_p = ky_1 \wedge \cdots \wedge y_p$ then

$$\wedge^p(T)(x_1 \wedge \cdots \wedge x_p) = k\wedge^p(T)(y_1 \wedge \cdots \wedge y_p) \qquad \text{for any } T \in GL(n).$$

Thus (in the same vague sense as the arbitrary sign before) a p-vector determines an oriented p-plane and something more, two different p-vectors of the same p-plane differing by a well-determined scalar. In the case $p = 1$ this is simply the observation that a vector gives more than a directed line, two vectors, x and y on the same directed line have a well-determined positive ratio k, if $x = ky$. One can not talk about the *length* of a vector until a metric is given. If a Euclidean metric is given, then the vector x can be determined on the oriented line by its length, $\|x\|$. (The metric fixes the arbitrary scale.) Thus the factor k is the ratio of the length of x to the length of y. The point is, that although the actual lengths depend on the choice of metric, the ratio of length (of two vectors *lying on the same line*) does not. We shall see later that a choice of Euclidean metric on V induces a metric on $\wedge^p(V)$ and therefore attaches a magnitude, $\|x_1 \wedge \cdots \wedge x_p\|$ to $x_1 \wedge \cdots \wedge x_p$. This magnitude is the volume of the parallelopiped spanned by $x_1, \ldots, x_p$. Again, whereas the volume, $\|x_1 \wedge \cdots \wedge x_p\|$, depends on the choice of metric, the ratio of two volumes does not. In any event, relative to some Euclidean metric, choosing $x_1 \wedge \cdots \wedge x_p$ amounts to choosing a p-plane, an orientation on this p-plane, and volume. Of course, choosing $x_1 \wedge \cdots \wedge x_p$ involves less than choosing $x_1, \ldots, x_p$. Picking $x_1, \ldots, x_p$ involves in addition to determining a parallelopiped whose volume is $\|x_1 \wedge \cdots \wedge x_p\|$ and an ordering of its edges.

While we are on the subject of independence we will state an elementary theorem with important applications later on.

THEOREM 4.4. (Cartan's lemma) *Let $x_1, \ldots, x_p$ be linearly independent and suppose that*

$$x_1 \wedge y_1 + \cdots + x_p \wedge y_p = 0. \tag{4.5}$$

Then

$$y_i = \sum_{j=1}^{p} A_{ij} x_j \qquad \textit{with } A_{ij} = A_{ji}.$$

Proof. Choose $x_{p+1}, \ldots, x_n$ so that $x_1, \ldots, x_p, x_{p+1}, \ldots, x_n$ is a basis of V. Then we can write

$$y_i = \sum_{j=1}^{p} A_{ij} x_j + \sum_{j=p+1}^{n} B_{ij} x_j.$$

Substituting this expression into (4.5) we obtain

$$\sum_{i<j\leqslant p} (A_{ij} - A_{ji}) x_i \wedge x_j + \sum_{i\leqslant p,\, j>p} B_{ij} x_i \wedge x_j = 0. \tag{4.6}$$

However, the expressions $x_i \wedge x_j$ $(i < j)$ are linearly independent by Theorem 4.1. Thus $B_{ij} = 0$ and $A_{ij} = A_{ji}$. Q.E.D.

Let f be a linear map of $V \to W$. Then for any $\sigma \in \pi_p$, $\sigma C^p(f) = C^p(f)\sigma$ so that

$$\mathscr{A} C^p(f) = C^p(f)\mathscr{A}. \tag{4.7}$$

Therefore $C^p(f)$ carries $\wedge^p(V)$ into $\wedge^p(W)$. We shall denote the restriction of $C^p(f)$ to $\wedge^p(V)$ by $\wedge^p(f)$. Equation (4.7) also shows that if we denote by $\wedge(f)$ the map of $\wedge(V) \to \wedge(W)$ which is defined to be $\wedge^p(f)$ on $\wedge^p(V)$ then $\wedge(f)$ is a homomorphism of $\wedge(V) \to \wedge(W)$. Furthermore, for $x \in C^p(V)$, $y^* \in C^p(V^*)$, and $\sigma \in \pi_p$

$$\langle \sigma(x), y^* \rangle = \langle x, \sigma^{-1}(y^*) \rangle.$$

Thus

$$\langle \mathscr{A} x, y^* \rangle = \langle x, \mathscr{A} y^* \rangle.$$

Thus if $x \in \wedge^p(V)$, $x = \mathscr{A} x$ so

$$\langle x, y^* \rangle = \langle x, \mathscr{A} y^* \rangle. \tag{4.8}$$

In view of the previous identification of $(C^p(V))^*$ with $C^p(V^*)$, we could identify $(\wedge^p(V))^*$ with $\wedge^p(V^*)$ in a natural manner using the pairing $\langle\ ,\ \rangle$, restricted to $\wedge^p(V)$ and $\wedge^p(V^*)$. However, it is more convenient to use the pairing $\langle\ |\ \rangle$ (we replace a comma by a $|$) defined by

$$\langle x \mid y^* \rangle = p!\langle x, y^* \rangle \tag{4.9}$$

for $x \in \wedge^p(V)$, $y^* \in \wedge^p(V^*)$. This pairing is, of course, also natural (i.e., behaves the right way under maps).

Let us make this duality between $\wedge^p(V)$ and $\wedge^p(V^*)$ more explicit by computing $\langle x | y^* \rangle$ for decomposable p-vectors. If $x = x_1 \wedge \cdots \wedge x_p$

$= \mathscr{A}(x_1 \otimes \cdots \otimes x_p)$ and $y = y^{1^*} \wedge \cdots \wedge y^{p^*} = \mathscr{A}(y^{1^*} \otimes \cdots \otimes y^{p^*})$ then by Exercise 3.2, Equation (4.9) and since $\mathscr{A}^2 = \mathscr{A}$,

$$\begin{aligned} \langle x \,|\, y^* \rangle &= p!\langle \mathscr{A}(x_1 \otimes \cdots \otimes x_p), \mathscr{A}(y^{1^*} \otimes \cdots \otimes y^{p^*})\rangle \\ &= p!\langle \mathscr{A}(x_1 \otimes \cdots \otimes x_p), y^{1^*} \otimes \cdots \otimes y^{p^*}\rangle \\ &= \sum_{\sigma \in \pi_p} \operatorname{sgn} \sigma \langle x_{\sigma(1)} \otimes \cdots \otimes x_{\sigma(p)}, y^{1^*} \otimes \cdots \otimes y^{p^*}\rangle \\ &= \sum_{\sigma \in \pi_p} \operatorname{sgn} \sigma \langle x_{\sigma(1)}, y^{1^*}\rangle \cdots \langle x_{\sigma(p)}, y^{p^*}\rangle \end{aligned}$$

or

$$\langle x \,|\, y^* \rangle = \det \langle x_i, y^{j^*} \rangle . \tag{4.10}$$

The vector space $\wedge^p(V^*) = (\wedge^p(V))^*$ is called the *space of exterior p-forms over V*. The algebra $\wedge(V^*)$ is called the *algebra exterior forms over V*.

We will define the pairing $\langle \;|\; \rangle$ of $\wedge(V)$ with $\wedge(V^*)$ by

$$\langle x \,|\, y^* \rangle = 0 \qquad \text{if} \quad x \in \wedge^p(V), \wedge^q(V) \quad (p \neq q)$$

and by (4.9) if $p = q$.

Exercise 4.3. Let $e_1, \ldots, e_n$ be a basis of V and $e^{1^*}, \ldots, e^{n^*}$ the dual basis of V^*. Show that relative to the pairing $\langle \;|\; \rangle$, the dual basis to

$$\{e_{i_1} \wedge \cdots \wedge e_{i_r}\} \qquad (1 < i_1 < \cdots < i_r) \quad (r = 0, \ldots, n)$$

is

$$\{e^{i_1^*} \wedge \cdots \wedge e^{i_r^*}\}.$$

The pairing of $\wedge(V)$ and $\wedge(V^*)$ allows us to define the adjoint of any linear operation on $\wedge(V)$ as a linear operator on $\wedge(V^*)$. In particular, we can define the adjoint of (left) exterior multiplication. It is called (right) interior multiplication. More precisely, for any $u \in \wedge(V)$ we define the linear operator $u \,\lrcorner\, \cdot$ on $\wedge(V^*)$ by

$$\langle x \,|\, u \,\lrcorner\, y^* \rangle = \langle u \wedge x \,|\, y^* \rangle, \qquad \text{for all} \quad x \in \wedge(V). \tag{4.11}$$

Exercise 4.4. Show that $u_1 \,\lrcorner\, (u_2 \,\lrcorner\, y^*) = (u_2 \wedge u_1) \,\lrcorner\, y^*$.

In view of (4.11), it is clear that if $u \in \wedge^p(V)$ and $y \in \wedge^r(V^*)$ that

$u \rfloor y^* \in \wedge^{r-p}(V^*)$ if $r \geqslant p$ and $= 0$ for $r < p$. In particular, if $p = r$, $u \rfloor y^* = \langle u | y^* \rangle$. Thus

$$(4.12)\quad e_{i_1} \wedge \cdots \wedge e_{i_p} \rfloor e^{j_1^*} \wedge \cdots \wedge e^{j_r^*} = \begin{cases} 0 \text{ unless } \{i_1, \ldots, i_p\} \subset \{j_1, \ldots, j_r\} \\ \pm e^{\alpha_1^*} \wedge \cdots \wedge e^{\alpha_{r-p}^*} \end{cases}$$

where $\alpha_1, \ldots, \alpha_{r-p}$ are the complementary set of indices in $j_1, \ldots, j_r$ to $i_1, \cdots i_p$, and the sign is the sign of the permutation $(j_1, \ldots, j_r) \to (i_1, \ldots, i_p, \alpha_1, \ldots, \alpha_{r-p})$. In particular,

$$(4.13)\quad e_i \rfloor e^{j_1^*} \wedge \cdots \wedge e^{j_r^*} = \begin{cases} 0 & \text{if } i \neq j_k \text{ for any } k, \\ (-1)^{k-1} e^{j_1^*} \wedge \cdots \wedge e^{j_{k-1}} \wedge e^{j_{k+1}} \cdots \wedge e^{j_r} & \text{if } i = j_k. \end{cases}$$

Formula (4.13) shows that interior multiplication by e_i can be viewed as the analogue in the exterior algebra of partial differentiation with respect to the ith variable. As to be expected, successive "partial differentiations" depends on the order: By Exercise 4.4 $e_i \rfloor (e_j \rfloor y^*) = - e_j \rfloor (e_i \rfloor y^*)$. Furthermore, if we multiply (4.13) by e_i and sum over i, we obtain (first for the bases and then by linearity) for any $y^* \in \wedge^r(V^*)$ the analogue of Euler's theorem on homogeneous functions

$$(4.14)\qquad \sum_{i=1}^{n} e_i^* \wedge (e_i \rfloor y^*) = ry^*.$$

Exercise 4.5. Show that interior multiplication by an $X \in V$ is actually an antiderivation of $\wedge(V^*)$. That is, show that for any $X \in V, y_1^* \in \wedge^p(V^*)$ and $y_2^* \in \wedge^q(V^*)$

$$X \rfloor (y_1^* \wedge y_2^*) = (X \rfloor y_1^*) \wedge y_2^* + (-1)^p y_1^* \wedge (X \rfloor y_1^*).$$

A frequent use of the interior product is to relate $\wedge(V)$ and $\wedge(V^*)$ as follows:

Theorem 4.5. *Let e and e^* be dual bases of the (one-dimensional) spaces $\wedge^n(V)$ and $\wedge^n(V^*)$ respectively. The linear map ${}_e\!\perp : \wedge^r(V) \to \wedge^{n-r}(V^*)$ defined by*

$${}_e\!\perp(x) = x \rfloor e^*$$

is an isomorphism of $\wedge^r(V)$ onto $\wedge^{n-r}(V^*)$ and carries decomposable elements into decomposable elements. If $w \neq 0$ is a decomposable r-vector corresponding *to the linear subspace W, then $\perp(w)$ corresponds to the subspace W^0 (of those linear functions vanishing on W).*

Proof. If we choose dual basis so that $e_1 \wedge \cdots \wedge e_n = e$ and $e^{1*} \wedge \cdots \wedge e^{n*} = e^*$ then (4.13) shows the map is an isomorphism. A decomposable element of degree n goes into $\wedge^0(V)$ so there is nothing to prove. Any decomposable element of degree $r < n$ can be written as $e_1 \wedge \cdots \wedge e_r$ where we can complete $(e_1, \ldots, e_r)$ to a basis $\{e_1, \ldots, e_n\}$ such that $e_1 \wedge \cdots \wedge e_n = e$. Then (4.13) shows that

$$\perp(e_1 \wedge \cdots \wedge e_r) = \pm e^{(r+1)*} \wedge \cdots \wedge e^{n*},$$

which is decomposable and which corresponds to the space spanned by $n - r$ independent forms $e^{(r+1)*}, \ldots, e^{n*}$ which vanish on W. Q.E.D.

EXERCISE 4.6. Show that every $n - 1$ vector is decomposable.

We can introduce the right interior multiplication on $\wedge(V)$ by elements of $\wedge(V^*)$ defined by

$$\langle x \llcorner u^* | y^* \rangle = \langle x | u^* \wedge y^* \rangle. \tag{4.15}$$

The analogues of the previous statements about $\lrcorner$ are then true for $\llcorner$.

Suppose that V carries a Euclidean metric. That is, assume that there is a scalar product $(\ ,\ )$ defined on V satisfying

1) $(x, y) = (y, x)$
2) $(a_1x_1 + a_2x_2, y) = a_1(x_1, y) + a_2(x_2, y)$
3) $(x, x) > 0$ unless $x = 0$.

Then we can identify $y \in V$ with $y^* \in V^*$ by

$$\langle x, y^* \rangle = (x, y).$$

We then define an inner product on $C^p(V)$ by setting

$$(x_1 \otimes \cdots \otimes x_r, y_1 \otimes \cdots \otimes y_r) = (x_1, y_1) \cdots (x_r, y_r)$$

for decomposable elements and extending by linearity. If we define $C^p(V)$ to be orthogonal to $C^q(V)$ when $p \neq q$, we then have an inner product defined on $C(V)$. Observe that $\mathscr{S}$ and $\mathscr{A}$ are self-adjoint operators relative to this inner product.

If we restrict $(\ ,\ )$ to $S(V)$ or $\wedge(V)$ we obtain inner products on these spaces. As before, we define an inner product on $\wedge^p(V)$, by changing $(\, , \,)$ by a factor:

$$(x|y) = p!(x, y) \qquad \text{for } x, y \in \wedge^p(V) \tag{4.16}$$

then

$$(x_1 \wedge \cdots \wedge x_p | y_1 \wedge \cdots \wedge y_p) = (\det(x_i, y_j)). \tag{4.17}$$

In particular,

$$\begin{aligned} \|x_1 \wedge \cdots \wedge x_p\| &= (x_1 \wedge \cdots \wedge x_p | x_1 \wedge \cdots \wedge x_p)^{\frac{1}{2}} \\ &= (\det(x_i, x_j))^{1/2} \end{aligned} \tag{4.18}$$

is the volume of the parallelopiped spanned by $x_1, \ldots, x_p$.

Since $\wedge^n(V)$ is one-dimensional, there are exactly two elements, e and $-e$, in $\wedge^n(V)$ such that $\|e\| = 1$. A choice between them is an orientation of V.

By Theorem 4.5 a choice of e (say) defines an isomorphism, φ, of $\wedge^r(V) \to \wedge^{n-r}(V^*)$. On the other hand, the inner product defines an isomorphism of $\wedge^{n-r}(V^*)$ with $\wedge^{n-r}(V)$. Combining these isomorphisms, we obtain an isomorphism, $*$,

$$* : \wedge^r(V) \to \wedge^{n-r}(V). \tag{4.19}$$

Exercise 4.7. Let $e_1, \ldots, e_n$ be an orthonormal basis of V with respect to $(\, , \,)$. Show that $\{e_{i_1} \wedge \cdots \wedge e_{i_p}\}$ form an orthonormal basis for $\wedge(V)$ with respect to $(\, , \,)$.

Exercise 4.8. In terms of this basis, what is $*(e_{i_1} \wedge \cdots \wedge e_{i_p})$?

Exercise 4.9. Show that $x \wedge *x = e$.

Exercise 4.10. Show that $**x = (-1)^{r(n-r)}x$ if $x \in \wedge^r(V)$.

Exercise 4.11. Show that if x and y are vectors in three-dimensional Euclidean space, $*(x \wedge y)$ is the "vector product" of x and y.

Exercise 4.12. Show that $*$ is an isometry; i.e., $(*x | *y) = (x|y)$.

Exercise 4.13. Show that $x \wedge *y = (x|y)e$ for $x, y \in \wedge^p(V)$.

If S.S.'s * operation is everybody else's this, is wrong it should be $(-1)^{(n+1)(n-r-1)}$ so that $** = -1$ for 2nd rank tensors in 4 dimensions

5. Exterior equations. We first establish a normal form for exterior 2-forms analogous to the (diagonal) canonical form for symmetric quadratic forms.

THEOREM 5.1. *For any* $q^* \in \bigwedge^2(V^*)$ *there exists a basis* $\{f^{1*}, \ldots, f^{n*}\}$ *of* V^* *so that*

$$q^* = f^{1*} \wedge f^{2*} + \cdots + f^{2(p-1)*} \wedge f^{2p*}. \tag{5.1}$$

The integer p *depends only on* q^*.

Proof. Suppose that q^* is given by

$$q^* = \sum_{i<j} a_{ij}\, e^{i*} \wedge e^{j*}$$

in terms of the basis $\{e^{i*}\}$. If $q^* \neq 0$ then we can (by rearranging the basis if necessary) assume that $a_{12} \neq 0$. Then

$$q^* = \left(e^{1*} - \frac{a_{23}}{a_{12}} e^{3*} - \cdots - \frac{a_{2n}}{a_{12}} e^{n*}\right) \wedge \left(a_{12}\, e^{2*} + a_{13}\, e^{3*} + \cdots + a_{1n}\, e^{n*}\right) + q^{1*}$$

where q^{1*} does not contain any expression involving e^{1*} or e^{2*}. If we set

$$f_1 = e^{1*} - \frac{a_{23}}{a_{1n}} f^{3*} - \cdots - \frac{a_{2n}}{a_{12}} e^{n*}$$

and

$$f_2 = a_{12}\, e^{2*} + a_{13}\, e^{3*} + \cdots + a_{1n}\, e^{n*},$$

then $f^{1*}, f^{2*}, e^{3*}, \ldots, e^{n*}$ are linearly independent and

$$q^* = f^{1*} \wedge f^{2*} + q^{1*}.$$

If $q^{1*} = 0$ we are done. Otherwise we continue the process until we achieve the desired normal form.

The integer p can be described solely in terms of q^*. Suppose we consider the linear subspace $E(q^*) \subset V^*$ spanned by those elements of the form

$$x \,\lrcorner\, q^* \tag{5.2}$$

for all $x \in V$. This subspace (called the associated system of q^*) is clearly also spanned by

$$g_i \,\lrcorner\, q^* \tag{5.3}$$

if $g_1, \ldots, g_n$ is any basis of V. If we take $g_i = f_i$ where $\{f_1, \ldots, f_n\}$ is the dual basis of $f^{1^*}, \ldots, f^{n^*}$, then the forms (5.3) are just $f^{1^*}, f^{2^*}, \ldots, f^{2p^*}$. Thus $2p = \dim E(q^*)$. Q.E.D.

The associated system can be defined for a $q^* \in \bigwedge^p(V^*)$ for any $p \geq 1$. A subspace $W^* \subset V^*$ will be said to *envelop* q^* if $q^* \in \bigwedge^p(W^*) \subset \bigwedge^p(V^*)$. We first show that W^* envelops q^* if and only if

$$x \rfloor q^* \in W^* \qquad \text{for all } x \in \textstyle\bigwedge^{p-1}(V). \tag{5.4}$$

To see this choose a basis $e^{1^*}, \ldots, e^{s^*}$ of W^*, extend it to a basis of V^*, and let $e_1, \ldots, e_n$ be the dual basis of V. As before the expression "all $x \in \bigwedge^{p-1}(V)$" in (5.4) can be replaced by all basis elements of $\bigwedge^{p-1}(V)$. If $q^* \in \bigwedge^p(V)$ then

$$q^* = \sum_{i_1 < \cdots < i_p} a_{i_1, \ldots, i_p} e^{i_1^*} \wedge \cdots \wedge e^{i_p^*}.$$

For q^* to lie in $\bigwedge^p(W^*)$ means that $a_{i_1, \ldots, i_p} = 0$ for $i_p > s$. This is the same as (5.4).

Condition (5.4) shows that there is a minimal subspace, $E(q^*)$, enveloping q^*, the space generated by $x \rfloor q^*$ for all $x \in \bigwedge^{p-1}(V)$. The dimension of $E(q^*)$ is called the rank of q^*. If q^* is decomposable, $q^* = y^{1^*} \wedge \cdots \wedge y^{p^*}$, then $E(q^*)$ is spanned by $y^{1^*}, \ldots, y^{p^*}$. Thus

THEOREM 5.2. *A p-form is decomposable if and only if its rank is p.*

EXERCISE 5.1. Show that a 2-form $\sum a_{ij} e^{i^*} \wedge e^{j^*}$ is decomposable if and only if

$$a_{ij} a_{kn} - a_{ik} a_{jn} + a_{in} a_{jk} = 0.$$

EXERCISE 5.2. Show that a 3-form is decomposable if and only if

$$a_{ijk} a_{nem} - a_{ijn} a_{kem} + a_{ije} a_{knm} - a_{ijm} a_{kne} = 0.$$

What is the general theorem?

It should be observed that $(E(q^*))^0$ [the set of vectors on which the forms of $E(q^*)$ vanish] has a simple characterization. In fact, if $x \in (E(q^*))^0$, then

$$x \rfloor q^* = 0. \tag{5.5}$$

On the other hand, the dimension of the solutions of (5.5) cannot exceed

$n - r$ (if r is the rank of q^*) for otherwise, q^* would be expressible in terms of less than r independent variables. Thus

$$(5.6) \qquad x \in (E(q^*))^0 \quad \textit{if and only if} \quad x \rfloor q^* = 0.$$

Let W be a subspace of V. If i denotes the injection mapping of $W \to V$ then the dual map of $V^* \to W^*$ induces a map $\bigwedge(i^*)$ of $\bigwedge(V^*) \to \bigwedge(w^*)$. If $y \in \bigwedge(w^*)$ the form $\bigwedge(i^*)y^* \in \bigwedge(w^*)$ is called the restriction of y^* to W. The subspace W is said to annihilate y^* if $\bigwedge(i^*)y^* = 0$. For linear y^*, this coincides with the usual terminology. The fundamental problem of the theory of exterior equations is: Given a set of forms $\{\Omega_\alpha\}$, find those subspaces which annihilate these forms. Some preliminary observations should be made. First of all, since $\bigwedge(i^*)$ is a homomorphism, if $\bigwedge(i^*)$ sends y^* into zero, it will do the same to any multiple of y^*. Thus, if $\bigwedge(i^*)$ annihilates $\{y_\alpha^*\}$ it will annihilate the ideal generated by $\{y_\alpha^*\}$. This ideal is homogeneous, i.e., is generated by homogeneous elements. Therefore, we can assume that the set $\{y_\alpha^*\}$ forms an ideal. Secondly, suppose that Ω^* is a p-form

$$(5.7) \qquad y^* = \sum a_{i_1,\ldots,i_j} e^{i_1^*} \wedge \cdots \wedge e^{i_p^*}$$

for some choice of basis $\{e_1, \ldots, e_n\}$. If $\Omega \neq 0$, we may well assume that $a_{12,\ldots,p} \neq 0$. If we consider the p-plane spanned by $e_1, \ldots, e_p$, then the restriction of y^* to W is $a_{12,\ldots,p} e^{1^*} \wedge \cdots \wedge e^{p^*} \neq 0$. Thus given a non-zero p-form, there is at least one p-plane which does not annihilate it. To put it another way, if p-form is annihilated by all p-planes it is zero. If $W_1 \subset W \subset V$, the restriction of a form to W_1 can be obtained by first restricting to W and then to W_1. Applying the previous remark we can assert:

Given a p-dimensional subspace W, with $i: W \to V$. A q-form y^* $(q \leqslant p)$ is annihilated by W, if and only if $\bigwedge(i^*)y^*$ is annihilated by every q-dimensional subspace of W.

We can now assert:

THEOREM 5.3. *Let $\mathscr{I}$ be a homogeneous*[3] *ideal of $\bigwedge(V^*)$. A p-dimensional subspace W of V annihilates $\mathscr{I}$, if and only if*

$$(5.8) \qquad \langle w | y^* \rangle = 0$$

[3] Homogeneous means that $\mathscr{I} = \sum \mathscr{I} \cap \bigwedge^p(V)$.

for all $y \in \mathscr{I} \cap \bigwedge^p(V^*)$, *where* w *is a* p*-vector corresponding to* W.

Proof. A p-dimensional subspace annihilates any form whose degree is greater than p, so we have only to look at forms of degree $\leqslant p$. Let us choose a basis $e_1, \ldots, e_n$ so that $w = e_1 \wedge \cdots \wedge e_p$. For any p-form given by (5.7)

$$\begin{aligned} \bigwedge(i^*)y^* &= a_{12,\ldots,p}\, e^{1^*} \wedge \cdots \wedge e^{p^*} \\ &= \langle w \mid y^* \rangle e^{1^*} \wedge \cdots \wedge e^{p^*}. \end{aligned} \tag{5.9}$$

Thus $\bigwedge(i^*)y^* = 0$ if and only if $\langle w \mid y^* \rangle = 0$. We have proved that (5.8) is a necessary condition. We know that (5.8) implies that W annihilates all the p-forms of $\mathscr{I}$. What about the q-forms where $q \leqslant p$? If

$$y^* = \sum_{i_1 < \cdots < i_q} a_{i_1,\ldots,i_q} e^{i_1^*} \wedge \cdots \wedge e^{i_q^*}$$

is such that $\bigwedge(i^*)y^* \neq 0$, then $a_{i_1,\ldots,i_q} \neq 0$ for some set of indices with $i_q \leqslant p$. By rearranging the basis, if necessary, we can assume that $a_{12,\ldots,q} \neq 0$. Then $z^* = y^* \wedge e^{q^*+1} \wedge \cdots \wedge e^{p^*}$ is a p-form with $\langle w \mid z^* \rangle \neq 0$. If $y^* \in \mathscr{I}$ then $z^* \in \mathscr{I}$ since $\mathscr{I}$ is an ideal. Thus if (5.8) hold, $\bigwedge(i^*)y^* = 0$ for any $y^* \in \mathscr{I}$. Q.E.D.

Now (5.8) is a system of linear equations in $\bigwedge^p(V)$. It would appear at first glance, therefore, that we have reduced the problem of solving exterior equations to the simple business of solving linear equations. This is not so, however, because we have to adjoin to (5.8) the conditions expressing that w is decomposable. These conditions are not linear but quadratic in the coordinates of W (cf. Exercises 5.1 and 5.2).

The one case where the problem becomes trivial is where the ideal, $\mathscr{I}$, is generated by its elements of degree one. Then W annihilates $\mathscr{I}$ if and only if $W \subset (\mathscr{I} \cap \bigwedge^1(V^*))^\perp$.

Exercise 5.3. As an illustration of the complications caused by the quadratic conditions for an exterior vector to be decomposable, it turns out that the set of forms annihilated by all subspaces annihilating $\mathscr{I}$, may be larger than $\mathscr{I}$. For example, let $\mathscr{I}$ be generated by

$$e^{1^*} \wedge e^{3^*},\ e^{1^*} \wedge e^{4^*} \quad \text{and} \quad e^{1^*} \wedge e^{2^*} - e^{3^*} \wedge e^{4^*}.$$

Show that $e^{1^*} \wedge e^{2^*}$ is annihilated by every subspace annihilating $\mathscr{I}$, although $e^{1^*} \wedge e^{2^*} \notin \mathscr{I}$.

There are some simplifications which can be made. Let us suppose that $\mathscr{I}$ has a system of generators belonging to $\bigwedge(U^*)$ for some subspace $U^* \subset V^*$. Then $(U^*)^0$ obviously annihilates $\mathscr{I}$. Furthermore, given any subspace W that annihilates $\mathscr{I}$, so will $W + (U^*)^0$. As in the case of forms, we would like to be able to find a minimal subspace U^* so that $\mathscr{I}$ has a set of generators belonging $\bigwedge(U^*)$. We can find such a subspace by the following:

Let $\mathscr{I}$ be a homogeneous ideal of $\bigwedge(V^*)$ and U a subspace of V. Then $\mathscr{I}$ has generators lying in $\bigwedge(U^0)$, if and only if

$$x \lrcorner y^* \subset \mathscr{I} \qquad \textit{for all } y^* \in \mathscr{I} \tag{5.10}$$

for all $x \in U$.

Proof. If the generators $\{y_\alpha^*\}$ of $\mathscr{I}$ lie in $\bigwedge(U^0)$, then

$$x \lrcorner y_\alpha^* = 0$$

for all $x \in U$. Thus

$$x \lrcorner (z^* \wedge y_\alpha^*) = (x \lrcorner z^*) \wedge y_\alpha^* \in \mathscr{I},$$

for an arbitrary form $z^* \in \bigwedge(V^*)$. Since every element of $\mathscr{I}$ is a sum of terms of the form $z^* \wedge y_\alpha^*$, this establishes (5.10).

We prove the converse, step by step on the degree of the generators of $\mathscr{I}$. For $y^* \in \mathscr{I} \cap \bigwedge^1(V^*)$, (5.10) implies $x \lrcorner y^* = 0$ or $y^* \in U^0$. Let $e_1, \ldots, e_p, \ldots, e_n$ be a basis of V where $e_1, \ldots, e_p$ is a basis of U. Then $e^{(p+1)^*}, \ldots, e^{n^*}$ is a basis of U^0. For $y^* \in \mathscr{I} \cap \bigwedge^2(V^*)$,

$$y^* = \sum_{i<j} a_{ij} e^{i^*} \wedge e^{j^*}.$$

Then by (5.10), $e_1^* \wedge (e_1 \lrcorner y^*) \in \mathscr{I}$ and $y_1^* = y^* - e^{1^*} \wedge (e_1 \lrcorner y^*)$ does not involve e^{1^*}. Then $y_2^* = y_1^* - e^{2^*} \wedge (e_2 \lrcorner y_1^*) \in \mathscr{I}$ and does not involve e^{1^*} or e^{2^*}. Proceeding in this way, we can write

$$y^* = y_p^* - w^*$$

where $y_p^* \in \bigwedge(U^0)$ and w^* is a sum of multiples of the generators of degree one. This procedure works inductively. Let us assume that all of $\mathscr{I} \cap \bigwedge^1(V^*), \ldots, \mathscr{I} \cap \bigwedge^{q-1}(V^*)$ is generated by elements of $\bigwedge(U^0)$. For any $y^* \in \mathscr{I} \cap \bigwedge^q(V^*)$, $y^* - e^{1^*} \wedge (e_1 \lrcorner y^*) \in \mathscr{I}$ and does not involve e^{1^*}. Proceeding as before, we obtain a system of generators for $\mathscr{I} \cap \bigwedge^q(V^*)$, lying in $\bigwedge(U^0)$. Q.E.D.

If we are interested in minimizing U^0, we maximize U. Thus we have

THEOREM 5.4. *Let $\mathscr{I}$ be a homogeneous ideal of $\bigwedge(V^*)$. There exists a unique minimal subspace $U^* \subset V^*$, called the adjoint system of $\mathscr{I}$, such that $\mathscr{I}$ has a system of generators lying in (U^*). The space $(U^*)^0$ is given by all $x \in V$ satisfying* (5.10).

It should be remarked that if $\mathscr{I}$ is generated by a single form y^*, then (5.10) reduces to (5.5).

In case the associated system U^0 is not all of V^* (the system is not trivial), it is sometimes convenient to make a reduction by eliminating the "extra variables." Let U' be a subspace of V supplementing to U so that $U + U' = V$, $U \cap U' = \{0\}$. If W annihilates $\mathscr{I}$, so does $W + U$. But $W + U = W' + U$ where $W' = (W + U) \wedge U'$. Thus, to every solution W we can associate a solution $W' \subset U'$. If we restrict attention to maximal solutions (those subspaces annihilating $\mathscr{I}$ contained in no larger subspace annihilating $\mathscr{I}$) then this correspondence is one-one: To every maximal solution, W, there corresponds a unique solution $W' = W \cap U'$ lying in U'. Now a subspace of U' will be a solution if and only if it annihilates $\bigwedge(i^*)$ where i is the injection of $U' \to V$. This is so because i^* is an isomorphism of U^0 onto $(U')^*$ and therefore sends no generator of $\mathscr{I}$ into zero. Thus we can always reduce an exterior system to the case where the associated system is trivial.

The construction of solutions of an exterior system can proceed by an inductive "building up" process. First we look for those lines which are solutions. A line l is a solution if it is a solution of the system of linear equations

$$\langle \cdot \mid y^* \rangle = 0 \qquad y^* \in \bigwedge^1(V^*) \cap \mathscr{I}. \tag{5.11}$$

We will call s_0 the number of linearly independent equations of (5.11), that is, the dimension of $\bigwedge^1(V) \cap \mathscr{I}$. Let l_1, be a solution (5.11) and v_1 some vector on l. A line l_2 will, together with l_1, span a two-dimensional solution if l_2 is a solution of

$$\langle \cdot \mid v_1 \lrcorner\, y^* \rangle = 0 \qquad y^* \in \bigwedge^2(V^*) \cap \mathscr{I}. \tag{5.12}$$

Let $\mathscr{I}$ be generated by $\{y^*_{j\alpha}\}$ where $y^*_{j\alpha} \in \bigwedge^j(V^*)$. Then the Equation (5.12) is equivalent to the system (5.11) together with the new system of equations

$$\langle \cdot \mid v_1 \lrcorner\, y^*_{2\alpha} \rangle = 0. \tag{5.13}$$

Thus a line, l_2, will span a solution space with l_1, if it satisfies the conditions (5.11) plus the set of conditions (5.13) which depends on the choice of l_1. Thus the dimension of the system (5.12) can be written as $s_0 + s_1(l_1)$. Now the number of linearly independent equations of the system (5.12) is given by the rank of the appropriate matrix whose entries depend (linearly) on the coefficients of l_1. Thus if $s_1 = \max_1 s_1(l_1)$, then the set of l_1 with $s_1(l_1) < s_1$ is given by the zeros of non-vanishing polynomials. Those lines l_1 for which $s_1(l_1) = s_1$ are called *regular*; if $s_1(l_1) < s_1$ then l_1 is called *singular*. If $s_0 + s_1(l_1) = n - 1$ then the only solution of (5.12) is l_1 itself so that there are no 2-dimensional solutions containing l_1. If $s_0 + s_1(l_1) < n - 1$ there will exist a two-dimensional solution containing l_1. We can continue this process. If W is a p-dimensional solution with p-vector $W = w_1 \wedge \cdots \wedge w_p$, then a line l_{1+1} will span a solution with W if l_{p+1} is a solution of

$$\langle \cdot, w \lrcorner y^* \rangle = 0 \qquad \text{for } y^* \in \Lambda^{p+1}(V^*) \cap \mathscr{I}. \tag{5.14}$$

This system of equations is called the *polar system of* w. The system (5.14) can be obtained by adjoining the previous equations used in defining W, the system given by the equations

$$\begin{gathered} \langle \cdot\, w_p \,|\, y^*_{2\alpha} \rangle,\ \langle \cdot \,|\, w_{i_1} \wedge w_p \lrcorner y^*_{3\alpha} \rangle, \cdots \\ \langle \cdot \,|\, w_1 \wedge \cdots \wedge w_p \lrcorner y^*_{(p+1)\alpha} \rangle \quad \text{where } 0 < i_1 < \cdots < p. \end{gathered} \tag{5.15}$$

Again, the number of the Equations (5.15), which are independent of the previous set is some function $s_p(w)$. We call $s_p = \max s_p(w)$ and the p-plane W regular if it contains a regular $p - 1$ plane and $s_p(w) = s_p$. Since W is a solution of the system (5.8)

$$s_0 + s_1 + \cdots + s_p \leqslant n - p. \tag{5.16}$$

If equality holds in (5.16), then there is no $p + 1$ solution plane containing a regular p-plane. Eventually we will hit an integer p for which equality holds. It is called the *genus* of the system.

CHAPTER II

Differentiable Manifolds

Differential geometry is the study of those geometrical problems that can be treated using analysis. Therefore, the underlying objects of study should be spaces on which such notions as differentiation and integration have a sense. In this chapter, we will introduce such a class of spaces, called *differentiable manifolds*, and study their most elementary properties. Before making the formal definitions, it may be instructive to look at the difficulties in an elementary example. Suppose we are interested in studying the circle; a reasonable question to ask is how to describe (i.e., name or parameterize) the individual points on the circle. If the circle is given as the locus

$$x^2 + y^2 = 1 \tag{1}$$

in the x, y plane, then each point could be described by giving its coordinates x and y. This description is unsatisfactory, however, for two reasons. First of all, giving both x and y is somewhat redundant. If x is given, then (1) can be solved to yield (at most) two possibilities for y. Furthermore, as long as we stay away from the bad points $x = \pm 1$, choosing a value of y_0 at x_0, (i.e., choosing between $\pm\sqrt{1 - x_0^2}$) uniquely determines y in a neighborhood of x_0 by continuity. In short, the circle is a "one-dimensional" object and should not require two parameters. A second objection is that these coordinates depend on the position of the circle in the plane. If we are interested in studying the circle in its own right, and not as a locus of points in the plane, we should be able to find an invariant description of it. A moment's reflection will show that there is really no way out of the first of these difficulties. No single one parameter can decently describe the whole circle. Perhaps the best attempt would be to use the angle from a fixed direction. But here at zero angle things go bad: The function assigning to each point its angle is not continuous at zero. The way out of the difficulty is to give up the idea of having a single parameter describe the whole circle but be satisfied with having several parametrizations for different pieces. We do want these parameters to be consistent with the topology of the circle (defined, say, as the quotient space of the real numbers modulo the integers). The topology tells us when functions on open subsets of the circle are continuous. The given quotient space representation of the circle allows us also to decide when a function is differentiable. In fact, in any proper open subset of the circle we *can* use a real (angular) parameter. A function f will be called differentiable on this subset if it is differentiable in terms of this parameter. It is easy to see that this definition is independent of the (additive constant involved in the) choice of angular parameter. Therefore, we can distinguish a class of functions by smoothness properties; for example, we can talk about the class of functions with continuous derivatives of all order. On an open sub-interval of the circle, we could use any C^∞ function defined there as parameter, providing that its derivative with respect to the angular parameter does not vanish. Thus the circle is covered by intervals, each having a parameter or coordinate. When two intervals overlap, the change of coordinates is given by a C^∞ function with non-vanishing derivative. Thus the abstract circle, S^1, is a topological space together with a class of functions and coordinate intervals with coordinates interrelated properly. We view the unit circle, (1), of the x, y plane as S^1, together with a map of S^1 into the x, y plane given by $x(\cdot) = \cos(\cdot)$, $y(\cdot) = \sin(\cdot)$. We now proceed to make our definitions.

1. Definitions. The topological spaces we wish to study are those which look, locally, like Euclidean space.

DEFINITION 1.1. *An n-manifold M is a Hausdorff space with a countable basis which is locally homeomorphic to n-dimensional Euclidean space,* R^n.

By locally homeomorphic, we mean that every point $x \in M$ has a neighborhood which is homeomorphic to R^n (or an open subset of R^n). We now adjoin to the topology of M a distinguished class of functions.

DEFINITION 1.2. *A differentiable structure of class C^k (analytic) on an n-manifold M is a collection, $\mathscr{F}$, of real valued functions, each defined on an open subset of M such that*

1) *if $U \subset V$ and $f \in \mathscr{F}$ is defined on V then this restriction of f to $U, f \mid U \in \mathscr{F}$;*
2) *if $U = \bigcup U_\alpha$, and f is defined on U with $f \mid U_\alpha \in \mathscr{F}$ for all α, then $f \in \mathscr{F}$;*
3) *every point $p \in M$ has a neighborhood U and a homeomorphism of U onto an open subset of R^n such that a function f defined on a subset of U lies in $\mathscr{F}$, if and only if, $f \circ h^{-1}$ is a function of class C^k (analytic), as a function of n real variables.*

The functions of $\mathscr{F}$ will be called functions of class C^k (or analytic) on the manifold. For the rest of this chapter we will use the word "differentiable" to mean "of class C^k" or "analytic" according to the case considered, in those situations where the argument is the same for all differentiability conditions. The first two conditions say that piecing together or restricting functions of $\mathscr{F}$ is an admissible operation. The third condition says that $f \in \mathscr{F}$ if it is differentiable in some "coordinate system." Let us make this notion more precise. Suppose the homeomorphism h of condition 3) is given by the n real functions $x^1(\cdot), \ldots, x^n(\cdot)$ defined on U.[1] If f is a function defined on $U, f \circ h^{-1}$ can be written as

$$f \circ h^{-1} = f(x^1, \ldots, x^n). \tag{1.1}$$

Condition 3) says that $f \in \mathscr{F}$ if and only if the right-hand side of (1.1) is a differentiable function of its arguments. In particular, the functions $x_i(\cdot) \in \mathscr{F}$. Let U' be some other neighborhood of p and h' a homeomorphism of it onto R^n. Let $V = U \cap U'$ and denote the restrictions of h and h' to V by the same letters (a slight abuse of language). Then $h' \circ h^{-1}$ is a homeomorphism of $h(V)$ onto $h'(V)$ and is given by

$$x'^i = x'^i(x^1, \ldots, x^n) \tag{1.2}$$

where the functions in (1.2) are differentiable. For $f \in \mathscr{F}$, we can write

$$f \circ h'^{-1} = f(x^{1\prime}, \ldots, x^{n\prime}). \tag{1.3}$$

---

[1] Here $x^1, \ldots, x^n$ are the standard n coordinate functions on R^n. Thus for any $p \in U$, $h(p) = x^1\ (h(p)), \ldots, x^n(h(p))$. Therefore, we should write $x^1(h(\cdot))$ instead of $x^1(\cdot)$. We prefer to omit the h when there is no possibility of confusion.

Composing (1.3) and (1.2) we have

$$(1.4)\qquad f(x^1,\ldots,x^n) = f'(x^{1'}(x^1,\ldots,x^n),\ldots,x^{n'}(x^1,\ldots,x^n)).$$

Formula (1.4) gives the relation between the "representations" of f in terms of two systems of coordinates. In particular, we can express the functions $x^i \circ h'^{-1}$ as

$$(1.5)\qquad x^i \circ h'^{-1} = x^i(x^{1'},\ldots,x^{n'})$$

and substitute into (1.4) to obtain the identity

$$(1.6)\qquad x^i = x^i(x^{1'}(x^1,\ldots,x^n),\ \ldots,x^{n'}(x^1,\ldots,x^n)).$$

If we take the Jacobian matrix of both sides of (1.5) we obtain

$$(1.7)\qquad \left(\frac{\partial x^i}{\partial x^{j'}}\right)\left(\frac{\partial x^{j'}}{\partial x^k}\right) = (\delta^i_k).$$

Thus

$$(1.8)\qquad \left|\left(\frac{\partial x^{i'}}{\partial x^j}\right)\right| \neq 0.$$

On the other hand, suppose there is given a set of differentiable functions, (1.2), which satisfy (1.8) at the point $x^1(p),\ldots,x^n(p)$. The implicit function theorem (cf. Appendix 1) then guarantees the existence of n differentiable functions (1.5) which satisfy (1.6), in some neighborhood, W, of $x^1(p),\ldots,x^n(p)$. If we compare these functions with h, we obtain a homeomorphism h' of $W = h^{-1}(W)$ onto an open subset of R^n; and h' has the properties required in condition 3) of Definition 1.2.

Definition 1.3. *An open set W and a homeomorphism h having the properties* 3) *of Definition* 1.2 (*relative to the differentiable structure $\mathcal{F}$*) *are called coordinate neighborhood and coordinate map respectively. If $h(p) = (x^1(p),\ldots,x^n(p))$, the functions $x^i(\cdot)$ are called local coordinates.*

According to Definition 1.2, every point has a coordinate neighborhood, U, with a coordinate map h. The sets $\{U\}$ form an open covering of M. By the countability assumption we can choose a countable subcovering $\{U_i\}$ with coordinate maps $\{h_i\}$. By the previous discussion (choosing $p \in U_i \cap U_j$ and $V = U_i \cap U_j$) the mapping $h_i \circ h_j^{-1}$ is given by n differentiable functions whose Jacobian does not vanish.

We can use the coordinates to define the differentiable structure. Suppose we are given a countable open covering $\{U_j\}$ of M and a homeomorphism, h_j, of each U_j into Euclidean n-space such that $h_i \circ h_j^{-1}$ is given by n differentiable functions with non-vanishing Jacobian. We can then consider the collection, $\mathscr{G}$, of all functions, f, defined on open subsets of any U_j and such that $f \circ h_j^{-1}$ is differentiable. If f is defined on a subset of $U_i \cap U_j$ and $f \circ h_j^{-1}$ is differentiable then so is

$$f \circ h_i^{-1} = f \circ h_j^{-1} \circ h_j \circ h_i^{-1},$$

since $h_j \circ h_i^{-1}$ is a differentiable transformation. The collection $\mathscr{G}$ satisfies condition 3) of Definition 1.2. Any differentiable structure containing $\mathscr{G}$, must, by condition 1) contain every function f whose restriction to all U_j belongs to G. Let $\mathscr{F}$ be the collection of all such functions. Then it is easy to check that $\mathscr{F}$ is a differentiable structure. It is the minimal differentiable structure containing all of $\mathscr{G}$. We can summarize this discussion in

THEOREM 1.1. *Let M be an n-manifold and $\mathscr{F}$ a differentiable structure on M. Then there is countable open covering $\{U_j\}$ of M and a set of homeomorphisms h_j of U_j onto an open subset of Euclidean space such that $h_i \circ h_j^{-1}$ is given by n differentiable functions with non-vanishing Jacobian. Conversely, given such an open covering and maps h_j there exists a unique minimal differentiable structure $\mathscr{F}$ for which the U_j are coordinate neighborhoods and the h_j are coordinate maps.*

DEFINITION 1.4. *A manifold M together with a differentiable structure $\mathscr{F}$ is called a differentiable manifold.*

When there is no risk of confusion we will omit the $\mathscr{F}$ and write the differentiable manifold as M.

DEFINITION 1.5. *Let U be an open subset of a manifold M and let h be a 1-1 map of U into E^n. The pair (U, h) is called a coordinate chart if $h_i \circ h^{-1}$ is a differentiable map of $h(U \cap U_i)$ onto $h_i(U \cap U_i)$ whose Jacobian does not vanish anywhere.*

(Here the h_i and U_i are as in Theorem 1.1.)

It is clear that the condition that (U, h) be a chart is independent of the (U_i, h_i) and depends only on the differentiable structure of M.

The functions $x^i \circ h$ [where $(x^1, \ldots, x^n)$ are coordinates on R^n] are called the *coordinates* of the chart (U, h). When there is no risk of confusion we will simply write these functions as x^i, and write $h = (x^1, \ldots, x^n)$ to indicate that $h(p) = (x^1(p), \ldots, x^n(p)) \in R^n$ for all $p \in U$.

DEFINITION 1.6. *A collection of charts (U_i, h_i) such that $M = \bigcup U_i$ is called an atlas of M.*

EXERCISE 1.1. Let $\mathscr{F}^k$ be a differentiable structure of class C^k on a manifold M. Show that there is a unique minimal, $\mathscr{F}^l$, structure of class C^l such that $\mathscr{F}^l \supset \mathscr{F}^k$, if $l < k$. (The converse of this statement is true: Given a structure, $\mathscr{F}^l$, of class C^l, there exists a structure of class $\mathscr{F}^k$ such that $\mathscr{F}^l$ is the unique minimal structure containing $\mathscr{F}^k$, if $l < k$. We can even take $k = \infty$ in this statement. This result is due to Whitney [22]. It depends on an embedding theorem part of which we will prove later on, cf. Theorem 4.1.)

If M_1 and M_2 are manifolds of dimension n_1 and n_2 respectively, then $M_1 \times M_2$ is a manifold. In fact, if $U_1 \subset M_1$, and $U_2 \subset M_2$ are homeomorphic to open subsets of E^{n_1} and E^{n_2}, then $U_1 \times U_2$ is homeomorphic to an open subset of $E^{n_1+n_2}$. Suppose that $\mathscr{F}_1$ and $\mathscr{F}_2$ are differentiable structures on M_1 and M_2. Let $\{U_{1j}\}$ and $\{U_{2j}\}$ be coverings of M_1 and M_2 by coordinate neighborhoods with coordinate maps h_{1j} and h_{2j}. Then the sets $\{U_{1i} \times U_{2j}\}$ form an open covering of $M_1 \times M_2$. The map h_{ij} of $U_{1i} \times U_{2j}$ into $E^{n_1+n_2}$ defined by

$$h_{ij}(p, q) = h_{1i}(p) \times h_{2j}(q)$$

is a homeomorphism. Furthermore, $h_{ij} \circ h_{kl}^{-1} = h_{1i} \circ h_{1k}^{-1} \times h_{2j} \circ h_{2l}^{-1}$ is obviously a differentiable map with non-vanishing Jacobian. Thus the coordinate system $\{U_{1i} \times U_{2j}\}$ and maps h_{ij} define a differentiable structure on $M_1 \times M_2$. This structure is called the product of the structures $\mathscr{F}_1$ and $\mathscr{F}_2$. The manifold $M_1 \times M_2$ with this differentiable structure is called the *product of the differentiable manifolds M_1 and M_2.*

EXAMPLES OF DIFFERENTIABLE MANIFOLDS

1. The Euclidean n-space R^n (or any open subset thereof) is obviously a differentiable manifold. It is covered by a single coordinate neighborhood.

2. Let S^n (the n-dimensional sphere) denote the set of all points in R^{n+1} satisfying

$$(x^1)^2 + \cdots + (x^{n+1})^2 = 1.$$

Let U_i^+ denote the subset of S^n where $x_i > 0$ and U_i^- the subset where $x_i < 0$. We then take as coordinates on $U_i^\pm$ the remaining x's. It is easy to check that the U's cover S^n and the coordinate maps satisfy the required condition.

3. Let $T^n = S^1 \times S^1 \times \cdots \times S^1$ (n times). It is called the *n-dimensional torus*.

4. Let $T(m, n)$ be the space of all $m \times n$ matrices. Then $T(m, n)$ can be regarded as E^{mn} and therefore is an analytic manifold. Let $T(m, n; k)$ denote the space of all $m \times n$ matrices of rank k (where $0 < k \leqslant \min(m, n)$), with the induced topology of $T(m, n)$. Then $T(m, n; k)$ is an analytic manifold of dimension $k(m + n - k)$. In fact, let $X_0 \in T(m, n)$. If rank $X_0 \geqslant k$, there are permutation matrices P and Q such that

$$PX_0Q = \begin{pmatrix} A_0 & B_0 \\ C_0 & D_0 \end{pmatrix}$$

where A_0 is a non-singular $k \times k$ matrix. There is an $\epsilon > 0$ such that if all the entries of $A - A_0$ are less than ϵ, then A is non-singular. Let U^* denote the set of those $X \in T(m, n)$ such that

$$PXQ = \begin{pmatrix} A & B \\ C & D \end{pmatrix} \tag{1.9}$$

where all the entries of $A - A_0$ are $< \epsilon$ in absolute value. Then U^* is an open set of $T(m, n)$. If $X \in U^*$ then $X \in T(m, n; k)$ if and only if $D = CA^{-1}B$. To see this, observe that the matrix

$$\begin{pmatrix} I_k & 0 \\ -CA^{-1} & I_{m-k} \end{pmatrix}\begin{pmatrix} A & B \\ C & D \end{pmatrix} = \begin{pmatrix} A & B \\ 0 & -CA^{-1}B + D \end{pmatrix} \tag{1.10}$$

(where I_j is the $j \times j$ identity matrix) has the same rank as

$$\begin{pmatrix} A & B \\ C & D \end{pmatrix} \quad \text{since} \quad \begin{pmatrix} I_k & 0 \\ -CA^{-1} & I_{m-k} \end{pmatrix}$$

is non-singular; but the matrix on the right of (1.10) has rank k if and only if $D = CA^{-1}B$. We will take as our coordinate neighborhood of

$X_0 \in T(m, n; k)$, the set $U = U^* \cap T(m, n; k)$, which is open in the induced topology. We shall define the coordinate map h by

$$h(X) = \begin{pmatrix} A & B \\ C & 0 \end{pmatrix} \tag{1.11}$$

where we view the Euclidean space $E^{(k^2 + k(m-k) + k(n-k))} = E^{k(m+n-k)}$ as the space of all matrices of the form

$$\begin{pmatrix} A & B \\ C & 0 \end{pmatrix}.$$

The mapping h^{-1} is given by

$$h^{-1}\begin{pmatrix} A & B \\ C & 0 \end{pmatrix} = P^{-1}\begin{pmatrix} A & B \\ C & CA^{-1}B \end{pmatrix} Q^{-1}. \tag{1.12}$$

The transition function $h' \circ h^{-1}$ for two overlapping neighborhoods U and U' is thus given by

$$h' \circ h^{-1}\begin{pmatrix} A & B \\ C & 0 \end{pmatrix} = \begin{pmatrix} A' & B' \\ C' & 0 \end{pmatrix} \tag{1.13}$$

where

$$P'P^{-1}\begin{pmatrix} A & B \\ C & CA^{-1}B \end{pmatrix} Q^{-1}Q' = \begin{pmatrix} A' & B' \\ C' & D' \end{pmatrix}. \tag{1.14}$$

As all the matrix entries in (1.13) are analytic (indeed rational) functions of their arguments (since A is non-singular) the coordinate neighborhoods and maps define an analytic structure on $T(m, n; k)$ of dimension $k(m + n - k)$.

5. Let $\mathscr{F}(m, V)$ denote the set of all ordered m-tuplets of linearly independent vectors in an n-dimensional real vector space V. ($\mathscr{F}(m, V)$ is called the *space of m-frames of V*). Then $\mathscr{F}(m, V)$ is a differentiable manifold of dimension mn. In fact, relative to a choice of basis of V, $\mathscr{F}(m, V)$ can be identified with $T(m, n; m)$.

6. Let V be an n-dimensional vector space and let $\mathscr{G}(p, V)$ denote the set of p-dimensional subspaces of V. We claim that $\mathscr{G}(p, V)$ can be made into an analytic manifold (the Grassmann manifold of V) of dimension

$p(n-p)$. In fact, let $v^{1^*}, \ldots, v^{n^*}$ be a basis of V^*. If $v^* \in V^*$ and $E \in \mathscr{G}(p, V)$ we will denote the restriction of v^* to E by v^*_E. Denote by $U^{i_1, \ldots, i_p}$ $(1 \leqslant i_1 < i_2 < \cdots < i_p \leqslant n)$ the set of those p-planes, E, on which $v_E^{i_1*}, \ldots, v_E^{i_p*}$ are linearly independent. Let $(j_1, \ldots, j_{n-p})$ be the complementary set of indices to $(i_1, \ldots, i_p)$ with $j_1 < \cdots < j_{n-p}$. Then

$$v_E^{j_k*} = \sum_1^p h_l^k \, v_E^{i_l*}$$

for $E \in U^{i_1, \ldots, i_p}$. Let $\Phi^{i_1, \ldots, i_p}$ be the map of $U^{i_1, \ldots, i_p} \rightarrow E^{p(n-p)}$ assigning to each E, the coefficients (h_l^k). The $\Phi^{i_1, \ldots, i_p}$ are one to one. They define a manifold structure on $\mathscr{G}(p, V)$ by pulling back the open sets of $E^{p(n-p)}$. They also define an analytic structure in $\mathscr{G}(p, V)$. The proof of these facts will be left as an exercise to the reader.

2. Differentiable maps. Let φ be a map of $M_1 \rightarrow M_2$. For any function f defined on a subset $W \subset M_2$ we define the function $\varphi^*(f)$ on $\varphi^{-1}(W)$ by

$$\varphi^*(f)(\cdot) = f \circ \varphi(\cdot). \tag{2.1}$$

Thus φ^* maps the space of functions defined on subsets of M_2 into the space of functions on subsets of M_1.

Definition 2.1. *Let $\mathscr{F}_1$ and $\mathscr{F}_2$ be differentiable structures of class C^k (analytic) defined on the manifolds M_1 and M_2. A continuous map, $\varphi: M_1 \rightarrow M_2$, is said to be differentiable of class C^k (analytic) if*

$$\varphi^*(\mathscr{F}_2) \subset \mathscr{F}_1. \tag{2.2}$$

Let $\psi: M_2 \rightarrow M_3$. Then it follows from (2.1) that $(\psi \circ \varphi)^* = \varphi^* \circ \psi^*$. Therefore we can assert:

If $\varphi: M_1 \rightarrow M_2$, and $\psi: M_2 \rightarrow M_3$ *are differentiable, so is* $\psi \circ \varphi: M_1 \rightarrow M_3$.

Let n_i be the dimension of M_i. Let (U, h) be a coordinate chart about $\varphi(p)$, where $p \in M_1$. Its coordinates are given by n_2 differentiable functions, $y_i(\cdot)$, on U. If φ is differentiable, the $y_i \circ \varphi$ are differentiable functions on M_1. In terms of (V, g), $g = (x^1, \ldots, x^{n_1})$, about p, φ (or rather $h \circ \varphi \circ g^{-1}$) is given by

$$y^i = \varphi^i(x^1, \ldots, x^{n_1}) \qquad (i = 1, \ldots, n_2), \tag{2.3}$$

where the φ^i are differentiable functions. Conversely, suppose we are given a continuous map, φ, which on all pairs of charts is given by (2.3). A function f on M_2 is differentiable if and only if its restriction to every coordinate neighborhood is differentiable in terms of the y's. But in terms of the chart (V, g),

$$(\varphi^* f) \circ g^{-1}(x^1, \ldots, x^{n_1}) = f(y^1(x^1, \ldots, x^{n_1}), \ldots, y^{n_2}(x^1, \ldots, x^{n_1})). \tag{2.4}$$

Thus $\varphi^* f$ is differentiable in terms of every chart. The map φ is therefore differentiable.

The form (2.3) of $h \circ \varphi \circ g^{-1}$, depends on the choice of local charts M_2 and g in M_1. To obtain the form of φ relative to any other pair of charts involves effecting a non-singular transformation on the variables y and x. Therefore, we are interested in those expressions derived from (2.3) which are invariant under such transformations. One such object is the rank of the $n_1 \times n_2$ Jacobian matrix

$$\left(\frac{\partial \varphi^i}{\partial x^j}\right)_{(x^1, \ldots, x^{n_1})}. \tag{2.5}$$

Suppose the matrix (2.5) has rank k at a point $p \in M_1$. We can, by renumbering the coordinates if necessary, assume that

$$\left|\left(\frac{\partial \varphi^\alpha}{\partial x^\beta}\right)\right| \neq 0 \qquad (\alpha, \beta = 1, \ldots, k) \tag{2.6}$$

holds at p. Consider the functions u^i defined on a neighborhood of p by

$$\begin{aligned} u^\alpha(x^1, \ldots, x^n) &= \varphi^\alpha(x^1, \ldots, x^n), \qquad (\alpha = 1, \ldots, k). \\ u^r &= x^r, \qquad (r = k + 1, \ldots, n_1). \end{aligned} \tag{2.7}$$

By (2.6), the Jacobian determinant $(\partial u^i/\partial x^j)$ does not vanish at p. Therefore (2.7) defines a local coordinate system at p. In terms of this chart the mapping φ is given by $(\psi^1, \ldots, \psi^{n_2})$ where

$$\psi^\alpha(u^1, \ldots, u^n) = u^\alpha \qquad (\alpha = 1, \ldots, k). \tag{2.8}$$

Now if the rank of φ is k at p, it must be $\geqslant k$ in some neighborhood of p.

If it actually is equal to k in some neighborhood of p, we can say somewhat more. Equation (2.8) implies that $\partial\psi^r/\partial u^s \equiv 0$ for $r, s > k$. In fact, the Jacobian matrix has the form,

$$
\begin{array}{l}
k\left\{\begin{array}{c} \\ \\ \\ \\ \\ \end{array}\right.
\end{array}
\left(\begin{array}{cccccccc}
\multicolumn{2}{c}{\overbrace{\hphantom{1 \quad 0 \quad 0}}^{k}} & & & & & & \\
1 & 0 & \text{---} & \text{---} & \text{---} & \text{---} & \text{---} & 0\\
0 & 1 & 0 & \text{---} & \text{---} & \text{---} & \text{---} & 0\\
\cdot & 0 & & 0 & \text{---} & \text{---} & \text{---} & 0\\
\cdot & \cdot & \cdot & & & & & \\
0 & \cdot & \cdot & 0 & 1 & 0 & \text{---} & 0\\
 & & & & & \frac{\partial\psi^{k+1}}{\partial u^{k+1}} & & \vdots\\
 & & & & & \vdots & & \frac{\partial\psi^{n_2}}{\partial u^{n_1}}
\end{array}\right).
$$

If any term in the lower right-hand corner did not vanish, then the rank would be $>k$. We can thus write $\psi^r = \psi^r(u^1, \ldots, u^k)$. Suppose we introduce the function $z^1, \ldots, z^{n_2}$ defined in some neighborhood of $\varphi(p)$ by

$$
\begin{aligned}
&z^\alpha = y^\alpha && (\alpha = 1, \ldots, k)\\
(2.9)\qquad &z^r = y^r - \psi^r(y^1, \ldots, y^k) && (r = k + 1, \ldots, n_2).
\end{aligned}
$$

An immediate verification shows that (2.9) is a change of coordinates at $\varphi(p)$ and in terms of the new coordinates, φ is given by $z^i = \eta^i(u^1, \ldots, u^{n_1})$ where

$$
\begin{aligned}
&\eta^\alpha(u^1, \ldots, u^{n_1}) = u^\alpha && (\alpha = 1, \ldots, k)\\
(2.10)\qquad &\eta^r(u^1, \ldots, u^{n_1}) = 0 && (r > k).
\end{aligned}
$$

Thus *if the rank of a differentiable map* φ *is* k *at* p *we can find* (V, g) *with* $g = (x^1, \ldots, x^{n_1})$ *at* p *and* (u, h) *with* $h = (y^1, \ldots, y^{n_2})$ *at* $\varphi(p)$ *so that* $y^i \circ \varphi = (\varphi^1, \ldots, \varphi^{n_2})$ *where*

$$
(2.11)\qquad \varphi^\alpha(x^1, \ldots, x^{n_1}) = x^\alpha \qquad (\alpha = 1, \ldots, k).
$$

If the rank of φ *is* k *in some neighborhood of* p *we can further arrange that*

$$
(2.12)\qquad \varphi^r(x^1, \ldots, x^{n_1}) = 0 \qquad \textit{for } r > k.
$$

There are various kinds of maps which are of particular importance to us.

DEFINITION 2.2. *A differentiable map* φ *of* $M^1 \to M^2$ *is called an immersion if the rank of* φ = *dimension of* M^1 *at all points of* M^1.

Note that an immersion is locally one to one, but not necessarily globally.

DEFINITION 2.3. *An immersion* φ *is called an imbedding if it is a homeomorphism onto its image (in the induced topology).*

Thus an imbedding is a one to one immersion, but not every one to one immersion is an imbedding. In Figure 2.1, $M_1 = E^1$ = real line and $M_2 = E^2$ = the plane. Then (a) is not an immersion (because of the

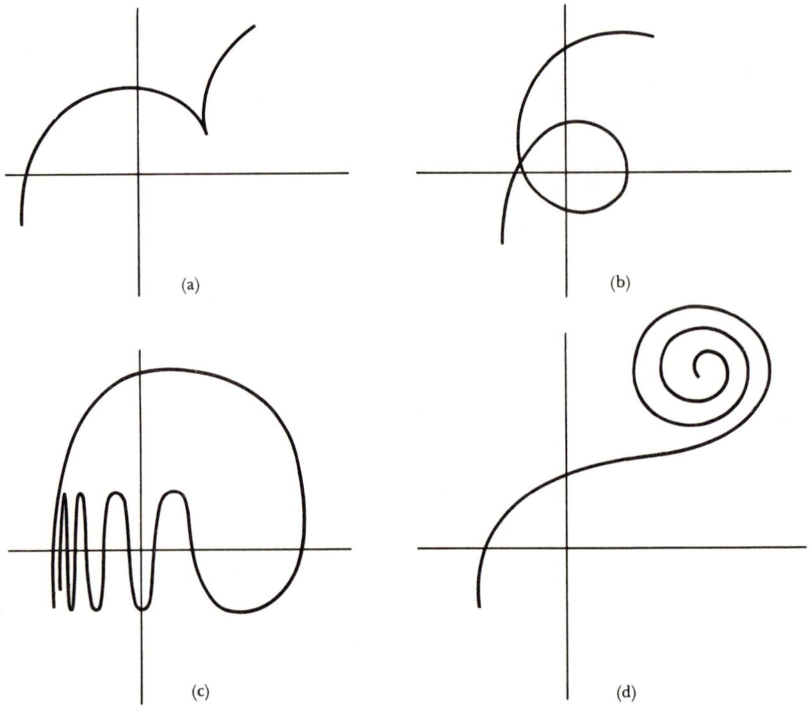

FIGURE 2.1

cusp), (b) is an immersion which is not one to one, (c) is a one to one immersion that is not an imbedding, and (d) is an imbedding.

Definition 2.4. *A map φ is called proper if $\varphi^{-1}(K)$ is compact for any compact subset, K, of M_2. An imbedding which is proper is called a proper imbedding.*

Thus none of the maps pictured in Figure 2.1 is proper. If M_1 is compact, then every differentiable map is proper.

Exercise 2.1. Show that a proper one to one immersion is a proper imbedding.

Corresponding to the various types of maps we introduce various notions of "submanifold."

Definition 2.5. *Let M be a differentiable manifold and N a subset of M which is the underlying set of a differentiable manifold, N. Let ι denote the injection map of $N \to M$. (The identity map of M restricted to N.) Then N is called an immersed submanifold if ι is an immersion; N is called a submanifold if ι is an imbedding; N is called a closed submanifold if ι is a proper imbedding.*

Notice, that in the case of an immersed submanifold, the induced topology on N does not necessarily coincide with the topology of N as a differentiable manifold.

Definition 2.6. *A differentiable map of M_1 onto M_2 is called a diffeomorphism if it and its inverse are imbeddings. Two manifolds M_1 and M_2 are said to be diffeomorphic if there exists a diffeomorphism of M_1 onto M_2.*

Exercise 2.2. Let M_1 be the real line with the differentiable structure consisting of all functions of the form $f(x^3)$ where f is C^∞. Let M_2 be the real line with the usual C^∞ structure. Show that the identity map of $R \to R$ is a C^∞ homeomorphism of $M_2 \to M_1$ which is not a diffeomorphism.

Exercise 2.3. Let N be an immersed submanifold of a differentiable manifold M. Show that every point of N has a neighborhood U such that the set of differentiable functions on U of N coincide with the set of restrictions to U of differentiable functions of M. Thus the differentiable structure of an immersed submanifold is determined by that of M.

EXERCISE 2.4. Let $\bar{N}$ be a subset of M. Show that $\bar{N}$ is the underlying set of a submanifold, if and only if every point p of $\bar{N}$ has a neighborhood U in M with coordinates $x^1, \ldots, x^n$ such that $\bar{N} \cap U$ coincides with the set of points where $x^p = x^{p+1} = \cdots, x^n = 0$.

EXERCISE 2.5. Let $N \subset M$. Show that a necessary and sufficient condition that N be a closed submanifold of M is that there exist a covering of M by coordinate neighborhoods $\{U_\alpha\}$ such that $p \in N \cap U_\alpha$ if and only if $x_\alpha^1(p) = \cdots = x_\alpha^k(p) = 0$, where $x_\alpha^1, \ldots, x_\alpha^n$ are local coordinates in U_α.

EXERCISE 2.6. Show that $T(m, n; k)$ is a submanifold of $T(m, n)$, [where the $T(m, n; k)$ are defined in Exercise 4 of Section 1].

EXERCISE 2.7. Show that the map $\pi: \mathscr{F}(m, V) \to \mathscr{G}(m, V)$ which assigns to $(v_1, \ldots, v_n)$ the subspace spanned by $v_1, \ldots, v_n$ is differentiable.

EXERCISE 2.8. Show that the map of $T(m, n) \times T(n, p) \to T(m, p)$ given by matrix multiplication is differentiable.

A useful result in the construction of differentiable manifolds is the following:

THEOREM 2.1. *Let M be a set, $\{U_i\}$ a denumerable covering of M, and $\{\varphi_i\}$ a collection of one to one maps of U_i onto differentiable manifolds M_i such that*

1) *For every pair of points $p, q \in M$ either p and q are both contained in some U_i or there are sets U_i and U_j with $U_i \cap U_j = \varphi, p \in U_i$, and $q \in U_j$;*
2) *$\varphi_i(U_i \cap U_j)$ is an open subset of M_i for all i and j;*
3) *$\varphi_j \circ \varphi_i^{-1}$ is a diffeomorphism of $\varphi_i(U_i \cap U_j)$ into M_j, for all i and j.*

Then there is a unique (topology and) differentiable structure on M such that the U_i are open and the φ_i are diffeomorphisms of U_i onto M_i.

Proof. A basis for the topology of M are the sets of the form $\varphi_i^{-1}(V)$ where V is an open subset of M_i. Conditions 2) and 3) guarantee that the sets of this form do indeed constitute a basis for a topology. Since the topology of the M_i has a countable bases so does the topology of M. By condition 1), this topology is Hausdorff. In fact if $p \neq q \in U_i$ then $\varphi_i(p)$ and $\varphi_i(q)$ have disjoint neighborhoods V_p and V_q; thus $\varphi_i^{-1}(V_p)$ and $\varphi_i^{-1}(V_q)$ are disjoint neighborhoods of p and q. If $p \in U_i$, $q \in U_j$ with

$U_i \cap U_j = \varphi$ then U_i and U_j are disjoint neighborhoods of p and q. If $p \in U_i$, choose a coordinate neighborhood, V, of $\varphi_i(p)$. Then $\varphi_i^{-1}(V)$ is a neighborhood of p homeomorphic to an open subset of E^n. Thus M is a manifold. A function f on M is called differentiable if $f \circ \varphi_i^{-1}$ is a differentiable function on M_i for all i. The class $\mathscr{F}$ of all such functions defines a differentiable structure [again in virtue of 2) and 3)]. The maps φ_i are differentiable: If f is a differentiable function on M_i, $f \circ \varphi_i \circ \varphi_j^{-1}$ is a differentiable function of M_j by 2). The map φ_i^{-1} is differentiable by definition. The uniqueness of the topology and differentiable structure is clear from the previous discussion.

3. Sard's theorem. Given a smooth map, φ, of $M^{n_1} \to M^{n_2}$, the maximum rank that φ can have is the minimum of n_1 and n_2. If $n_1 < n_2$, we would expect the image of M^{n_1} to be a "lower dimensional" object in M^{n_2}; "in general," a point of M^{n_2} will not lie on $\varphi(M^{n_1})$. It turns out that no matter what values n_1 and n_2 are, "in general" a point of M^{n_2} is not an image of a point of M^{n_1} where the rank of φ is less than n_2. In order to formulate the notion of "generality" we will introduce the concept of a set of measure zero.

DEFINITION 3.1. *A set A of E^n is said to be of measure zero (meas $A = 0$) if for any $\epsilon > 0$, there exists a sequence of balls B_r^ϵ such that*

$$1) \quad A \subset \bigcup_{r=1}^{\infty} B_r^\epsilon$$

$$2) \quad \sum \text{vol } B_r^\epsilon < \epsilon.$$

(A ball is a set of points of the form $\{x \mid \|x - y\| \leqslant \delta\}$ for some y and δ, where $\|\ \|$ is the Euclidean distance. By vol B_r^ϵ, we mean the ordinary Euclidean volume.) We make some obvious remarks

(3.1) If $A \subset \bigcup A_k$ and meas $A_k = 0$, then meas $A = 0$.

In fact, for any $\epsilon > 0$, choose balls $B_{k,r}$ covering A_k such that

$$\sum_r \text{vol } B_{k,r} \leqslant \epsilon/2^k.$$

Then the $B_{k,r}$ form a denumerable covering of A with $\sum_{k,r} \text{vol } B_{k,r} < \epsilon$.

(3.2) If φ is a C^1 map of a neighborhood of A into E^n, and meas $A = 0$, then meas $\varphi(A) = 0$.

To prove this, we may assume that A is contained in some compact set. If A is not we could write it as a denumerable union of such (say of $A \cap \{x \mid \|x\| < n\}$) and apply (3.1). For such A, φ will satisfy a uniform Lipschitz condition

$$\|\varphi(x) - \varphi(y)\| < \mu\|x - y\|. \tag{3.3}$$

Then for any $\epsilon > 0$, cover A by sets B_k such that $\sum \text{vol } B_k < \epsilon/\mu^n$. By (3.3), $\varphi(B_k)$ lies in some ball C_k whose radius is μ times that of B_k, thus $\varphi(A) \subset \bigcup C_k$ and $\sum \text{vol } C_k < \epsilon$.

Exercise 3.1. Any countable set has measure zero.

Exercise 3.2. Show that in the definition of measure zero we could use a covering by an arbitrary sequence of open sets, not necessarily balls. That is, if, for any $\epsilon > 0$, A can be covered by a sequence of open U_i with $\sum \text{vol } U_i < \epsilon$ then meas $(A) = 0$.

Definition 3.2. *A subset A of a differentiable manifold is said to be of measure zero (meas $A = 0$) if $A = \bigcup A_n$, each A_n lies in some coordinate neighborhood U_n with coordinate maps φ_n, and $\varphi_n(A)$ is of measure zero.*

In view of (3.1) and (3.2), the definition does not really depend on the decomposition of A into A_n or on U_n; i.e., for any coordinate neighborhood V, and maps ψ, $\psi(V \cap A)$ will have measure zero. Furthermore, (3.2) and Definition 3.2, imply that:

(3.4) *If φ is a C^1 map of $M_1^n \to M_2^n$, then* meas $\varphi(A) = 0$ *whenever* meas $A = 0$.

If $A \subset \bigcup A_n$ where meas $(A_n) = 0$, then meas $(A) = 0$. We can use this remark to prove

Lemma 3.1. *Let A be a subset of a manifold M. If every point, p, of M has a neighborhood U_p such that* meas $(U_p \cap A) = 0$ *then* meas $A = 0$.

In fact, since the topology of M has a countable base, we can replace the sets $\{U_p\}$ by a countable collection, $\{U_i\}$. Then $A = \bigcup(A \cap U_i)$ and meas $(A \cap U_i) = 0$.

Exercise 3.3. Show that the set of points where $x_{k+1} = \cdots = x_n = 0$ in E^n has measure zero (if $k < n$).

EXERCISE 3.4. Let f be a C^1 map of a manifold $M_1 \to M_2$. Suppose that $\operatorname{rank} f = k < \dim M_2$. Then $\operatorname{meas} f(M_1) = 0$. [Use (2.12) and Exercise 3.3.]

DEFINITION 3.3. *Let f be a C^1 map of $M_1 \to M_2$ where* $\dim (M_i) = n_i$. *The points of M_1 where rank $f < n_2$ are called the critical points of f. All other points of M_1 are called regular. A point $q \in M_2$ such that $f^{-1}(q)$ contains at least one critical point is called a critical value. All other points of M_2 are called regular values.*

Remarks. Notice that if $\dim (M_1) < \dim (M_2)$ all points of M_1 are critical. Notice also that if $x \in M_2$ does not lie in $f(M_1)$ it is a regular value; that is, in our notation, a point which is not a value is always a regular value.

EXERCISE 3.5. Show that if y is a regular point of M_2 then either $f^{-1}(y)$ is a submanifold of M_1 of dimension $n_1 - n_2$ or $f^{-1}(y)$ is empty.

We now state Sard's theorem.

THEOREM 3.1.[2] *Let M_1 and M_2 be C^k manifolds of dimension n_1 and n_2 respectively. Let f be a map of class C^k of $M_1 \to M_2$. The critical values of f form a set of measure zero if*

$$k - 1 \geqslant \max (n_1 - n_2, 0). \tag{3.5}$$

Remarks. (a) The theorem is trivial for $n_1 < n_2$ (cf. Exercise 3.2); (b) Although the theorem is stated for arbitrary differentiable manifolds, it is clear from Definition 3.2 that we can reduce the theorem to the case that M_2 is Euclidean space, M_1 is a subset of the unit cube

$$C: \{x \in E^{n_1} \mid 0 \leqslant x_i \leqslant 1\}$$

and f is a C^k map of some neighborhood of $C \to E^{n_2}$. The proof for the case $n_1 = n_2$ is easy; we shall examine it first. What we must prove is

LEMMA 3.2.[2] *Let $\varphi = (\varphi^1, \ldots, \varphi^n)$ be a map of $C \to E^n$ where each of*

[2] For many applications, the case $n_1 = n_2$ of Sard's theorem, i.e., Lemma 3.2 suffices. For a first reading, the reader may wish to skip the "hard" part ($n_1 > n_2$) of Sard's theorem and proceed directly to Section 4. Sard's theorem for ($n_1 > n_2$) will only be used in Theorems 4.5 and 4.6 in this book.

the functions φ^i is defined and C^1 in some neighborhood of C. Then the set of critical values of φ has measure zero.

Proof. For two points $x = (x_1, \ldots, x_n)$ and $y = (y_1, \ldots, y_n)$ of C, the mean value theorem asserts that

$$\varphi^i(y) - \varphi^i(x) = \sum_{j=1}^{n} \frac{\partial \varphi^i}{\partial x^j}(z_i)(y^j - x^j), \tag{3.6}$$

where z_i is some point on the line segment joining x to y. From this we deduce two facts. First of all, since $\partial \varphi^i / \partial x^j$ is bounded on C, we can find a uniform Lipschitz constant, a, such that

$$\|\varphi(y) - \varphi(x)\| < a\|x - y\|. \tag{3.7}$$

Secondly, let T_x denote the affine map tangent to φ at x, i.e., if $T_x(y) = (T_x^1(y), \ldots, T_x^n(y))$,

$$T_x^i(y) = \varphi^i(x) + \sum_{j=1}^{n} \frac{\partial \varphi^i}{\partial x^j}(x)(y^j - x^j). \tag{3.8}$$

By (3.6)

$$\varphi^i(y) - T_x^i(y) = \sum_{j=1}^{n} \left(\frac{\partial \varphi^i}{\partial x^j}(z_i) - \frac{\partial \varphi^i}{\partial x^j}(x) \right)(y^j - x^j). \tag{3.9}$$

Since $\varphi \in C^1$, the functions $\partial \varphi^i / \partial x^j$ are continuous and hence uniformly continuous on C. Therefore, we can find a function $b(\epsilon)$ such that $b(\epsilon) \to 0$ as $\epsilon \to 0$ and

$$\|\varphi(y) - T_x(y)\| \leqslant b(\|x - y\|)\|x - y\|. \tag{3.10}$$

If x is a critical point, the determinant of T_x vanishes so T_x maps E^n into some plane P_x of dim $n - 1$. If $\|y - x\| < \epsilon$ then by (3.7) $\|\varphi(x) - \varphi(y)\| < a\epsilon$, and by (3.10) the distance of $\varphi(y)$ from P_x is less than $\epsilon b(\epsilon)$. Thus $\varphi(y)$ lies in the intersection of any cube of edge $a\epsilon$ centered at $\varphi(x)$ and the region between two parallel planes of distance $\epsilon b(\epsilon)$ from P_x. Taking the cube so that one of its faces is parallel to P_x, we conclude that the volume of this intersection is $2^n a^{n-1} \epsilon b(\epsilon)$.

Now divide C into p^n cubes of edge $1/p$. Any cube containing a critical point, x, is contained in the ball of radius $\sqrt{n}/p$ about x. The image of this ball has volume $\leqslant 2^n a^n (\sqrt{n}/p)^n b(\sqrt{n}/p)$. The image of all cubes containing at least one critical point contains the set of critical values of φ. Since there are at most p^n such cubes the total volume is at most

$$(3.11)\qquad p^n 2^n a^n (\sqrt{n}/p)^n b(\sqrt{n}/p) = 2^n a^n n^{n/2} b(\sqrt{n}/p).$$

As we let p tend to infinity this expression tends to zero. This proves Lemma 3.2.

Let us examine what happens if $n_1 > n_2$. Everything we said carries over up to the last step where we would have to replace (3.11) by

$$(3.12)\qquad p^{n_1} 2^{n_2} a^{n_2} (\sqrt{n_1}/p)^{n_2} b(\sqrt{n_1}/p) = 2^{n_2} a^{n_2} n_1^{n_2/2} p^{n_1 - n_2} b(\sqrt{n_1}/p),$$

since there are now p^{n_1} cubes of edge $1/p$ in C. We no longer know, and it is generally false, that $b(\epsilon)/\epsilon^{n_1 - n_2} \to 0$, as $\epsilon \to 0$. The proof relies on a more careful estimate than that given by the mean value theorem. What we want is to be able to conclude that a function vanishes with high enough order at its set of zeros. This is accomplished by the following lemma due to A. P. Morse [14].

LEMMA 3.3. *Let A be a subset of E^n and q a non-negative integer. Then $A \subset \bigcup_{i=0}^{\infty} A_i$ where A_0 is countable and the $A_i (i \geqslant 1)$ have the following property:*

Let f be a function of class C^q defined on a neighborhood of A such that every point of A is a critical point of f. Then there exist functions $b_i(\epsilon)$ (depending on f) such that $b_i(\epsilon)$ are monotone decreasing, $b_i(\epsilon) \to 0$ as $\epsilon \to 0$ and

$$(3.13)\qquad |f(x) - f(y)| < b_i(\|x - y\|)\|x - y\|^q$$

for any points $x, y \in A_i$.

For the case $n = 1$, the lemma is obvious. In fact, the set of critical points of a function $f(x)$ is the set where $f'(x) = 0$. Let A_0 be the set of discrete points of A. At any other points $x \in A$, $x \notin A_0$, $f'(x) = f^{(2)}(x) = \cdots = f^{(q)}(x) = 0$, since x is a limit point of a sequence of points x_i with $f'(x_i) = 0$. Thus $f^{(q)}(x) = 0$ for $x \in A - A_0$, and (3.13) is a consequence of Taylor's formula. (The sets A_i should be taken to be $A \cap K_i$ where the K_i form a sequence of compact sets covering A). For dimension $n > 1$, the proof is a little more complicated since not all higher-order partial derivatives need vanish at limit points of A. However, if some higher-order partial derivative does not vanish at a point p, we will be able to use the implicit function theorem to conclude that there is a neighborhood

N_p such that $N_p \cap A$ lies on a lower-dimensional submanifold. In this way we reduce the problem by double induction of q and n.

In any event, before giving the formal proof of Lemma 3.3, we will use it to prove Theorem 3.1. We first prove

LEMMA 3.4. *Let* $\varphi = (\varphi^1, \ldots, \varphi^{n_2})$ *be a map of the cube* $C: \{x \in E^{n_1} \mid 0 \leqslant x^i \leqslant 1\}$ *where the* φ^i *are functions of class* C^q *defined on some neighborhood of* C. *Let* A *be the set of points of* C *where* φ *has rank zero. Then* $\varphi(A)$ *is of measure zero if* $q \geqslant n_1/n_2$.

Proof. Decompose A using Lemma 3.2: $A \subset \bigcup A_r$. We need only prove that $\varphi(A_r)$ has measure zero. For $r = 0$ this is trivial since it is countable, let us consider $r \geqslant 1$. The map φ is of rank zero at x if and only if x is a critical point for all the functions φ^i. By (3.11), if x and $y \in A_r$ and $\|x - y\| < \epsilon$

$$|\varphi^i(x) - \varphi^i(y)| < b(\epsilon)\epsilon^q,$$

or

$$\|\varphi(x) - \varphi(y)\| < \sqrt{n_2}\, b(\epsilon)\epsilon^q. \tag{3.14}$$

We now proceed as in the proof of Lemma 3.2. Divide C into p^{n_1} cubes C_α of side $1/p$. By (3.14), $\varphi(A_r \cap C_\alpha)$ lies in a ball of radius

$$\sqrt{n_1 n_2}\, b\left(\frac{\sqrt{n_1}}{p}\right)\left(\frac{\sqrt{n_1}}{p}\right)^q.$$

Thus the total volume of $\varphi(A_r)$ is less than

$$Kb\left(\frac{\sqrt{n_1}}{p}\right)^{n_2} p^{n_1 - qn_2} \tag{3.15}$$

with $K = \sqrt{n_1 n_2}\,(\sqrt{n_1})^{qn_2}\omega_{n_2}$ where ω_{n_2} is the volume of the unit ball in E^{n_2}. We can make (3.15) arbitrarily small by making p large providing $q \geqslant n_1/n_2$. This proves Lemma 3.4.

LEMMA 3.5. *Let* φ *be as in Lemma* 3.4. *Let* A *be the set of points where* φ *has rank* r. *Then* $\varphi(A)$ *has measure zero if* $q > (n_1 - r)/(n_2 - r)$, $(0 \leqslant r < n_2)$.

Proof. For $r = 0$ this is Lemma 3.4. By Lemma 3.1 it suffices to prove that every point of A has a neighborhood N such that $\varphi(N \cap A)$ has measure zero. Let $x_0 \in A$. By (2.11), we can find a neighborhood N of x_0 and coordinates $u^1, \ldots, u^{n_1}$ about x_0 and $y^1, \ldots, y^{n_2}$ about $\varphi(x_0)$ such that the functions $y^i \circ \varphi = y^i(u^1, \ldots, u^n)$ are given by

$$\begin{aligned} y^\alpha(u^1, \ldots, u^{n_1}) &= u^\alpha & (\alpha = 1, \ldots, r) \\ y^\beta(u^1, \ldots, u^{n_1}) &= F^\beta(u^1, \ldots, u^{n_1}) & (\beta = r + 1, \ldots, n_2). \end{aligned} \tag{3.16}$$

For each fixed value of $u^1, \ldots, u^r$, the map η of $E^{n_1 - r} \to E^{n_2 - r}$ given by

$$\eta(u^{r+1}, \ldots, u^{n_1}) = F(u^1, \ldots, u^r, u^{r+1}, \ldots, u^n), \tag{3.17}$$

has rank zero whenever $(u^1, \ldots, u^{n_1}) \in N \cap A$ since φ has rank r. By Lemma 3.4, the set $\varphi(N \cap A) \cap P_{(u^1, \ldots, u^r)}$ has measure zero in $P_{(u^1, \ldots, u^r)}$, where $P_{(u^1, \ldots, u^r)}$ is the $n_2 - r$ dimensional plane in E^{n_2} given by $y^\alpha = u^\alpha$, $\alpha = 1, \ldots, r$. Thus we have proved that the set $\varphi(N \cap A)$ has the property that its intersection with each of the planes $P_{(u^1, \ldots, u^r)}$ has measure zero. For those readers who know some measure theory, it is now an immediate consequence of Fubini's theorem that $\varphi(N \cap A)$ is of measure zero. We can, however, give a direct proof of this fact. First we observe that we can take N to be a compact neighborhood. Since A is closed, the set $N \cap A$ will then be compact. Since φ is continuous, $\varphi(N \cap A)$ is compact. What must be shown, then, is the following:

Let B be a compact subset of E^n whose intersection with every plane $P_{(u^1, \ldots, u^r)}$ has measure zero in $P_{(u^1, \ldots, u^r)}$. Then B has measure zero in E^n.

It is clear that we can use induction to reduce the statement to the case $r = n - 1$.[3]

We preface our proof by the following remark: Let $[a, b]$ be a closed interval of the real line. We know that from every open covering we can extract a finite subcovering. We want a slightly stronger statement.

From any covering of $[a, b]$ by intervals, we can extract a subcovering by intervals whose total length is $\leqslant 2|b - a|$.

Choose a finite *minimal* subcovering; that is choose intervals $I_1, \ldots, I_p$ of the covering such that $[a, b] \subset \bigcup_{j=1}^p I_j$ but

$$[a, b] \not\subset \bigcup_{i=1,\, i \neq q}^{p} I_i \qquad (1 \leqslant q \leqslant p).$$

(This can be done by starting with a finite subcovering and throwing

[3] I am indebted for the following proof to Professor H. Furstenberg.

away superfluous intervals.) Number the intervals according to the order of their left-hand endpoints; i.e., if $I_j = (a_j, b_j)$, $i < j$ if $a_i < a_j$ (if $a_i = a_j$ the covering is not minimal). Then $a_i < a_{i+1} < b_i < b_{i+1}$ and the total overlap is at most the length of $[a, b]$. Thus the total length of the intervals is at most $2|b - a|$.

Now suppose that $B \subset E^{n-1} \times [a, b]$ and meas $(B \cap P_{u^n}) = 0$. For any $\epsilon > 0$, we can find a covering of $B \cap P_{u^n}$ by open sets $R^i_{u^n}$ in P_{u^n} of total volume $< \epsilon$. For sufficiently small α, the sets $R^i_{u^n} \times I^{u^n}_\alpha$ cover

$$B \cap \bigcup_{z^n \in I^{u^n}_\alpha} P_{z^n}$$

where $I^{u^n}_\alpha$ is the interval of length α centered at u^n. (This follows from the compactness of B.) The sets $I^{u^n}_\alpha$ form a covering of $[a, b]$ of which we can extract a finite subcovering, $\{I_j\}$ of length $\leqslant 2|b - a|$. Let R^L_j denote the set $R^i_{u^n}$ if $I_j = I^{u^n}_\alpha$. Then the $R^i_j \times I_j$ form a cover of B whose total volume is $\leqslant 2|b - a|$. Since we can make this arbitrarily small, B is of measure zero. This proves Lemma 3.5.

Theorem 3.1 now follows from Lemma 3.4. If A is the set of critical points of φ, then $A = \bigcup_{r=0}^{n_2-1} A_r$, where A_r is the set of points where φ has rank r. The set of critical values is thus contained in $\bigcup_{r=0}^{n_2-1} f(A_r)$. If $q > n_1 - n_2 + 1$ then $q > (n_1 - r)/(n_2 - r)$, $(0 \leqslant r < n_2)$ so that each of the sets $f(A_r)$ is of measure zero.

We now proceed to the proof of Lemma 3.3. We have stated that its proof will be essentially induction on n and q. To formalize the argument we state

Lemma 3.6. *Let A be a subset of E^n and k a non-negative integer. There exists a sequence of sets $A_i (i \geqslant 0)$ and maps $\varphi_i (i \geqslant 1)$ such that A_0 is countable, $A \subset \bigcup_{r=0}^{\infty} A_r$, and, for $i \geqslant 1$*

(3.18) *φ_i is a C^1 homeomorphism of the ball $B^{m_i}_{\epsilon_i}$ into E^n with*

$A_i \subset \varphi_i(B^{m_i}_{\epsilon_i})$ (where B^m_ϵ is the set of $x \in E^m$ with $\|x\| < \epsilon$).

(3.19) $$\|\varphi_r(x) - \varphi_r(y)\| \geqslant \|x - y\|$$

(3.20) *for any $f \in C^k$ vanishing on A there exist monotone functions $b_i(\epsilon)$ with $\lim_{\epsilon \to 0} b_i(\epsilon) = 0$ such that*

$$|f(\varphi_i(x))| < b_i(\|x - y\|)\,\|x - y\|^k$$

for all $x, y \in B^{m_i}_{\epsilon_i}$ with $\varphi_i(y) \in A_i$.

The lemma is trivial for $k = 0$, just take the $A_i = A \cap K_i$ where K_i is a set of closed balls constituting a basis for E^n. Take the φ_i to be the translation of the ball centered at 0 to K_i. Then (3.20) is just uniform continuity. We have essentially proved the lemma for $n = 1$ in our discussion on page 49. We proceed by induction on $n + k$. Our induction hypothesis is that Lemma 3.6 is true for $n + k < p$.

Assume that $n + k = p$, $n > 1$, $k > 0$, and let $A \subset E^n$. Let A^1 be the set of all $x \in A$ such that

$$\sum_{j=1}^{n} \left(\frac{\partial f}{\partial x_j}\right)^2 (x) = 0 \tag{3.21}$$

for *all* $f \in C^k$ vanishing on A; let $A^2 = A - A^1$.

By induction, there is a decomposition $A^1 = \bigcup A_r^1$ and maps φ_r^1 of $B_\epsilon^{m_r}$ such that $\varphi_r^1(B_\epsilon^{m_r}) \supset A_r^1$, (3.19) holds, and

$$|g(\varphi_r^1(x))| < b_r^1(\|x - y\|)\|x - y\|^{k-1} \tag{3.22}$$

for all $g \in C^{k-1}$ vanishing on A^1. In particular (3.22) holds for $g = \partial f/\partial x_i$ if $f \in C^k$ and vanishes on A. We now show that the sets A_r^1 and φ_r^1 actually satisfy (3.20). This follows immediately from

LEMMA 3.7. *Let φ be a C^1 map of $B^m \to E^n$ satisfying* (3.19). *If,* $f \in C^k$, $(k \geq 1)$

$$\left|\frac{\partial f}{\partial x_j}(\varphi(x))\right| < b(\|x - y\|)\|x - y\|^{k-1}$$

with b monotone and $\lim b(\epsilon) = 0$, $(j = 1, \ldots, n)$. *Then*

$$|f(\varphi(x)) - f(\varphi(y))| < Kb(\|x - y\|)\|x - y\|^k$$

where K depends only on φ.

Proof. Set $F(t) = f(\varphi[y + (x - y)t])$. Then $f(\varphi(x)) = F(1)$ and $F(0) = f(\varphi(y))$. By the mean value theorem $f(\varphi(x)) - f(\varphi(y)) = F'(t)$ for some $0 < t < 1$. But

$$F'(t) = \sum_{j,\alpha} \frac{\partial f}{\partial x_j}(\varphi(y + (x - y)t)) \times \frac{\partial \varphi^j}{\partial x^\alpha} \times (x^\alpha - y^\alpha)$$

if $\varphi = (\varphi^1, \ldots, \varphi^n)$. Thus

$$|F'(t)| < K_1\|x - y\|\left\{\sum \left|\frac{\partial f}{\partial x_j}(\varphi(y + (x - y)t))\right|\right\}$$

where $K_1 = n \max |\partial\varphi^j/\partial x^\alpha|$. Then

$$|F'(t)| < K_1 n b(\|x - y\|)\|x - y\|^k,$$

proving the lemma with $K = n_1 K_1$.

Applying Lemma 3.7 to f vanishing on A (thus $f(\varphi(y)) = 0$) yields (3.20).

Let us return to Lemma 3.6. We have found sets A_r^1 and maps φ_r^1 which satisfy (3.18), (3.19), and (3.20) and cover A^1. Now we want to do the same for A^2. Let $p \in A^2$. This means that there is some function g vanishing on A with some $\partial g/\partial x_i(p) \neq 0$. We may assume that $\partial g/\partial x_n(p) \neq 0$. By the implicit function theorem, we can find a neighborhood N of p and a function $\varphi^n(x_1, \ldots, x_{n-1}) \in C^k$ defined in some B_ϵ^{n-1} such that the only solution of

$$g(x^1, \ldots, x^n) = 0$$

is given by the points $(x^1, \ldots, x^{n-1}, \varphi^n(x^1, \ldots, x^{n-1}))$. If we consider the map φ of $B_\epsilon^{n-1} \to E^n$ given by $\varphi = (\varphi^1, \ldots, \varphi^n)$ with

$$\varphi^\alpha (x^1, \ldots, x^{n-1}) = x^\alpha \qquad 1 \leqslant \alpha \leqslant n - 1$$
$$\varphi^n (x^1, \ldots, x^{n-1})$$

then $\varphi \in C^k$, $\|\varphi(x) - \varphi(y)\| > \|x - y\|$ and $N \cap A \subset \varphi(B_\epsilon^{n-1})$. We can now apply the induction hypothesis for $n - 1$ to $\varphi^{-1}(N \cap A)$. We can write $\varphi^{-1}(N \cap A) = \bigcup D_r$, with D_0 countable, and maps ψ_r satisfying (3.18) and (3.19) such that

$$|h(\psi_r(x))| < b_r(\|x - y\|)\|x - y\|^k$$

if $\psi_r(y) \in \varphi^{-1}(N \cap A)$, for any $h \in C^k$ vanishing on $\varphi^{-1}(N \cap A)$. But if $f \in C^k$ vanishes on A, then $h = f \circ \varphi$ vanishes on $\varphi^{-1}(N \cap A)$. If we set $\varphi_r = \varphi \circ \psi_r$, we have a decomposition of $N \cap A$ of the type desired in Lemma 3.5. We cover all of A_2 by neighborhoods N and select a countable subcovering. Putting together all the decompositions (and calling the union of the countable sets A_0) we obtain Lemma 3.6.

Lemma 3.3 now follows from Lemma 3.6. In fact we just use Lemma 3.7, since all the derivatives of a function vanish at its critical points. This concludes the proof of Sard's theorem.

It should be noted that the smoothness condition

$$k - 1 \geqslant \max (0, n_1 - n_2)$$

in Theorem 3.1 is necessary. A counter example in case this inequality doesn't hold is given by Whitney [21].

4. Partitions of unity, approximation theorems. In this section we obtain a number of approximation theorems which will be useful later on. The proof of each of them splits into two parts: a "local" result usually using Sard's theorem, and a method of piecing together the local information.

First we review some results from topology.

DEFINITION 4.1. *A family of sets* $\{U_\alpha\}$ *of a topological space* X *is called locally finite if every point* $p \in X$ *has a neighborhood* N, *which meets only a finite number of the* U_α; i.e., $N_p \cap U_\alpha = \varnothing$ *except for a finite number of* α. *On a differentiable manifold* M, *an atlas* $\{U_\alpha, h_\alpha\}$ *is called locally finite if the collection of sets* $\{U_\alpha\}$ *is locally finite.*

DEFINITION 4.2. *A covering* $\{V_\beta\}$ *is called a refinement of a covering* $\{U_\alpha\}$ *if for every* β, *there is (at least) one* α *such that* $V_\beta \subset U_\alpha$. *A space* X *is called paracompact if it is Hausdorff and if every open covering has a locally finite refinement.*

LEMMA 4.1. *If* X *is locally compact, has a countable basis, and is Hausdorff then* X *is paracompact. In particular every manifold is paracompact.*

Proof. Let $\{U_i\}$ be a basis for X such that $\overline{U}_i$ is compact. Define the sequence of compact sets A_i inductively as follows: Let $A_1 = \overline{U}_1$; assuming that A_i has been defined, let j be the smallest integer $>i$ such that $A_i \subset U_1 \cup \cdots \cup U_j = V_j$. Set $A_{i+1} = \overline{V}_j$. Then the A_i are compact, $A_i \subset \operatorname{Int} A_{i+1}$ and $\bigcup_{i=0}^{\infty} A_i = X$. Now let $\{W_\alpha\}$ be any open covering of X. For each i, consider the open sets $(\operatorname{Int} A_{i+2} - A_{i-1}) \cap W_\alpha$. These sets cover the compact set $A_{i+1} - \operatorname{Int} A_i$. Therefore, we can choose a finite subcovering $V_1^i, \ldots, V_{p(i)}^i$. The collection of all the sets V_j^i cover X since the sets $A_{i+1} - \operatorname{Int} A_i$ cover X. The covering $\{V_k^i\}$ is a refinement of $\{W_\alpha\}$. For $x \in A_i$, then $\operatorname{Int} A_{i+1}$ is a neighborhood of x which meets no V_k^j for $j > i + 1$ so $\{V_k^i\}$ is locally finite.

For a differentiable manifold we can improve on the preceding lemma by choosing the refinement to be coordinate neighborhoods:

LEMMA 4.2. *Let M be an n-dimensional differentiable manifold and $\{W_\alpha\}$ an open covering of M. There exists an atlas (V_i, h_i)*

1) *$\{V_i\}$ is a locally finite refinement of $\{W_\alpha\}$*
2) $h_i(V_i) = B_3^n \quad (\{x \in E^n \mid \|x\| < 3\})$
3) *The sets $O_i = h_i^{-1}(B_1^n)$ cover X.*

To see this, we go back to the proof of Lemma 4.1 and choose our covering of $A_{i+1} - \text{Int } A_i$ a little more carefully. Each point, p, of the open set $(\text{Int } A_{i+2} - A_{i-1}) \cap W_\alpha$ has a coordinate neighborhood $N_{p,\alpha}$ and map $h_{p,\alpha}$ such that $N_{p,\alpha} \subset (\text{Int } A_{i+2} - A_{i-1}) \cap W_\alpha$ and $h_{p,\alpha}(N_{p,\alpha}) \supset B_3^n$. Let $V_{p,\alpha} = h_{p,\alpha}^{-1}(B_3^n)$ and $O_{p,\alpha} = h_{p,\alpha}^{-1}(B_1^n)$. Now choose a finite number of the $O_{p,\alpha}$ which cover $A_{i+1} - \text{Int } A_i$ and proceed as before.

LEMMA 4.3. *There is a C^∞ function φ defined on E^n such that $\varphi(x) \geqslant 0$ and*

$$\varphi(x) = \begin{cases} 1 & \text{for } \|x\|^2 \leqslant 3/2 \\ 0 & \text{for } \|x\|^2 \geqslant 2. \end{cases}$$

Proof. Let

$$f(r) = \begin{cases} e^{-1/r^2} & r > 0 \\ 0 & r \leqslant 0. \end{cases}$$

Then f is C^∞. The function

$$g(r) = \frac{f(2-r)}{f(2-r) + f(r-3/2)}$$

is a C^∞ function which vanishes for $r > 2$ and for $r \leqslant 3/2$. Define $\varphi(x) = g(\|x\|)$; it satisfies the requirements of Lemma 4.3.

DEFINITION 4.3. *Given a function ψ on a space X, the support of ψ, denoted by* supp ψ, *is the closure of the set of points, p, where $\psi(p) \neq 0$.*

DEFINITION 4.4. *A C^k partition of unity on a differentiable manifold, M, is a collection of C^k functions $\varphi_i \geqslant 0$ such that:*

1) *the collection* $\{\text{supp } (\varphi_i)\}$ *is locally finite*
2) supp φ_i *is compact*

and

3) $\sum_1^\infty \varphi_i(p) = 1$ *for all $p \in M$.*

It should be noted that in view of 1), the sum in 3) has, for any fixed p, only a finite number of non-zero terms.

DEFINITION 4.5. *A partition of unity, $\{\varphi_i\}$, is said to be subordinate to an open covering $\{W_\alpha\}$ if for every i there is an α such that* supp $\varphi_i \subset W_\alpha$.

THEOREM 4.1. *Let $\{W_\alpha\}$ be a covering of a C^k differentiable manifold, M. There exists a C^k partition of unity $\{\varphi_i\}$ subordinate to $\{W_\alpha\}$.*

Proof. Choose a covering $\{V_i\}$ refining $\{W_\alpha\}$ as in Lemma 4.2. Let $\psi_i = \varphi \circ h_i$ where φ is the function of Lemma 4.3. Then ψ_i is a C^k function with supp $(\psi_i) \subset V_i$, and $\psi_i(p) = 1$ for $p \in O_i$. Since the covering $\{V_i\}$ is locally finite the collection $\{\psi_i\}$ is. Set

$$\varphi_i(p) = \frac{\psi_i(p)}{\sum_{j=1}^{\infty} \psi_j(p)}. \tag{4.1}$$

Since the ψ_j are non-negative, $\psi_i(p) = 1$ for at least one i (the O_i cover M) and all but a finite number of the $\psi_i(p)$ vanish, the expression (4.1) makes sense. The $\{\varphi_i\}$ form a partition of unity. Since supp φ_i = supp ψ_i, this partition of unity is subordinate to $\{W_\alpha\}$.

We now use the preceding results to state a number of approximation theorems for C^k mappings. Before stating these theorems, we must first discuss our measurement of approximation. What we want is some topology on the space of C^k maps of $M_1 \to M_2$ where the M_i are differentiable manifolds. In the case that M_1 and M_2 are Euclidean spaces, then every map is given by n_2 functions $f^1, \ldots, f^{n_2}$ of n_1 variables $x^1, \ldots, x^{n_1}$. On the space of C^k functions there is an obvious metric, that of uniform C^k convergence. That is,

$$\|f - g\|_k = \sum \sup_{x \in E^n} \left| \frac{\partial^\alpha f(x)}{(\partial x^1)^{\alpha_1}, \ldots, (\partial x^n)^{\alpha_n}} - \frac{\partial^\alpha g(x)}{(\partial x^1)^{\alpha_1}, \ldots, (\partial x^n)^{\alpha_n}} \right|$$

where the sum is over all partial derivatives of order $\leqslant k$. Thus g is close to f if all the partial derivatives of g are uniformly close to those of f (including the zero-th order) up to order k. This metric also extends to the case of n-functions in the obvious manner, that is defined $\|f - g\|_k = \sup_i \|f^i - g^i\|_k$. In the case of manifolds, however, a map is represented by functions only locally, and only after local coordinates have been chosen in M_1 and M_2. Furthermore, the partial derivatives also make sense only in terms of local coordinates. Therefore, we have to define closeness in terms of a fixed choice of coordinate systems in M_1 and M_2.

DEFINITION 4.6. *Let f be a C^k map of $M_1 \to M_2$. Let $Z = \{Z_i, H_i\}$ be a locally finite atlas on M_2 with $H_i = (z_i^1(\cdot), \ldots, z_i^{n_2}(\cdot))$. Let $W = \{W_i, h_i\}$ be a locally finite atlas on M_1 with $h_i = (w_i^1, \ldots, w_i^{n_1})$ such that $\overline{W}_i$ is compact and $f(\overline{W}_i) \subset Z_{j(i)}$* for some $j(i)$. *(Given Z there always exists such a W by Lemma 4.2.) Let $\epsilon(\cdot)$ be a positive function on M_1 and k a non-negative integer. Then $\mathcal{N}_{(W, Z, \epsilon, k)f}$ is the set of all C^k maps g of $M_1 \to M_2$ such that the following two conditions hold:*

$$g(\overline{W}_i) \subset Z_j \text{ for all } i \text{ and } j \text{ such that } f(\overline{W}_i) \subset Z_j. \tag{4.2}$$

For these values of i and j let $H_j \circ f \circ h_i^{-1}$ be given by $(f_{ij}^1, \ldots, f_{ij}^{n_2})$ and $H_j \circ g \circ h_i^{-1}$ be given by $(g_{ij}^1, \ldots, g_{ij}^{n_2})$. Then

$$|Df_{ij}^r(w) - Dg_{ij}^r(w)| < \epsilon(h_i^{-1}(w)) \quad \text{for all } w \in h_i(W_i) \text{ and } r = 1, \ldots, n_2, \tag{4.3}$$

where D is any differential operator $\partial^\alpha/(\partial w^1)^{\alpha_1} \cdots (\partial w^{n_1})^{\alpha_{n_1}}$ with $\alpha_1 + \cdots + \alpha_{n_1} \leqslant k$. (This includes the case $\alpha = 0$).

In view of the local finiteness assumption, conditions (4.2) and (4.3) reduce to a finite number of "closeness" conditions at every point of M_1. We leave to the reader the verification of the fact that the sets $\mathcal{N}_{(W, Z, \epsilon, k)f}$ form a system of neighborhoods of f for a topology on the space of C^k maps of $M_1 \to M_2$.

We now state the first of our approximation theorems.

THEOREM 4.2. *Let f be a C^k map of an n_1-dimensional manifold M_1, into an n_2-dimensional manifold M_2, with $k \geqslant 2$. If $n_2 \geqslant 2n_1$ then for any (W, Z, ϵ, k), there is a $g \in \mathcal{N}_{(W, Z, \epsilon, k)f}$ which is an immersion. If rank $f = n_1$ on a closed set K, we can choose g so that $g(x) = f(x)$ for $x \in K$.*

The proof splits into three parts. We first prove a local version of the theorem; i.e., any map can be locally approximated by an immersion. We show next that a map which is close to an immersion is an immersion. We finally use Lemma 4.2 to put together the local approximations.

LEMMA 4.4. *Let F be a C^2 map of an open set, U, of $E^{n_1} \to E^{n_2}$ where $n_2 \geqslant 2n_1$. For any $\delta > 0$ there is an $n_1 \times n_2$ matrix $A = (a_j^i)$ with*

$$|a_j^i| < \delta \tag{4.4}$$

such that the map

$$x \to F(x) + A \cdot x \tag{4.5}$$

is an immersion.

(The expression $A \cdot x$ is the matrix product of A with the $1 \times n_1$ matrix $x = (x^1, \ldots, x^{n_1})$.)

Proof. Let $J(x)$ denote the Jacobian matrix of F at x. The Jacobian of (4.5) at x is $J(x) + A$. We wish to choose A so that this matrix has rank n_1 at all x. Another way of saying the same thing is that we wish to find an A which is not of the form $B - J(x)$ for any B of rank $< n$ for any $x \in U$. In order to do this, we examine the set of all matrices of the form $B - J(x)$ where rank $B = k < n$ and $x \in U$.

Let $T(n_1, n_2; k)$ denote the manifold of all $n_1 \times n_2$ matrices of rank k (cf. Example 4 of Section 1). Let G_k be the map of $U \times T(n_1, n_2; k) \to T(n_1, n_2)$ given by

$$G_k(x, B) = B - J(x).$$

The map G_k is clearly C^1. The dimension of $U \times T(n_1, n_2; k)$ is $n_1 + k(n_1 + n_2 - k)$. Now the derivative of $n_1 + t(n_1 + n_2 - t)$ is $n_1 + n_2 - 2t$ which is >0 if $t < n_1 < n_2$. Thus $n_1 + k(n_1 + n_2 - k)$ is monotone increasing with k for $k < n_1$; i.e., $n_1 + k(n_1 + n_2 - k) \leqslant n_1 + (n_1 - 1)(n_1 + n_2 - n_1 + 1) = n_1 n_2 - (n_2 - 2n_1) - 1$ which is $< n_1 n_2$ if $n_2 \geqslant 2n_1$. Since the dimension of $T(n_1, n_2)$ is $n_1 n_2$, (the trivial case of) Sard's theorem implies that the image of G_k has measure zero in $T(n_1, n_2)$. Therefore, we can find an $A \in T(n_1, n_2)$ as close to the zero matrix as we please which does not lie in the image of any of the G_k. This proves Lemma 4.4.

LEMMA 4.5. *Let F be a C^1 map of a subset U of $E^{n_1} \to E^{n_2}$. Let F have rank n_1 on a compact subset K of U. Then there exists an $\epsilon > 0$ such that any map G of $U \to E^{n_2}$ satisfying $\|G - F\|_1 < \epsilon$ has rank n_1 on K.*[4]

Let $\delta(x)$ be the maximum of the absolute values of the determinants of the $n_1 \times n_1$ submatrices of the Jacobian matrix of F at x. Then δ is a

[4] Recall that

$$\| F \|_1 = \sup_{i, x} |f^i(x)| + \sup_{i, j, x} \left|\frac{\partial f^i}{\partial x^j}(x)\right|.$$

continuous function which is positive on the compact set K. Thus $\delta(x) > \delta > 0$ on K. Therefore, we can find an ϵ so that changing the entries of $J(x)$ by less than ϵ changes all the determinants by less than $\delta/2$. This ϵ works for the lemma.

We proceed to the proof of Theorem 4.2. We are given the locally finite covering $\{W_i\}$. Every point, p, has a neighborhood N_p meeting only a finite number of the W_i. Since the set where rank $(f) = n_1$ is open, we can find an open set $W \supset K$ such that rank $(f) = n_1$ on W. Let $N_p^1 = N_p \cap W$ and $N_p^2 = N_p \cap (M_1 - K)$. The sets $\{N_p^1, N_p^2\}$ $p \in M$ form an open covering of M_1. We can find a denumerable locally finite refinement of this covering by coordinate neighborhoods V_i satisfying 1), 2), and 3) of Lemma 4.2. By the definition of the N_p^α, each V_i is contained in either W or $M - K$. Let us reindex the V_i by positive and negative integer subscripts so that $V_i \subset W$ if and only if $i \leqslant 0$.

According to Lemma 4.2 the sets $O_i = h_i^{-1}(B_1^n)$ cover M. Let $P_i = \overline{h_i^{-1}(B_2^n)}$ so that $O_i \subset P_i \subset V_i$ and P_i is compact. We will construct, inductively, a sequence of C^k maps $\{f_p\}$ $p \geqslant 0$ such that

i) $f_0 = f$

ii) $f_s \in \mathcal{N}_{(W,\, Z,\, \epsilon/2^s,\, k)f_{s-1}}$

iii) f_s has rank n on $\bigcup_{i \leqslant s} \bar{O}_s = Q_s$

iv) $f_s(x) = f_{s-1}(x)$ for $x \in M - V_s$.

Once we have constructed such a sequence we will have proved the theorem. In fact, by iv) (and since the V_i are locally finite) $f_s(x) = f_{s+1}(x)$ for sufficiently large s, for all x in the neighborhood of any fixed $p \in M_1$. Thus the sequence $\{f_s(x)\}$ converges to a C^k map $g(x)$ which has rank n_1 everywhere by iii) and satisfies (4.2) and (4.3) by i) and ii).

Now we proceed to the construction of the $\{f_s\}$. We take $f_0 = f$ and assume that we have constructed f_{s-1}. We wish to choose an f_s satisfying ii), iii), and iv). Condition iv) determines f_s outside of V_s. We must define f_s in V_s so as to satisfy ii) and iii) and to "match up" smoothly with f_{s-1} outside V_s. We shall guarantee the "matching up" by actually choosing $f_s = f_{s-1}$ in $V_s - P_s$. The whole problem of defining f_s now lies entirely within V_s so we can transform the problem to Euclidean space: Let $g_{s-1} = H_\rho \circ f_{s-1} \circ h_s^{-1}$ where (Z_ρ, H_ρ) is a chart of the atlas Z with $f_{s-1}(V_s) \subset Z_\rho$. Let $R_{s-1} = h_s(Q_{s-1} \cap V_s)$. We wish to find a mapping g_s of $B_3^{n_1} \to H_\rho(Z_\rho) \subset E^{n_2}$ so that setting $f_s = H_\rho^{-1} \circ g_s \circ h_s$ on

V_s and $f_s = f_{s-1}$ on $M_1 - V_s$, the map f_s satisfies conditions ii) and iii). We are going to define g_s by

$$g_s(x) = g_{s-1}(x) + \varphi(x)A \cdot x \tag{4.6}$$

where φ is the function of Lemma 4.3 and A is an $n_1 \times n_2$ matrix to be chosen momentarily. Let us examine what restrictions conditions ii) and iii) impose on g_s, that is, on A. First of all let us look at condition ii). There are a finite number of W_i containing V_s and a finite number of Z_j containing $f_{s-1}(\overline{W}_i)$ for $W_i \supset V_s$. The map f_s is to map V_s into each of these Z_j. If h_i' denote the coordinate maps on the W_i and H_j on the z_j, we can use them to translate the condition to g_s: We let $\Phi_j = H_j \circ H_\rho^{-1}$ and $\varphi_i = h_s \circ h_i'^{-1}$ then the maps $\Phi_j \circ g_{s-1} \circ \varphi_i$ take $B_3^{n_1}$ into compact subsets of the open sets $H_j(Z_j)$. The first requirement of condition ii) is that the $\Phi_j \circ g_s \circ \varphi_i$ do the same. This can certainly be achieved by taking the entries of A sufficiently small. All that condition ii) requires in addition to this is that the partial derivatives of the $\Phi_j \circ g_s \circ \varphi_i$ up to order k be close to those of $\Phi_j \circ g_{s-1} \circ \varphi_i$. Since there are only a finite number of i and j and since the function φ, and hence its derivatives has been chosen in advance, this also can be achieved by taking A small.

Next, let us look at condition iii). This requires that g_s be an immersion on $\bar{B}_1^{n_1} \cup R_{s-1}$, where g_{s-1} is an immersion on R_{s-1}. Since $\varphi = 0$ outside $B_{3/2}^{n_1}$, g_s will be of rank n_1 at all points of $R_{s-1} - R_{s-1} \cap \bar{B}_2^{n_1}$ regardless of how we choose A. On the other hand, Lemma 4.5 says that g_s will have rank n_1 on $R_{s-1} \cap B_2^{n_1}$ so long as A is sufficiently small. By Lemma 4.4 we can choose A to satisfy all of these requirements and such that the map $x \to g_{s-1}(x) + A \cdot x$ has rank n_1 on $B_3^{n_1}$. Since $\varphi(x) \equiv 1$ on $B_{3/2}^{n_1}$, g_s coincides with this map on $B_{3/2}^{n_1}$. Thus g_s has rank n_1 on $\bar{B}_1^{n_1} \cup R_{s-1}$. We can thus construct the desired sequence $\{f_s\}$ proving Theorem 4.2.

We have shown that any smooth map of $M_1 \to M_2$ can be approximated by an immersion if $m \geqslant 2n$. In particular we obtain as a corollary,

Any C^k manifold ($k > 1$) of dimension n_1 can be immersed in E^{2n_1}.

In fact choose f to be any C^k map into E^{2n} and apply Theorem 4.2. We shall now show that any immersion can be approximated by a 1–1 immersion, if $n_2 > 2n_1$.

THEOREM 4.3. *Let f be a C^k ($k \geqslant 1$) immersion of an n_1-dimensional manifold M into E^{n_2}. Let Z be E^{n_2} with its usual coordinates and $\{W_i\}$ as* *in Definition* 4.6. *Then if* $n_2 > 2n_1$, *for any* $\epsilon(x)$ *and* $k \geqslant 0$ *there is a* 1–1 *immersion* g *of* $M \to R^n$ *such that* $g \in \mathcal{N}_{(W, Z, \epsilon, k)f}$. *If* f *is* 1–1 *in a neighborhood,* W, *of a closed set* K, *we can choose* g *so that* $g(x) = f(x)$ *for* $x \in K$.

Proof. The proof proceeds as before. The local aspect is trivial in this case: Every point p has a neighborhood, N_p, such that f is 1–1 on N_p by the implicit function theorem. Let $N_p^1 = W \cap N_p$ and $N_p^2 = M - K \cap N_p$. Choose a partition of unity $\{\varphi_i\}$ subordinate to this covering. Number the φ_i by positive and negative integers so that supp $\varphi_i \subset W$ if and only if $i \leqslant 0$. We will obtain g as

$$g = f + \sum_1^{\infty} b_i \varphi_i \tag{4.7}$$

where the coefficients b_i will be chosen inductively: Assume that $b_1, \ldots, b_p$ have been defined so that

$$f_p = f + \sum_1^{p} b_p \varphi_p$$

is an immersion, and $f_s - f_{s-1}$ satisfies condition ii) on page 60, for $s \leqslant p$. We can arrange that ii) holds for $s = p + 1$, merely by choosing $|b_{p+1}|$ sufficiently small. The requirement that $f_p + b_{p+1}\varphi_{p+1}$ be an immersion can also be met simply by choosing $|b_{p+1}|$ small by Lemma 4.5. Finally, let D_{p+1} be the open subset of $M \times M$ consisting of those pairs (y_1, y_2) such that $\varphi_{p+1}(y_1) \neq \varphi_{p+1}(y_2)$. Let G_{p+1} be the map of $D_{p+1} \to E^{n_2}$ defined by

$$G_{p+1}(y_1, y_2) = \frac{f_p(y_2) - f_p(y_1)}{\varphi_{p+1}(y_1) - \varphi_{p+1}(y_2)}. \tag{4.8}$$

Since G_{p+1} is differentiable and D_{p+1} is of dimension $2n_1$, the set $G_{p+1}(D_{p+1})$ has measure zero. Therefore, we can (and do) choose b_{p+1} arbitrarily small so that $b_{p+1} \notin G_{p+1}(D_{p+1})$. Then $f_{p+1}(y_1) = f_{p+1}(y_2)$ implies

$$f_p(y_1) + b_{p+1}\varphi_{p+1}(y_1) = f_p(y_2) + b_{p+1}\varphi_{p+1}(y_2)$$

or

$$b_{p+1}(\varphi_{p+1}(y_1) - \varphi_{p+1}(y_2)) = f_p(y_2) - f_p(y_1).$$

Since $b_{p+1} \notin G_{p+1}(D_{p+1})$ this can only happen if $\varphi_{p+1}(y_1) = \varphi_{p+1}(y_2)$ and $f_p(y_2) = f_p(y_1)$.

Now define g by (4.7). In view of the way we constructed g it is clear that g is an immersion approximating f, i.e., $g \in \mathcal{N}_{(W, Z, \epsilon, k)f}$. We shall show that it is 1–1. Suppose $g(y_1) = g(y_2)$. Since the φ_i form a partition of unity, for some sufficiently large r, $\varphi_p(y_1) = \varphi_p(y_2) = 0$ for $p > r$. Thus $g(y_1) = f_r(y_1) = f_r(y_2) = g(y_2)$. By our construction of b_r this implies that $\varphi_r(y_1) = \varphi_r(y_2) = 0$ and $f_{r-1}(y_1) = f_{r-1}(y_2)$. Proceeding in this way we conclude that $f(y_1) = f(y_2)$ and

$$\varphi_i(y_1) - \varphi_i(y_2) = 0 \qquad \text{for all } i > 0. \tag{4.9}$$

But (4.9) implies that y_1 and y_2 lie in W, where f is 1–1. Thus $y_1 = y_2$.

Theorems 4.2 and 4.3 combine to give Whitney's theorem:

THEOREM 4.4. *Any manifold, M, of class C^k $(k \geqslant 2)$ of dimension n can be C^k imbedded in E^{2n+1}.*

Proof. Given any C^k map of M into E^{2n+1}, it can be approximated by an immersion (Theorem 4.2) and this immersion can be approximated by a 1–1 immersion (Theorem 4.3). Thus starting with any C^k map, f, (even the constant map!) we end up with a 1–1 immersion, g. By Exercise 2.1, we know that the 1–1 immersion, g, is an imbedding if $g^{-1}(Q)$ is compact for any compact $Q \subset E^{2n+1}$, or, what amounts to the same thing,

$$g^{-1}(\bar{B}_p^n) \text{ is compact for all } p. \tag{4.10}$$

We wish to choose f so that g satisfies (4.10). Choose a partition of unity $\{\varphi_i\}$ and define f by

$$f(x) = \sum k\varphi_k(x). \tag{4.11}$$

(This is well defined since $\{\text{supp } \varphi_i\}$ is locally finite.) Take $\delta(x) = 1/2$ in Theorem 4.2 and 4.3 and let g be the 1–1 immersion obtained by applying these theorems. Then $\|\varphi(x) - f(x)\| < 1$ so that $\|\varphi(x)\| \leqslant p$ implies that $\|f(x)\| \leqslant p + 1$, so that

$$x \in \bigcup_{i=1}^{p+1} \text{supp } \varphi_i$$

which is compact, i.e., (4.9) holds. Thus g is an imbedding of $M \to E^{2n+1}$.

Remarks. a) Whitney proved a stronger theorem in [22]. He showed if M is a C^1 manifold it has a C^1 imbedding, f, into E^{2n+1} where $f(M)$ is a real analytic submanifold of E^{2n+1}. This shows that any C^1 manifold carries a real analytic structure.

b) There exist manifolds of dimension n which cannot be imbedded in E^{2n-1} (for example, projective spaces of suitable dimension.)

c) If topological restrictions are placed on M, then it can be imbedded in Euclidean spaces of dim $< 2n$. The conditions can be expressed in terms of the vanishing of various homotopy groups, of [9].

d) In the case of real analytic manifolds, the corresponding theorem, that there is a real analytic embedding, lies much deeper. It is due to Morrey [13].

Our last approximation theorems have to do with regularity of intersection of two submanifolds. For example, given two curves in the plane we would expect that "in general" they are not tangent to one another at points of intersection. We formulate the condition of regularity of intersection as follows:

Definition 4.7. *Let f be a differentiable map of an n_1-manifold M_1 into an n_2-manifold M_2. Let N be an $(n_2\text{-}p)$-dimensional submanifold of M_2. The map f is said to be transverse regular* (TR) *to N at x_0 if either $f(x_0) \notin N$ or if there are coordinates $x^1, \ldots, x^{n_1}$ at x_0 and $y^1, \ldots, y^{n_2}$ at $f(x_0)$ such that*

1) *In some neighborhood U, of $f(x_0)$, $N \cap U$ is the set of points where $y^1 = \cdots = y^p = 0$, and*

2) $$\left(\frac{\partial y^i}{\partial x^j}\right) \quad \begin{matrix} i = 1, \ldots, p \\ j = 1, \ldots, n_1 \end{matrix} \quad \textit{has rank } p \textit{ at } x_0.$$

Note that 2) can occur only if $p \leqslant n_1$. If $n_1 < p$, transverse regularity means that $f(x_0) \notin N$.

Thus in Figure 4.1 ($m = 2$, $n = p = 1$) (a) is a TR intersection and (b) is not.

Definition 4.8. *The map f is said to be transverse regular on N if it is transverse regular to N at every point of M_1.*

Definition 4.9. *Let f be an immersion of M_1 into M_2. Let x_1 and x_2 be points of M_1 such that $f(x_1) = f(x_2)$. We can find disjoint neighborhoods U_1 and U_2 of x_1 and x_2 such that $f|U_1$ and $f|U_2$ are imbeddings.*

The immersion f will be called regular at (x_1, x_2) if $f|U_1$ is transverse regular to $f(U_2)$ at x_1. The immersion f will be called regular if it is transverse regular at all pairs (x_1, x_2) such that $f(x_1) = f(x_2)$.

EXERCISE 4.1. Show that if f is regular at (x_1, x_2) it is regular at (x_2, x_1).

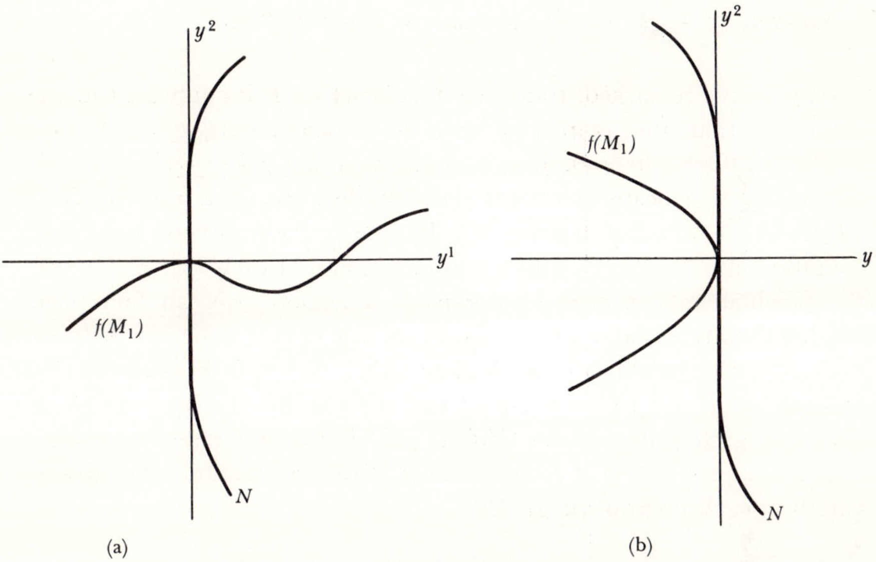

FIGURE 4.1

THEOREM 4.5. (Thom).[5] *Let f be a C^k map of $M_1 \to M_2$ and let N be a submanifold of M_2. For any (W, Z, ϵ, k) there is a $g \in \mathcal{N}_{(W, Z, \epsilon, k)}f$ which is transverse regular to N. Furthermore, if A is a closed subset of M_1 such that f is transverse regular for all $x \in A$ then we can choose g so that $g(x) = f(x)$ for $x \in A$.*

Before proceeding to the proof of Theorem 4.5 we will rewrite condition 1) of Definition 4.7. Let (U, h) be the chart at $f(x)$ and let $\pi: E^m \to E^p$ be defined by

$$\pi(y^1, \ldots, y^m) = (y^1, \ldots, y^p). \tag{4.12}$$

[5] The proof of this theorem depends on the "hard" part of Sard's theorem. It may be omitted on first reading as Theorems 4.5 and 4.6 will not be used in any serious way in the rest of this book. The reader may skip to Section 5.

Then condition 2) says that the origin, $(0, \ldots, 0)$ be a regular value of the map $\pi \circ h \circ f$ restricted to some neighborhood of x_0.

We now state the analogue of Lemma 4.5.

LEMMA 4.6. *Let f be a C^1 map of an open set $U \subset E^{n_1}$ into E^{n_2}. Suppose that f is transverse regular to the plane $P = \{y \mid y^1 = \cdots = y^p = 0\}$ at all points of a compact set K. Then there is an $\epsilon > 0$ such that any map g satisfying $\|f - g\|_1 < \epsilon$ is also transverse regular to P on K.*

As we have remarked, transverse regularity on K is equivalent to the condition that the map $\pi \circ f$ have no critical points, p, on K with $\pi \circ f(p) = (0, \ldots, 0) = 0$. If $n_1 < p$, this says that $0 \notin \pi \circ f(K)$. Since K is compact, this means that 0 is at a positive distance, say ϵ, from $\pi \circ f(K)$. Then if $\|g - f\|_1 < \epsilon$, $0 \notin \pi \circ g(K)$. If $n_1 \geqslant p$, the transverse regularity condition says that $(\pi \circ f)^{-1}(0)$ is a finite set, and at each $x_i \in (\pi \circ f)^{-1}(0)$ the Jacobian matrix of $\pi \circ f$ has rank p. Therefore, we can find open neighborhoods, U_i, about each of these points so that some $p \times p$ determinant of the Jacobian is of absolute value $> \delta > 0$ for all $x \in \bigcup U_i$. Since $K' = K - \bigcup U_i$ is compact and $0 \notin \pi \circ f(K')$ a map g which is close enough to f must have $g^{-1}(0) \in \bigcup U_i$. But then if g is close enough to f, the appropriate $p \times p$ determinant of the Jacobian of $\pi \circ g$ will not vanish at each point of $(\pi \circ g)^{-1}(0)$.

Proof of Theorem 4.5. Choose an atlas $\{Y_\alpha, H_\alpha\}$ on M_2 with $H_\alpha = (y^1_\alpha, \ldots, y^{n_2}_\alpha)$ such that either $N \cap Y_\alpha = \varnothing$ or $N \cap Y_\alpha$ is given by the set where $y^1_\alpha = \cdots = y^p_\alpha = 0$. Let $\{X_i\}$ be a locally finite refinement of $\{Y_\alpha\}$. The sets $f^{-1}(X_i)$ form a covering of M_1. Let W be an open set containing A on which f is transverse regular to N. The covering $\{W_i\}$ is given. Then the sets $\{W \cap f^{-1}(X_i) \cap W_j,\ M_1 - \overline{A \cap f(X_i) \cap W_j}\}$ form a covering of M_1. Let $\{V_s\}$ be a refinement of this covering given in Lemma 4.2, reindexed so that $V_i \subset W$ if and only if $i \leqslant 0$.

We will construct a sequence satisfying conditions i), ii), and iv) of page 60 and [instead of iii)]

v) f_s is transverse regular to N on $Q_s = \bigcup \bar{O}_s$

and

vi) for each i there is a $j(i)$ such that $f_s(P_i) \subset X_{j(i)}$ for all s.

The theorem is an immediate consequence of the existence of such a sequence.

We proceed by induction as before. Assume f_{s-1} has been defined and satisfies ii), iv), v), and vi). We set $f_s = f_{s-1}$ on $V_s - P_s$. By induction, $f_{s-1}(P_s) \subset X_{j(s)}$. Let H be a coordinate map on $X_{j(s)}$. We would like to choose f_s so that the map $\pi \circ H \circ f_s$ does not have any critical points on $Q_s \cap V_s$ which are mapped into the origin. The set of critical values has measure zero in E^p by Sard's theorem. Therefore we can choose an $a \in E^{n_2}$ as close to the origin as we please such that $\pi(a)$ is a regular value of $\pi \circ H \circ f_{s-1}$. Now define f_s on V_s by

$$f_s(x) = H^{-1} \circ [H \circ f_{s-1}(x) - a\,\varphi(h_s(x))]. \tag{4.13}$$

For $x \in \bar{O}_s$, this is $\pi \circ H \circ f_{s-1}(x) - a$. Since a is a regular value of $\pi \circ H \circ f_{s-1}$, the origin is a regular value of $(\pi \circ H \circ f_{s-1} - a)$ so there is no critical $x \in \bar{O}_s$ with $\pi \circ H \circ f_s(x) = 0$. For $x \in Q_s \cap P_s$ we can apply Lemma 4.6 if we choose a small enough. All the remaining conditions can also be achieved by taking a small. For condition ii) this is obvious. As to condition vi), let us remark that only a finite number of the P_i (those for which $P_i \cap V_s \neq \varnothing$) really enter. On all others, $f_s = f_{s-1}$. For these, we have a finite number of compact sets $P_i \subset M_1$ and open sets $X_{j(i)} \subset M_2$ with $f_{s-1}(P_i) \subset X_{j(i)}$. It is thus clear that if f_s is close enough to f_{s-1} (i.e., if a is small) we will have $f_s(P_i) \subset Y_{j(i)}$. This proves Theorem 4.5.

THEOREM 4.6. *Let f be a C^k immersion of $M_1 \to M_2$. For any (W, Z, ϵ, k) there is a regular immersion, g, of $M_1 \to M_2$ with $g \in \mathcal{N}_{(W, Z, \epsilon, k)f}$.*

Note that Theorem 4.6 is a generalization of Theorem 4.2. If $\dim M_2 > 2 \dim M_1$, regular means one to one.

Let $\{V_i\}$ be a refinement of $\{W_i\}$ as given in Lemma 4.2, such that f is one to one on each V_i. We will construct a sequence of functions $\{f_s\}$ satisfying i), ii), and iv) of page 60 and also

vii) f_s is regular at all pairs $(x_1, x_2) \in Q_s \times Q_s$,
viii) f_s is an immersion and is one to one on P_i for all i.

We construct the sequence f_s by induction, defining $f_s = f_{s-1}$ on $V_s - P_s$. Let Z be a coordinate neighborhood in M_2 containing $f_{s-1}(V_s)$ and H a coordinate map of Z. Let $N_i = H(f_{s-1}(V_i) \cap Z)$, $i = 1, \ldots, s-1$. By the induction assumption, the map $g_{s-1} = H \circ f_{s-1} \circ h_s^{-1}$ of $B_3^{n_1} \to E^{n_2}$ is a one to one immersion. We will construct a g_s which is a one to one immersion of $B_3^{n_1} \to E^{n_2}$ coinciding with g_{s-1} on $\bar{B}_3^{n_1} - B_2^{n_1}$

which is transverse regular to the N_i. First observe that the condition that g_s be one to one will be fulfilled if g_s is close enough to g_{s-1}. In fact, let

$$m = \min \frac{\|g_{s-1}(x) - g_{s-1}(y)\|}{\|x - y\|} \qquad \text{for } x, y \in P_s.$$

Then arrange that the function $d_s = g_s - g_{s-1}$ have its derivatives so small that $\|d_s(x) - d_s(y)\| < m\|x - y\|$. Then

$$\begin{aligned}\|g_s(x) - g_s(y)\| &= \|d_s(x) - d_s(y) - (g_s(x) - g_s(y))\| \\ &\geqslant \|g_s(x) - g_s(y)\| - m\|x - y\| > 0.\end{aligned}$$

This guarantees that f_s is one to one on P_s. To get f_s one to one on all the P_i, we need only worry about the finite number of P_i with $P_i \cap V_s \neq 0$. But on each of the corresponding V_i this can be achieved by taking f_s close to f_{s-1}, and this reduces to a finite number of smallness conditions on $g_s - g_{s-1}$.

Next observe, that Lemma 4.6 (or a slight modification thereof) says that if we have a map g_s^k which is transverse regular to $N_1, N_2, \ldots, N_k$ on a compact set A then any map close enough to g_s^k will also be transverse regular to these submanifolds on A.

By assumption g_{s-1} is transverse regular to $N_1, \ldots, N_{s-1}$ on $h(Q_{s-1})$. Now choose $G_s^1 : B_3^{n_1} \to E^{n_2}$ by Theorem 4.5 to be transverse regular to N_1 and so close to g_{s-1} that the map g_s^1 given by

$$g_s^1(x) = G_s^1(x) + \varphi(x)(g_{s-1}(x) - G_s^1(x))$$

is transverse regular to all the N_i on $h(Q_s \cap V_s)$. Next choose $G_s^2 : B_3^{n_1} \to E^{n_2}$ transverse regular to N_2 on $B_3^{n_1}$ and so close to g_s^1 that the map g_s^2 given by

$$g_s^2(x) = G_s^2(x) + \varphi(x)(G_s^2(x) - g_s^1(x))$$

is transverse regular to all the N_i on $h(Q_s \cap V_s)$ and to N_1 on $\bar{B}_1^{n_1}$. Continue inductively in this way, until we have a map $g_s = g_s^s$ which is transverse regular to $N_1, \ldots, N_k$ on $h(Q_s \cap V_s) \cup \bar{B}_1^{n_1}$. Also $g_s(x) - g_{s-1}(x)$ can be made as small as we like by imposing additional "closeness" conditions at each stage. Take $f_s = H^{-1} \circ g_s \circ h$, the claim is that f satisfies conditions ii), iv), vii), and viii). Conditions ii), iv), and viii) are obviously satisfied. Let us examine condition vii). Let $(x_1, x_2) \in Q_s \times Q_s$. If $x_1 \in V_s$ and $x_2 \in V_s$ then the condition is trivially verified since f_s is 1–1 on V_s. If $x_1 \in V_s$ and $x_2 \notin V_s$, then $f_s(x) = f_{s-1}(x)$ for some neighborhood of x_2. By our construction f_s is transverse regular to $f_{s-1}(V_i)$ at all $x \in Q_s \cap V_s$. If $x_1 \notin V_s$ and $x_2 \notin V_s$ then $f_s(x) = f_{s-1}(x)$ for all x in some neighborhood of x_1 and x_2 so that vii) is satisfied by the induction hypothesis.

The mapping $g = \lim f_s$ provides the desired approximation in Theorem 4.6.

5. The tangent space. In our study of differentiable manifolds it is useful to introduce various "infinitesimal" objects. We will give a coherent treatment of such objects in the next section. In order to motivate that treatment we first study the notion of a "tangent vector" on a manifold. In classical vector analysis, the notion of the tangent vector to a curve is straightforward. If $\varphi(t) = (\varphi^1(t), \ldots, \varphi^n(t))$ is a curve in E^n with $\varphi(0) = p$, then the tangent vector to φ at p is given by $(\varphi^{1\prime}(0), \ldots, \varphi^{n\prime}(0))$. Two curves, φ and ψ, have the same tangent vector if $\psi^{i\prime}(0) - \varphi^{i\prime}(0) = 0$ for all i. Now in a manifold, we cannot write every curve as an n-tuplet of functions of one variable unless we choose a local coordinate system. Nevertheless, we can define a tangent vector at p in a more abstract way as an equivalence class of curves through p; two curves being equivalent if they have the same derivatives in one, and hence all, coordinate systems about p.

More precisely, let $Cu(M)$ be the set of all pairs (p, φ) where $p \in M$ and φ is a differentiable map of the interval $(-\epsilon, \epsilon) \to M$ with $\varphi(0) = p$. That is, $Cu(M)$ is the space of all curves of M. There is an obvious map $\pi: Cu(M) \to M$ defined by $\pi(p, \varphi) = p$. On the space $Cu(M)$ we introduce the following equivalence relation:

$$(p, \varphi) \sim (p', \varphi') \quad \text{if} \quad p = p' \quad \text{and} \quad \left.\frac{d(x^i \circ \varphi)}{dt}\right|_{t=0} = \left.\frac{d(x^i \circ \varphi')}{dt}\right|_{t=0} = 0,$$

where $(x^1, \ldots, x^n)$ is some coordinate system about p.

We observe that if $y^i = y^i(x^1, \ldots, x^n)$ form a second coordinate system about p, that

$$\frac{dy^i \circ \varphi}{dt} = \sum_{j=1}^{n} \frac{\partial y_i}{\partial x_j} \frac{dx^j \circ \varphi}{dt}. \tag{5.1}$$

Thus the equivalence relation is independent of the coordinate system. We will denote the space of cosets of $Cu(M)$ by $T(M)$; an element, X, of $T(M)$ [i.e., a coset of $Cu(M)$] is called a *tangent vector*. An element $(p, \varphi) \in X$ is said to be tangent to X. It is immediate that the map π is constant on cosets; therefore, it induces a map, which we continue to denote by π, of $T(M) \to M$. If $X \in T(M)$ and $\pi(X) = p$, we say that X is a tangent vector at p. The set $\pi^{-1}(p)$ is denoted by $T_p(M)$ and called the *tangent space to M at p*. We observe that a choice of coordinates about p induces a map of $T_p(M) \to E^n$. In fact, let $h = (x^1, \ldots, x^n)$ be coordinates at p and $X \in T_p(M)$. Let (p, φ) be tangent to X and set

$$\bar{h}(X) = (X^1, \ldots, X^n), \; X^i = \left.\frac{dx^i \circ \varphi}{dt}\right|_{t=0}. \tag{5.2}$$

It is clear that (5.2) does not depend on the choice of φ. The map $\bar{h}$ of $T_p(M) \to E^n$ is clearly one to one. We claim that $\bar{h}(T_p(M)) = E^n$. In fact, if $(X^1, \ldots, X^n) \in E^n$, define the curve $\varphi(t)$ by

$$\varphi(t) = h^{-1}(h(p) + t(X^1, \ldots, X^n)).$$

A direct computation shows that if Y is tangent to (p, φ) then $\bar{h}(Y) = (X^1, \ldots, X^n)$. If $h' = (y^1, \ldots, y^n)$ is a second coordinate system with $h' \circ h^{-1}$ given by $y^1(x^1, \ldots, x^n), \ldots, y^n(x^1, \ldots, x^n)$, then $\bar{h}'(t) = (Y^1, \ldots, Y^n)$ where

$$Y^i = \sum \frac{\partial y^i}{\partial x^j} X^j, \tag{5.3}$$

(the partial derivatives being evaluated at $h(p)$). Thus the map $\bar{h}' \circ \bar{h}^{-1}$ is a linear map of E^n onto E^n. *This implies that we have a canonical vector space structure on* $T_p(M)$. In fact if a and b are real numbers and $Y, X \in T_p(M)$, define

$$aY + bX = \bar{h}^{-1}(a\bar{h}(Y) + b\bar{h}(X)). \tag{5.4}$$

Since $\bar{h}' \circ \bar{h}^{-1}$ is linear, (5.4) is independent of h. The map $\bar{h}$ is then an isomorphism of $T_p(M)$ onto E^n.

Let $\mathscr{F}_U$ denote the (linear) space of all differentiable functions of M defined on U.

We now introduce the notion of directional derivative of a function f with respect to a tangent vector X_p at the point p.

THEOREM 5.1. *To every tangent vector,* X_p, *of a manifold* M, *there corresponds a unique linear function* L_{X_p} *defined on* $\mathscr{F}_U$ *for any open set* U *containing* p. *This linear function satisfies*

$$L(fg) = f(p)L(g) + g(p)L(f) \tag{5.5}$$

for all $f, g \in \mathscr{F}_U$. *If* $L_{X_p} = L_{Y_p}$ *then* $X_p = Y_p$. *Furthermore,*

$L_{aX_p+bY_p} = aL_{X_p} + bL_{Y_p}$. *If M is analytic or C^∞ and L is a linear function defined on $\mathscr{F}_U$ satisfying* (5.5) *for some coordinate neighborhood, U, of p, there exists a unique $X_p \in T_p(M)$ such that $L = L_{X_p}$.*

Proof. Let (p, φ) be tangent to X_p. Then define L_{X_p} by

$$L_{X_p}(f) = \left.\frac{df[\varphi(t)]}{dt}\right|_{t=0}. \tag{5.6}$$

It is immediate that L_{X_p} satisfies (5.5). If $(p, \varphi) \sim (p, \psi)$ then

$$\left.\frac{df[\varphi(t)]}{dt}\right|_{t=0} = \left.\frac{df[\psi(t)]}{dt}\right|_{t=0} \tag{5.7}$$

for all f so that L_{X_p} does not depend on the choice of (p, φ). If (5.7) holds, then (taking $f = x^i$), $(p, \varphi) \sim (p, \psi)$ so that $L_{X_p} = L_{Y_p}$ implies that $X_p = Y_p$. It is also clear that the map $X_p \to L_{X_p}$ is linear.

Now assume M is C^∞ or analytic. Let L be a linear functional satisfying (5.6) on $\mathscr{F}_U$ where (U, h) is a coordinate system with $h = (x^1, \ldots, x^n)$ and $h(p) = (0, \ldots, 0)$. By (5.5), $L(1) = 2L(1) = 0$ (if 1 denotes the function assuming the constant value 1). Thus L vanishes on the constants. For any f,

$$f \circ h^{-1}(x^1, \ldots, x^n) = f(p) + \sum a_i x^i + \sum x^i x^j f_{ij}(x^1, \ldots, x^n)$$

where f_{ij} is C^∞ or analytic. (Note that if f were C^k then f_{ij} would be C^{k-2} and the proof would break down.) Then $L(f) = \sum a_i L(x^i)$ since

$$\begin{aligned} L(x^i x^j f_{ij}) &= x^i(p)L(x^j) f_{ij} \circ h(p) + x^j(p)L(x^i) f_{ij} \circ h(p) \\ &\qquad + x^i(p)x^j(p)L(f_{ij} \circ h) \\ &= 0 + 0 + 0 = 0. \end{aligned}$$

Thus

$$L(f) = \left.\frac{df[\varphi(t)]}{dt}\right|_{t=0} \qquad \text{if} \quad h \circ \varphi(t) = (tL(x^1), \ldots, tL(x^n)).$$

That is, $L = L_{X_p}$ where X_p is tangent to (p, φ). In any event, no matter what the differentiability class of M, any linear functional of the form (5.6) can be identified with the tangent vector tangent to (p, φ). Thus if φ^i is the curve given by

$$x^j(\varphi^i(t)) = \delta_i^j t \tag{5.8}$$

then the tangent to (p, φ^i) can be identified with the linear functional

$$\left(\frac{\partial}{\partial x^i}\right)_p$$

where

$$\left(\frac{\partial}{\partial x^i}\right)_p (f) = \left.\frac{\partial f}{\partial x^i}\right|_p .$$

It follows from the definitions that $\bar{h}(\partial/\partial x^i)_p = (\delta_i^1, \ldots, \delta_i^n)$ so that the vectors $(\partial/\partial x^i)_p$ form a basis of $T_p(M)$.

Let f be a differentiable function defined at p. Then f determines a linear function on $T_p(M)$ via (5.6). We will denote this function by $(df)_p$; thus

$$\langle X_p, (df)_p \rangle = L_{X_p}(f). \tag{5.9}$$

In view of the definitions, we have

$$\left\langle \left(\frac{\partial}{\partial x^i}\right)_p, (dx^j)_p \right\rangle = \delta_i^j \tag{5.10}$$

for any coordinate system $(x^1, \ldots, x^n)$. Thus the $(dx^j)_p$ constitute the basis of $(T_p(M))^*$ dual to $(\partial/\partial x^i)_p$. We will denote $(T_p(M))^*$ by $T_p^*(M)$ and set $T^*(M) = \bigcup_{p \in M} T_p^*(M)$. An element of $T_p^*(M)$ will be called a cotangent vector at p. If f is a differentiable function then (5.9) and (5.10) imply that

$$df_p = \sum \left(\frac{\partial f}{\partial x_i}\right)_p dx_p^i . \tag{5.11}$$

If g is a differentiable map of a manifold $M_1 \to M_2$, g induces, in the obvious way, a map, g_*, of $Cu(M_1) \to Cu(M_2)$. In fact, if $(p, \varphi) \in M$, let

$$g_*(p, \varphi) = (g(p), g \circ \varphi) \in Cu(M_2). \tag{5.12}$$

It is apparent that if $(p, \varphi) \sim (p', \varphi')$ then $g_*(p, \varphi) \sim g_*(p', \varphi')$. That is, g_* induces a map of $T(M_1) \to T(M_2)$. We will also call this map g_*. Let $h = (x^1, \ldots, x^{n_1})$ be local coordinates at $p \in M_1$ and $H = (y^1, \ldots, y^{n_2})$ be local coordinates at $g(p)$, with g given by the functions $y^i(x^1, \ldots, x^{n_1})$. Then if φ^i is the curve given by (5.8) the curve $g \circ \varphi^i$ has, in local coordinates, the form $(y^1(\delta_1^i t, \ldots, \delta_{n_1}^i t), \ldots, y^{n_2}(\delta_1^i t, \ldots, \delta_{n_1}^i t)$ so that

$$g_* \left(\frac{\partial}{\partial x^i}\right)_p = \sum \frac{\partial y^j}{\partial x^i} \left(\frac{\partial}{\partial y^j}\right)_{g(p)} . \tag{5.13}$$

From (5.13) we see that the map g_* is linear on each $T_p(M)$.

EXERCISE 5.1. Show that g is an immersion if and only if g_* is one to one on each $T_p(M)$.

We will see in the next section that $T(M)$ and $T^*(M)$ can be made into differentiable manifolds, and that the map g_* will be differentiable. For the present let us verify this for the case that M is a vector space, say E^n. On E^n there is a canonical coordinate system. Relative to this coordinate system a tangent vector is given by an n-tuplet, which we can regard as an element of E^n. That is we have a map, l, of $T_p(E^n)$ onto E^n. Thus via l, we can identify $T(E^n)$ with $E^n \times E^n$.

EXERCISE 5.2. Let $g_a: E^n \to E^n$ be translation by a, i.e., $g_a(x) = x + a$. Show that for any $t \in T(E^n)$, $l((g_a)_*(t)) = l(t)$.

EXERCISE 5.3. Let $\underline{t} \in T_p(E^n)$, and f be a differentiable function defined at p. Show that

$$L_{\underline{t}}(f) = \lim_{s=0} \frac{1}{s} (f(p + sl(\underline{t})) - f(p)).$$

It is this identification which is used when one talks, in classical analysis, of a vector at a point of E^n. If g is an immersion of a manifold $M \to E^n$ then $l(g_*(T_p(M)))$, $p \in M$ is a subspace of E^n. It is convenient, pictorially, to regard the hyperplane parallel to this subspace passing through $g(p)$. It is "tangent" to $g(M)$ in the obvious sense of Figure 5.1.

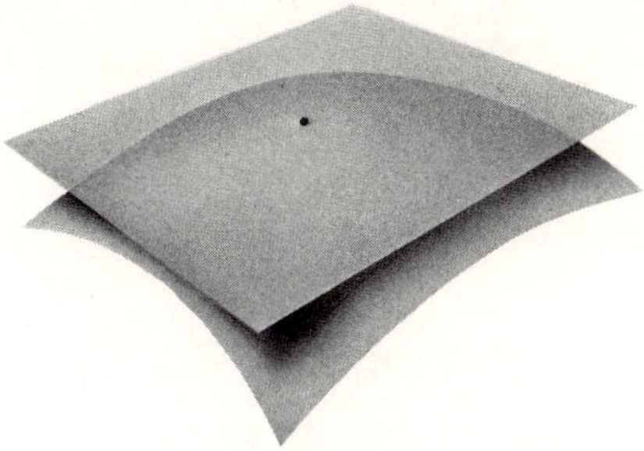

FIGURE 5.1

6. The principal bundle. In the study of differentiable manifolds, it is useful to introduce "geometrical objects" analogues to the quantities of

Chapter I, Section 2. Here, the analogue of a basis of the vector space V would be a chart (U, h) of a manifold M. A geometrical object would then be an element $q(U, h)$ of some space Q attached to each chart (U, h), together with a rule of transformation of q corresponding to every local change of coordinates. An example of such an object would be a tangent vector: to each local coordinate system a tangent vector determined an element of E^n; under a change of coordinates, this element changed via (5.3). In order to make this definition precise, we would have to find some analogue for the group $GL(n)$ and discuss its action on the set of all local coordinates. Such a program can be carried out (cf. for example [16]). In most cases that arise in practice, however (such as tangent and cotangent vectors), the law of transformation of the q's depend only on the Jacobian matrix of the change of coordinates. This makes things much simpler, and allows us to set up definitions very analogous to those of Section 2 of Chapter I. If the laws of transformation of the objects under consideration depend only on the linear part of the change of coordinates, it is natural to identify two coordinate systems at a point if they have the same "linear part." With this in mind we proceed to the definitions.

Let M be an n-dimensional differentiable manifold. Denote by $\underline{U}(M)$ the space of all pairs $(p, (U, h))$ where $p \in M$ and (U, h) is a chart about p. The group $GL(n)$ acts on $\underline{U}(M)$ as follows: If $A \in GL(n)$ and $h = (x^1, \ldots, x^n)$ set $(p, (U, h))A = (p, (U, g))$ where $g = (y^1, \ldots, y^n)$ with

$$y^i = \sum \beta^i_j x^j \tag{6.1}$$

where $A^{-1} = (\beta^i_j)$. Then $(p, (U, h))(AA') = ((p, (U, h))A)A'$. Let π denote the map of $\underline{U}(M)$ onto M which sends $(p, (U, h))$ into p. Then $\pi(\underline{u}A) = \pi(\underline{u})$ for $\underline{u} \in \underline{U}(M)$.

On $\underline{U}(M)$ introduce the equivalence relation, $\sim$, defined by

$$(p, (U, h)) \sim (p', (U', h')) \quad \text{if} \quad p = p' \quad \text{and} \quad J_{h(p)}(h' \circ h^{-1}) = I.$$

(Here, if φ is a map of $U \subset E^n \to E^n$, $J_x(\varphi)$ denotes the Jacobian matrix of φ at x; I is the unit matrix). It follows immediately from the multiplicative law for Jacobians that $\sim$ is an equivalence relation. We will denote by $\mathscr{F}^*(M)$ the space of cosets of $\underline{U}(M)$ under this equivalence relation. If $\underline{u}_1 \sim \underline{u}_2$ then $\pi(\underline{u}_1) = \pi(\underline{u}_2)$ so that π induces a map, π, of $\mathscr{F}^*(M)$ onto M.

EXERCISE 6.1. We can give an alternative description of $\mathscr{F}^*(M)$ in terms of the space $T^*(M)$. In fact, every coordinate system $(x^1, \ldots, x^n)$ at p

determines the bases $dx_p^1, \ldots, dx_p^n$ of $T_p^*(M)$. Show that if $(p, (U, h)) \sim (p', (U', h'))$ they determine the same basis of $T_p^*(M)$. Conversely, show that every basis of $T_p^*(M)$ is of the form $dx_p^1, \ldots, dx_p^n$ for some coordinate system $(x^1, \ldots, x^n)$. Thus $\mathscr{F}^*(M)$ can be identified with the set of all frames (i.e., bases) of all spaces $T_p^*(M)$ for $p \in M$.

If $u_1 \sim u_2$ then $u_1 A \sim u_2 A$ so that we have a well defined action of $GL(n)$ on $\mathscr{F}^*(M)$. Also $\pi(zA) = \pi(z)$ for $z \in \mathscr{F}^*(M)$. We can make a stronger statement:

LEMMA 6.1. *If $zA = z$ for any $z \in \mathscr{F}^*(M)$ then $A = I$. If $z_1 A = z_2$ for some $A \in GL(n)$ then $\pi(z_1) = \pi(z_2)$. Conversely, if $\pi(z_1) = \pi(z_2)$ there is an $A \in GL(n)$ such that $z_1 A = z_2$.*

The first two statements are immediate consequences of the definition of the action of $GL(n)$ on $\mathscr{F}^*(M)$. We prove the last statement: Let $(p, (u_1, h_1))$ and $(p, (u_2, h_2))$ be elements of z_1 and z_2. If

$$A = J_{h_2(p)}(h_1 \circ h_2^{-1}) \qquad \text{then} \quad z_1 A = z_2.$$

We now make $\mathscr{F}^*(M)$ into a differentiable manifold (of class C^{k-1} if M is of class C^k or analytic if M is analytic).

Let $\{(U_i, h_i)\}$ be a denumerable family of charts such that the U_i form a basis for the topology of M. Define the mapping ψ_i of $\pi^{-1}(U_i) \to GL(n)$ as follows: Let $z \in \pi^{-1}(U_i)$ and $(\pi(z), (U', h'))$ be a chart in the coset of z. Set $\psi_i(z) = J_{h_i(\pi(z))}(h' \circ h_i^{-1})$. This map is well defined since the equivalence relation defining $\mathscr{F}^*(M)$ guarantees that $J_{h_i(\pi(z))}(h' \circ h_i^{-1})$ depends only on z. A direct calculation shows that

$$\psi_i(zA) = A^{-1}\,\psi_i(z). \tag{6.2}$$

On $U_i \cap U_j$ we have

$$\psi_i(z) = \psi_j(z) J_{h_i(\pi(z))}(h_j \circ h_i^{-1}). \tag{6.3}$$

Now define the map $\varphi_i : \pi^{-1}(U_i) \to U_i \times GL(n)$ by

$$\varphi_i(z) = (\pi(z), \psi_i(z)). \tag{6.4}$$

By Lemma 6.1, φ_i is one to one and onto. Now $GL(n) = T(n, n; n)$ is a differentiable manifold. Furthermore, $\varphi_i(\pi^{-1}(U_j)) = U_i \cap U_j \times GL(n)$ and

$$\varphi_i \circ \varphi_j^{-1}(x, A) = (x, AJ_{h_i(x)}(h_j \circ h_i^{-1})) \tag{6.5}$$

by (6.3). Thus $\varphi_i \circ \varphi_j^{-1}$ is a diffeomorphism (of class C^{k-1} or analytic if M is C^k or analytic). Finally, the sets $\pi^{-1}(U_i)$ satisfy condition 1) of Theorem 2.1. In fact if $\pi(z_1) = \pi(z_2) = p$ then z_1 and z_2 lie in $\pi^{-1}(U_i)$ for any U_i containing p. If $\pi(z_1) = p \neq q = \pi(z_2)$, since the $\{U_i\}$ form a basis for the (Hausdorff) topology of M, we can find an i and j such that $p \in U_i$, $q \in U_j$ and $U_i \cap U_j = \varnothing$. Thus, by Theorem 2.1 there is a unique differentiable structure on $\mathscr{F}^*(M)$ such that the φ_i's are diffeomorphisms.

EXERCISE 6.2. Show that this differentiable structure is independent of the choice of the atlas $\{(U_i, h_i)\}$.

DEFINITION 6.1. *The space $\mathscr{F}^*(M)$ with the above topology and differentiable structure will be called the bundle of coframes of M.*

Since the maps φ_i are diffeomorphisms, the map π is differentiable. In fact, in $\pi^{-1}(U_i)$ it is given by $\pi = p_1 \circ \varphi_i$ where $p_1(x, A) = x$. We will call π the projection of $\mathscr{F}^*(M)$ onto M. The action of $GL(n)$ on $\mathscr{F}^*(M)$ is also nice. It is clear from (5.1) that the map $z \to z \cdot A$ is a diffeomorphism. We can say somewhat more.

DEFINITION 6.2. *Let Q be a differentiable manifold. $GL(n)$ is said to act differentiably on Q to the right (or to the left) if there is a differentiable map, φ, of $GL(n) \times Q \to Q$ such that $\varphi(\varphi(x, A), B) = \varphi(x, AB)$ (or $= \varphi(x, BA)$), and $\varphi(x, I) = x$ if I is the identity of $GL(n)$.*

It follows from the definitions that $GL(n)$ acts differentiably to the right on $\mathscr{F}^*(M)$. In fact, in terms of local coordinates, the map of $GL(n) \times \pi^{-1}(U_i) \to \pi^{-1}(U_i)$ is given by

$$\varphi_i^{-1} \circ (L \times \mathrm{id}) \circ (\mathrm{id} \times \varphi_i) \tag{6.6}$$

where $L: GL(n) \times GL(n) \to GL(n)$ is defined by $L(A, B) = A^{-1}B$. [In (6.6) id $\times\ \varphi_i$ maps $GL(n) \times \pi^{-1}(U_i) \to GL(n) \times GL(n) \times d_i$ and $L \times$ id maps $GL(n) \times GL(n) \times U_i \to GL(n) \times U_i$.] By Exercise 2.8 the map L (group multiplication) is differentiable and therefore so is (6.6).

We are now in a position to define differential geometric quantities similar to the quantities of Chapter I, Section 2. Let Q be a differentiable manifold on which $GL(n)$ acts differentiably to the left. Let us denote the action by $\varphi(A, q) = \rho(A) \cdot q$ so that $\rho(AB) \cdot x = \rho(A) \cdot (\rho(B) \cdot x)$.

On the space $\mathscr{F}^*(M) \times Q$ introduce the equivalence relation

$$\begin{aligned} (z, q) \sim (z', q') \quad & \text{if there is an } A \in GL(M) \text{ such that} \\ z' = zA \quad & \text{and} \quad q = \rho(A)q'. \end{aligned}$$

Let $E(M, Q)$ denote the quotient space of $\mathscr{F}^*(M) \times Q$ by this equivalence relation. On $E(M, Q)$ we can introduce a differentiable structure. In fact, there is a map, Φ_i^{-1}, of $U_i \times Q$ onto the quotient of $\pi^{-1}(U_i) \times Q$ by $\sim$ given by

$$\Phi_i^{-1}(x, q) = \{\varphi_i^{-1}(x, I), q\} \tag{6.7}$$

where $\{z, q\}$ denotes the coset of (z, q) and I is the identity of $GL(n)$. It is easy to check that Φ_i^{-1} is one to one and onto. Let us denote by Φ_i the inverse of Φ_i^{-1}. Then

$$\Phi_i \circ \Phi_j^{-1}(x, q) = (x, q')$$

where q' is defined by $\{\varphi_i^{-1}(x, I), q'\} = \{\varphi_j^{-1}(x, I), q\}$.

Now

$$\begin{aligned} \varphi_j^{-1}(x, I) &= \varphi_i^{-1}(x, J_{h_i(x)}(h_j \circ h_i^{-1})) && \text{by (6.5)} \\ &= \varphi_i^{-1}(x, I)\ (J_{h_i(x)}(h_j \circ h_i^{-1}))^{-1} && \text{by (6.2) and (6.4).} \end{aligned}$$

Thus

$$q' = \rho((J_{h_i(x)}(h_j \circ h_i^{-1}))^{-1})q = \rho(J_{h_j(x)}(h_i \circ h_j^{-1}))q.$$

In other words,

$$\Phi_i \circ \Phi_j^{-1}(x, q) = (x, \rho(J_{h_j(x)}(h_i \circ h_j^{-1})) \cdot q). \tag{6.8}$$

The map of $h_j(U_i \cap U_j) \to G$ sending $x \to J_{h_j(x)}(h_i \circ h_j^{-1})$ is differentiable as is the map of $G \times Q \to Q$ given by $(A, q) \to \rho(A) \cdot q$. Their composite, which occurs in (6.8) is differentiable and thus $\Phi_i \circ \Phi_j^{-1}$ is differentiable also. It is now immediate that the conditions of Theorem 2.1 are satisfied. As in Exercise 6.2, it is easy to verify that this differentiable structure does not depend on the choice of U_i. If we define the map $\bar{\pi}_Q$ of $\mathscr{F}^*(M) \times Q$ onto M by $\bar{\pi}_Q(z, q) = \pi(z)$, then $\bar{\pi}_Q$ is constant on cosets and defines a map, π_Q, of $E(M, Q)$ onto M. It is easy to check that π_Q is differentiable. We will call it the projection of $E(M, Q)$ onto M; sometimes, when there is no danger of confusion, we will drop the subscript Q and denote π_Q by π.

EXERCISE 6.2. Show that for any $x \in M$, $\pi_Q^{-1}(x)$ is a submanifold of $E(M, Q)$ which is diffeomorphic to Q. The diffeomorphism depends on a choice of coordinates about x. Show that any two such diffeomorphisms differ by the action of $GL(n)$ on Q.

DEFINITION 6.3. *The differentiable manifold $E(M, Q)$ is called the space of quantities of type Q. A differentiable map, s, $M \to E(M, Q)$ satisfying*

$$\pi_Q \circ s = \text{id} \tag{6.9}$$

is called a field of quantities of type Q. The field will be called of class C^k or analytic if the map s is C^k or analytic.

For any chart (U, h), which we can take as a (U_i, h_i), the map $s \mid U_i$ induces a map $s_i : U_i \to Q$ defined by

$$(x, s_i(x)) = \Phi_i(s(x)). \tag{6.10}$$

In $U_i \cap U_j$ we have $(x, s_i(x)) = \Phi_i \circ \Phi_j^{-1}(x, s_j(x))$ or, by (6.8),

$$s_i(x) = \rho(J_{h_j(x)}(h_i \circ h_j^{-1}))\, s_j(x). \tag{6.11}$$

Conversely, let $\{(U_i, h_i)\}$ be an atlas of M and to each (U_i, h_i) a map $s_i : U_i \to Q$ satisfying (6.11) on $U_i \cap U_j$. Then

$$\Phi_i^{-1}(x, s_i(x)) = \Phi_j^{-1}(x, s_j(x))$$

on $U_i \cap U_j$ so the maps $x \to \Phi_i^{-1}(x, s_i(x))$ piece together to form a well defined differentiable field $s : M \to E(M, Q)$. If U is an open subset of M, we can view $E(U, Q)$ in the obvious way, as an open submanifold of $E(M, Q)$. If s is a field of $E(U, Q)$, we will sometimes say (by abuse of language) that s is a field of type Q (of M) defined on U. By the previous discussion, if s^i are fields of type Q defined on open sets U_i such that $s^i = s^j$ on $U_i \cap U_j$ then there is a unique field on $\bigcup U_i$ which coincides with the s^i on U_i.

Before proceeding, let us give some examples.

1. Let Q be a differentiable manifold on which $GL(n)$ acts trivially; i.e., $\rho(A)q = q$ for all A and Q. Then $\Phi_i \circ \Phi_j^{-1} = \text{id}$ by (5.7) and $E(M, Q)$ can be identified as $M \times Q$. A field of type Q is then just given by a map of $M \to Q$. In the special case where Q is the real line, we will denote the space $E(M, Q)$ by $R(M)$.

2. Let $Q_T = E^n =$ the space of all n-tuplets $X = (X^1, \ldots, X^n)$. If $A = (\alpha_j^i)$ let $\rho(A)X = Y = (Y^1, \ldots, Y^n)$ where

$$Y^i = \sum \alpha_j^i X^j.$$

The space $E(M, Q_T)$, can be identified with $T(M)$, in view of the discussion about Equation (5.3). *With this identification we have introduced a differentiable structure on* $T(M)$. The space $T(M)$ with this differentiable structure is called the *tangent bundle of* M. A field of type Q_T is called a *(contravariant) vector field.* In terms of a local coordinate system, (U, h) with $h = (x^1, \ldots, x^n)$, a vector field is given by a differentiable map of $U \to Q_T$; i.e., by n functions $X^1(\cdot), \ldots, X^n(\cdot)$. In view of the discussion in Section 5, the vector field in U corresponding to the constant function $(\delta_i^1, \ldots, \delta_i^n)$ assigns to each point $p \in u$ the tangent vector $(\partial/\partial x^i)p$. We will call this vector field $\partial/\partial x^i$. By linearity, the vector field given by $(X^1, \ldots, X^n)$ can be written as

$$\sum_{i=1}^{n} X^i \frac{\partial}{\partial x^i}.$$

If $y^1, \ldots, y^n$ is a second coordinate system we can write (5.3) as

$$\sum Y^i \frac{\partial}{\partial y^i} = \sum \frac{\partial y^i}{\partial x^j} X^j \frac{\partial}{\partial y^i} = \sum X^j \frac{\partial}{\partial x^j} \tag{6.12}$$

or

$$\frac{\partial}{\partial x^j} = \sum \frac{\partial y^i}{\partial x^j} \frac{\partial}{\partial y^i}. \tag{6.13}$$

Thus the vector fields transform as "differential operators."

3. Let Q_{T^*} = the space of all n-tuplets $X = (X_1, \ldots, X_n)$. If $A = (\alpha_j^i)$ and $A^{-1} = (\beta_j^i)$ let $\rho(A)X = Y = (Y_1, \ldots, Y_n)$ where

$$Y_j = \sum \beta_j^i X_i.$$

The space $E(M, Q_{T^*})$ can be identified as before, with $T^*(M)$. The space $T^*(M)$, with this differentiable structure is called the *cotangent bundle.* The field corresponding to $(\delta_1^i, \ldots, \delta_n^i)$ in the coordinate system $(x^1, \ldots, x^n)$ will be denoted by dx^i. Thus $dx^i(p) = dx_p^i$. A field of type Q_{T^*} is called a *covariant vector field* or a *(linear) differential form.* Thus the most general linear differential form can be written, in a coordinate neighborhood, as $\sum X_i\, dx^i$ where the X_i are differentiable functions. If f is a differentiable function then by (5.11), the map $p \to (df)_p$ is a differential form which we shall denote by df. Thus, in a local coordinate system,

$$df = \sum \frac{\partial f}{\partial x^i} dx^i.$$

4. More generally, if Q is a finite-dimensional real vector space and ρ is any linear representation of $GL(n)$ on Q, then it is clear that $GL(n)$ acts

differentiably on Q. Thus we can talk about linear quantities. In particular, if ρ is a tensor representation of G we will call $E(M, Q)$ the space of tensors of type ρ. A field of such quantities will be called a *tensor field.* For example, a tensor field of type ${}^{1}{}_{23}$ is given, in (u, h) the n^3 functions $s^{i_1}{}_{i_2 i_3}$. If $t^{i_1}{}_{i_2 i_3}$ are the functions in the (v, h') system, the change is given by

$$t^{i_1}{}_{i_2 i_3}(p) = \sum \left(\frac{\partial y^{i_1}}{\partial x^{j_1}}\right)_{h(p)} \left(\frac{\partial x^{j_2}}{\partial y^{i_2}}\right)_{h'(p)} \left(\frac{\partial x^{j_3}}{\partial y^{i_3}}\right)_{h'(p)} s^{j_1}{}_{j_2 j_3}(p).$$

EXERCISE 6.3. Let $\mathscr{F}(M)$ be the set of all pairs $(p; X_1, \ldots, X_n)$ where $p \in M$ and $X_1, \ldots, X_n$ is a basis of $T_p(M)$. Show that $\mathscr{F}(M)$ carries a natural differential structure. (What is the space Q in this case?) The space $\mathscr{F}(M)$ with this structure is called the *bundle of frames* of M.

If $Q_1, \ldots, Q_k$ are spaces on which $GL(n)$ acts differentiably, we can define $GL(n)$ to act on $Q_1 \times \cdots \times Q_k$, by $\rho(A)(q_1, \ldots, q_k) = (\rho_1(A)q_1, \ldots, \rho_k(A)q_k)$. To see that this is a differentiable action we note that the map of $GL(n) \times Q_1 \times \cdots \times Q_k \to Q_1 \times \cdots \times Q_k$ is the composite of the maps

$$\overbrace{GL(n) \times \cdots \times GL(n)}^{k\text{-times}} \times Q_1, \ldots, Q_k \to Q_1 \times \cdots \times Q_k$$

defined by sending $(A_1, \ldots, A_k, q_1, \ldots, q_k) \to (\rho_1(A_1)q_1, \ldots, \rho_k(A_k)q_k)$ and the diagonal map $A \to (A, \ldots, A)$. Both of these are differentiable. We thus can define $E(M, Q_1 \times \cdots \times Q_k)$. Let $s^1, \ldots, s^k$ be fields of type $Q^1, \ldots, Q^k$ which define the maps s_i^r of $u_i \to Q_i$ by (6.10). The maps $s_i : u_i \to Q_1 \times \cdots \times Q_k$ given by

$$s_i(x) = (s_i^1(x), \ldots, s_i^k(x))$$

define a field, s, of $E(M, Q_1 \times \cdots \times Q_k)$. To see this we have to check (6.11). This says that

$$\begin{aligned} s_i(x) &= (s_i^1(x), \ldots, s_i^k(x)) = \rho(J_{h_j(x)}(h_i \circ h_j^{-1}))s_j(x) \\ &= (\rho_1(J_{h_j(x)}(h_i \circ h_j^{-1}))s_j^1(x), \ldots, \rho_k(J_{h_j(x)}(h_i \circ h_j^{-1}))s_j^k(x)) \end{aligned}$$

which is true since the s^r are fields of type Q_r. The field s will be called the product of $s^1, \ldots, s^k$ and will be denoted by $(s^1, \ldots, s^k)$.

Let Q_1 and Q_2 be spaces on which $GL(n)$ acts differentiably. A map $f: Q_1 \to Q_2$ is said to be equivariant if $f(\rho_1(A)q) = \rho_2(A)f(q)$ for all

$A \in GL(n)$ (where ρ_1 and ρ_2 denote the action on Q_1 and Q_2). For example, if $Q_1 = Q \times Q$ and $Q_2 = Q$ where Q is some *linear* representation of $GL(n)$, then the map of $Q_1 \to Q_2$ sending $(a, b) \to a + b$ is clearly equivariant. If the map f of $Q_1 \to Q_2$ is equivariant then the map id $\times f$ of $\mathscr{F}^*(M) \times Q_1 \to \mathscr{F}^*(M) \times Q_2$ takes cosets into cosets and thus defines a mapping, $f_\#$, of $E(M, Q_1)$ into $E(M, Q_2)$. In terms of the chart (U_i, h_i), the map $f_\#$ restricted to $\pi^{-1}(U_i)$ is given by

$$\Phi_{2i}^{-1} \circ (\text{id} \times f) \circ \Phi_{1i}$$

where Φ_{1i} and Φ_{2i} are the coordinate maps, (6.7), of $E(M, Q_1)$ and $E(M, Q_2)$ respectively. Thus if f is differentiable so is $f_\#$.

EXERCISE 6.4. Let f be an equivariant map of $Q_1 \to Q_2$, and let s be a field of type Q_1 which is given in local coordinates by the maps $s_i : u_i \to Q_1$. Show that $f_\# \circ s$ is a field of type Q_2 which is given, in local coordinates, by $f \circ s_i$.

We can use the previous remarks to introduce algebraic operations on fields of linear quantities, in particular on tensor fields. For instance, if f is the equivariant map of $Q \times Q \to Q$ giving addition, then $f_\#(s^1, s^2)$ will be a field of type Q whenever s^1 and s^2 are fields of type Q. We will denote this field by $s^1 + s^2$. If R is the real numbers with the trivial action of $GL(n)$, let $f: R \times Q \to Q$ send $(\lambda, v) \to \lambda v$. Then $f_\#(u, s)$ is a field of type Q if u is a differentiable function (i.e., a field of type R) and s a field of type Q; we denote this field by us. It is easy to check that this makes the set of fields of type Q into a module over the ring of differentiable functions. That is, we have to check that $s^1 + s^2 = s^2 + s^1$, $u(s^1 + s^2) = us^1 + us^2$, etc. The proof of all these assertions can be subsumed under the following observation: if f_1 and f_2 are equivariant maps, so is $f_1 \circ f_2$ and $(f_1 \circ f_2)_\# = f_{1\#} \circ f_{2\#}$. For example, let us prove that $s^1 + s^2 = s^2 + s^1$. To do this, let $f_1: Q \times Q \to Q \times Q$ send $(v_1, v_2) \to (v_2, v_1)$ and $f_2: Q \times Q \to Q$ send $(v_1, v_2) \to v_1 + v_2$. Then $f_{1\#}(s^1, s^2) = (s^2, s^1)$ and $f_{2\#}(s^1, s^2) = s^1 + s^2$. In view of the commutability of addition in Q, $f_2 \circ f_1 = f_1$. Thus $f_{2\#} \circ f_{1\#} = f_{1\#}$, i.e., $s^2 + s^1 = s^1 + s^2$. If $Q_1, \ldots, Q_n$ are tensor spaces, the map of $(Q_1, \ldots, Q_n) \to Q_1 \otimes \cdots \otimes Q_n$ is equivariant and therefore defines the tensor product $s^1 \otimes \cdots \otimes s^k$ of the tensor fields $s^1, \ldots, s^k$. Similarly symmetrization and antisymmetrization are equivariant maps as are exterior product, interior product, etc. In short, all of the algebraic operations of Chapter I induce the corresponding algebraic operations on the tensor fields. Thus every tensor field can be written (locally) as a sum of terms consisting of products of functions with tensor products of various $\partial/\partial x^i$ and dx^j. Thus, for example, let $Q_{\wedge^p T^*}$ be the pth antisymmetric product of the space Q_{T^*} with itself (and with the corresponding representation of $GL(n)$). A field of type $Q_{\wedge^p T^*}$ will be called an *exterior differential form of degree p*. If ω is an exterior differential form of degree p, then in terms of local coordinates we can write

$$\omega = \sum_{i_1 < \cdots < i_p} a_{i_1, \ldots, i_p} dx^{i_1} \wedge \cdots \wedge dx^{i_p}$$

(where the a's are differentiable functions).

Before proceeding to a more detailed study of tensor bundles and fields, we examine the behavior of the coframe bundle under diffeomorphisms. Let g be a diffeomorphism of M_1 onto an open subset of M_2. Then there is a map g^* of the subset $\pi^{-1}(g(M_1)) \subset \mathscr{F}^*(M_2)$ onto $\mathscr{F}^*(M_1)$, defined as follows: If $(y, (U, h)) \in \underline{U}(M_2)$ with $y \in g(M_1)$ then $(g^{-1}(U), h \circ g)$ is a chart about $g^{-1}(y)$. Let $\underline{g}: \bar{\pi}^{-1}(g(M_1)) \subset \underline{U}(M_2) \to \underline{U}(M_1)$ be given by $\underline{g}((y, (U, h))) = (g^{-1}(y), (g^{-1}(U), h \circ g))$. It is immediate to verify that if $z_1 \sim z_2$ then $\underline{g}(z_1) \sim \underline{g}(z_2)$. The induced map on the coset space will be g^*. Let (U_i^2, h_i^2) be a chart on M_2 and (U_j^1, h_j^1) on M_1, with φ_i^2 and φ_j^1 corresponding maps of $\pi^{-1}(U_i^2) \to U_i^2 \times GL(n)$ and $\pi^{-1}(U_i^1) \to U_j^1 \times GL(n)$. Then the map g^* on $\pi^{-1}(U_i^2 \cap g(U_j^1))$ is given by

$$(6.14) \qquad g^*(z) = (\varphi_j^1)^{-1}(g^{-1}(q), J_{h_j^1(g^{-1}(q))}(h \circ g \circ (h_j^1)^{-1}))$$

if $z \in \{(q, (U, h))\}$. Thus g^* is differentiable (of degree one less than that of g). A direct verification shows that the mapping g^* has the two further properties:

$$(6.15) \qquad \pi = g \circ \pi \circ g^*$$

and

$$(6.16) \qquad g^*(zA) = g^*(z)A.$$

Of course, the π's in (6.15) and multiplication by A in (6.16) have different meanings on the different sides of the equation. On the left they refer to $\mathscr{F}^*(M_2)$ and on the right to $\mathscr{F}^*(M_1)$.

Exercise 6.5. Let g_1 be a diffeomorphism of M_1 onto an open subset of M_2 and g_2 a diffeomorphism of M_2 onto an open subset of M_3. Show that $(g_2 \circ g_1)^* = g_1^* \circ g_2^*$.

Let Q be a space on which $GL(n)$ acts differentiably. The map $g^* \times \text{id}$

of $\pi^{-1}(g(M_1)) \times Q \subset \mathscr{F}^*(M_2) \times Q \to \mathscr{F}^*(M_1) \times Q$ takes cosets into cosets by (6.15). Therefore, it defines a map (which we shall also denote by g^*) of $\pi_Q^{-1}(g(M_1)) \subset E(M_2, Q) \to E(M_1, Q)$. In terms of charts, for $x \in U_i^2 \cap g(U_j^1)$

$$\Phi_j^1 \circ g^* \circ (\Phi_i^2)^{-1}(x, q) = (g^{-1}(x), q^1)$$

where

$$\{(\varphi_j^1)^{-1}(g^{-1}(x), I), q^1\} = \{g^* \circ (\varphi_j^2)^{-1}(x, I), q\}.$$

But

$$g^* \circ (\varphi_i^2)^{-1}(x, I) = (\varphi_j^1)^{-1}(g^{-1}(x), J_{h_j^1(g^{-1}(x))}(h_i^2 \circ g \circ (h_j^1)^{-1})) \quad \text{by (6.14)}$$

$$= (\varphi_j^1)^{-1}((g^{-1}(x), I); (J_{h_j^1(g^{-1}(x))}(h_i^2 \circ g \circ (h_j^1)^{-1}))^{-1}) \quad \text{by (6.1).}$$

Thus

$$q^1 = \rho(J_{h_j^1(g^{-1}(x))}(g \circ (h_j^1)^{-1}))q^{-1}$$
$$= \rho(J_{h_i^2(x)}(h_j^1 \circ g^{-1} \circ (h_i^2)^{-1})q.$$

In other words,

$$(6.17) \quad \Phi_j^1 \circ g^* \circ (\Phi_i^2)^{-1}(x, q) = (g^{-1}(x), \rho(J_{h_i^2(x)}(h_j^1 \circ g^{-1} \circ (h_i^2)^{-1}))q).$$

Thus the map g^* is differentiable. It follows from the definitions that (6.15) holds for the g^* defined on $E(M_2, Q)$. Let f be an equivariant map of $Q_1 \to Q_2$. Then it is immediate that

$$(6.18) \qquad g^* f_\# = f_\# g^*.$$

If s is a field of type Q on M_2, $s: M_2 \to E(M_2, Q)$ then (6.15) implies that $g^* \circ s \circ g: M_1 \to E(M_1, Q)$ defines a field of type Q on M_1. In fact $\pi \circ g^* \circ s \circ g = g^{-1} \circ \pi \circ s \circ g = \text{id}$. We thus have a map from fields of type Q on M_2 to fields of type Q on M_1. We shall also denote this map by g^*. Thus

$$(6.19) \qquad g^*(s) = g^* \circ s \circ g.$$

Let $s^1, \ldots, s^k$ be fields of type $Q_1, \ldots, Q_k$. We let the reader verify that

$$(6.20) \qquad g^*((s^1, \ldots, s^k)) = (g^*(s^1), \ldots, g^*(s^k))$$

and

$$(6.21) \qquad g^*(f_\#(s)) = f_\#(g^*(s)).$$

We resume part of the previous discussion in

THEOREM 6.1. *On a differentiable manifold the space of all tensor fields has the structure of an algebra with an operation of contraction. There is also the usual algebraic structure on the space of symmetric and antisymmetric contravariant (and covariant) tensor fields. A diffeomorphism g of M_1 onto an open subset of M_2 induces a map g^* of tensor fields of M_2 into tensor fields of M_1 which preserves all algebraic operations. If $g_1 : M_1 \to M_2$ and $g_2 : M_2 \to M_3$ then*

$$(g_2 \circ g_1)^* = g_1^* \circ g_2^*. \tag{6.22}$$

7. The tensor bundles. We have seen in Section 6 that the tangent bundle, which is the space of quantities of type Q_T can be viewed, as a set, as the collection $\bigcup T_p(M)$. That is the submanifold $\pi^{-1}(p)$ is a vector space and, indeed, can be viewed as the vector space introduced in Section 5. Furthermore, the vector space structure is consistent with the algebraic structure on the tangent space; i.e., if X and Y are two vector fields then

$$(X + Y)(p) = X(p) + Y(p)$$

[where + on the left is the addition of vector fields and + on the right is vector addition in $T_p(M)$]. This can be seen, for example, by looking at the expressions of both sides in a local coordinate system. We can make similar statements, of course, for any space of *linear* quantities $E(M, Q)$. In fact, by Exercise 6.3, $\pi^{-1}(p)$ is diffeomorphic to Q and two distinct diffeomorphisms differ by a *linear* transformation of Q. Thus if X_p and Y_p are elements of $\pi^{-1}(p)$, we can set $aX_p + bY_p = h^{-1}(ah(X_p) + bh(Y_p))$ where h is any one of the diffeomorphisms of Exercise 6.2. If h' is a second such diffeomorphism we can write $h' = Lh$ where L is a linear transformation of Q. Then $(h')^{-1}(ah'(X_p) + bh'(Y_p)) = h^{-1}L^{-1}(aLh(X_p) + bLh(Y_p)) = h^{-1}(ah(X_p) + bh(Y_p))$. Thus the linear combination $aX_p + bY_p$ is well defined independently of the choice of h. It is also immediate that this vector space structure is consistent with the algebraic structure on the fields of type Q.

In the case of tensor quantities we can make a further identification: Let Q be a tensor representation of $GL(n)$. Then Q is obtained by taking various tensor products of copies of Q_T and Q_{T^*} and applying various operations such as symmetrization, antisymmetrization, direct sum, and so on. In short, Q is obtained by a sequence of operations on Q_T and

Q_{T^*}; let us denote this sequence of operations by $\mathcal{Q}$. Consider the tangent $T_p(M)$ and the cotangent space $T_p^*(M)$. For each choice of chart (U, h) about p we have an isomorphism h of $T_p(M)$ and $T_p^*(M)$ onto Q_T and Q_{T^*}. Therefore, we have an induced isomorphism of $\mathcal{Q}(T_p(M), T_p^*(M))$ onto $\mathcal{Q}(Q_T; Q_{T^*}) = Q$, depending on h. If (U', h') is a second chart with $(U, h') \sim (U', h')$ it is clear that $\bar{h}' = \bar{h}$, thus the isomorphism depends only on the coset $z \in \mathcal{F}^*(M)$ of (U, h). Thus to every $z \in \pi^{-1}(p) \subset \mathcal{F}^*(M)$, we have an isomorphism, $\bar{z}$, of $\mathcal{Q}(T_p(M), T_p^*(M))$ onto Q. If z' is another element of $\pi^{-1}(p)$, with $z' = zA$ then the definition of tensor product and the facts already established for the tangent and cotangent bundle imply that $\bar{z}' = \rho(A) \circ \bar{z}$ where $\rho(A): Q \to Q$ is given by the representation, ρ, of $GL(n)$ on Q. We can thus assign to each $\underline{t} \in \mathcal{Q}(T_p(M), T_p^*(M))$ the coset $\{z, \bar{z}(t)\}$ in $E(M, Q)$. Conversely, to each coset $\{z, q\}$ we can assign the unique element, $\bar{z}^{-1}(q)$ of $\mathcal{Q}(T_p(M), T_p^*(M))$.

We can thus identify $\mathcal{Q}(T_p(M), T_p^*(M))$ *with* $\pi^{-1}(p) \subset E(M, Q)$.

As an illustration let Q be $Q_T^* \circledcirc Q_T^*$. Then $E(M, Q)$ can be identified as $\bigcup_{p \in M} T_p^*(M) \circledcirc T_p^*(M)$. A field of type Q assigns to each p an element of $V^* \circledcirc V^*$, that is, a metric on $T_p(M)$. In terms of local coordinates such a field is given by $\sum g_{ij}\, dx^i \circledcirc dx^j$.

DEFINITION 7.1. *A field of type $Q_T^* \circledcirc Q_T^*$ is called a Riemann metric on M if it induces a positive definite metric on $T_p(M)$ for all $p \in M$. If, in terms of local coordinates, the field is given by $\sum g_{ij}\, dx^i \circledcirc dx^j$, this amounts to the assertion that the matrix (g_{ij}) is positive definite. A manifold M together with a Riemann metric is called a Riemann manifold.*

In a Riemann manifold it makes sense to talk of the length of a tangent vector. Since a tangent vector is an "infinitesimal curve," we should expect to be able to talk about the length of a curve on M by "integrating" its infinitesimal parts. We will see later on in this section that this can be done.

Let Q be a tensor space which is obtained solely from operations, $\mathcal{Q}$ involving Q_T (but not Q_T^*). Then Q is called a *contravariant tensor space* and $E(M, Q)$ is said to be a *contravariant tensor bundle*, and will be denoted by $\mathcal{Q}T(M)$. Similarly if Q is obtained by $\mathcal{Q}$ from Q_{T^*} the space $E(M, Q)$ is called a *covariant tensor bundle* and will be denoted by $\mathcal{Q}T^*(M)$. The important difference between contravariant and covariant tensors lie in their behavior under mappings.

Let g be a differentiable map of a manifold M_1 into a manifold M_2. We saw in Section 5 that g induces a map g_* of $T(M_1) \to T(M_2)$ which is given by (5.13) in terms of local coordinate systems (U, h) on M_1 and (V, H) on M_2. In view of the identification of $E(M, Q_T)$ with $T(M)$, we see from (5.1) that the map g_* is indeed a differentiable map of $T(M_1) \to T(M_2)$. Since g_* restricted to each $T_p(M_1)$ is a linear map of $T_p(M_1) \to T_{g(p)}(M_2)$, it induces a linear map of any contravariant tensor space over $T_p(M_1)$ into a space of the same type over $T_{g(p)}(M_2)$. Thus we have a map of $\mathscr{Q}T(M_1) \to \mathscr{Q}T(M_2)$ for any (contravariant) tensor product operation $\mathscr{Q}$; we will denote this map also by g^*. It is an immediate verification that g_* is also differentiable. Thus we have

THEOREM 7.1. *Let $g: M_1 \to M_2$ be a differentiable map of $M_1 \to M_2$ (of class at least C^2). Then for any (contravariant) operation $\mathscr{Q}$ there is a differentiable map $g_*: \mathscr{Q}T(M_1) \to \mathscr{Q}T(M_2)$. In particular, the map $g_*: T(M_1) \to T(M_2)$ is given in terms of local coordinates by* (5.13). *If $g_1: M_1 \to M_2$ and $g_2: M_2 \to M_3$ then*

$$(g_2 \circ g_1)_* = g_{2*} \circ g_{1*}. \tag{7.1}$$

The only thing we haven't proved is (7.1), but this follows immediately from the definition of g_* on $Cu(M_1) \to Cu(M_2)$.

The map g_*, in contrast to the g^* defined for diffeomorphisms in Section 5, does not take vector fields into vector fields. Given a vector field $X: M_1 \to T(M_1)$ the only obvious candidate for $g_*(X)$ would be the "map" of $M_2 \to T(M_2)$ which takes $g(x) \to g_*(X(x))$. This is not well defined for two reasons. First of all, if g is not one to one we can have $y = g(x_1) = g(x_2)$ and $g_*(X(x_1)) \neq g_*(X(x_2))$. Secondly, even if g is one to one, our proposed map of $M_2 \to T(M_2)$ would be defined only on $g(M_1)$ and not on all of M_2. In fact, the only case where g_* defines a map of vector fields into vector fields is if g is a diffeomorphism of M_1 onto M_2. This follows from

EXERCISE 7.1. Let g be a diffeomorphism of M_1 onto M_2. Show that for any contravariant tensor space Q, the map g_* and $(g^{-1})^*$ (defined in Section 6) of $E(M_1, Q) \to E(M_2, Q)$ coincide.

Since g_* is a linear map of $T_x(M_1)$ into $T_{g(x)}(M_2)$ it induces a linear map, $(g_*)^*_x$ of $(T_{g(x)}(M_2))^*$ onto $(T_x(M_1))^*$. If $\mathscr{Q}$ is a covariant operation, then $(g_*)^*$ induces a map of $\mathscr{Q}(T_{g(x)}(M))$ onto $\mathscr{Q}(T_x(M))$. This time we can not define a $(g_*)^*$ as a map of $E(M_2, Q) \to E(M_1, Q)$ because of the same trouble as before: The rule assigning x to $g(x)$ is not well defined over all M_2. On the other hand, let $s: M_2 \to E(M_2, Q)$ be a field of type Q. Then the map $g^*(s): M_1 \to E(M, Q)$ defined by

$$g^*(s)(x) = (g_*)^* \circ s \circ (g(x))$$

satisfies $\pi \circ g^*(s) = \text{id}$.

EXERCISE 7.2. Show that $g^*(s)$ is a differentiable map of $M_1 \to E(M_1, Q)$.

EXERCISE 7.3. Let Q be a space on which $GL(n)$ acts trivially. By example (1) of Section 6, a section, s, of $E(M_2, Q)$ can be regarded as a map $s: M_2 \to Q$. Show that $g^*(s) = s \circ g$.

In view of Exercise 7.2, $g^*(s)$ is a field of type Q on M_1, where $\mathcal{Q}T^*(M_1) = E(M_1, Q)$. In fact, we have

THEOREM 7.2. *Let g be a differentiable map of $M_1 \to M_2$. If $\mathcal{Q}$ is any covariant tensor operation there is a map g^* of the tensor fields of $\mathcal{Q}T^*(M_2)$ to the tensor fields of $\mathcal{Q}T^*(M_1)$. The map g^* preserves all algebraic operations on tensor fields. In particular, if ω is a field of $T^*(M_2)$ i.e., a linear differential form, whose expression in local coordinates is $\omega = \sum Y_i dy^i$ and if, locally, the map g is given by functions $y^i(x^1, \ldots, x^{n_1})$ then*

$$g^*(\omega) = \sum_{i,j} Y_i \frac{\partial y^i}{\partial x^j} dx^j. \tag{7.2}$$

If $g_1: M_1 \to M_2$ and $g_2: M_2 \to M_3$ then

$$(g_2 \circ g_1)^* = g_1^* \circ g_2^*. \tag{7.3}$$

We leave those assertions not already proved as exercises for the reader.

EXERCISE 7.4. Show that if g is a diffeomorphism of $M_1 \to M_2$, the g^* defined here coincides with the g^* defined in Section 6.

As an illustration, let us compute $g^*(b)$ where b is a symmetric bilinear form, given, in local coordinates by

$$\underline{b} = \sum g_{ij}\, dy^i \circledcirc dy^j. \tag{7.4}$$

Then

$$
\begin{aligned}
g^*(\underline{b}) &= \sum_{i,j,k,l} [g_{ij}\circ g] \left(\frac{\partial y^i}{\partial x^k}\, dx^k\right) \circledcirc \left(\frac{\partial y^j}{\partial x^l}\, dx^l\right) \\
&= \frac{1}{2} \sum_{i,j,k,l} [g_{ij}\circ g] \left[\frac{\partial y^i}{\partial x^k}\frac{\partial y^j}{\partial x^l} + \frac{\partial y^i}{\partial x^l}\frac{\partial y^j}{\partial x^k}\right] dx^k \circledcirc dx^l.
\end{aligned} \tag{7.5}
$$

In particular let $C\colon [a, b] \to M$ be a curve where M is a Riemann manifold with Riemann metric $\underline{b}$. We will assume, for simplicity, that C is a differentiable map which can be extended to some neighborhood of the interval $[a, b]$, and let t be a coordinate on $[a, b]$. Then $C^*(\underline{b})$ is a Riemann metric on (a, b). Let us assume for the moment that $C([a, b])$ is contained in a coordinate neighborhood U where $\underline{b}$ has the form (7.4). Then $C^*(\underline{b})$ by (7.5) can be written as

$$
C^*(\underline{b}) = \sum [g_{ij}\circ C] \left[\frac{\partial y^i}{\partial t}\frac{\partial y^j}{\partial t}\right] dt \circledcirc dt. \tag{7.6}
$$

We then define the length $L(C)$ of C to be

$$
L(C) = \int_a^b \sqrt{\sum g_{ij}\circ C \frac{\partial y^i}{\partial t}\frac{\partial y^j}{\partial t}}\, dt. \tag{7.7}
$$

This length is independent of the parametrization of C. More precisely, let φ be a differentiable monotone map of the interval $[c, d]$ onto the interval $[a, b]$ and s be a coordinate on $[c, d]$. Then $C\circ\varphi$ is also a curve lying in U. (It is what is called the "same curve with another parametrization.") By (7.3),

$$
\begin{aligned}
(C\circ\varphi)^*b &= \varphi^*(C^*(\underline{b})) \\
&= \sum [g_{ij}\circ C\circ\varphi] \left(\frac{\partial y^i}{\partial t}\circ\varphi\right)\left(\frac{\partial y^j}{\partial t}\circ\varphi\right)\varphi^*\,(dt \circledcirc dt) \\
&= \sum [g_{ij}\circ C\circ\varphi] \left(\frac{\partial y^i}{\partial t}\circ\varphi\right)\left(\frac{\partial y^j}{\partial t}\circ\varphi\right)\left(\frac{dt}{ds}\right)^2 ds \circledcirc ds.
\end{aligned}
$$

By (7.7) $L(C\circ\varphi)$ is given by

$$
\int_c^d \sqrt{\sum (g_{ij}\circ C\circ\varphi)\left(\frac{\partial y^i}{\partial t}\circ\varphi\right)\left(\frac{\partial y^j}{\partial t}\circ\varphi\right)}\, \left|\frac{dt}{ds}\right| ds
$$

which is the same as $L(C)$ in view of the formula for change of variables in a definite integral. A similar argument shows that (7.7) does not depend on the choice of coordinates $(y^1, \ldots, y^n)$.

EXERCISE 7.5. Let $a < a' < b$ and let C_1 be C restricted to $[a, a']$ and C_2 be C restricted to $[a', b]$. Show that $L(C) = L(C_1) + L(C_2)$.

If C is a curve that does not lie entirely in one coordinate neighborhood, we can break up the interval $[a, b]$ into a finite number of subintervals whose image under C does lie in a coordinate neighborhood. We then define the length of C to be the sum of the lengths of each of the subintervals. In view of the previous discussion and Exercise 7.5, this definition of length is independent of the subdivision, the choice of coordinates and of the parametrization of C.

8. Vector fields and Lie derivatives. In this section we show that vector fields can be regarded as "infinitesimal transformations" and we will study their properties.

DEFINITION 8.1. *A family φ_t of diffeomorphisms of $M \to M$ is called a one parameter group of differentiable transformations of M if the map $\varphi : R \times M \to M$ sending $(t, p) \to \varphi_t(p)$ is a differentiable action of the additive group of real numbers on M, i.e., if*

1) *The map φ is differentiable,*
2) *$\varphi_{t+s} = \varphi_t \circ \varphi_s$ for all t and s,*

and

3) *φ_0 is the identity.*

A one parameter group, φ_t, induces a vector field on M. In fact, at every $p \in M$, $t \to \varphi_t(p)$ is a curve passing through p. Define X_p to be the tangent to this curve. For any differentiable function, f, on M, it follows from the definitions that

$$\langle X_p, (df)_p \rangle = \lim \left(\frac{1}{t}\right)(f(\varphi_t(p)) - f(p)). \tag{8.1}$$

In a coordinate system x, taken $f = x^1, \ldots, x^n$ yields

$$X_p = \sum \lim \left(\frac{1}{t}\right)(x^i(\varphi_t(p)) - x^i(p)) \frac{\partial}{\partial x^i}.$$

Since φ is differentiable the map $p \to X_p$ defines a vector field, X, on M. The vector field X is called the *infinitesimal generator of the one parameter group* φ_t.

The converse is not quite true. Not every vector field on M need generate a one parameter family of transformations of M. However, locally, it is true. More precisely,

THEOREM 8.1. *Let X be a vector field on M. For every $p \in M$ there exists a neighborhood U of p, an $\epsilon > 0$ and a unique family of differentiable maps φ_t of $U \to M$, defined for $|t| < \epsilon$ such that*

1) *the map φ of $(-\epsilon, \epsilon) \times U \to M$ sending $(t, p) \to \varphi_t(p)$ is differentiable*
2) *if $|t|$, $|s|$, $|t + s|$ are all $< \epsilon$ and if $q \in U$ and $\varphi_t(q) \in U$ then*
$$\varphi_{s+t}(q) = \varphi_s \circ \varphi_t(q).$$
3) *X_q is tangent to (q, φ_q) for $q \in U$ where φ_q is the curve $t \to \varphi_t(q)$.*

Proof. Let (V, h) be a chart about p so that in V we can write
$$X = \sum X^i \frac{\partial}{\partial x^i}.$$

Consider the system of ordinary differential equations

$$\frac{d\varphi^i}{dt} = X^i(\varphi^1, \ldots, \varphi^n). \tag{8.2}$$

By the existence theorem for ordinary differential equations (cf. appendix) there exist δ_1 and $\epsilon_1 > 0$ such that (8.2) has a unique solution, $\varphi^i(t; x^1, \ldots, x^n)$ satisfying

$$\varphi^i(0, x^1, \ldots, x^n) = x^i \quad \text{for} \quad |t| < \epsilon_1, x \in B^n_{\delta_1}. \tag{8.3}$$

If we choose $\epsilon \leqslant \epsilon_1$ and $\delta \leqslant \delta_1$ sufficiently small, then

$$\varphi^i(t, x^1, \ldots, x^n) \in h(V) \quad \text{for} \quad |t| < \epsilon, (x^1, \ldots, x^n) \in B^n_\delta.$$

Let $U = h^{-1}(B^n)$. The map φ_t of $U \to M$ defined, in local coordinates, by

$$\varphi^i_t(x^1, \ldots, x^n) = \varphi^i(t; x^1, \ldots, x^n)$$

is clearly a differentiable map $(-\epsilon, \epsilon)$, $U \to M$. Now the functions $\psi^i(t; x^1, \ldots, x^n) = \varphi^i(s + t; x^1, \ldots, x^n)$ are solutions of (8.2) satisfying the initial conditions

$$\psi^i(0; x^1, \ldots, x^n) = \varphi^i(s; x^1, \ldots, x^n).$$

By uniqueness, we must have

$$\varphi^i(s + t; x^1, \ldots, x^n) = \varphi^i(t; \varphi^1(s; x^1, \ldots, x^n), \ldots, \varphi^n(s; x^1, \ldots, x^n)).$$

Thus condition 2) of the theorem is fulfilled. Condition 3) is fulfilled by construction.

We cannot define φ_t for all t or on all of M since the existence theorem on differential equations guarantees only local solutions. Even on Euclidean space, there are examples of vector fields which do not generate one parameter groups on the whole space. If the manifold M is compact, it is not too difficult to show that every vector field generates a one-parameter group. We will not prove this result here.

Let X be a vector field on M, and η a field of type Q where Q is a *linear* quantity. If U is a neighborhood and φ_t a family of maps given by Theorem 8.1, then $\varphi_t^*(\eta)$ is a field of type Q over U. We can then consider the derivative

$$\mathscr{L}_X(\eta)_p = \lim_{t=0} \left(\frac{1}{t}\right) (\varphi_t^*(\eta)_p - \eta_p). \tag{8.4}$$

Since the φ_t are essentially unique, the left-hand side of (8.4) is independent of U. We let the reader verify that the map $p \to \mathscr{L}_X(\eta)_p$ is a field of type Q in the general case. We will denote it by $\mathscr{L}_X(\eta)$, and call it the *Lie derivative* of η with respect to X. We shall be content with proving that $\mathscr{L}_X(\eta)$ is a tensor field when η is a tensor. First observe, that since φ_t^* is linear,

$$\mathscr{L}_X(a\eta + b\xi) = a\,\mathscr{L}_X(\eta) + b\,\mathscr{L}_X(\xi) \tag{8.5}$$

where a and b are *constants*. Also, if $f: Q_1 \to Q_2$ is a *linear* map, (such as contraction of tensor, for example) $\varphi_t^* f_\#(\eta) = f_\# \varphi_t^*(\eta)$ by formula (6.21), for any field, η, of type Q_1. Therefore,

$$f_\#\,\mathscr{L}_X(\eta) = \mathscr{L}_X(f_\#\,\eta). \tag{8.6}$$

Thus (8.6) says, for instance, that the Lie derivative of a contraction of a tensor field is the contraction of a Lie derivative. Now let f be a *bilinear* map of $Q_1 \times Q_2 \to Q_3$. Then again $\varphi_t^* \circ f_\# = f_\# \circ \varphi_t^*$ so

$$\begin{aligned}\varphi_t^* f_\#(\eta, \xi) - f_\#(\eta, \xi) &= f_\#(\varphi_t^*\eta, \varphi_t^*\xi) - f_\#(\eta, \xi)\\ &= f_\#(\varphi_t^*\eta, \varphi_t^*\xi) - f_\#(\varphi_t^*\eta, \xi)\\ &\quad + f_\#(\varphi_t^*\eta, \xi) - f_\#(\eta, \xi).\end{aligned}$$

Dividing by t and letting t tend to zero we obtain (since f commutes with the limiting operation)

$$\mathscr{L}_X(f_\#(\eta, \xi)) = f_\#(\mathscr{L}_X\eta, \xi) + f_\#(\eta, \mathscr{L}_X\xi). \tag{8.7}$$

We can take as $f_\#$ the various operations of tensor product, exterior product, and so on. This procedure shows that in order to obtain the formula for the Lie derivative of a tensor field, it suffices to know the Lie derivative of 1) functions, 2) linear differential forms, and 3) vector fields. All the others can be obtained from these using (8.5), (8.6), and (8.7).

1) *Functions.* If u is a differentiable function on M,

$$\varphi_t^* u = u \circ \varphi_t : (\mathscr{L}_X u)_p = \lim \frac{1}{t} (u(\varphi_t(p)) - u(p)).$$

In terms of local coordinates

$$u = u(x^1, \ldots, x^n)$$
$$(\varphi_t^* u) = u(\varphi_t^1(x^1, \ldots, x^n), \ldots, \varphi_t^n(x^1, \ldots, x^n))$$

so that

$$(\mathscr{L}_X u) = \sum \frac{\partial x^u}{\partial x^i} \frac{d\varphi^i}{dt} = \sum X^i \frac{\partial x^u}{\partial x^i}.$$

Thus, if $X = \sum X^i (\partial/\partial x^i)$,

$$\mathscr{L}_X(u) = \sum X^i \frac{\partial u}{\partial x^i} = \langle X, du\rangle. \tag{8.8}$$

Since this is, symbolically, just the action of the "differential operator," X, on u we will usually write

$$\mathscr{L}_X(u) = X(u) = \langle X | du\rangle. \tag{8.9}$$

2) *Linear differential forms.* According to (7.2) $\varphi_t^*(dx^i)$ is given by

$$\varphi_t^* (dx^i) = d(\varphi_t^* x^i) = d(\varphi_t^i) = \sum \frac{\partial \varphi_t^i}{\partial x^j} dx^j.$$

Now $\partial\varphi_0^i/\partial x^j = \delta_j^i$, so that

$$\lim \frac{1}{t} \left(\frac{\partial \varphi_t^i}{\partial x^j} - \delta_j^i \right) = \frac{d}{dt} \frac{\partial \varphi_t^i}{\partial x^j}\bigg|_{t=0} = \frac{\partial X^i}{\partial x^j}.$$

Thus

$$\mathscr{L}_X\,(dx^i) = \sum \frac{\partial X^i}{\partial x^j}\,dx^j. \tag{8.10}$$

If a_i is a function on u, then, by (8.7),

$$\begin{aligned}\mathscr{L}_X(a_i\,dx^i) &= \mathscr{L}_X(a^i)\,dx^i + a_i\,\mathscr{L}_X\,(dx^i)\\ &= \sum X^j \frac{\partial a_i}{\partial x^j}\,dx^i + a_i \sum \frac{\partial x^i}{\partial x^j}\,dx^j.\end{aligned}$$

Summing the last equation over i yields

$$\mathscr{L}_X\Big(\sum a_i\,dx^i\Big) = \sum_{i,j}\left(X^j \frac{\partial a_i}{\partial x^j} + a_j \frac{\partial X^j}{\partial x^i}\right) dx^i. \tag{8.11}$$

3) *Vector fields.* We could proceed as in 2) using (7.2). It is more convenient, however, to proceed as follows: Given a vector field, Y, and a differential form ω, the evaluation map $(Y, \omega) \to \langle Y, \omega\rangle$ is bilinear. We can thus apply (8.7). Take $Y = \partial/\partial x^j$ and $\omega = dx^i$. Then

$$\left\langle \mathscr{L}_X\left(\frac{\partial}{\partial x^j}\right), dx^i \right\rangle + \left\langle \frac{\partial}{\partial x^j}, \mathscr{L}_X\,dx^i \right\rangle = \mathscr{L}_X\,(\delta^j_i) = 0.$$

Using (8.10),

$$\mathscr{L}_X\left(\frac{\partial}{\partial x^j}\right) = \sum_i -\frac{\partial X^i}{\partial x^j}\frac{\partial}{\partial x^i}.$$

By (8.7), if Y^j is a function on u,

$$\begin{aligned}\mathscr{L}_X\left(Y^j \frac{\partial}{\partial x^j}\right) &= \mathscr{L}_X(Y^j)\frac{\partial}{\partial x^j} + Y^j\,\mathscr{L}_X\left(\frac{\partial}{\partial x^j}\right)\\ &= \sum_i \frac{\partial Y^j}{\partial x^i} X^i \frac{\partial}{\partial x^j} - \sum Y^j \frac{\partial X^i}{\partial x^j}\frac{\partial}{\partial x^i}.\end{aligned}$$

If Y is a vector field given by $Y = \sum Y^j\,(\partial/\partial x^j)$ then summing the last equation over j, yields

$$\mathscr{L}_X(Y) = \sum_j\left(\sum_i\left(\frac{\partial Y^j}{\partial x^i} X^i - \frac{\partial X^j}{\partial x^i} Y^i\right)\frac{\partial}{\partial x^j}\right). \tag{8.12}$$

It follows from (8.8), (8.11), and (8.12) that if X is a vector field of class C^k the Lie derivative of a differentiable function, differential form, or vector field of class C^k is a function, form, or vector field of class C^{k-1}. By (8.7) this then is true for an arbitrary tensor field.

We will follow the standard usage and define

$$[Y, X] = \mathscr{L}_X(Y). \tag{8.13}$$

The left-hand side of (8.13) is called the *Lie bracket of Y and X.* It follows immediately from (8.12) that

$$[X, Y] = -[Y, X]. \tag{8.14}$$

EXERCISE 8.1. Show that $X(Y(u)) - Y(X(u)) = [X, Y]u$ for any function u.

Let Z be a third vector field. Then for any function u, by Exercise 8.1,

$$[X, [Y, Z]](u) = XYZ(u) - XZY(u) - YZX(u) + ZYX(u),$$
$$[Z, [X, Y]](u) = ZXY(u) - ZYX(u) - XYZ(u) + YXZ(u),$$

and

$$[Y, [Z, X]](u) = YZX(u) - YXZ(u) - ZXY(u) + XZY(u).$$

If we add these three equations we obtain

$$([X, [Y, Z]] + [Z, [X, Y]] + [Y, [Z, X]])(u) = 0$$

for any function u. The only vector field which takes every function into zero is the zero vector field. Thus

$$[X, [Y, Z]] + [Z, [X, Y]] + [Y, [Z, X]] = 0. \quad \text{(Jacobi's identity)} \tag{8.15}$$

DEFINITION 8.2. *A (non-associative) algebra which is anticommutative and whose elements satisfy Jacobi's identity is called a Lie algebra.*

Then we can resume the previous discussion in

THEOREM 8.2. *The space of all C^∞ (analytic) vector fields on a C^∞ (analytic) manifold is a Lie algebra if multiplication is taken to be the Lie bracket.*

EXERCISE 8.2. Let A be an associative algebra. Define $L(A)$ to be the space A the new multiplication given by

$$[a, b] = ab - ba.$$

Show that $L(A)$ is a Lie algebra.

EXERCISE 8.3. Show that

$$[uX, vY] = uv[X, Y] + vY(u)X - uX(v)Y, \tag{8.16}$$

if u, v are functions and X, Y are vector fields.

For later use (in Chapter V) we state the following results:

Let M_1 and M_2 be differentiable manifolds and f a differentiable map of $M_1 \to M_2$. As we have seen, there is no way of moving vector fields from M_1 to M_2 via f. Nevertheless, it is convenient to introduce the following definition.

DEFINITION 8.3. *Let X be a vector field on M_1 and Y a vector field on M_2. We shall say that X and Y are f-related if $Y_{f(p)} = f_*(X_p)$ for all $p \in M_1$.*

THEOREM 8.3. *Let X_1 and Y_1, X_2 and Y_2 be f-related. Then $[X_1, X_2]$ and $[Y_1, Y_2]$ are f-related.*

Proof. Let X be a vector field on M_1 and Y a vector field on M_2. Then X and Y are f-related means that for any differentiable function g defined on an open set U of M_2,

$$(X(g \circ f))_p = (Yg)_{f(p)}$$

for all $p \in f^{-1}(U)$. If X_i and Y_i are f-related ($i = 1, 2$), then for any g

$$Y_2(Y_1 g)_{f(p)} = (X_2(Y_1 g \circ f))_p = (X_2 X_1(g \circ f))_p.$$

Thus $([Y_1, Y_2]g)_{f(p)} = ([X_1, X_2](g \circ f))_p$ or $[X_1, X_2]$ and $[Y_1, Y_2]$ are f-related.

THEOREM 8.4. *Let f be a differentiable map of $M_1 \to M_2$. If f maps M_1 onto M_2, then to every X on M_1 there is at most one Y on M_2 so that X and Y are f-related. If f is an immersion of $M_1 \to M_2$ then to every Y on M_2 there is at most one X on M_1 so that X and Y are f-related. In this latter case, there will exist such an X if and only if for every $p \in M_1$, $Y_{f(p)} \in f_*(T_p(M_1))$.*

The proof will be left as an exercise for the reader.

EXERCISE 8.4. Let φ_{1t} be a one parameter family of transformations of M_1 with infinitesimal generator X, and φ_{2t} a one parameter family of transformations of M_2 with generator Y. Let $f: M_1 \to M_2$ satisfy $f \circ \varphi_{1t} = \varphi_{2t} \circ f$ for all t. Show that X and Y are f-related. State and prove the appropriate converse.

As an illustration of the notion of f-relatedness let us consider any space of quantities $\mathcal{Q}(M)$ over a manifold M. Any diffeomorphism, α, of M onto itself induces a diffeomorphism, α_*, of $\mathcal{Q}(M)$ onto itself. Furthermore, $\pi \circ \alpha_* = \alpha$ where π is the projection of $\mathcal{Q}(M)$ onto M. If α_t is a one parameter group of transformations of M the α_{t*} is a one parameter group of transformations of $\mathcal{Q}(M)$ and $\pi \circ \alpha_{t*} = \alpha_t$. In this way, if X is any vector field on M we get a vector field on $\mathcal{Q}(M)$ which we shall denote by $\mathcal{Q}(X)$. By Exercise 8.4, X and $\mathcal{Q}(X)$ are π-related. The most important case of this construction is when $\mathcal{Q}(M) = T(M)$. For any vector field X and M we get a vector field, $T(X)$, on $T(M)$. The vector field $T(X)$ is known as the *variational vector field* or "the equations of variation" of X. It describes how tangent vectors are moved under the flow generated by X. It will prove to be of fundamental importance in Chapter IV.

CHAPTER III

Integral Calculus On Manifolds

In this chapter we continue the study of analysis on differentiable manifolds. Our first order of business is to introduce a theory of integration. In order to do this, we must have something that plays the role of an "integrand" and something that plays the role of a "domain of integration." To see what they should be, let us examine the classical theory of line integrals in the plane. A line integral is usually written as

$$\text{(i)} \qquad \int_C P\,dx + Q\,dy$$

where C is a curve and P and Q are continuous functions in the plane. If the curve C is made up of a finite number of pieces C_i, where each C_i is given by differentiable maps $C_i = (x_i, y_i)$ of $[0, 1]$ into E^2, then the value of (i) is given by

$$\text{(ii)} \qquad \int_C P\,dx + Q\,dy = \sum_i \int_0^1 \left(P\,\frac{dx_i}{dt} + Q\,\frac{dy_i}{dt}\right) dt.$$

Of course, (ii) is taken as the definition of (i). If C is a simple closed curve bounding a region D, then Green's theorem asserts that

$$\text{(iii)} \qquad \int_C P\,dx + Q\,dy = \int_D\!\!\int \left(\frac{\partial P}{\partial y} - \frac{\partial Q}{\partial x}\right) dxdy.$$

In (iii) the curve C must be transcribed in the counter-clockwise direction.

traversed

Let us translate some of this into our language. An expression such as $P\,dx + Q\,dy$ has an independent meaning for us. It is just a linear differential form. If C_i is a map of [0, 1] into E^2, then

$$C_i^*(P\,dx + Q\,dy) = \left(P\frac{dx}{dt} + Q\frac{dy}{dt}\right)dt.$$

Thus if $\omega = P\,dx + Q\,dy$, we could rewrite (ii), symbolically as

(iv) $$\int_C \omega = \sum_i \int_0^1 C_i^*(\omega).$$

In this case, our "integrand" is a linear differential form. Our "domain of integration" is a finite sum of maps of [0, 1] into E^2. The generalization of this to manifolds is straightforward. The "integrands" of a p-dimensional theory of integration on a manifold M will be exterior differential forms. The "domain of integration" will be linear combinations of maps of a standard p-dimensional object into M.

Before discussing its generalizations, let us observe that there is something unsatisfactory about the form of Equation (iii). In order for it to be true, the curve C must be traversed in a clockwise direction. Let us interchange x and y; more precisely, let us map the (x, y) plane into the (u, v) plane by, φ, sending $x \to v$, $y \to u$. If $P'(u, v) = Q(x, y)$ and $Q'(u, v) = P(x, y)$, then $\varphi^*(P'\,du + Q'\,dv) = P\,dx + Q\,dy$. If $\varphi(C)$ is the image curve in the (u, v) plane, then (iv) and (ii) show that

$$\int_{\varphi(C)} P'\,du + Q'\,dv = \int_C P\,dx + Q\,dy.$$

Now

$$\frac{\partial P}{\partial y} = \frac{\partial Q'}{\partial u}, \quad \frac{\partial Q}{\partial x} = \frac{\partial P'}{\partial v} \quad \text{and} \quad dudv = dxdy.$$

Thus

$$\int_D \left(\frac{\partial P}{\partial y} - \frac{\partial Q}{\partial x}\right) dxdy = \int_{\varphi(D)} \left(\frac{\partial Q'}{\partial u} - \frac{\partial P'}{\partial v}\right) dudv$$
$$= -\int_{\varphi(D)} \left(\frac{\partial P'}{\partial v} - \frac{\partial Q'}{\partial u}\right) dudv.$$

In short, while the left-hand side of (iii) is invariant under the interchange of axes, the right-hand side changes sign. Of course the reason is that counterclockwise becomes clockwise under this interchange; but this condition of orientation does not explicitly occur in the formula. We can remedy the situation if we replace the symbol $dxdy$ by $dx \wedge dy$. Then $du \wedge dv = -dx \wedge dy$ and both sides are invariant. Now

(v) $$\left(\frac{\partial P}{\partial y} - \frac{\partial Q}{\partial x}\right) dx \wedge dy = dP \wedge dx + dQ \wedge dy.$$

If we call the exterior 2-form of (v), $d\omega$, then (iii) has the simple expression

$$\int_C \omega = \int_D d\omega.$$

In general, we would like to have an equation of the form

$$\text{(vi)} \qquad \int_{\partial(D)} \omega = \int_D d\omega$$

where d is some operation taking p forms into $p + 1$ forms and ∂, when applied to a $p + 1$ dimensional "domain of integration" yields its "boundary," which should be a p-dimensional "domain of integration."[1]

In Section 1, we introduce the operator d and study its properties. In Section 2, we introduce the differentiable chains and define integration and establish (vi). The rest of the chapter is devoted to a study of various useful theorems of analysis on manifolds.

1. The operator *d*.

THEOREM 1.1. *Let M be a differentiable manifold. There is one and only one map, d, of the exterior algebra of forms of class C^k into the algebra of forms of class C^{k-1} such that*

1) $d(\omega_1 + \omega_2) = d\omega_1 + \omega_2.$
2) *If ω_1 is a form of degree r,*

$$\text{(1.1)} \qquad d(\omega_1 \wedge \omega_2) = d\omega_1 \wedge \omega_2 + (-1)^r \omega_1 \wedge d\omega_2.$$

3) *If f is a function df is the form given in Example* 3 *of Section* 6 *of Chapter II.*
4) $d(df) = 0.$[2]

---

[1] It should be observed that for the case of functions (i.e., forms of degree zero) (vi) reduces to the fundamental theorem of the calculus. In this case, D is a curve and $\partial(D)$ its end points (with appropriate signs) and (vi) becomes $\int_C df = f(C(1)) - f(C(0))$.

[2] In order to motivate 4), consider any "two-dimensional domain of integration" say a triangle

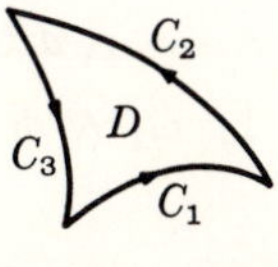

If (vi) is to hold, we must have $\int_D d(df) = \int_{\partial D} df = f(C_1(1)) - f(C_1(0)) + f(C_2(1)) - f(C_2(0)) + f(C_3(1)) - f(C_3(0)) = 0$. Thus $d(df)$ yields zero when integrated over anything and thus should be $= 0$.

Proof. It is sufficient to prove the theorem in a coordinate system. In fact, if we have existence and uniqueness of $d\omega|_U$ for each restriction $\omega|_U$ of ω to neighborhood U when (U, h) is a chart, then by uniqueness $(d\omega|_U)_{U\cap V} = (d\omega|_V)_{U\cap V} = d\omega|_{U\cap V}$ so that the $d\omega|$'s agree on the overlap and define a global differential form.

In a chart, every exterior form can be written as a linear combination (with constant coefficients) of forms

$$f dx^{i_1} \wedge \cdots \wedge dx^{i_p}.$$

By 2), 3), and 4),

$$d(f dx^{i_1} \wedge \cdots \wedge dx^{i_p}) = df \wedge dx^{i_1} \wedge \cdots \wedge dx^{i_p}.$$

This proves uniqueness. If

$$\omega = \sum_{i_1 < \cdots < i_p} a_{i_1 \cdots i_p} dx^{i_1} \wedge \cdots \wedge dx^{i_p},$$

then

$$(d\omega) = \sum da_{i_1 \cdots i_p} \wedge dx^{i_1} \wedge \cdots \wedge dx^{i_p}. \tag{1.2}$$

For instance, if $\omega = \sum a_i\, dx^i$,

$$d\omega = \sum_{i<j} \left(\frac{\partial a_i}{\partial x_j} - \frac{\partial a_j}{\partial x_i}\right) dx^i \wedge dx^j. \tag{1.3}$$

To prove existence, we define d by (1.2). We have to verify that this definition satisfies conditions 1) through 4). Conditions 1) and 3) are immediate. Condition 4) is also easy: If f is a function then $df = \sum (\partial f/\partial x^i)\, dx^i$. By (1.3) this amounts to the assertion that $\partial^2 f/\partial x^i\, \partial x^j = \partial^2 f/\partial x^j\, \partial x^i$. All that remains to be checked is 2). Since both the d given by (1.2) and exterior multiplication are linear, it suffices to verify 2) in the case that $\omega_1 = f dx^{i_1} \wedge \cdots \wedge dx^{i_r}$ and $\omega_2 = g\, dx^{j_1} \wedge \cdots \wedge dx^{j_s}$. Then

$$\begin{aligned}
d(\omega_1 \wedge \omega_2) &= d(fg) \wedge dx^{i_1} \wedge \cdots \wedge dx^{i_r} \wedge dx^{j_1} \wedge \cdots \wedge dx^{j_s} \\
&= (f\, dg + g\, df) \wedge dx^{i_1} \wedge \cdots \wedge dx^{i_r} \wedge dx^{j_1} \wedge \cdots \wedge dx^{j_s} \\
&= (df \wedge dx^{i_1} \wedge \cdots \wedge dx^{i_r}) \wedge (g\, dx^{j_1} \wedge \cdots \wedge dx^{j_s}) \\
&\quad + (-1)^r (f\, dx^{i_1} \wedge \cdots \wedge dx^{i_r}) \wedge (dg \wedge dx^{j_1} \wedge \cdots \wedge dx^{j_s}) \\
&= d\omega_1 \wedge \omega_2 + (-1)^r\, \omega_1 \wedge d\omega_2.
\end{aligned}$$

This proves Theorem 1.1.

DEFINITION 1.1. *The operator, d, given by Theorem* 1.1 *is called exterior differentiation.*

THEOREM 1.2. *For any* ω,

$$d(d\omega) = 0. \tag{1.4}$$

It suffices to verify (1.4) in every coordinate system. By 1) of Theorem 1.1, it suffices to check it for $\omega = f\,dx^{i_1} \wedge \cdots \wedge dx^{i_p}$. But $d(d\omega) = d(df \wedge dx^{i_1} \wedge \cdots \wedge dx^{i_p}) = (ddf) \wedge dx^{i_1} \wedge \cdots \wedge dx^{i_p} - df \wedge d(dx^{i_1} \wedge \cdots \wedge dx^{i_p}) = -df \wedge d(dx^{i_1} \wedge \cdots \wedge dx^{i_p})$ by 4). Therefore, we are reduced to showing that $d(dx^{i_1} \wedge \cdots \wedge dx^{i_p}) = 0$, i.e., we can prove (1.4) by induction on p.

The converse of Theorem 1.2 is not true in general. For example, if M is the circle, S^1, then there is no non-zero form of degree 2 and therefore every form ω of degree 1 satisfies $d\omega = 0$. Now the form $d\theta$ is well defined but there is no function θ defined on all of S^1. We will discuss these matters further in Sections 2 and 3.

THEOREM 1.3. *Let* φ *be a differentiable map*[3] *of a manifold* $M_1 \to M_2$. *Then*

$$d\varphi^*(\omega) = \varphi^*(d\omega). \tag{1.5}$$

Again, it is sufficient to prove (1.5) for coordinate neighborhoods in M_1 and M_2 and for the case that $\omega = f\,dx^{i_1} \wedge \cdots \wedge dx^{i_p}$. Proof by induction:

If $\omega = f\,dx^{i_1} \wedge \cdots \wedge dx^{i_p}$ then

$$\begin{aligned} d\varphi^*\omega &= d\varphi^*(f\,dx^{i_1} \wedge \cdots \wedge dx^{i_p}) = d\varphi^*((f\,dx^{i_1} \wedge \cdots \wedge dx^{i_{p-1}}) \wedge dx^{i_p}) \\ &= d(\varphi^*(f\,dx^{i_1} \wedge \cdots \wedge dx^{i_{p-1}}) \wedge \varphi^*(dx^{i_p})) \\ &= \varphi^*(df \wedge dx^{i_1} \wedge \cdots \wedge dx^{i_{p-1}}) \wedge \varphi^*(dx^{i_p}) + 0, \qquad \text{by (1.1)} \\ &= \varphi^*(df \wedge dx^{i_1} \wedge \cdots \wedge dx^{i_{p-1}} \wedge dx^{i_p}) \\ &= \varphi^*(d\omega). \end{aligned}$$

Thus, we see that the operator d is invariant under differentiable maps.

As an immediate corollary to Theorem 1.3, in view of the definition of Lie derivative, we have

for any vector field X *and form* ω

$$d(\mathscr{L}_X\omega) = \mathscr{L}_X(d\omega). \tag{1.6}$$

---

[3] of class C^2 at least.

There are some other important relations between Lie derivatives and exterior derivatives. According to Theorem 1.1, d is an antiderivation of the algebra of differential forms: i.e., it is a linear map, θ, of the algebra of forms into itself satisfying

$$\theta(\omega_1 \wedge \omega_2) = \theta(\omega_1) \wedge \omega_2 + (-1)^k \omega_1 \wedge \theta(\omega_2) \tag{1.7}$$

if ω_1 is a form of degree k. Another such antiderivation is interior multiplication by a vector field, cf. Exercise 4.5 of Chapter I: If $\theta(\omega) = X \rfloor \omega$ then θ also satisfies (1.7). On the other hand $\mathscr{L}_X$ is, by [II, (8.7)] a derivation, i.e., satisfies

$$\theta(\omega_1 \wedge \omega_2) = \theta(\omega_1) \wedge \omega_2 + \omega_1 \wedge \theta(\omega_2). \tag{1.8}$$

The importance of these observations is that any derivation or antiderivation is completely determined by what it does on functions and linear differential forms. The proof is obvious (it is the same as the proofs we have been giving earlier in this section). If θ_1 and θ_2 are antiderivations such that $\deg \theta_i(\omega) = \deg(\omega) + r_i$ where r_i is odd ($i = 1, 2$) then $\theta_1\theta_2 + \theta_2\theta_1$ is a derivation. In fact,

$$\begin{aligned}\theta_1\theta_2(\omega_1 \wedge \omega_2) = \theta_1\theta_2\,\omega_1 \wedge \omega_2 &+ (-1)^k\theta_1\omega_1 \wedge \theta_2\omega_2 \\ &+ (-1)^k\omega_1 \wedge \theta_1\theta_2\omega_2 \\ &+ (-1)^{k+r_1}\theta_2\omega_1 \wedge \theta_1\omega_2\end{aligned}$$

$$\begin{aligned}\theta_2\theta_1(\omega_1 \wedge \omega_2) = \theta_2\theta_1\omega_1 \wedge \omega_2 &+ (-1)^k\theta_2\omega_1 \wedge \theta_1\omega_2 \\ &+ (-1)^k\omega_1 \wedge \theta_2\theta_1\omega_2 + (-1)^{k+r_2}\theta_1\omega_1 \wedge \theta_2\omega_2.\end{aligned}$$

Since r_1 and r_2 are odd,

$$\begin{aligned}(\theta_1\theta_2 + \theta_2\theta_1)\omega_1 \wedge \omega_2 = (\theta_1\theta_2 + \theta_2\theta_1)\omega_1 \wedge \omega_2 \\ + \omega_1 \wedge (\theta_1\theta_2 + \theta_2\theta_1)\omega_2\end{aligned}$$

i.e., $\theta_1\theta_2 + \theta_2\theta_1$ is a derivation. We use these remarks to prove

$$\mathscr{L}_X(\omega) = X \rfloor (d\omega) + d(X \rfloor \omega). \tag{1.9}$$

The operation, left interior multiplication by X is an antiderivation and so is d. Therefore the operator $\omega \to X \rfloor (d\omega) + d(X \rfloor \omega)$ is a derivation as is $\mathscr{L}_X$. Since the operators on both sides of (1.4) are derivations, it suffices to check (1.9) for functions and linear forms. Now $X \rfloor f = 0$ so (1.9), for functions, reduces to [II, (8.8)]. If $\omega = df$ for some function, f, then (1.9) reduces to $\mathscr{L}_X(df) = d(X \rfloor df)$. But $X \rfloor df = \mathscr{L}_X(f)$ by [II, (8.8)] so (1.9) follows from (1.6). We can thus verify (1.9) locally since every linear form is a linear combination of $f\,dg$'s. But this implies (1.9) in the large.

Let us take the interior product of (1.9) with a vector field Y. We obtain

$$Y \rfloor \mathscr{L}_X(\omega) = Y \rfloor (X \rfloor d\omega) + Y \rfloor d(X \rfloor \omega).$$

Now

$$Y \rfloor (X \rfloor d\omega) = (X \wedge Y) \rfloor d\omega$$

and

$$\begin{aligned}\mathscr{L}_X(Y \rfloor \omega) &= \mathscr{L}_X(Y) \rfloor \omega + Y \rfloor \mathscr{L}_X(\omega)\\ &= [Y, X] \rfloor \omega + Y \rfloor \mathscr{L}_X(\omega).\end{aligned}$$

Thus

$$(X \wedge Y) \rfloor d\omega = \mathscr{L}_X(Y \rfloor \omega) - Y \rfloor d(X \rfloor \omega) - [Y, X] \rfloor \omega.$$

If we take ω to be a linear form then $X \wedge Y \rfloor d\omega = \langle X \wedge Y | d\omega \rangle$, $\mathscr{L}_X(Y \rfloor \omega) = X\langle Y | \omega \rangle$, and $Y \rfloor d(X \rfloor \omega) = Y\langle X | \omega \rangle$. Thus

$$(1.10) \qquad \langle X \wedge Y | d\omega \rangle = X\langle Y | \omega \rangle - Y\langle X | \omega \rangle - \langle [X, Y] | \omega \rangle.$$

Equation (1.10), which we will be using frequently further on, is, in a sense, an infinitesimal analogue of Stokes' theorem. In fact, let us first consider the case where $[X, Y] = 0$. Consider an "infinitesimal parallelogram" whose sides are parallel to X and Y. That is, let $\varphi_p(s)$ denote the solution curve of X passing through p and $\psi_p(s)$ the solution curve of Y. Consider the "rectangle" $abcd$ where (a, b) is the curve $\varphi_a(s)$ $(0 \leqslant s \leqslant \epsilon)$ and $b = \varphi_a(\epsilon)$ and (b, c) is the curve $\psi_b(s)$ $(0 \leqslant s \leqslant \epsilon)$, $c = \psi_b(\epsilon)$. Let ad be the curve $\psi_a(s)$ $(0 \leqslant s \leqslant \epsilon)$ and dc' be the curve $\varphi_d(s)$ $(0 \leqslant s \leqslant \epsilon)$.

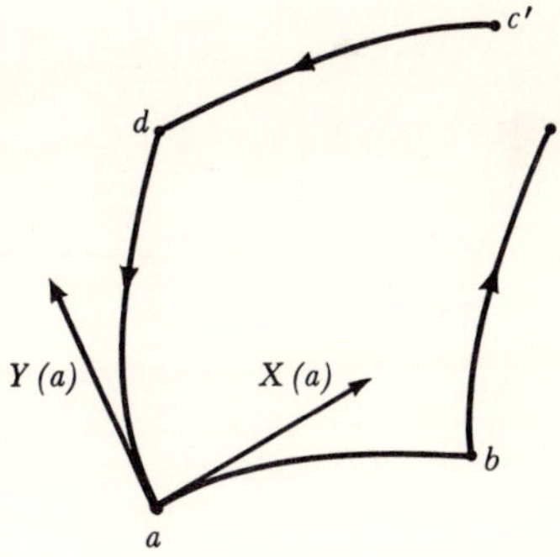

If we consider ϵ as small we can regard the "value" of ω on the curves as "approximately equal" to the value of ω on the tangent vectors to the curves. Then the infinitesimal value of ω around the "parallelogram" is given by the right-hand side of (1.10). The first term measures $\omega(bc) + \omega(da)$, the second term measures $\omega(ab) + \omega(dc)$ and the third term takes into account the fact that infinitesimally the "parallelogram" need not be closed, i.e., $c \neq c'$. Thus $d\omega$ evaluated on an infinitesimal parallelogram measures the value obtained by evaluating ω around its boundary.

EXERCISE 1.1. What is the formula generalizing (1.10) for the case that ω is a differential form of degree k and $X_1, \ldots, X_{k+1}$ are vector fields?

The importance of the operator d is the role that it plays in the theory of integration and the fact that it is a differential operator which is defined purely by the differential structure of the differentiable manifold and is invariant under differentiable maps (Theorem 1.3). In view of this invariance property, many of the equations of physics and geometry take a simple form when expressed in terms of d. We will see some of them later on.

2. Chains and integration.[4] In this section we will present the first of two theories of integration of differential forms. They can be shown to be essentially the same.

We first choose a standard p-dimensional object.

DEFINITION 2.1. *We will denote by* Δ_p *the simplex in* E^p *defined by* $0 \leqslant x^i \leqslant 1$, $\sum_{i=1}^p x^i \leqslant 1$.

Thus Δ_0 is a point, Δ_1 is the unit interval, Δ_2 is the triangle

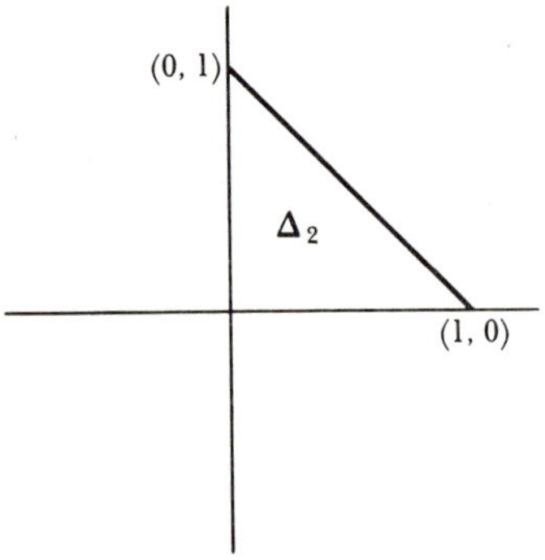

FIGURE 2.1

etc. On the simplex Δ_p it is convenient to introduce the so-called barycentric "coordinates" defined by

$$(2.1) \qquad y^0 = 1 - \sum_{i=1}^{p} x^i, \qquad y^i = x^i, \qquad \text{for } i = 1, \ldots, p,$$

[4] Except for Definitions 2.5 and 2.7 the results of this section will not be used elsewhere in this book.

so that $y_0 + \cdots + y_p = 1$ and $0 \leqslant y_i \leqslant 1$ $(i = 0, \ldots, p)$.

DEFINITION 2.2. *The map* $\delta_{p,i}$ $(i = 0, \ldots, p + 1)$ of Δ_p *into* Δ_{p+1} *is defined by*

$$\bar{y}^j[\delta_{p,i}(y^0, \ldots, y^p)] = \begin{cases} y^j & \text{if} \quad j < i \\ 0 & \text{if} \quad j = i \\ y^{j-1} & \text{if} \quad j > i \end{cases} \tag{2.2}$$

where $\bar{y}^j$ *are the barycentric coordinates of* Δ_{p+1}.

Since $\sum_0^{p+1} \bar{y}^j(\delta_{p,i}) = \sum_0^p y^j = 1$, (2.2) does indeed define a map of Δ_p into Δ_{p+1}. This map can clearly be extended to a differentiable map of a neighborhood of Δ_p in E^p into E^{p+1}. In Figure 2.2 we represent the images of Δ_1 in Δ_2 under $\delta_{p,0}$, $\delta_{p,1}$ and $\delta_{p,2}$.

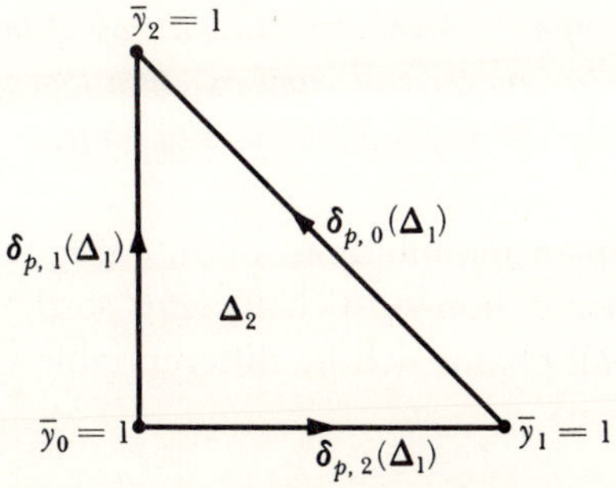

FIGURE 2.2

Let us compare $\delta_{p+1,i} \circ \delta_{p,l}$ with $\delta_{p+1,l} \circ \delta_{p,i}$. If $\bar{\bar{y}}^k$ are the barycentric coordinates of Δ_{p+2}, then

$$\bar{\bar{y}}^k \circ \delta_{p+1,i} \circ \delta_{p,l}(y^0, \ldots, y^p) = \begin{cases} \bar{y}^k \circ \delta_{pl}(y^0, \ldots, y^p) & \text{if} \quad k < i \\ \quad 0 & \text{if} \quad k = i \\ \bar{y}^{k-1} \circ \delta_{pl}(y^0, \ldots, y^p) & \text{if} \quad k > i. \end{cases}$$

Now if $l \leqslant i$, we can write this as

$$y^k \circ \delta_{p+1,i} \circ \delta_{p,l}(y^0, \ldots, y^p) = \begin{cases} y^k & \text{if} \quad k < l \\ 0 & \text{if} \quad k = l \\ y^{k-1} & \text{if} \quad l < k < i \\ 0 & \text{if} \quad k = i \\ y^{k-2} & \text{if} \quad k > i. \end{cases}$$

On the other hand

$$\bar{y}^k \circ \delta_{p+1,l} \circ \delta_{p,i} = \begin{cases} \bar{y}^k \circ \delta_{pi}(y^0, \ldots, y^p) & \text{if } k < l \\ 0 & \text{if } k = l = \\ \bar{y}^{k-1} \circ \delta_{pi}(y^0, \ldots, y^p) & \text{if } k > l \end{cases} \begin{cases} y^k & \text{if } k < l \\ 0 & \text{if } k = l \\ y^{k-1} & \text{if } l < k < i+1 \\ 0 & \text{if } 1 = i+1 \\ y^{k-2} & \text{if } l > i+1. \end{cases}$$

Thus

$$\delta_{p+1,i} \circ \delta_{p,l} = \delta_{p+1,l} \circ \delta_{p,i+1} \qquad \text{if } l \leqslant i. \tag{2.3}$$

DEFINITION 2.3. *A differentiable singular p-simplex on a manifold M is a map of $\Delta_p \to M$ which can be extended to a differentiable map of a neighborhood of Δ_p in E^p to M. A differentiable p-chain is a finite formal linear combination (with real coefficients) of singular p-simplices. An infinite differentiable p-chain is an infinite formal sum (i.e., a map of the set of differentiable singular p-simplices into the real numbers) such that if $\{s_\alpha\}$ denotes the set of s with $c_s \neq 0$, the sets $\{s_\alpha(\Delta_p)\}$ are locally finite.*

Remarks. On a compact manifold, an infinite singular chain can have at most a finite number of non-zero coefficients so the two notions coincide. In this chapter all chains will be differentiable-singular so we just say chain for short.

The set of all p-chains (resp. infinite p-chains) form a vector space in the obvious way; we will denote it by $C_p(M)$ [resp. $IC_p(M)$]. If f is a differentiable map of $M_1 \to M_2$, define

$$f_*(s) = s \circ f \tag{2.4}$$

for simplices and extend by linearity to obtain a linear map, f_* of $C_p(M_1) \to C_p(M_2)$. For the infinite chains we have to be a little more careful. A map f is called *proper*, if $f^{-1}(K)$ is compact for any compact set, K, in M_2. If f is a proper map, and $c = c_s$ is a p-chain of M_1, define $f_*(s) = d_t$ where

$$d_t = \sum_{t = f_*(s)} c_s \tag{2.5}$$

it being understood that $d_t = 0$ if $t \neq f_*(s)$ for any s. Since f is proper, the sum in (2.15) is finite and the set of t with $d_t \neq 0$ is locally finite so (2.5) does define an infinite p-chain of M_2.

If s is a p-simplex, $s \circ \delta_{p-1,i}$ is a $(p-1)$ simplex. We define the boundary of s by

(2.6) $$\partial(s) = \sum_{i=0}^{p} (-1)^i s \circ \delta_{p-1,i}.$$

Thus in Figure 2.3 s is a singular 2-simplex and $\partial(s)$ its boundary.

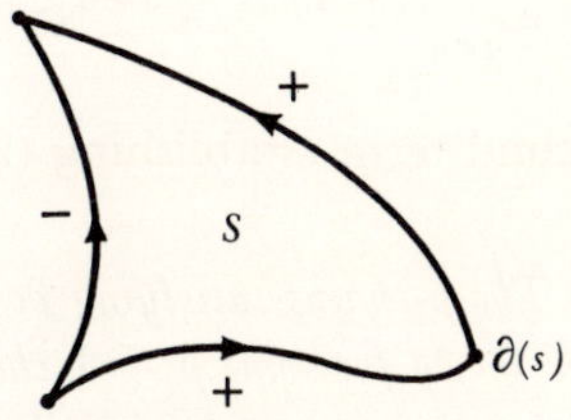

FIGURE 2.3

We extend ∂ by linearity to a map of $C_p(M) \to C_{p-1}(M)$. Similarly, for infinite chains, we define

(2.7) $$\partial(c_s) = \sum_{i=0}^{p} (-1)^i d_t$$
$$d_t = \sum_{s \mid t = s \circ \delta_{p-1,i}} c_s.$$

Notice that in view of the local finiteness condition this sum is always finite.

In both cases we have the fundamental relation

(2.8) $$\partial \circ \partial = 0.$$

We will prove that $\partial(\partial s) = 0$ for a simplex. It will then follow for any chain.

Let s be a q-dimensional simplex. If $q \leqslant 1$ there is nothing to prove. If $q \geqslant 2$

$$\begin{aligned}\partial(\partial s) &= \sum_{i=0}^{q} \sum_{j=0}^{q-1} (-1)^i(-1)^j s \circ \delta_{q-1,i} \circ \delta_{q-2,j} \\ &= \sum_{0 \leqslant j \leqslant i \leqslant q} (-1)^{i+j} s \circ \delta_{q-1,i} \circ \delta_{q-2,j} \\ &\quad + \sum_{0 \leqslant i < j \leqslant q} (-1)^{i+j} s \circ \delta_{q-1,i} \circ \delta_{q-2,j}.\end{aligned}$$

By (2.3), $s \circ \delta_{q-1,i} \circ \delta_{q-2,j} = s \circ \delta_{q-1,j} \circ \delta_{q-2,i+1}$ if $j \leqslant i$. Therefore, we can rewrite the first sum as

$$\sum_{0 \leqslant j \leqslant i \leqslant q} (-1)^{i+j} s \circ \delta_{q-1,j} \circ \delta_{q-2,i+1}.$$

If we put $k = i + 1$ and $l = j$ this becomes

$$\sum_{0 \leqslant l < k \leqslant q} (-1)^{k+l-1} s \circ \delta_{q-1,l} \circ \delta_{q-2,k}$$

which cancels the second term establishing (2.8).

Definition 2.4. *The p-chains satisfying $\partial c = 0$ are called cycles. The p-chains of the form $c = \partial d$ for some $p + 1$ chains d are called boundaries. Equation (2.8) says that the space of boundaries is a subspace of the space of cycles. The quotient space is called the p-dimensional homology group and denoted by $H_p(M)$. The same definitions hold for infinite chains giving $IH_p(M)$.*[5]

In view of its definition, if $g: M_1 \to M_2$, g_* commutes with ∂, i.e., $\partial g_* = g_* \partial$. Thus g_* maps cycles into cycles and boundaries into boundaries and hence induces a map of $H_p(M_1) \to H_p(M_2)$ which we will continue to denote by g_*.

Now Equation (1.4) appears analogous to (2.8).

Definition 2.5. *A differential form, ω, is called closed if $d\omega = 0$. It is called exact, if $\omega = d\Omega$. We will sometimes call a closed form a cocycle and an exact form a coboundary. The quotient of the space of closed p-forms by the exact p-forms is called the p-dimensional (De Rham) cohomology group of M and will be denoted by $H^p(M)$.*

The relation between the two is given by the theory of integration.

Definition 2.6. *Let S be a p-simplex and ω a differential form of degree p. The form $S^*(\omega)$ is defined in some neighborhood of the Euclidean*

[5] We could have gone through this whole procedure using continuous singular chains and continuous maps instead of differentiable. It is not hard to show that the homology groups so obtained are the same as the corresponding groups $H^p(M)$ and $IH^p(M)$. Thus $H^p(M)$ and $IH^p(M)$ depends only on the topological nature of M and not on its differentiable structures.

p-simplex Δ_p. Suppose that $S^*(\omega) = \sum a_{i_1,\ldots,i_p}\, dx^{i_1} \wedge \cdots \wedge dx^{i_p}$ where $x^1, \ldots, x^n$ are the standard coordinates on E^n. Then we define

$$(2.9) \qquad \int_s \omega = \sum_{i_1 < \cdots < i_p} \int_p\!\int \cdots \int a_{i_1,\ldots,i_p}\, dx^{i_1}, \ldots, dx^{i_p}$$

where the summands on the right are ordinary multiple integrals.

For any finite p-chain we extend (2.9) by linearity. We cannot integrate arbitrary p-forms over infinite p-chains since this will lead to infinite series which do not converge in general. Therefore, we restrict the class of forms which we integrate.

Definition 2.7. *The support of a form ω is the smallest closed set outside of which $\omega = 0$. The quotient of the space of closed p-forms with compact support over the space of forms $d\Omega$ where Ω is a $p + 1$ form of compact support is called $H_c^p(M)$.*

If ω is a form with compact support, then for any infinite chain, c_s, the sum

$$(2.10) \qquad \int_{c_s} \omega = \sum_s c_s \int_s \omega$$

has only a finite number of non-zero terms, and is therefore well defined.

If $g: M_1 \to M_2$ is a proper differentiable map, it again follows immediately from the definitions that

$$(2.11) \qquad \int_{g_*(c)} \omega = \int_c g^*\omega.$$

The fundamental theorem of the subject is

Theorem 2.1. (Stokes' theorem first version). *For any p-chain c and $(p - 1)$ form ω (respectively, infinite p-chain c and $(p - 1)$ form ω with compact support).*

$$(2.12) \qquad \int_c d\omega = \int_{\partial c} \omega.$$

Since both sides of (2.12) are linear and so is ∂ we can restrict attention to the case $c = s$.

Now we can write $s = s_*(i)$ where i is the identity map of $\Delta_p \to \Delta_p$. We can thus apply s^* to ω and (2.12) reduces to

$$\int_i d\Omega = \sum (-1)^i \int_{\delta_{p-1,i}} \tag{2.13}$$

where we have written $\Omega = s^*(\omega)$. In this expression, $\delta_{p-1,i}$ (a map from $\Delta_{p-1} \to \Delta_p$) is regarded as a singular $(p-1)$ simplex of E^p. Now $\Omega = \sum a_{i_1,\ldots,i_{p-1}}(x^1, \ldots, x^p)\, dx^{i_1} \wedge \cdots \wedge dx^{i_p}$. It suffices to prove (2.13) for each of these terms. What we must show is

$$\begin{aligned}(-1)^{j-1}\int_i \frac{\partial a}{\partial x^j}\, dx^1 \wedge \cdots \wedge dx^p \\ = \sum_{i=0}^{p} (-1)^i \int_{\delta_{p-1,i}} a\,dx^1 \wedge \cdots \wedge \widehat{dx^j} \wedge \cdots \wedge dx^p.\end{aligned} \tag{2.14}$$

Now $\delta^*_{p-1,i}(x^i) = 0$ so the only non-zero terms are those with $i = 0$ and $i = j$. If $z^1, \ldots, z^{p-1}$ are the Euclidean coordinates of Δ_{p-1}, we have

$$\begin{aligned}\delta^*_{p-1,j}(a\, dx^1 \wedge \cdots \wedge \widehat{dx^j} \wedge \cdots \wedge dx^p) \\ = a \circ \delta_{p-1,j}(z^1, \ldots, z^p)\, dz^1 \wedge \cdots \wedge dz^{p-1}\end{aligned}$$

and

$$\begin{aligned}&\delta^*_{p-1,0}(a\, dx^1 \wedge \cdots \wedge \widehat{dx^j} \wedge \cdots \wedge dx^p) \\ &= a \circ \delta_{p-1,j}(d(1 - z^1 - z^2 - z^p) \wedge dz^1 \wedge \cdots \wedge \widehat{dz^j} \wedge \cdots \wedge dz^{p-2}) \\ &= (-1)^{j-1} a \circ \delta_{p-1,j}\, dz^1 \wedge \cdots \wedge dz^{p-1}.\end{aligned}$$

In view of the definitions, (2.14) reduces to

$$\begin{aligned}&\int_{\Delta_p} \frac{\partial a}{\partial x^j}\, dx^1, \ldots, dx^p \\ &= \int a\left(x^1, \ldots, x^{j-1}, 1 - \sum_{i \neq j} x^i, \ldots, x^p\right) dx^1 \wedge \cdots \wedge dx^j \wedge \cdots \wedge dx^p \\ &\qquad - \int a(x^1, \ldots, x^{j-1}, 0, \ldots, x^p)\, dx^1 \wedge \cdots \wedge dx^j \wedge \cdots \wedge dx^p\end{aligned}$$

which is true by partial integration. Thus we have proved Theorem 2.1.

From (2.12) it follows that the integral of an exact form over a cycle vanishes, and that the integral of a closed form over a boundary vanishes. Thus the integral of a closed form over a cycle depends only on the cohomology class of the form and the homology class of the cycle. Thus

COROLLARY 2.1. *The bilinear map of $C_p(M) \times \wedge^p(M) \to R$ given by $\int_C \omega$ induces a bilinear map of $H_p(M) \times H^p(M) \to R$.*

We have

COROLLARY 2.2. *The bilinear map of $IC_p(M) \times \wedge_C^p(M) \to R$ given by $\int_C \omega$ induces a bilinear map of $IH_p(M) \times H_C^p(M) \to R$.*

A fundamental theorem in algebraic topology (De Rham's theorem) asserts that this bilinear pairing is non-singular; that is, we can regard $H^p(M)$ as the dual space on $H_p(M)$, cf. De Rham [19]. In particular, the spaces H^p and IH^p, which were defined in terms of differential forms, and hence would appear to depend on the differential structure of the manifold, are actually topological invariants.

3. Integration of densities. Although the theory of integration given above, i.e., using singular chains, proves to be very convenient in the study of differential topology we cannot effectively compute with it. Therefore, we shall introduce an alternate approach, of more geometric applicability, which will allow us to draw some immediate conclusions without the use of any complicated algebraic machinery. It is a consequence of De Rham's theorem that the two theories are essentially the same.

We wish to define, on a manifold M, objects which are to be integrated over a suitable collection of domains of M. We know how to integrate reasonably well behaved functions on Euclidean space. On a manifold, it cannot make any sense to integrate a function. Even on Euclidean space, if we change coordinates, the form of the expression of the integral changes with change of coordinates: $x^1, \ldots, x^n$ are Euclidean coordinates on E^n and $y^1, \ldots, y^n$ is some other coordinate system, then for any function f_1 (cf. Appendix II)

$$\int_{E^n} f(x^1, \ldots, x^n)\, dx^1, \ldots, dx^n$$
$$= \int_{E^n} f(x^1(y^1, \ldots, y^n), \ldots, x^n(y^1, \ldots, y^n)) \left|\det\left(\frac{\partial x^i}{\partial y^j}\right)\right| dy^1, \ldots, dy^n.$$

Thus in terms of the y coordinates, we integrate $f|\det(\partial x^i/\partial y^j)|$ and not f. We are therefore interested in an object ρ on a manifold which is presented as follows: For any chart (U_i, h_i) of the atlas $\{(U_i, h_i)\}$ we are given a function ρ_i defined on U_i so that

$$\rho_i(x) = \rho_j(x)|\det J_{h_i(x)}(h_j \circ h_i^{-1})|. \tag{3.1}$$

If we compare with Equation (6.11) of Chapter II, we see that ρ is a field of type Q where Q is a one-dimensional vector space and

$$\rho(A)q = |\det A|^{-1} q \qquad \text{for any } A \in GL(n). \tag{3.2}$$

(Thus Q is a space of densities, compare with Example 4, of Section 2, of Chapter I.)

Definition 3.1. *Let Q be the real numbers and let $GL(n)$ act on Q according to* (3.2). *A field of type Q defined on a manifold M is called a density.*

In terms of a chart (U_i, h_i), ρ is given by a function ρ_i, the rule for change of charts being (3.1).

Exercise 3.1. Let M be a Riemann manifold. On each chart (U_i, h_i), $h_i = (x^1, \ldots, x^n)$ define the function ρ_i as follows:

$$\rho_i(x) = \text{volume of parallelopiped spanned by}$$

$$\left.\frac{\partial}{\partial x^1}\right|_x, \ldots, \left.\frac{\partial}{\partial x^n}\right|_x.$$

(Here $T_p(M)$ has a scalar product, i.e., is a Euclidean space and so we can compute volumes.) Show that this defines a density on M. It is called the *volume density* of the Riemann metric.

It is clear that the space of densities is a vector space, and, in fact, a module over the ring of differentiable functions. That is, we can add two densities and multiply a density by a function. The value of a density at a point doesn't make sense, since a density is not a function. However, it does make sense to ask whether a density is zero or not at a point (and in fact to talk about the sign of a density at a point) since this is independent of the representation of the density in terms of a chart.

Definition 3.2. *The support of a density ρ (denoted by* supp ρ*) is the closure of the set of points where $\rho \neq 0$.*

Similarly, it makes sense to speak of density being continuous or discontinuous at a point.

Definition 3.3. *A density ρ is called locally bounded if for any chart (U_i, h_i) and any compact subset $K \subset U$ the function ρ_i is bounded on K.*

(Here ρ_i is the function representing ρ on (U_i, h_i).)

It is easy to verify that if we have boundedness of the ρ_i for all $K \subset (U_i, h_i)$ where $\{(U_i, h_i)\}$ form an atlas then we have boundedness of ρ.

We now turn to our definition of domains of integration. Recall[6] that a subset $B \subset E^n$ is said to be of content zero if, for any $\epsilon > 0$, there are a finite number of rectangles $\square_1, \ldots, \square_k$ so that $B \subset \bigcup \square_i$ and $\sum \text{vol}(\square_i) < \epsilon$.

Definition 3.4. *A subset B of a manifold M is said to have content zero if there is an atlas $\{(U_i, h_i)\}$ with $B \subset \bigcup B_i$ where*

1) *all but a finite number of the B_i are empty*
2) *B_i is compact*
3) *$B_i \subset U_i$ and $h_i(B_i)$ has content zero in E^n.*

We let the reader verify that the union of a finite number of sets of content zero has content zero.

Definition 3.5. *A set $D \subset M$ is said to have content if*

1) *D is compact*
2) *the boundary of D has measure zero.*

We denote the set of all domains with content by $\mathscr{D}$.

We let the reader verify that the set of domains with content is closed under intersection and finite union.

Definition 3.6. *A density, ρ, is said to be almost continuous if it is continuous except at a set of content zero. We denote the space of locally bounded almost continuous densities by* P.

Exercise 3.2. Show that P is a vector space. Let us call a function almost continuous if it is continuous except at a set of content zero. Show that the set of bounded almost continuous functions form a ring and P is a module over this ring.

Exercise 3.3. For any set $D \subset M$, we denote by φ_D the characteristic function of D; i.e., $\varphi_D(x) = 1$ if $x \in D$ and $\varphi_D(x) = 0$ if $x \notin D$. Show that $D \in \mathscr{D}$ if and only if supp φ is compact and φ is almost continuous.

[6] Cf. Appendix II for a review of the theory of Riemann integration on E^n.

We are now in a position to present our theory of integration.

THEOREM 3.1. *There is a unique rule,* $\int$, *which assigns to* $D \in \mathscr{D}$ *and* $\rho \in \mathrm{P}$ *a number* $\int_D \rho$ *(which we call the integral of* ρ *over* D*) satisfying*

1) $\int_D \rho$ *is a linear functional on* P *for each fixed* D,
2) *if* $D_1 \cap D_2 = \varnothing$ *then* $\int_{D_1 \cup D_2} \rho = \int_{D_1} \rho + \int_{D_2} \rho$, *and*
3) *if* $\operatorname{supp} \rho \subset U$ *where* (U, h) *is a chart and* f *is a function representing* ρ *in* (U, h) *then*

$$\int_D \rho = \int_{h(U)} (\varphi_D \circ h^{-1})(f \circ h^{-1})\, dx^1, \ldots, dx^n.$$

In proving Theorem 3.1, we remark that we can reformulate the theory so as to avoid explicit mention of the domain of integration. In fact, let P_c denote the space of densities of compact support. We will prove:

THEOREM 3.2. *There is a unique linear functional,* $\int$, *on* P_c *such that if* $\operatorname{supp} \rho \subset U$ *for any chart* (U, h) *then*

$$\int \rho = \int_{h(U)} f \circ h^{-1}\, dx^1, \ldots, dx^n \tag{3.3}$$

where f *is the function representing* ρ *on* (U, h).

If we then define

$$\int_D \rho = \int \varphi_D \rho \tag{3.4}$$

for any $D \in \mathscr{D}$ and any $\rho \in \mathrm{P}$, we see that Theorem 3.2 is equivalent to Theorem 3.1; we need only check condition 2) of Theorem 3.1. But if $D_1 \cap D_2 = \varnothing$ then $\varphi_{D_1 \cup D_2} = \varphi_{D_1} + \varphi_{D_2}$ so 2) is a consequence of the linearity of $\int$.

We now proceed to the proof of Theorem 3.2. Let φ_i be a partition of unity subordinate to some atlas and let (U_i, h_i) be a chart containing $\operatorname{supp} \varphi_i$ where $h_i(x_i^1, \ldots, x_i^h)$. For any $\rho \in \mathrm{P}_c$ only a finite number of the densities $\varphi_i \rho$ are different from zero and so

$$\int \rho = \sum \int \varphi_i \rho = \sum \int (\varphi_i \circ h_i^{-1})(f_i \circ h_i^{-1})\, dx^1, \ldots, dx^n \tag{3.5}$$

when f_i is the function representing ρ on (U_i, h_i). This proves uniqueness. To prove existence we must check that the $\int$ defined by (3.5) satisfies (3.3). For that it suffices to check (3.3) for each of the summands $\varphi_i\, \rho$, i.e., that

$$\int_{h(U)} \varphi_i f \circ h^{-1}\, dx^1, \ldots, dx^n = \int_{h_i(U_i)} \varphi_i f_i \circ h_i^{-1}\, dx_i^1, \ldots, dx_i^n.$$

In view of the relation between f and f_i, this is exactly the change of variables formula for the multiple integral.

Exercise 3.4. For any Riemann manifold, M, define the volume of a $D \in \mathscr{D}$ to be the integral of the volume density over D. Compute the volume of the n-sphere relative to the metric induced by the standard imbedding in E^{n+1}.

Exercise 3.5. Let φ_1 and φ_2 be two immersions of M into the Riemann manifold N and let g_1 and g_2 denote the induced Riemann metric. Let V_1 and V_2 denote the volumes of M with respect to these metrics. Show that for any $\eta > 0$ there is a $(W, Z, \epsilon, 1)$ so that if $\varphi_1 \in \mathscr{N}_{(W,\, Z,\, \epsilon,\, 1)}\varphi_2$ then $|V_1 - V_2| < \eta$.

Exercise 3.6. Show that the above definition of the volume of any subset of E^n regarded as a Riemann manifold coincides with the usual definition.

Exercise 3.7. Let φ be an immersion of a compact manifold M into E^n. Let V denote its volume with respect to the induced Riemann metric. Using Exercises 3.6 and 3.7 show that V is the limit of the surface volume (area) of approximating polyhedra P_i where the P_i approach $\varphi(M)$ in position and direction.

If ω is an exterior n-form on an n-dimensional manifold, then for any chart (U, h), $h = (x^1, \ldots, x^n)$, we can write $\omega = f\, dx^1 \wedge \cdots \wedge dx^n$. If (V, g), $g = (y^1, \ldots, y^n)$ is another chart with $\omega = g\, dy^1 \wedge \cdots \wedge dy^n$, then

$$f = g(\det J_{g \circ h^{-1}}).$$

This looks almost like (3.1) except that we have $\det J_{g \circ h^{-1}}$ instead of $|\det J_{g \circ h^{-1}}|$. We would be able to associate a density to an n-form, and conversely, if we could choose an atlas for which all the Jacobian determinants are positive. For this reason, we introduce the following:

Definition 3.7. *An n-dimensional differentiable manifold, M, is called orientable if there exists an exterior differential form ω of degree n on M which is never zero, i.e., $\omega_p \neq 0$ for all $p \in M$. Two such forms are said to define* *the same orientation if they differ by a factor which is positive everywhere. A manifold with a choice of orientation is said to be oriented.*

Since $\wedge^n(T_p^*)$ is one-dimensional, any two n-forms differ by a (differentiable) function factor. Thus, on a connected orientable manifold there are exactly two orientations.

Let ω be a form defining an orientation of M. Since any other form, Ω, defined on an open subset, U, of M can be written as $\Omega = f\omega$, it makes sense to say that $\Omega > 0$ or $\Omega < 0$ according as $f > 0$ or $f < 0$ and this depends only on the orientation and not the specific choice of ω. If x is a coordinate system on U, $dx^1 \wedge \cdots \wedge dx^n \neq 0$ anywhere on U. Thus either $dx^1 \wedge \cdots \wedge dx^n > 0$ or $dx^1 \wedge \cdots \wedge dx^n < 0$. In the latter case, the functions $-x^1, x^2, \ldots, x^n$ give a coordinate system for which the first alternative holds. If we cover M by coordinate neighborhoods, we can thus choose coordinates $(x^1, \ldots, x^n)$ in these neighborhoods so that $dx^1 \wedge \cdots \wedge dx^n > 0$ in each. This means that in the overlap of two coordinate systems, x and y, we have

$$dx^1 \wedge \cdots \wedge dx^n = f\, dy^1 \wedge \cdots \wedge dy^n \qquad \text{with } f > 0 \tag{3.6}$$

i.e.,

$$\left| \left(\frac{\partial x^i}{\partial y^j} \right) \right| > 0. \tag{3.7}$$

Conversely, suppose that we can cover M by a countable number of coordinate neighborhoods, U_i, in such a way that (3.7) holds. If we write $\omega_i = dx_i^1 \wedge \cdots \wedge dx_i^n$ where $(x_i^1, \ldots, x_i^n)$ are the coordinates on U_i, and if φ_i is a partition of unity subordinate to U_i, then

$$\omega = \sum \varphi_i \omega_i \neq 0$$

since $\omega_i \neq 0$ in U_i, $\varphi_i \geqslant 0$. Thus

THEOREM 3.3. *A necessary and sufficient condition that a manifold be orientable is that there exist an atlas so that in the overlap of any two charts* (3.7) *holds. A choice of orientation picks out preferred charts which we call positive relative to the orientation.*

DEFINITION 3.8. *Let M be a differentiable manifold. A subset D of M is called a domain with regular boundary (domain for short) if every point $p \in \bar{D}$ either*

1) *has a neighborhood belonging entirely to D*

2) *has a neighborhood U, with coordinates $x^1, \ldots, x^n$ such that $U \cap \overline{D}$ is the set of all points of U with $x^n \geqslant 0$, and $x^n(p) = 0$.*[7]

(a)

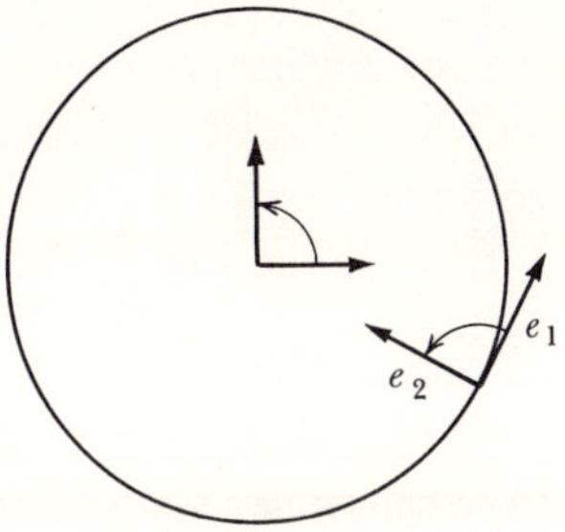

(b)

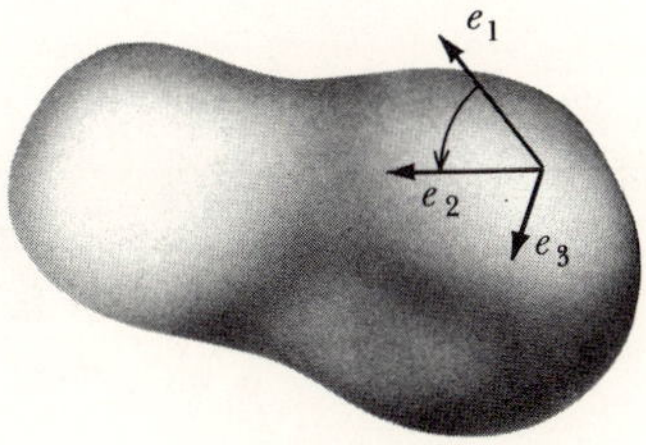

(c)

These two properties are obviously mutually exclusive. Points with property 1) are called *interior points*; those satisfying 2) are called *boundary points*. It follows from Exercise 2.5 of Chapter II that the set of boundary points of D can be regarded as a submanifold, $\partial(D)$, of M. If M is orientable, so is $\partial(D)$. In fact, choose coordinates in the neighborhoods of Definition 3.8 such that $dx^1 \wedge \cdots \wedge dx^n > 0$ relative to some fixed orientation of M. (This can always be done if $n > 1$. If $n = 1$, $\partial(D)$ is a collection of points and is trivially orientable.) Now by definition, the sets $U_i \cap \partial(D)$ form a covering of $\partial(D)$ by coordinate neighborhoods,

[7] Thus the manifold M itself is a domain with regular boundary.

where $\{U_i\}$ is a countable collection of sets of type 2) in Definition 3.8; on each such set, U, the functions $x^1, \ldots, x^{n-1}$ form a coordinate system in $U \cap \partial(D)$. If $y^1, \ldots, y^n$ is a second such coordinate system then $y^n = 0$ on $\partial(D)$ so that $\partial y^n/\partial x^i = 0$ at all points of $\partial(D)$. Since $y^n > 0$ when $x^n > 0$, $\partial y^n/\partial x^n \geqslant 0$. Thus we can write

$$\left.\frac{\partial(y^1, \ldots, y^n)}{\partial(x^1, \ldots, x^n)}\right|_{\partial(D)} = \frac{\partial(y^1, \ldots, y^{n-1})}{\partial(x^1, \ldots, x^{n-1})} \frac{\partial y^n}{\partial x^n} > 0.$$

Thus $\partial(y^1, \ldots, y^{n-1})/\partial(x^1, \ldots, x^{n-1}) > 0$ and $\partial(D)$ is orientable. We have in fact a definite orientation induced on $\partial(D)$ by that of M. We shall consider the form $dx^1 \wedge \cdots \wedge dx^{n-1}$ corresponding to a positive coordinate system of M to have the sign $(-1)^n$.

Thus in one, two, and three dimensions the orientations are as in the figures on page 117.

THEOREM 3.4. *Let M be an oriented n-dimensional manifold, and let D be a domain of M. There is a unique functional which assigns to every n-form, ω, with compact support a real number, called its integral over D, and denoted by $\int_D \omega$, such that*

1) $$\int_D (\omega_1 + \omega_2) + \int_D \omega_1 + \int_D \omega_2.$$

2) *If the support of ω is contained in some coordinate neighborhood U and $h = (x^1, \ldots, x^n)$ is a coordinate system in U with $dx^1 \wedge \cdots \wedge dx^n > 0$ and $\omega = f\, dx^1 \wedge \cdots \wedge dx^n$, then*

$$\int_D \omega = \int \varphi f(x^1, \ldots, x^n)\, dx^1, \ldots, dx^n \tag{3.8}$$

where φ is the characteristic function of $D(\varphi(x) = 1$ if $x \in D$ and $\varphi(x) = 0$ if $x \notin D)$ and the integral on the right of (3.8) is the ordinary Riemann integral.

The proof is an immediate consequence of Theorem 3.1.

Let M be an oriented manifold and D_i a finite number of disjoint domains of M. For any ω we have

$$\int_{\cup D_i} \omega = \sum \int_D \omega. \tag{3.9}$$

THEOREM 3.5. (Stokes' theorem, second version). *Let* ω *be a form of degree* $n - 1$ *of compact support on an oriented manifold* M. *For any domain* D *in* M

$$\int_D d\omega = \int_{\partial(D)} i^*(\omega) \tag{3.10}$$

where $\partial(D)$ *is considered as an oriented manifold with the orientation induced by that of* M *and* i *is the injection map of* $\partial(D) \to M$.

Choose a countable atlas $\{(U_i, h_i)\}$ of M such that either $U_i \cap \partial(D) = \varnothing$ or U_i is a neighborhood of the type required in 2) of Definition 3.8. Let g_j be a partition of unity subordinate to this covering. Then

$$\int_{\partial(D)} i^*(\omega) = \sum \int_{\partial(D)} i^*(\varphi_j \omega)$$

and

$$\int_D d\omega = \sum \int_D d(\varphi_i \omega).$$

Therefore, it suffices to verify (3.10) for each of the $\varphi_j\,\omega$. That is to establish (3.10) for the case that $\operatorname{supp}(\omega) \subset U_i$ for some i. Let $x^1, \ldots, x^n$ be coordinates in U_i with $dx^1 \wedge \cdots \wedge dx^n > 0$. Let

$$\omega = \sum (-1)^{j-1} a_j\, dx^1 \wedge \cdots \wedge dx^{j-1} \wedge dx^{j+1} \wedge \cdots \wedge dx^n.$$

Then

$$d\omega = \left(\sum \frac{\partial a_j}{\partial x^j}\right) dx^1 \wedge \cdots \wedge dx^n.$$

If $U_i \cap \partial(D) = \varnothing$ then $\int_{\partial(D)} i^*(\varphi_i\,\omega) = 0$. On the other hand the function φ of Theorem 3.4 is either $\equiv 0$ or $\equiv 1$ in U_i. In the first case there is nothing to prove. In the other case, we must evaluate

$$\int_{h_i(U_i)} \left(\sum \frac{\partial a_j}{\partial x_j}\right) dx^1, \ldots, dx^n$$

where we write (by abuse of language) $a_j = a_j \circ h_i^{-1}$ where $\operatorname{supp}(a_j) \subset h_i(U_i)$. We may extend the a_i to be defined throughout E^n by setting $a_i = 0$ outside $h_i(u_i)$. Now let R be so large that $|x^j| < R$ for $(x^1, \ldots, x^n) \in h_i(U_i)$, and let B denote the cube $|x_i| < R$. Then

$$\int_{h_i(U_i)} \left(\sum \frac{\partial a_j}{\partial x_j}\right) dx^1, \ldots, dx^n = \sum_j \int_B \frac{\partial a_j}{\partial x^j}\, dx^1, \ldots, dx^n.$$

On the other hand, integration by parts gives

$$\int_B \frac{\partial a_j}{\partial x^j} dx^1, \ldots, dx^n = \pm \int \Big\{ a_j(x^1, \ldots, x^{j-1}, R, x^{j+1}, \ldots, x^n)$$
$$- a_j(x^1, \ldots, x^{j-1}, -R, x^{j+1}, \ldots, x^n) \Big\} dx^1, \ldots, dx^n = 0.$$

Thus (3.10) is verified for supp $\omega \subset U_i$ where $U_i \cap \partial(D) = \varnothing$. Next assume $U_i \cap \partial D \neq \varnothing$ and $x^1, \ldots, x^n$ are coordinates in U_i as prescribed in 2) of Definition 3.8. Then if $\omega = \sum (-1)^{j-1} a_j \, dx^1 \wedge \cdots \wedge dx^{j-1} \wedge dx^{j+1} \wedge \cdots \wedge dx^n$, $i^*(\omega) = (-1)^{n-1} a_n \, i^*(dx^1 \wedge \cdots \wedge dx^{n-1})$ and

$$\int_{\partial(D)} i^*(\omega) = (-1)^{n-1} \int a_n \, dx^1, \ldots, dx^{n-1}.$$

Extending the a's as before

$$\int_D d\omega = \sum \int_B \psi \frac{\partial a_j}{\partial x^j} dx^1, \ldots, dx^n,$$

where $\psi = 1$ for $x_n \geqslant 0$ and $\psi = 0$ for $x_n < 0$. Again, integration by parts gives

$$\int_B \psi \frac{\partial a_j}{\partial x^j} dx^1, \ldots, dx^n = 0 \qquad \text{for } j < n$$

and

$$\int_B \psi \frac{\partial a_n}{\partial x^n} dx^1, \ldots, dx^n = (-1)^n \int_{E^{n-1}} a_n \, dx^1, \ldots, dx^{n-1},$$

proving Theorem 3.5.

4. 0 and *n*-dimensional cohomology, degree. Although we will not present general techniques for computation of the various homology and cohomology groups we study some special cases here.

By definition, there is no zero-dimensional coboundary. Therefore, the zero cohomology groups coincide with the cocycles. Now a zero form is just a function. A zero cocycle is thus a function f such that $df \equiv 0$. That is, f is locally constant. Thus $H^0(M)$ is the space of all locally constant real valued functions on M. Since every locally constant function is constant on the connected components of M, we can write $H^0(M) \cong R^{\mathrm{Co}(M)}$ where $\mathrm{Co}(M)$ is the set of components of M. In particular, if M is connected $H^0(M) \cong R$.

In order for f to be a compact cocycle, f must be locally constant and supp (f) compact. This means that $f \equiv 0$ on the non-compact components of M. Thus $H_C^0(M) = R^{CCo(M)}$ where $CCo(M)$ is the set of compact components of M. In particular, if M is connected, $H_C^0(M) = R$ if M is compact and $H_C^0(M) = 0$ if M is not compact.

We can also compute $H^p(B_r^n)$ for any p.[8] In fact, we have the following

THEOREM 4.1. (Poincaré's lemma). *Let ω be a form of degree $k \geqslant 1$ defined on B_r^n with $d\omega = 0$. Then there exists a form, Ω, defined on B_r^n with $d\Omega = \omega$.*

We will prove this by induction on the dimension n. For $n = 1$, we can assume $k = 1$, all other cases being trivial. If $\omega = f(x)\,dx$ just set $\Omega = g$ where $g(x) = \int_0^x f(t)\,dt$.

We next make the following observation:

LEMMA 4.1. *If $\partial/\partial x^n \;\lrcorner\; \omega = 0$ and $d\omega = 0$ then $\omega = \sum a_{i_1,\ldots,i_k} dx^{i_1} \wedge \cdots \wedge dx^{i_k}$ where $i_k < n$ and $\partial a_{i_1,\ldots,i_k}/\partial x^n = 0$.*

In fact,

$$d\omega = (-1)^k \sum \frac{\partial a_{i_1,\ldots,i_k}}{\partial x^n}\, dx^{i_1} \wedge \cdots \wedge dx^{i_k} \wedge dx^n + \cdots$$

where the additional terms do not involve dx^n.

Now to the proof of the theorem.

Let $y^1, \ldots, y^{n-1}$ be the standard coordinates on B_r^{n-1}. Let i be the map of $B_r^{n-1} \to B_r^n$ given by

$$x^i(y^1, \ldots, y^{n-1}) = y^i \qquad 1 \leqslant i \leqslant n-1$$
$$x^n(y^1, \ldots, y^{n-1}) = 0$$

where $y^1, \ldots, y^{n-1}$ are Euclidean coordinates on B_r^{n-1}. Thus $i^*(dx^j) = dy^j$ for $1 \leqslant j \leqslant n-1$ and $i^*(dx^n) = 0$. Let p denote the map of B_n^r onto B_r^{n-1} defined by

$$y^i(x^1, \ldots, x^n) = x^i \qquad \text{for } 1 \leqslant i \leqslant n-1.$$

[8] Recall that B_r^n is the ball of radius r in E^n.

Now $d(i^*(\omega)) = i^*(d\omega) = 0$. By induction, $i^*(\omega) = d(\Omega')$ for some Ω' on B_r^{n-1}. Let us write

$$\omega = \omega_1 + dx^n \wedge \omega_2$$

where ω_1 and ω_2 are forms in $dx^1, \ldots, dx^{n-1}$. Define Ω by the following conditions:

$$1) \qquad i^*(\Omega) = \Omega'$$

$$2) \qquad \frac{\partial}{\partial x^n} \rfloor \Omega = 0$$

$$3) \qquad \frac{\partial}{\partial x^n} \rfloor d\Omega = \omega_2.$$

In terms of the coordinate system $(x^1, \ldots, x^n)$ condition 2) says that Ω can be written as

$$\Omega = \sum_{i_1 < \cdots < i_{k-1}} a_{i_1, \ldots, i_{k-1}} dx^{i_1} \wedge \ldots \wedge dx^{i_{k-1}}, \quad i_{k-1} \leqslant n - 1.$$

Condition 1) says that $\sum a_{i_1, \ldots, i_{k-1}}(y^1, \ldots, y^{n-1}, 0)\, dy^{i_1} \wedge \cdots \wedge dy^{i_n} = \Omega'$. Condition 3) says that

$$\frac{\partial a_{i_1, \ldots, i_{k-1}}}{\partial x^n} = b_{i_1, \ldots, i_{k-1}}$$

if

$$\omega_2 = \sum b_{i_1, \ldots, i_{k-1}} dx^{i_1} \wedge \cdots \wedge dx^{i_{k-1}}.$$

Thus, conditions 1), 2), and 3) can be solved uniquely by integration, with respect to x^n. We assert that $d\Omega = \omega$. In fact, $d\Omega - \omega = d\Omega - dx_n \wedge \omega_2 - \omega_1$. Now

$$d\Omega - dx^n \wedge \left(\frac{\partial}{\partial x^n} \rfloor d\Omega\right)$$

is a differential form which does not involve dx^n. Thus

$$d\Omega - \omega = dx^n \wedge \left(\frac{\partial}{\partial x^n} \rfloor d\Omega\right) - dx^n \wedge \omega_2 + \Omega = \Omega$$

where Ω does not depend on dx^n. Thus since $d(d\Omega - \omega) = 0$ we can apply Lemma 4.1 to conclude that $d\Omega - \omega = \sum d_{i_1, \ldots, i_k} dx^{i_1} \wedge \cdots \wedge dx^{i_k}$ where $i_k \leqslant n - 1$ and $\partial d_{i_1 \cdots i_k}/\partial x^n = 0$.

On the other hand $\sum d_{i_1,\ldots,i_k}(y^1,\ldots, y^{n-1}, 0)\ dy^{i_1} \wedge \cdots \wedge dy^{i_n} = i^*(d\Omega - \omega) = d(i^*(\Omega)) - \omega = d\Omega' - \omega = 0$. Thus

$$d_{i_1,\ldots,i_k}(x^1,\ldots,x^{n-1},0) = 0$$

and therefore $d_{i_1,\ldots,i_k}(x^1,\ldots,x^n) = 0$. In other words, $d\Omega - \omega = 0$, proving Theorem 3.1.

Corollary 4.1.

$$H^p(B^n_r) = 0 \qquad \textit{for } p < 1.$$

In the proof of Theorem 3.1, the coefficients of Ω were obtained from those of ω by a process of integration. Thus if the coefficients of ω depended differentiably on some additional parameters, Ω depends differentiably on these parameters.

Corollary 4.2. *If in Theorem* 3.1, *the form* ω *depends differentiably on some parameters, i.e.,*

$$\omega = \sum a_{i_1,\ldots,i_k}(x^1,\ldots,x^n; z^1,\ldots,z^l)\ dx^{i_1} \wedge \cdots \wedge dx^{i_k}$$

(where the a's are differentiable functions of $x^1,\ldots,x^n, z^1,\ldots,z^l$*) and* $d\omega = 0$ *(the d being taken with respect to the x's) then* $\omega = d\Omega$ *where*

$$\Omega = \sum b_{i_1,\ldots,i_{k-1}}(x^1,\ldots,x^n; z^1,\ldots,z^l)\ dx^{i_1} \wedge \cdots \wedge dx^{i_{k-1}}$$

where the b's are differentiable functions of all $n + l$ *variables.*

We now wish to compute $H^n_c(M)$, for any manifold M.

Lemma 4.2. *Let* ω *be an n-form defined on* E^n *such that*

1) *supp* $(\omega) \subset D^n_r$ *where* D^n_r *is the cube given by* $\{x \mid |x_i| < r\}$

and

2) $\int_{E^n} \omega = 0.$

Then we can find an $(n - 1)$ *form* Ω *such that*

3) supp $(\Omega) \subset D_r$

and

4) $d\Omega = \omega.$

The proof will proceed by induction. We will have to make the induction hypothesis somewhat stronger in order to proceed. For any k-form, π, let $|\pi|_r^n$ denote the maximum of the absolute value coefficients of π in D_r^n if we express π in terms of $dx^{i_1} \wedge \cdots \wedge dx^{i_k}$. We will say that a family of forms π_t depends differentiably on t if each of the coefficients is a differentiable function of $x^1, \ldots, x^n t$. Here t stands for some s-dimensional parameter, $t = (t_1, \ldots, t_s)$. Our induction assertion now is:

Let ω_t be a family of forms on E^n depending differentiably on t. Suppose that for each value of t the form ω_t satisfies 1) *and* 2). *Then there is a differentiable family of forms Ω_t satisfying* 3) *and $d\Omega_t = \omega_t$ and also*

$$5) \qquad |\Omega_t|_r^n < K_r^n |\omega_t|_r^n,$$

where K_r^n depends only on n and r.

For $n = 1$, we can write $\omega_t = f_t(\cdot)\, dx^1$. Let Ω_t be the function given by

$$\Omega_t(x) = \int_{-r}^{x} f_t(y^1)\, dy^1.$$

Then Ω_t is clearly a differentiable family of 0-forms satisfying conditions 3), 4), and 5).

Assume the induction hypothesis for $n - 1$. Let

$$\omega_t = f_t\, dx^1 \wedge \cdots \wedge dx^n$$

and let

$$\pi = \varphi(y^1, \ldots, y^{n-1})\, dy \wedge \cdots \wedge dy^{n-1}$$

be a positive form on E^{n-1} with supp $\pi \subset D_r^{n-1}$ and $\int_{E^n} \pi = 1$. Consider the forms $\pi_{x^n,t}$ on E^{n-1} (depending on the parameters x^n, t) given by

$$\begin{aligned}
\pi_{x^n,t} &= f_t(\ldots, x^n)\, dy^1 \ldots dy^{n-1} \\
&\qquad - \left(\int \cdots \int_{E^{n-1}} f_t(y^1, \ldots, y^{n-1}, x^n)\, dy^1 \ldots dy^{n-1}\right)\pi \\
&= \left[f_t(\cdot, x^n) - \left(\int_{E^{n-1}} f_t(y^1, \ldots, y^{n-1}, x^n)\right.\right. \\
&\qquad \left.\left. dy^1 \ldots dy^{n-1}\right)\varphi\right] dy^1 \ldots dy^{n-1}.
\end{aligned} \tag{4.1}$$

They constitute a differentiable family of $(n - 1)$ forms which satisfy the conditions of the induction hypothesis. Furthermore $\pi_{x^n,t} = 0$ for $|x^n| > r$. Therefore, we can write

$$\pi_{x^n,t} = d\Omega_{x^n,t}. \tag{4.2}$$

Let us write

$$\Omega_{x^n,t} = \sum_{i=1}^{n=1} (-1)^i g_t(\cdot, x^n)\, dy^1 \wedge \cdots \wedge \widehat{dy^i} \wedge \cdots \wedge dy^{n-1}$$

so that the functions $g_t(\cdot, \circ)$ are differentiable functions on E^n. Define the form Ω_t' by

$$\Omega_t' = \sum (-1)^i g_{t,i}\, dx^1 \wedge \cdots \wedge \widehat{dx^i} \wedge \cdots \wedge dx^{n-1}$$

so that

$$\iota_{x^n}^* \Omega_t' = \Omega_{t,x^n}$$

where $\iota_{x_0^n}^*$ is the map of $E^{n-1} \to E^n$ given by

$$\begin{aligned} x^i &= y^i \qquad\qquad i \leqslant n-1 \\ x^n &= x_0^n. \end{aligned}$$

If we compute $d\Omega_t'$ there are two types of terms, one coming from partial derivatives with respect to $x^i (i \leqslant n-1)$ and one from taking partial derivatives with respect to x^n. If we multiply by dx^n, we eliminate the terms involving dx^n so that for each value of x^n, we have

$$\begin{aligned} d(\Omega_t' \wedge dx^n) &= (p^*\, d\Omega_{x^n,t}) \wedge dx^n \\ &= (p^*\, \pi_{x^n,t}) \wedge dx^n, \end{aligned}$$

where p is the projection of $E^n \to E^{n-1}$ along x^n. On the other hand, let

$$F(x^n) = \int_{-r}^{x_n} \left(\int_{E^{n-1}} f_t(y^1, \ldots, y^{n-1}, z)\, dy \right) dz$$

so

$$dF(x^n) = \left(\int_{E^{n-1}} f_t(y^1, \ldots, y^{n-1}, x^n)\, dy \right) dx^n.$$

Put

(4.3) $$\Omega_t = \Omega_t' \wedge dx^n + (-1)^n F p^*(\pi).$$

Then

$$d\Omega_t = d(\Omega_t' \wedge dx^n) + (-1)^{n-1} dF \wedge p^*(\pi) + F p^*(d\pi).$$

Now $d\pi = 0$ since π is an $n-1$ form on E^n. Thus

$$d\Omega_t = d(\Omega_t' \wedge dx^n) + (-1)^{n-1} dF \wedge p^*(\pi)$$

(4.4) $$= p^*(\pi_{x^n,t}) \wedge dx^n + p^*(\pi) \wedge \left(\int_{E^{n-1}} f_t(y^1, y^{n-1}, x^n)\, dy \right) dx^n$$

$$= f_t(x^1, \ldots, x^n)\, dx^1 \wedge \cdots \wedge dx^n = \omega_t.$$

Now the forms π and $\Omega_{x^n,t}$ vanish for $|x^i| > r$ $(i = 1, \ldots, n-1)$, and therefore Ω_t does too. For $x^n < -r$ the integral vanishes and for $x^n > r$ the integral also vanishes since

$$\int_{-r}^{r} \int_{E^n} \cdots \int f_{t_1,\ldots,t_s}(y^1, \ldots, y^{n-1}, z)\, dy^1, \ldots, dy^{n-1}\, dz = \int_{E^n} \omega_{t_1,\ldots,t_s} = 0.$$

The $\Omega_{x^n,t}$ vanishes for $|x^n| > r$, since, by induction assumption 5)

$$|\Omega_{x^n,t}|_r^{n-1} \leqslant K|f_{t_1,\ldots,t_s}(\cdots, x^n)|_r^{n-1} = 0 \qquad \text{for } |x^n| > r.$$

Finally, the forms occurring on the right-hand side are obviously a differentiable family of forms and satisfy a bound such as 5) where K_r^n depends on K_r^{n-1} and the choice of φ.

We will now use Lemma 4.1 to establish

LEMMA 4.2. *If M is a connected n-manifold, $H_c^n(M)$ is either R or $\{0\}$; (i.e., $H_c^n(M)$ is at most one-dimensional).*

Let $\{(U_i, h_i)\}$ be an atlas of M such that each U_i is of the form $h_i^{-1}(D_1^n)$. Let Ω be an n-form with supp $(\Omega) \subset U_1$ and $\int_{E^n} h_1^*(\Omega) \neq 0$. In order to prove the lemma it will suffice to show that given any n-form, ω, of compact support, there is a real number c such that

$$\omega - c\Omega = d\omega^* \tag{4.5}$$

where ω^* is an $(n-1)$ form of compact support. Let $\{\varphi_i\}$ be a partition of unity subordinate to $\{U_i\}$. Then $\omega = \sum \varphi_i\, \omega$ is a finite sum and it suffices to prove (4.5) for each of the summands. Thus we are reduced to proving (4.5) for the case that supp $\omega \subset U_j$ for some j. Let p be a point of U_1 and q a point of U_j. Let $C(t)$ be a curve with $C(0) = p$ and $C(1) = q$. Cover $C(t)$ by a finite number of sets U_i. By renumbering if necessary we can assume that we have a set $U_1, \ldots, U_j$ such that $U_i \cap U_{i+1} \neq \varnothing$. Let ω_i $(i = 1, \ldots, j-1)$ be forms such that

$$\text{supp } \omega_i \subset U_i \cap U_{i+1}$$

and

$$\int_{E^n} h_i^*(\omega_i) \neq 0.$$

Put $\Omega = \omega_0$ and $\omega = \omega_j$. Now $h_i^*(\omega_i)$ and $h_i^*(\omega_{i-1})$ are forms on E^n whose support lies in D_1^n. Since $\int_{E^n} h_i^*(\omega_{i-1}) \neq 0$, we can find constants

$$c_i = -\left(\int_{E^n} h_i^*(\omega_i) \Big/ \int_{E^n} h_1^*(\omega_{i-1})\right)$$

such that

$$\int_{E^n} h_i^*(\omega_i - c_i\,\omega_{i-1}) = 0.$$

By Lemma 4.2,

$$(4.6) \qquad h_i^*(\omega_i - c_i\,\omega_{i-1}) = d\omega_i^{**}$$

where supp $(\omega_i^{**}) \subset D_1^n$. The form $(h_i^{-1})^*\,\omega_i^{**}$ is defined in U_i and has its support in U_i. Let ω_i^* be the form defined on M by setting $\omega_i^* = 0$ outside U_i and $\omega_i^* = (h_i^{-1})^*\,\omega_i^{**}$ in U_i. Then (3.6) becomes

$$(4.7) \qquad \omega_i - c_i\,\omega_{i-1} = d\omega_i^*.$$

If we put together all the Equation (4.7) we obtain (4.5) with $c = c_j + c_j\,c_{j-1} + \cdots + c_j \cdots c_0$ and $\omega^* = \omega_j^* + c_{j-1}\,\omega_{j-1}^* + \cdots + c_{j-1} \cdots c_0\,\omega_0^*$.

THEOREM 4.2. *If M is a connected oriented n-manifold then $H_c^n(M) \cong R$. Let M_1 and M_2 be connected oriented n-manifolds and f a proper differentiable map of $M_1 \to M_2$. There exists an integer called the degree of f and denoted by* deg (f) *such that*

$$(4.8) \qquad \int_{M_1} f^*\,\omega = \deg(f) \int_{M_2} \omega,$$

for any n-form, of compact support ω, on M_2. For any regular value $y \in M_2$

$$(4.9) \qquad \deg(f) = \sum_{x \in f^{-1}(y)} \operatorname{sign} J_x(f) \qquad [=0 \quad \text{if} \quad f^{-1}(y) = \varnothing]$$

where sign $J_x(f)$ is the sign of the Jacobian determinant of f relative to positive coordinate systems at x and y (in terms of the given orientations).

Proof. By Lemma 3.2, $H_c^n(M)$ is either zero or one-dimensional. By Theorem 4.4, the map $\omega \to \int_M \omega$ sends exact forms into zero and therefore induces a homomorphism of $H_c^n(M)$ to the reals. Since $\int_M \omega > 0$ for a positive form ω, this homomorphism is non-trivial and therefore is an isomorphism.

The map f^* of $\wedge^n(M_2) \to \wedge^n(M_1)$ is a homomorphism sending exact forms into exact forms. Therefore $\omega \to \int_M f^*(\omega)$ is a homomorphism of $H_c^n(M_2)$ into the reals. Since $H_c^n(M_2)$ is one-dimensional, it must be some multiple of the isomorphism $\omega \to \int_{M_2} \omega$. We thus have (4.8) but we don't yet know that deg (f) is an integer. We will show that this is so by verifying (4.9). Note that in (4.9) the sum is finite: f is proper so $f^{-1}(y)$

is compact, on the other hand, $f^{-1}(y)$ contains no singular points and therefore must be discrete. Regular values exist, by Sard's theorem, since the set of critical values is of measure zero. Let y be a regular value, and W a compact neighborhood of y. For each $x \in f^{-1}(y)$ let U_x be an open neighborhood of x with $U_x \subset f^{-1}(W)$ such that f is regular at all points of U_x, and such that $U_{x_1} \cap U_{x_2} = \varnothing$ if $x_1 \neq x_2$ (this can be done since f is continuous and the set of regular points is open).

$$V = f^{-1}(W) - \bigcup_{x \in f^{-1}(y)} U_x$$

is a compact set with $f(V) \subset W$ and $y \notin f(V)$. Since the image of compact sets under a continuous map is compact, $f(V)$ is a compact subset of W. Let $U \subset W - f(V)$ be a coordinate neighborhood of y; thus $f^{-1}(U) \subset \bigcup U_x$, $(f^{-1}(U) = \varnothing$ if $f^{-1}(y) = \varnothing)$. Now let ω be a positive form on M_2 with supp $\omega \subset U$. Then supp $(f^*(\omega)) \subset \bigcup U_x$. Therefore we have

$$\int_{M_1} f^*(\omega) = \sum_{x \in f^{-1}(y)} \int_{U_x} f^*(\omega). \tag{4.10}$$

Now if $y^1, \ldots, y^n$ is a positive coordinate system on U then $y^1 \circ f, \ldots, y^n \circ f$ is a positive or negative coordinate system of $f^{-1}(U) \cap U_x$ according to the sign of the Jacobian of f at x_j. Thus

$$\int_{f^{-1}(U) \cap U_x} f^*(\omega) = \text{sign } J_x(f) \int_U \omega. \tag{4.11}$$

But (4.10) and (4.11) together give (4.9), proving the theorem.

The notion of degree is very important in obtaining various kinds of information about maps. We illustrate its use here by presenting some theorems that we will need later on.

THEOREM 4.3. *Let M_1 and M_2 be oriented n-manifolds, with M_2 connected. Let f be a proper differentiable map of $M_1 \to M_2$. Assume that*

$$J_x(f) \geqslant 0 \qquad \textit{for all } x \in M_1. \tag{4.12}$$

Then either $J_x(f) \equiv 0$ or $f(M_1) \supset M_2$.

Notice that (4.12) makes sense. The value of the Jacobian depends on the choice of local coordinates of M_1 and M_2; however the sign depends only on the orientations of M_1 and M_2. To prove Theorem 3.3 let us remark that we can assume that M_1 is connected. If not, we just prove the theorem for each of the connected components of M_1: Either $J_x(f) \equiv 0$ on all the components; i.e., on M_1, or $f(M) \supset f(M_2)$ for some component M of M_1. Now if M_1 is connected, we can apply Theorem 4.2. If $f(M_1) \not\supset M_2$ then deg $(f) = 0$ by (3.9) [choose $y \notin f(M_1)$]. However if deg $(f) = 0$ then $J_x(f) \equiv 0$. Otherwise $J_x(f) > 0$ for some $x \in U_1$ where U is an open set of M_1. Then $f(U)$ contains an open set V of M_2. By Sard's theorem, V must contain a regular point y. By (4.9) deg $(f) > 0$. Q.E.D.

EXERCISE 4.1. Let $z \to \rho(z)$ be a polynomial map of the complex numbers into themselves. Show that this map, regarded as a map of $E^2 \to E^2$ is proper and satisfies (4.12). Conclude the fundamental theorem of algebra: that every non-constant polynomial has a root. Show that the degree of this map coincides with the degree of the polynomial.

We now state a slightly different version of Theorem 3.3 for "relative manifolds."

THEOREM 4.4. *Let M_1 and M_2 be oriented n-manifolds and M a connected open submanifold of M_1 with $\overline{M}$ compact. Let f be a differentiable map of $M \to M_2$ satisfying* (4.12) *which extends to a continuous map, $\bar{f}$, of $\overline{M} \to M_2$. Let $M' = \overline{M} - M$. Then if*

(4.13) $$f(M) \cap f(M') = \varnothing$$

either

(4.14) *boundary of* $$f(M) \subset f(M')$$

or

$$J_x(f) \equiv 0 \textit{ in } M.$$

Proof. By (3.13), f maps M into the open submanifold $M_2 - \bar{f}(M')$ of M_2. Since M is connected, f maps M into a connected component, say M_2^* of $M_2 - \bar{f}(M')$. Let K be a compact subset of M_2^*. Then $\bar{f}^{-1}(K)$ is a closed subset of $\overline{M}$ and hence compact. On the other hand, $\bar{f}^{-1}(K) \cap M = 0$, thus $f^{-1}(K)$ is a compact subset of M. Thus f is a proper map of $M \to M_2'$. By Theorem 4.3, if $J_x(f) \not\equiv 0$ then $f(M) \supset M_2'$. But then $\overline{f(M)} - f(M) = M_2^* - M_2^* \subset f(M')$.

EXERCISE 4.2. On the n-dimensional torus, T^n, the forms $d\theta_1, \ldots, d\theta_n$ are well defined and independent. (Where $\theta_1, \ldots, \theta_n$ are angular coordinates defined mod 2π.) Thus any k-form can be written as

$$\sum a_{i_1,\ldots,i_k} d\theta^{i_1} \wedge \cdots \wedge d\theta^{i_k}$$

where the $a_{i_1,\ldots,i_k}$ are functions of $\theta_1, \ldots, \theta_n$ which are periodic with period 2π in all n-variables. Use this to show that the cohomology classes of the forms $d\theta^{i_1} \wedge \cdots \wedge d\theta^{i_k}$ are a basis for $H^r(T^n)$.

5. Frobenius' theorem. As we have seen in Chapter II, a vector field, X, on a manifold M determines, at least locally a flow, i.e., a one parameter group. If we replace the X by fX, where f is a differentiable function, the new flow has the same trajectories as the old one, the change is in the "speed" along each trajectory. Thus the trajectories are determined by a field of "lines" rather than a field of vectors. We can study a more general situation; a field of "p-dimensional subspaces." To make this notion precise, we observe that $GL(n)$ acts to the left on the manifold $\mathcal{G}(p, E^n)$ where $\mathcal{G}(p, E^n)$ = the set of all p-dimensional subspaces of E^n [cf. Example 6 of Chapter II, Section 1]. Let us denote the manifold of quantities of type $\mathcal{G}(p, E^n)$ over M by $\mathcal{G}^p(M)$. As in the discussion of Chapter II, we can view $\mathcal{G}^p(M)$ as the set of all p-dimensional subspaces of the vector spaces $T_q(M)$, $q \in M$. Thus we can talk about a field of p-planes, i.e., a field of type $\mathcal{G}^p(M)$. We shall call such a field a *differential system of dimension p.*

DEFINITION 5.1. *Let $\mathcal{D}$ be a differential system of dimension p on M. A submanifold N is called an integral manifold of $\mathcal{D}$ if*

$$\iota_*(T_q(N)) \subset \mathcal{D}[\iota(q)] \tag{5.1}$$

for all $q \in N$, where ι is the injection of $N \to M$.

Notice that $\mathcal{D}[\iota(q)]$ is a subspace of $T_{\iota(q)}(M)$ as is $\iota_*(T_q(N))$ so that (5.1) makes sense. Thus an integral manifold is at most p-dimensional.

DEFINITION 5.2. *A differential system $\mathcal{D}$ of dimension p is called completely integrable if about every point $p \in M$ there is a chart (U, h), $h = (x^1, \ldots, x^n)$ such that all the submanifolds of U given by x^{p+1} = const, x^{p+2} = const, $\ldots$, x^n = const are integral manifolds of $\mathcal{D}$.*

EXERCISE 5.1. Let $\mathscr{D}$ be completely integrable and U a neighborhood as in Definition 5.2. Show that any connected integral manifold of $\mathscr{D}$ lying in U is a sub-manifold of one of the manifolds given in Definition 5.2.

The following exercise will prove to be useful later on.

EXERCISE 5.2. Let $\mathscr{D}$ be a completely integrable differential system of dimension p on a manifold M. Let N be a p-dimensional integral manifold of $\mathscr{D}$. Suppose that φ is a differentiable map of a connected manifold K into M such that $\varphi(K) \subset N$. Show that φ is a differentiable map of K into N.

The purpose of this section is to state criteria for a differential system to be completely integrable. Before doing this we observe that there are various ways that a differential system may be given. If $X_1, \ldots, X_p$ are p vector fields linearly independent at each point of M, then $X_1 \wedge \cdots \wedge X_p$ determines a differential system (as can easily be verified by the reader). Similarly, if $\omega^{p+1}, \ldots, \omega^n$ are $n - p$ differential forms linearly independent at each point of M, then the $(\omega^1 \wedge \cdots \wedge \omega^{n-p})^\perp$ also determines a differential system. In practice, most differential systems arise in one of these two ways. Locally any differential system can be given in either way.

In fact, let $\mathscr{D}$ be a differential system of degree k. At a point p, choose coordinates $x^1, \ldots, x^n$ so that $(\partial/\partial x^1)_p, \ldots, (\partial/\partial x^k)_p$ span $\mathscr{D}(p)$. (This can always be arranged: Take an arbitrary coordinate system and apply the appropriate linear change of coordinates.) Then $(dx^1)_p \wedge \cdots \wedge (dx^k)_p$ restricted to $\mathscr{D}(p)$ is not zero. By continuity, this implies that $(dx^1)_q \wedge \cdots \wedge (dx^k)_q|_{\wedge^k \mathscr{D}(q)} \neq 0$ for q near enough to p. On this neighborhood, the linear functions $(dx^1)_q|_{\mathscr{D}(q)}, \ldots, (dx^k)_q|_{\mathscr{D}(q)}$, are linearly independent on $\mathscr{D}_q$. There exist, therefore, functions a^α_β (which are differentiable since $\mathscr{D}$ is) such that

$$(dx^\alpha)_q|_{\mathscr{D}(q)} = \sum_{\beta=1}^{k} a^\alpha_\beta (dx^\beta)_q|_{\mathscr{D}(q)} \quad \alpha = k+1, \ldots, n.$$

Thus the $(n - k)$ independent forms

$$dx^\alpha - \sum a^\alpha_\beta \, dx$$

vanish on $\mathscr{D}(q)$. Thus they determine $\mathscr{D}(q)$.

These forms all annihilate the k vector fields

$$\frac{\partial}{\partial x^\beta} + \sum_{a=k+1}^{n} a_\beta^\alpha \frac{\partial}{\partial x^\alpha}$$

which are linearly independent and hence span $\mathscr{D}$.

Thus $\mathscr{D}$ can always be given, locally, in either way.

DEFINITION 5.3. *Let $\mathscr{D}$ be a differential system. By $\mathscr{V}(\mathscr{D})$ we mean the set of all vector fields, X, such that $X(p) \in \mathscr{D}(p)$ for all $p \in M$. If $X \in \mathscr{V}(\mathscr{D})$ we say that X belongs to $\mathscr{D}$. By $\mathscr{I}(\mathscr{D})$ we mean the set of all differential forms Ω such that Ω_p vanishes on $\mathscr{D}(p)$ for all $p \in M$.*

If V is any linear space of vector fields, we can consider the subspace, $\mathscr{D}(p) = \{X(p) | X \in V\}$ for each $p \in V$. If $\dim \mathscr{D}(p) = \text{const}$ we then have a differential system $\mathscr{D}$ with $\mathscr{V} \subset \mathscr{V}(\mathscr{D})$. The previous discussion shows that every $\mathscr{D}$ comes from $\mathscr{V}(\mathscr{D})$ in this way. Thus, we can express our condition for complete integrability in terms of $V(\mathscr{D})$. If $\mathscr{D}$ is of dimension p, and $X_1, \ldots, X_p$ are vector fields defined on an open set U, we say that the X's span $\mathscr{V}(\mathscr{D})$ on U if $X_i \in \mathscr{V}(\mathscr{D})$ and any $X \in \mathscr{V}(\mathscr{D})$ can be written as $X = \sum_{i=1}^{p} f_i X_i$ where the X_i are differentiable functions.

THEOREM 5.1. (Frobenius' theorem, first version). *Let $\mathscr{D}$ be a differential system of dimension p on an n-dimensional manifold M. $\mathscr{D}$ is completely integrable if and only if $\mathscr{V}(\mathscr{D})$ is a Lie algebra. That is, $[X, Y] \in \mathscr{V}(\mathscr{D})$ wherever $X \in \mathscr{V}(\mathscr{D})$ and $Y \in \mathscr{V}(\mathscr{D})$. Thus if $X^1, \ldots, X^p$ span $\mathscr{V}(\mathscr{D})$ on U, then $\mathscr{D}$ is completely integrable if and only if there are functions C_{jk}^i defined on U so that*

$$[X_j, X_k] = \sum_{i=1}^{p} C_{jk}^i X_i. \tag{5.2}$$

The necessity is almost obvious. In fact, first observe that the functions C_{jk}^i are uniquely determined by (5.2) since the X_i are linearly independent at every point. Therefore, it suffices to show their existence in the neighborhood of every point of U. If $\mathscr{D}$ is completely integrable every point x has a neighborhood and coordinates as in Definition 5.2 and in terms of this coordinate system $\mathscr{D}$ is spanned at all of U by $\partial/\partial x^1, \ldots, \partial/\partial x^p$. Thus, we can find functions a_{ik} and b_{il} so that

$$X_i = \sum_{k=1}^{p} a_{ik} \frac{\partial}{\partial x^k} \quad \text{and} \quad \frac{\partial}{\partial x^i} = \sum b_{il} X_l.$$

Thus

$$[X_i, X_j] = \left[\sum a_{il} \frac{\partial}{\partial x^l}, \sum a_{jm} \frac{\partial}{\partial x^m}\right]$$

$$= \sum_h \left[\frac{\partial a_{ik}}{\partial x^m} a_{jm} - \frac{\partial a_{jh}}{\partial x^m} a_{im}\right] \frac{\partial}{\partial x^h}$$

$$= \sum b_{hk} \left[\frac{\partial a_{ih}}{\partial x^m} a_{jm} - \frac{\partial a_{jh}}{\partial x^m} a_{im}\right] X_k,$$

establishing the existence of the functions C^i_{jk}.

To prove the sufficiency we can again restrict attention to local matters. For the case $p = 1$, the theorem reduces to the existence theorem on ordinary differential equations. In fact, given a vector field X, by this existence theorem one can always introduce a local coordinate system $y^1, \ldots, y^n$ so that X is given by $\partial/\partial y^1$. We will prove Theorem 4.1 by induction on p. Let $x \in M$ and $X_1, \ldots, X_p$ vector fields defined in some neighborhood, V, of x spanning $\mathscr{V}(\mathscr{D})$. Introduce coordinates $y^1, \ldots, y^n$ so that $X_1 = \partial/\partial y^1$ and $y^1(x) = 0$. Let $f_i = X_i y^1$ and

$$Y_1 = X_1, Y_i = X_i - f_i X_1 \qquad i = 2, \ldots, p.$$

Then the vector fields Y_i are linearly independent, thus span $\mathscr{V}(\mathscr{D})$, and

$$(5.3) \qquad Y_i y^1 = 0 \qquad \text{for } i > 1.$$

Consider the submanifold, W, of V defined by $y^1 = 0$. Then (5.3) implies that the vector fields Y_i are tangent to W, i.e., $Y_j(q) = \iota_*(Z_j(q))$ for any $q \in W$ where the $Z_j (j = 2, \ldots, p)$ are vector fields on W. In fact, let $z^2, \ldots, z^n$ be the coordinates on W defined by $z^j = y^j | W$. Then $\iota_*((\partial/\partial z^i)(q)) = (\partial/\partial y^i)(q)$ and if $Y_j = \sum_2^n Y_j^k \, \partial/\partial y^k$ we have $Z_j = \sum_2^n Z_j^k \, \partial/\partial z^k$ where Z_j^k is the restriction of Y_j^k to W.

The vector fields Z_j satisfy (5.2) (*with p replaced by $p - 1$*).

In fact, if $[Z_i, Z_j](q)$ did not lie in the space spanned by $Z_2, \ldots, Z_p$ then $[Y_i, Y_j](q)$ would not lie in the space spanned by $Y_1, \ldots, Y_p$ [since none of the Y_j's $(j = 2, \ldots, n)$ has any component relative to $\partial/\partial y^1$]. Thus, by induction, we can find coordinates $w^2, \ldots, w^n$ in some neighborhood of p on W such that the differential system spanned by $Z_2, \ldots, Z_p$ has $w^{p+1} = \text{const}, \ldots, w^n = \text{const}$ as integral manifolds. Now define coordinates $x^1, \ldots, x^n$ about x as follows: Let $x^1 = y^1$ and $x^j(y^1, \ldots, y^n) = w^j(y^2, \ldots, y^n)$. The Jacobian determinant of the x's with respect to the y's doesn't vanish at x so that the x's do indeed form a coordinate system about x. Furthermore, since $\partial x^j/\partial y^1 = 0$ for $j \geqslant 2$, we have $Y_1 = \partial/\partial y^1 = \partial/\partial x^1$. Now

$$Y_1 x^{p+r} = 0 \tag{5.4}$$

so that

$$\begin{aligned}\frac{\partial}{\partial x^1}(Y_i x^{p+r}) &= Y_1(Y_i x^{p+r}) \\ &= Y_1(Y_i x^{p+r}) - Y_i(Y_1 x^{p+r}) \\ &= [Y_i, Y_1]\, x^{p+r}.\end{aligned}$$

But the Y's satisfy (5.2), i.e., there are functions C^i_{jk} so that

$$[Y_i, Y_1] = C^1_{i1}\, Y_1 + \sum_{j=2}^{p} C^j_{i1}\, Y_j.$$

Thus, by (5.4) again, we get

$$\frac{\partial}{\partial x^1}(Y_i x^{p+r}) = \sum_{j=2}^{p} C^j_{i1}(Y_j x^{p+r}) \qquad i = 2, \ldots, p. \tag{5.5}$$

We can regard (4.4) as saying that for a fixed value of r, the functions $Y_i x^{p+r}$ ($i = 2, \ldots, p$) are solutions of the homogeneous linear system of $p - 1$ differential equations with respect to the independent variable x^1. On the other hand, when $x^1 = 0$, we have $Y_i x^{p+r} = Z_i w^{p+r} = 0$, i.e., the initial values of this system are zero and therefore, by uniqueness the functions $Y_i x^{p+r}$ vanish identically. Thus the vector fields $Y_1, \ldots, Y_p$ are expressible in terms of $\partial/\partial x^1, \ldots, \partial/\partial x^p$. Thus $\partial/\partial x^1, \ldots, \partial/\partial x^p$ span $\mathscr{V}(\mathscr{D})$, i.e., $\mathscr{D}$ is completely integrable.

EXERCISE 5.3. Let $X_1, \ldots, X_p$ be vector fields on M, linearly independent everywhere. Show that if $[X_i, X_j] = 0$ then about every point $q \in M$ there is a local coordinate system $x^1, \ldots, x^n$ such that $X_i = \partial/\partial x^i$, $i = 1, \ldots, p$.

It is convenient to have a formulation of Theorem 4.1 for the case that the differential system is given by differential forms instead of vector fields. In Definition 5.3, it is clear that $\mathscr{I}(\mathscr{D})$ is an ideal in the ring of differential forms. Furthermore, it is clearly generated by its elements of degree one.

THEOREM 5.2. (Frobenius' theorem, second version). *Let $\mathscr{D}$ be a differential system of dimension p on an n-manifold M. Then $\mathscr{D}$ is completely* *integrable if and only if $d\mathscr{I}(\mathscr{D}) \subset \mathscr{I}(\mathscr{D})$, i.e., if $d\omega \in \mathscr{I}(\mathscr{D})$ whenever $\omega \in \mathscr{I}(\mathscr{D})$. Let $\omega^{p+1}, \ldots, \omega^n$ be $n - p$ forms defined on an open set U which generate $\mathscr{I}(\mathscr{D})$. Then the previous assertion can be rephrased as saying that $\mathscr{D}$ is completely integrable if and only if there exist functions c^i_{jk} such that*

$$d\omega^i = \sum_{j<k,\, p<j} c^i_{jk}\, \omega^j \wedge \omega^k. \tag{5.6}$$

Proof. Again, since everything is local, we will restrict attention to a neighborhood on which there are forms $\omega^1, \ldots, \omega^n$ and $\mathscr{I}(\mathscr{D})$ is generated by $\omega^{p+1}, \ldots, \omega^n$. Let $X_1, \ldots, X_n$ be the vector fields which at each point give a dual basis to $\omega^1, \ldots, \omega^n$. Then $X_1, \ldots, X_p$ span $\mathscr{D}$, and complete integrability is equivalent to (5.2) by Theorem 5.1. But

$$\langle X_i \wedge X_j \,|\, d\omega^\alpha \rangle = X_i\langle X_j \,|\, \omega^\alpha\rangle - X_j\langle X_i \,|\, \omega^\alpha\rangle - \langle [X_i, X_j] \,|\, \omega^\alpha\rangle.$$

If we take $i \leqslant p, j \leqslant p$, and $\alpha > p$, the first two terms on the right vanish. Thus the left-hand side vanishes if and only if $[X_i, X_j]$ is expressible in terms of $X_1, \ldots, X_p$, i.e., if (5.2) holds. But the vanishing of $\langle X_i \wedge X_j \,|\, d\omega^\alpha\rangle$ is precisely equivalent to (5.6), proving Theorem 5.2.

Suppose we start with a homogeneous ideal, $\mathscr{I}$, of differential forms on a differentiable manifold M, which is not necessarily generated by linear forms. At each $x \in M$, $\mathscr{I}$ determines an ideal $\mathscr{I}_x$ in the exterior algebra $\wedge T^*_x(M)$. By Theorem 5.4 of Chapter I there is a well determined subspace of $T_x(M)$ associated to $\mathscr{I}_x$. Thus, starting with an ideal $\mathscr{I}$ we assigned a subspace $\mathscr{D}(x)$ of $T_x(M)$ to each $x \in M$. It is spanned by the set of all vector fields satisfying $X \lrcorner \mathscr{I} \subset \mathscr{I}$. This does not necessarily define a differential system: the dimension of $\mathscr{D}(x)$ may vary from point to point. We will call the ideal $\mathscr{I}$ non-singular if the assignment $x \to \mathscr{D}(x)$ is a differential system. If $\mathscr{I}$ is non-singular, we will call this differential system the *characteristic system* of $\mathscr{I}$.

Theorem 5.3. (Cartan). *Let $\mathscr{I}$ be a non-singular ideal satisfying $d\mathscr{I} \subset \mathscr{I}$. Then its characteristic system is completely integrable.*

Let $X_1, \ldots, X_p$ be vector fields which span the characteristic system of $\mathscr{I}$. Then $X_i \lrcorner \mathscr{I} \subset \mathscr{I}$. If $\omega \in \mathscr{I}$, then (1.10) implies that

$$[X_i, X_j] \lrcorner \omega = X_i \lrcorner (X_j \lrcorner d\omega) - X_i \lrcorner d(X_j \lrcorner \omega) - X_j \lrcorner d(X_i \lrcorner \omega).$$

Since $d\mathscr{I} \subset \mathscr{I}$ all three terms on the right lie in $\mathscr{I}$. Thus $[X_i, X_j] \,\lrcorner\, \mathscr{I} \subset \mathscr{I}$ or $[X_i, X_j]$ is a linear combination of $X_1, \ldots, X_p$, i.e., (5.2) is satisfied. Thus Theorem 5.3 is verified by Theorem 5.1.

For the first three theorems of this section we have been concerned with existence theorems of a purely local nature. Let us now turn to the problem of finding global solutions. Let $\mathscr{D}$ be a completely integrable differential system of degree k on a manifold M. Then there is a covering of M by coordinate neighborhoods U_α such that in U_α, the k-dimensional integral manifolds are given by $x_\alpha^{k+1} = \text{const}, \ldots, x_\alpha^n = \text{const}$. Suppose N is a connected integral manifold of M. More precisely, N is a connected manifold, together with a map $f: N \to M$ such that $f(T_x(N)) = \mathscr{D}(f(x))$ for all $x \in N$. Any $x \in N$ has a coordinate V such that $f(V) \subset U_\alpha$ for some α. Thus $f(V)$ is contained in a subset of U_α given by $x_\alpha^{k+1} = \text{const}, \ldots, x_\alpha^n = \text{const}$. Let y be another point of N. If C is a curve joining x to y, we can cover C by a finite number of neighborhoods of the type V. *Thus x and y can be joined by a finite number of curves C_i (with $C_i(1) = C_{i+1}(0)$) such that each $f \circ C_i$ is a one-dimensional integral curve of $\mathscr{D}$.*

We now see how to construct a *maximal* integral manifold of $\mathscr{D}$. Let p be a point of M. Let K be the set of all $q \in M$ such that there exists a finite number of curves $C_i: [0, 1] \to M$ with $C_1(0) = p$, $C_n(1) = q$, $C_{i+1}(0) = C_i(1)$ and such that each C_i is a one-dimensional solution of $\mathscr{D}$. Let us make K into a differentiable manifold. Let q be a point of K. Then $q \in U_\alpha$ for some α. Then the set of points, z, in U_α given by $x_\alpha^{k+1}(z) = x_\alpha^{k+1}(q), \ldots, x_\alpha^n(z) = x_\alpha^n(q)$ also belong to K. We will take this as a coordinate neighborhood of q with coordinates $x_\alpha^1, \ldots, x_\alpha^k$. We leave, as an exercise to the reader, the verification that this does indeed define the structure of a differentiable manifold on K. (In particular, to verify the Hausdorff and countability axioms). From the previous discussion, it is clear that *any* connected integral manifold of $\mathscr{D}$ passing through p maps into a subset of K. Thus

THEOREM 5.4. *Let $\mathscr{D}$ be a completely integrable differential system on M. Through each point, p, of M there passes a maximal connected integral manifold K_p. Any other connected integral manifold is a submanifold of K_p.*

Remark. The manifold K_p is not, in general, an embedded submanifold of M. For instance, if M is the two-dimensional torus (regarded as the plane modulo the unit lattice) and $\mathscr{D}$ is given by a constant vector field with irrational angle, every maximal integral manifold is clearly dense on the torus.

6. Darboux's theorem.[9] In this section we will discuss the problem of finding a normal form for linear differential forms. First we need some preliminaries.

Let M_1 and M_2 be differentiable manifolds. There are relations between pairs of vector fields on M_1 and M_2 and vector fields on $M_1 \times M_2$ which we will now study. Let $\pi_1: M_1 \times M_2 \to M_1$ and $\pi_2: M_1 \times M_2 \to M_2$ be the obvious projections. The differential system defined by $\pi_2^*(\wedge^1(M_2))$ is obviously completely integrable as is that defined by $\pi_1^*(\wedge^1(M_1))$. In fact, according to the very definition of the manifold structure on $M_1 \times M_2$ there is a covering of $M_1 \times M_2$ by neighborhoods of the form $U_1 \times U_2$ with coordinates $(x^1, \ldots, x^{n_1}, y^1, \ldots, y^{n_2})$ where $x^1, \ldots, x^{n_1}$ are coordinates on M_1 and $y^1, \ldots, y^{n_2}$ are coordinates on M_2. It is then clear that $y^1 = \text{const}, \ldots, y^{n_2} = \text{const}$ is an integral manifold for $\pi_2^*(\wedge^1(M_2))$ and that $x^1 = \text{const}, \ldots, x^{n_1} = \text{const}$ is an integral manifold for $\pi_1^*(\wedge^1(M_1))$.

If ω_1 is a differential form on M_1 and ω_2 a form on M_2 then they define a differential form, $\pi_1^*(\omega_1) + \pi_2^*(\omega_2)$, on $M_1 \times M_2$. On the other hand, if X is a vector field on M_1 and Y is a vector field on M_2, we can define a vector field (X, Y) on $M_1 \times M_2$. If $X = (X^1, \ldots, X^{n_1})$ relative to the coordinate system $x^1, \ldots, x^{n_1}$ and $Y = (Y^1, \ldots, Y^{n_2})$ relative to $y^1, \ldots, y^{n_2}$, we set $(X, Y) = (X^1, \ldots, X^{n_1}, Y^1, \ldots, Y^{n_2})$ relative to $x^1, \ldots, x^{n_1}, y^1, \ldots, y^{n_2}$. In view of the law of change of coordinates, this does indeed define a vector field on $M_1 \times M_2$. It is clear that

$$(6.1) \quad (X, Y) \rfloor (\pi_1^*(\omega_1) + \pi_2^*(\omega_2)) = \pi_1^*(X \rfloor \omega_1) + \pi_2^*(Y \rfloor \omega_2).$$

More generally, if s and t are contravariant tensors on M_1 and M_2 of a given type, we can define the tensor (s, t) of the same type on $M_1 \times M_2$. From the algebraic point of view this simply reduces to the observation that there are, for each point $(x, y) \in M_1 \times M_2$, maps $\iota_{1*}: T_x(M_1) \to T_{(x,y)}(M_1 \times M_2)$ and $\iota_{2*}: T_y(M_2) \to T_{(x,y)}(M_1 \times M_2)$ such that $\pi_{1*} \circ \iota_1 = \text{id}$ and $\pi_{2*} \circ \iota_{2*} = \text{id}$. In other words, the tangent space at every point of $M_1 \times M_2$ splits into a direct sum of a space "tangent" to M_1 and a space "tangent" to M_2. If X_1 and X_2 are vector fields on M_1 and

[9] The results of this section will not be used elsewhere in the book.

Y_1 and Y_2 are vector fields on M_2, then a direct verification (in terms of local coordinates, say) shows that

$$(6.2) \qquad [(X_1, Y_1), (X_2, Y_2)] = ([X_1, X_2], [Y_1, Y_2]).$$

If $\mathscr{E}_1$ is a differential system on M_1 and $\mathscr{E}_2$ is a differential system on M_2, then setting $\mathscr{E}(x, y) = \iota_{1*}\,\mathscr{E}_1(x) + \iota_{2*}\,\mathscr{E}_2(y)$ we get a differential system, $\mathscr{E}$, on $M_1 \times M_2$. It follows from (5.2) and Theorem 4.2 that *if* $\mathscr{E}_1$ *and* $\mathscr{E}_2$ *are completely integrable so is* $\mathscr{E}$.

We can state a useful generalization of this remark. Let $\mathscr{E}$ be a differential system on $M_1 \times M_2$ which is consistent with the direct sum decomposition of the tangent space, i.e., that

$$(6.3) \quad \mathscr{E}(x, y) = \iota_{1*}\pi_{1*}\,\mathscr{E}(x, y) + \iota_{2*}\pi_{2*}\,\mathscr{E}(x, y) \text{ for all } (x, y) \in M_1 \times M_2.$$

For each $y \in M_2$, $\mathscr{E}$ induces a differential system $\mathscr{E}_y$ on M_1. In fact, $\mathscr{E}_y$ is defined by $\mathscr{E}_y(x) = \pi_{1*}\,\mathscr{E}(x, y)$; similarly we define $\mathscr{E}_x$ by $\mathscr{E}_x(y) = \pi_{2*}\,\mathscr{E}(x, y)$. To see that these are indeed differential systems, let us introduce local coordinates $(x^1, \ldots, x^{n_1}, y^1, \ldots, y^{n_2})$ in the neighborhood of (x, y). We can then describe $\mathscr{E}$, in some neighborhood of (x, y) by the vanishing of certain forms $\omega^1, \ldots, \omega^k$. If we write each $\omega^i = \sum_{j=1}^{n_1} a^i_j\, dx^j + \sum_{j=1}^{n_2} b^i_j\, dy^j = \omega^i_1 + \omega^i_2$ then (6.3) implies [since ω^i_1 vanishes on $\iota_* T_x(M)$ and ω^i_2 vanishes $\iota_* T_y(M)$] that both ω^i_1 and ω^i_2 vanish on $\mathscr{E}(x, y)$. Thus we can renumber the forms as $\omega^1, \ldots, \omega^l, \omega^{l+1}, \ldots, \omega^k$ so that

$$\omega^i = \sum a^i_j\, dx^j \qquad \text{if } i \leqslant l$$

and

$$\omega^i = \sum b^i_j\, dy^j \qquad \text{if } i > l$$

(where the a's and b's are functions of all the variables and the ω's are linearly independent). Then the differential system $\mathscr{E}_y$ is given by the forms $\sum a^i_j(x^1, \ldots, x^{n_1}, y^1, \ldots, y^{n_2})dx^j$ and $\mathscr{E}_x$ is given by $\sum b^i_j(x^1, \ldots, x^{n_1}, y^1, \ldots, y^{n_2})dy^j$. We can now assert

LEMMA 6.1. *Let* $\mathscr{E}$ *be a differential system on* $M_1 \times M_2$ *satisfying* (6.3). *If* $\mathscr{E}_y$ *is completely integrable for all* $y \in M_1$ *and* $\mathscr{E}_x = \{0\}$ *for all* $x \in M_1$ *then* $\mathscr{E}$ *is completely integrable.*

Proof. $\mathscr{E}_x$ is the trivial differential system assigning to each $y \in M_2$ the zero subspace; thus in local coordinates $\mathscr{I}(\mathscr{E})$ is spanned by $\omega^1, \ldots, \omega^l$ and $dy^1, \ldots, dy^{n_2}$ where

$$\omega^i = \sum a^i_j(x^1, \ldots, x^{n_1}, y^1, \ldots, y^{n_2})\, dx^j.$$

Now $\mathscr{E}_y$ is given by the forms $\iota_y^*(\omega^i)$ where $\iota_y: M_1 \to M_1 \times M_2$ sends $x \to (x, y)$. Since $\mathscr{E}_y$ is completely integrable, the forms $d(\iota_y^*(\omega^i))$ and linear combinations of the $\iota_y^*(\omega^j) \wedge \iota^*(\omega^k)$. But $\pi_1^*\,(d\iota_y^*(\omega^j))$ differs from $d\omega^j$ only by terms involving the dy's. Thus the differential system generated by $\omega^1, \ldots, \omega^l$ and $dy^1, \ldots, dy^{n_2}$ is completely integrable.

EXERCISE 6.1. Show by example that if $\mathscr{E}_x$ and $\mathscr{E}_y$ are completely integrable for all $x \in M_1$ and $y \in M_2$, $\mathscr{E}$ need not be completely integrable.

We will now apply the previous discussion to the problem of finding a normal form for a linear differential form. Let ω be a linear differential form. Then $d\omega$ is an exterior differential form of degree 2 and the ideal, $\mathscr{I}$, generated by $d\omega$ is closed under d. The characteristic system of this ideal therefore coincides with the map that assigns to each $x \in M$ the associated system of $(\mathscr{I})_x$; in this case, a subspace of dimension $n - 2p$ where p is the rank of $(d\omega)_x$, cf. Theorem 5.1 of Chapter I. Let us assume that this rank is constant over M. By Cartan's theorem (Theorem 5.3) the characteristic system of $\mathscr{I}$ is completely integrable. Thus about each $x \in M$ we can find a coordinate neighborhood U and coordinates $u^1, \ldots, u^{n-1}, y^1$ so that the characteristic system of $\mathscr{I}$ is given by $du^{n-2p+1} = \cdots = du^{n-1} = dy^1 = 0$. By Theorem 5.1 of Chapter I, this implies that $d\omega$ can be expressed in terms of $du^{n-2p+1}, \ldots, du^{n-1}$, and dy^1. Let us regard U as a product $U = U_1 \times U_2$ where $u^1, \ldots, u^{n-1}$ are coordinates on U_1 and y^1 on U_2. For each fixed value of y^1, the restriction of $d\omega$ to the submanifold $U_1 \times y^1$ is a closed two form, $\iota_{y1}^*(d\omega)$ of rank $p - 1$. By Lemma 5.1 the characteristic system of $\iota_{y1}^*(d\omega)$ and dy^1 generate a completely integrable differential system. Therefore, we can find coordinates $u^1, \ldots, u^{n-2p}, w^{n-2p+1}, \ldots, w^{n-2}, y^2, y^1$ so that *this* differential system is given by $dw^{n-2p+2} = \cdots = dy^2 = dy^1 = 0$. Relative to these coordinates the forms $\iota_{y1}^*(d\omega)$ are expressible in terms of $dw^{n-2p+2}, \ldots, dy^2$. We can then regard U as a product $U = U_3 \times U_4$ where y^1, y^2 are coordinates in U_4 and consider the differential system generated by the characteristic system of $\iota_{y_1 y_2}^*(d\omega)$ and dy^1, dy^2. Proceeding inductively in this way, we can finally reach a system of coordinates $z^1, \ldots, z^{n-p}, y^1, \ldots, y^p$ so that $\iota_{y^1,\ldots,y^p}^*(d\omega) = 0$ where $\iota_{y^1,\ldots,y^p}^*$ denotes the injection $(z^1, \ldots, z^{n-p}) \to (z^1, \ldots, z^{n-p}, y^1, \ldots, y^p)$. Thus by Poincaré's lemma (Theorem 4.1), we can write $\iota_{y^1,\ldots,y^p}^* = dS_{y^1,\ldots,y^p}$ where $S_{y^1,\ldots,y^p}$ is a differentiable function of $(z^1, \ldots, z^{n-p})$ depending differentiably on the variables $y^1, \ldots, y^p$. Therefore, it defines a function S on U. Furthermore, the form $\omega - dS$ has the property that $\iota^*_{y^1,\ldots,y^p}(\omega - dS) = 0$, for all $y^1, \ldots, y^p$. Thus we can write $\omega - dS = f_1\, dy^1 + \cdots + f_p\, dy^p$. Thus $d\omega = df_1 \wedge dy^1 + \cdots + df_p \wedge dy^p$. Since $d\omega$ is of rank p, $d\omega \wedge \cdots \wedge d\omega \neq 0$ (p-factors). Thus $df_1 \wedge \cdots \wedge df_p \wedge dy^1 \wedge \cdots \wedge dy^p \neq 0$; in other words, we can extend the functions $f_1, \ldots, f_p, y^1, \ldots, y^p$ to a coordinate system $x^1, \ldots, x^{n-p}, y^1, \ldots, y^p$, $(x^i = f_i, 1 \leqslant i \leqslant p)$. In terms of this coordinate system we can write

$$\omega = dS + x^1\, dy^1 + \cdots + x^p\, dy^p. \tag{6.4}$$

Suppose we started with a *closed* two form, Ω, of rank p, i.e., such that $\Omega \wedge \cdots \wedge \Omega \neq 0$ (p-factors) for all x, but $\Omega \wedge \cdots \wedge \Omega \equiv 0$ ($p + 1$ factors). According to Poincaré's lemma, about each $x \in M$ we can find a linear differential form ω such that $d\omega = \Omega$. By the discussion above, we can find coordinates $x^1, \ldots, x^{n-p}, y^1, \ldots, y^p$ such that ω has the form (6.4). By exterior differentiation we obtain

THEOREM 6.1. *Let Ω be a closed two form of rank exactly p everywhere on an n-dimensional manifold M. About every point of M we can introduce coordinates $x^1, \ldots, x^{n-p}, y^1, \ldots, y^p$ such that, in terms of these coordinates,*

$$\Omega = dx^1 \wedge dy^1 + \cdots + dx^p \wedge dy^p. \tag{6.5}$$

Let us return to the problem of finding a normal form for a linear differential form. If we examine (6.4), we don't know whether dS is a linear combination of the dx's and dy's or not. We will consider two cases: Either dS is everywhere a linear combination of the dx's and dy's or it is independent of them. These two cases can be distinguished as follows: By assumption $\bar{\omega} = d\omega \wedge \cdots \wedge d\omega$ (p-terms) is a non-zero $2p$ form. In terms of the local coordinates $\omega \wedge \bar{\omega} = dS \wedge dx^1 \wedge \cdots \wedge dx^p \wedge dy^1 \wedge \cdots \wedge dy^p$. If $(\omega \wedge \bar{\omega})_x = 0$ then dS is a linear combination of dx^i and dy^i at x, otherwise it is independent. Our two cases thus coincide with the cases $\omega \wedge \bar{\omega} \equiv 0$ or $\omega \wedge \bar{\omega} \neq 0$ for any x. In the second case, we can extend the set of functions $S, x^1, \ldots, x^p, y^1, \ldots, y^p$ to a coordinate system. That is, in the new coordinate system we can write $\omega = x^1\, dy^1 + x^2\, dy^2 + \cdots + x^p\, dy^p + dx^{p+1}$.

In the first case we leave as a (non-trivial!) exercise[10] to the reader, to

[10] Hint: Consider the differential system generated by all vector fields satisfying $\langle X \mid \omega \rangle = 0$ and $X \lrcorner (\omega \wedge d\omega) = 0$. Show that this defines a differential system which is completely integrable. Then proceed as above.

verify that if $\omega \neq 0$ we can choose coordinates so that $\omega = x^1\,dy^1 + \cdots + x^p\,dy^p$. We thus have

THEOREM 6.2. (Darboux). *Let ω be a linear differential form such that $d\omega$ has rank p everywhere. If $\omega \neq 0$ and*

$$\omega \wedge \overbrace{d\omega \wedge \cdots \wedge d\omega}^{p\text{-factors}} \equiv 0,$$

about every point we can find coordinates $x^1, \ldots, x^{n-p}, y^1, \ldots, y^p$ such that

$$\omega = x^1\,dy^1 + \cdots + x^p\,dy^p. \tag{6.6}$$

If

$$\omega \wedge \overbrace{d\omega \wedge \cdots \wedge d\omega}^{p\text{-factors}} \neq 0 \qquad \textit{everywhere},$$

about every point we can introduce coordinates $x^1, \ldots, x^{n-p}, y^1, \ldots, y^p$ so that

$$\omega = x^1\,dy^1 + \cdots + x^p\,dy^p + dx^{p+1}. \tag{6.7}$$

7. Hamiltonian structures. An important geometrical object that arises in the study of Riemann geometry and classical mechanics is a differentiable manifold carrying a distinguished exterior two form.

DEFINITION 7.1. *Let M be a $2n$-dimensional differentiable manifold. We will call M a Hamiltonian manifold (or say that M has a Hamiltonian structure) if there is given a closed two form, Ω, of rank n defined everywhere on M.*

On a Hamiltonian manifold, we have a one–one mapping from the space of vector fields onto the space of linear differential forms. In fact, if X is a vector field, define ω_X by

$$\omega_X = X \,\lrcorner\, \Omega. \tag{7.1}$$

The map $X \to \omega_X$ is one–one and onto. In fact at each $x \in M$, (7.1) just describes the map of $T_x(M)$ onto $T_x^*(M)$ given by the non-singular bilinear form Ω_x.

We shall denote the inverse map by $\omega \to X_\omega$. Thus

$$\omega = X_\omega \,\lrcorner\, \Omega. \tag{7.2}$$

We can now transfer the Lie bracket on vector fields to an operation on forms. This operation will also be denoted by [,] and is called the *Poisson bracket.* Thus, for two linear differential forms ω_1 and ω_2.

$$[\omega_1, \omega_2] = [X_{\omega_1}, X_{\omega_2}] \lrcorner \Omega. \tag{7.3}$$

Suppose ω_2 is closed. Then

$$\begin{aligned} \mathscr{L}_{X_{\omega_2}}\omega_1 &= \mathscr{L}_{X_{\omega_2}}(X_{\omega_1} \lrcorner \Omega) \\ &= (\mathscr{L}_{X_{\omega_2}}X_{\omega_1}) \lrcorner \Omega + X_{\omega_1} \lrcorner (\mathscr{L}_{X_{\omega_2}}\Omega) \\ &= [X_{\omega_1}, X_{\omega_2}] \lrcorner \Omega + X_{\omega_1} \lrcorner (X_{\omega_2} \lrcorner d\Omega + d\omega_2) \\ &= [\omega_1, \omega_2]. \end{aligned}$$

Thus, if ω_1 and ω_2 are closed

$$[\omega_1, \omega_2] = \mathscr{L}_{X_{\omega_2}}\omega_1 = - \mathscr{L}_{X_{\omega_1}}\omega_2. \tag{7.4}$$

For any closed form ω and any vector field X, $\mathscr{L}_X\omega = d(X \lrcorner \omega)$. Thus the Lie derivative of any closed form is exact. Thus we obtain:

The Poisson bracket of two closed forms is exact.

Suppose ω_1 and ω_2 are exact, $\omega_1 = df$ and $\omega_2 = dg$. Then

$$\begin{aligned} [df, dg] &= \mathscr{L}_{X_{df}}(dg) \\ &= d(X_{df}g). \end{aligned} \tag{7.5}$$

We define

$$[f, g] = X_{df}g = - X_{dg}f, \tag{7.6}$$

so that

$$[df, dg] = d[f, g].$$

In particular, if $[f, g] = 0$, then f is constant on the orbits of X_{dg} and conversely. Thus f is constant along the orbits of X_{dg} if and only if g is constant along the orbits of X_{df}.

According to Theorem 6.1, about every point we can introduce coordinates $x^1, \ldots, x^n, y^1, \ldots, y^n$ so that

$$\Omega = dy^1 \wedge dx^1 + \cdots + dy^n \wedge dx^n. \tag{7.7}$$

[For convenience, we have interchanged the x and y of (6.5).]

In terms of these coordinates, let us compute X_ω and ω_X. Suppose X is a vector field given by

$$X = \sum_{i=1}^{n} \left(a^i \frac{\partial}{\partial x^i} + b^i \frac{\partial}{\partial y^i}\right).$$

Then

$$\omega_X = X_\omega = \sum (b^i\, dx^i - a^i\, dy^i). \tag{7.8}$$

If ω is a differential form

$$\omega = \sum (c^i\, dx^i + d^i\, dy^i).$$

Then (7.7) gives

$$X_\omega = \sum_{i=1}^{n} \left((-d^i)\frac{\partial}{\partial x^i} + c^i \frac{\partial}{\partial y^i}\right). \tag{7.9}$$

In particular, for any function f, we have

$$X_{df} = \sum \left(-\frac{\partial f}{\partial y^i}\frac{\partial}{\partial x^i} + \frac{\partial f}{\partial x^i}\frac{\partial}{\partial y^i}\right).$$

Thus

$$[f, g] = \sum \left(\frac{\partial f}{\partial x^i}\frac{\partial g}{\partial y^i} - \frac{\partial f}{\partial y^i}\frac{\partial g}{\partial x^i}\right). \tag{7.10}$$

Of particular importance are the infinitesimal autcmorphisms of the Hamiltonian structure, i.e., the generators of one parameter groups leaving Ω invariant.

DEFINITION 7.2. *A vector field X is called Hamiltonian if*

$$\mathscr{L}_X\Omega = 0. \tag{7.11}$$

Since $\mathscr{L}_X\Omega = d(X \rfloor \Omega) + X \rfloor d\Omega$ and $d\Omega = 0$, we obtain:

A vector field, X, is Hamiltonian if and only if ω_X is closed.

From the discussion above it follows that the Lie bracket of two Hamiltonian vector fields is again Hamiltonian. Thus

The set of all Hamiltonian vector fields on a C^∞ Hamiltonian manifold is a Lie algebra.

There is an extremely important way of arriving at Hamiltonian manifolds that we will now examine.

THEOREM 7.1. *Let M be any differential manifold. On $T^*(M)$ there is a linear differential form θ [called the fundamental form of $T^*(M)$] defined as follows: For any tangent vector X_u at a point u of $T^*(M)$,*

$$\langle X_u, \theta_u\rangle = \langle \pi_* X_u, u\rangle \tag{7.12}$$

where π is the projection of $T^*(M)$ onto M. The form $d\theta$ is of rank n and therefore defines a Hamiltonian structure on $T^*(M)$.

Notice that $\pi_* X_u$ is a tangent vector at $\pi(u)$ and u is a linear function on $T_{\pi(u)}(M)$. Thus (7.12) makes sense. Let us check the rest of the theorem. Let $x^1, \ldots, x^n$ be a coordinate system on an open set U of M. We can introduce coordinates $(x^1, \ldots, x^n; y^1, \ldots, y^n)$ on $\pi^{-1}(U)$ as follows: $dx^1, \ldots, dx^n$ span the space of forms of U. Therefore, for any $u \in \pi^{-1}(U)$, we have

(7.13) $$u = \sum y^i(u)\, dx^i_{\pi(u)}.$$

Thus $x^1(\pi(u)), \ldots, x^n(\pi(u)), y^1(u), \ldots, y^n(u)$, give the coordinates of U. Any tangent vector X_u can be written as

$$X_u = \sum \left(X^i_u \left(\frac{\partial}{\partial x^i}\right)_u + Y^i_u \left(\frac{\partial}{\partial y^i}\right)_u \right).$$

Then

$$\pi_* X_u = \sum X^i_u \left(\frac{\partial}{\partial x^i}\right)_{\pi(u)}$$

and (7.12) becomes

$$\langle X_u, \theta_u \rangle = \sum y^i(u) X^i_u.$$

Thus

$$\theta_u = \sum y^i(u) (dx^i)_u$$

or

(7.14) $$\theta = \sum y^i\, dx^i,$$

and is thus a linear differential form. In particular, $d\theta = \sum dy^i \wedge dx^i$ and is thus of rank n everywhere, proving the theorem.

It should be remarked that in classical mechanics, if the manifold M is "configuration space," the manifold $T^*(M)$ is called "phase space." In classical particle mechanics, the manifold M represents the configuration space of n-particles. Thus $M = E^{3n}$. If we exclude the possibility of any two particles lying in the same point of space then M consists of all of those points $(x_1, y_1, z_1, \ldots, x_n, y_n, z_n)$ of E^{3n} for which

$$(x_i - x_j)^2 + (y_i - y_j)^2 + (z_i - z_j)^2 \neq 0 \quad \text{for all pairs } i, j, \text{ with } i \neq j.$$

In the laws of motion one assumes the existence of n positive numbers $m_1, \ldots, m_n$ (where m_i is called the mass of the ith particle) and a function

V on M (called the *potential energy*) such that the equations of motion take the form

$$
(7.15)\qquad \begin{aligned} m_i \frac{d^2x_i}{dt^2} &= -\frac{\partial V}{\partial x_i}, \\ m_i \frac{d^2y_i}{dt^2} &= -\frac{\partial V}{\partial y_i}, \\ m_i \frac{d^2z_i}{dt^2} &= -\frac{\partial V}{\partial z_i}. \end{aligned}
$$

For simplicity in notation, let us write

$$x_1 = q_1, y_1 = q_2, z_1 = q_3, x_2 = q_4, \ldots, z_n = q_{3n}$$

and

$$M_1 = M_2 = M_3 = m_1, \quad M_4 = M_5 = M_6 = m_2, \quad \text{etc.}$$

Then (7.15) becomes

$$
(7.16)\qquad M_i \frac{d^2q_i}{dt^2} = -\frac{\partial V}{\partial q^i}.
$$

Let us consider the Riemann metric given by $(v, v) = \sum M_i \dot{q}_i^2$ when $v = \sum \dot{q}_i(\partial/\partial q^i)$. Any Riemann metric induces a diffeomorphism, λ, of $T(M)$ onto $T^*(M)$. In terms of the given metric, it is easy to see that

$$\lambda\left(\sum \dot{q}_i \frac{\partial}{\partial q^i}\right) = \sum p_i\, dq_i$$

where

$$p_i = M_i \dot{q}_i.$$

The p_i is called *momentum conjugate to* q_i. [Notice that in view of (7.14), on any manifold, once we choose coordinates $q^1, \ldots, q^n$ the conjugate momenta are determined, *independently* of the choice of Riemann metric. What the Riemann metric (i.e., choice of mass) does is give a rule for passing from velocity to momentum.]

Equation (7.16) can now be written as

$$\frac{dq_i}{dt} = \frac{p_i}{M_i} \quad \frac{dp_i}{dt} = \frac{-\partial V}{\partial q_i}.$$

Thus we have a flow on $T^*(M)$ whose vector field is

$$X = -\sum \left(\frac{p_i}{M_i}\frac{\partial}{\partial q_i} + \frac{\partial V}{\partial q_i}\frac{\partial}{\partial p_i}\right).$$

Now $p_i/M_i = \partial T/\partial p_i$ where $T = \frac{1}{2}\sum p_i^2/M_i$ is the kinetic energy. Thus

(7.17) $$X \lrcorner\, d\theta = -dH$$

where

$$H = T + V$$

is the total energy.

Now

$$\begin{aligned} XH &= \langle X \mid dH \rangle \\ &= -\langle X \mid X \lrcorner\, \Omega \rangle = 0. \end{aligned}$$

Thus the function H is constant along the trajectories of X. This is known as the *law of conservation of energy*. More generally, let f be any function such that $X_{df}H = 0$. Then

$$\begin{aligned} Xf &= \langle X \mid df \rangle \\ &= \langle X \mid X_{df} \lrcorner\, \Omega \rangle \\ &= -\langle X_{df} \mid X \lrcorner\, \Omega \rangle \\ &= \langle X_{df} \mid dH \rangle = XH = 0. \end{aligned}$$

Thus

(7.18) $X_{df}H = 0$ implies that f *is constant along trajectories.*

There is a way of passing from vector fields on M to functions on $T^*(M)$. In fact, any one parameter group g_t of diffeomorphisms of M gives a one parameter group g_{t^*} diffeomorphisms of $T^*(M)$ satisfying $\pi\varphi_{t^*} = \varphi_t \circ \pi$. In fact, for any $(p, v^*) \in T^*(M)$ define

$$\varphi_{t^*}(p, v^*) = (\varphi_t(p), (\varphi_{-t})^*{}_{\varphi_t(p)}v^*).$$

This clearly gives a one parameter group of diffeomorphisms of $T^*(M)$. Furthermore,

$$(\varphi_{t^*})^*\,\theta = \theta.$$

In fact, for any $X \in T_{(p,v^*)}(T^*(M))$

$$\begin{aligned} \langle X, (\varphi_{t^*})^*\theta \rangle &= \langle (\varphi_{t^*})_* X, \theta_{\varphi_t{}^*(p,v^*)} \rangle \\ &= \langle \pi_* \circ (\varphi_{t^*})_* X, (\varphi_{-t})^*\, v^* \rangle && \text{by definition of } \theta \\ &= \langle \varphi_{t^*} \circ \pi_* X, (\varphi_{-t})^*\, v^* \rangle && \text{since } \pi \circ \varphi_{t^*} = \varphi_t \circ \pi \\ &= \langle \pi_* X, v^* \rangle \\ &= \langle X, \theta \rangle. \end{aligned}$$

Thus any vector field, Y, on M, gives rise to a vector field $\tilde{Y}$ on $T^*(M)$ such that $\pi_{*u}\tilde{Y}_u = Y_{\pi(u)}$ for any $u \in T^*(M)$ and $\mathscr{L}_{\tilde{Y}}\theta = 0$.

Now $$\mathscr{L}_{\tilde{Y}}\,\theta = \tilde{Y} \,\lrcorner\, d\theta + d\langle\tilde{Y}|\theta\rangle.$$

Thus $\tilde{Y} \,\lrcorner\, d\theta = -d\langle\tilde{Y}|\theta\rangle$. It is easy to evaluate the function $\langle\tilde{Y}|\theta\rangle$. In fact, at any point $u = (p, v^*)$,

$$\begin{aligned}\langle\tilde{Y}_u, \theta_u\rangle &= \langle\pi_* \tilde{Y}_u, \theta_u\rangle \\ &= \langle Y, v^*\rangle.\end{aligned}$$

In the case of mechanics, for example, if $Y = \partial/\partial q^i$ then $\langle\tilde{Y}|\theta\rangle = p^i$. Thus the linear momentum of a particle in given directions corresponds to the one parameter group of translation of that particle in the given direction. Similarly, the angular momenta come from the corresponding one parameter group of rotations. In the case of mechanics, if Y is any vector field on M with $\tilde{Y}H = 0$, then (7.18) says that $\langle\tilde{Y}|\theta\rangle$ is conserved under the flow generated by X.

As an example of such vector fields, suppose that V depends only on the distances between particles. Then any simultaneous Euclidean motion of all the particles leaves both V and T invariant and thus [induces a transformation of $T^*(M)$ which] leaves H invariant. Thus the vector fields, Y, corresponding to one parameter groups of translations or rotations of all the particles give functions invariant under the flow generated by X. This is known as the *laws of conservation of the total linear momenta and of total angular momenta.*

CHAPTER IV

The Calculus of Variations

In this chapter we will study the calculus of variations on manifolds. The calculus of variations deals with the following kind of question: Given a function, I, on the space of differentiable curves on a manifold M, find those curves, C, for which I achieves a minimum among all curves satisfying some auxiliary conditions. For instance, suppose M is a Riemann manifold, with a Riemann metric, g. For any differentiable curve $C: [a, b] \to M$ let $I[C]$ be the arc length of C, i.e.,

$$(1) \qquad I[C] = \int_a^b \sqrt{g\left(C_*\left(\frac{\partial}{\partial t}\right)_t, C_*\left(\frac{\partial}{\partial t}\right)_t\right)}\, dt.$$

We might ask for the "shortest" curve joining two fixed points p and q of M. We would then be looking for a curve C for which (1) achieves its minimum among all curves satisfying $C(a) = p$, $C(b) = q$. Similarly, we might ask for the shortest curve joining two submanifolds N_1 and N_2. Then the auxiliary conditions would read $C(a) \in N_1$ and $C(b) \in N_2$.

Our main purpose in studying the calculus is, indeed, to treat the special problem of arc length on Riemann manifolds. It is convenient, however, to make a slight detour. The integrand of (1) is just the length of the tangent vector to C. (It is the "speed" at which a particle moves along the curve C.) The length of a tangent vector is, unfortunately, not a differentiable function of the tangent vector; it misbehaves at vectors of length zero. Another difficulty with (1) is that arc length is independent of parametrization. That implies that we could not expect to find a unique solution curve to the minimization problem as such, since any reparametrization would give the same solution. On the other hand, arc length suggests itself as a natural parameter and it should, somehow, come out of the minimization problem itself.

It turns out, that instead of considering the integral of arc length, (1), it is more convenient to study the integral of (kinetic) energy

$$\frac{1}{2}\int_a^b g\left(C_*\left(\frac{\partial}{\partial t}\right)_t, C_*\left(\frac{\partial}{\partial t}\right)_t\right) dt. \tag{2}$$

Here the integrand is the square of the length of the tangent vector, which is indeed a smooth function. Furthermore, the integral (2) does depend on the parametrization. The important remark (to be finally established in Section 6) is that the problems of minimizing (1) and (2) are essentially the same. (This is known as the *principle of least action of Maupertius.*) Furthermore, the curves which minimize (2) are automatically parametrized proportionally to arc length. (This is a special case of the law of conservation of energy to be proved in Section 3.) In Sections 1–4 we will present some of the main theorems in the calculus of variations. The reader is advised to keep the special case of kinetic energy in mind throughout the discussion, as all of the (increasingly restrictive) assumptions we shall be making will always hold in this special case.

Since our main concern is the problem of minimizing the kinetic energy, we shall make two simplifying assumptions in the treatment of the general theory. Our first one is that everything that comes up (i.e., functions, forms, maps, etc.) be defined everywhere. This, of course, is *not* the case in many of the calculus of variations problems that arise in nature. The reason for our making this assumption is simply not to clutter up the geometrical ideas involved with various worries as to domains of definition. None of these worries are serious, however. The astute reader should have no difficulty in extending the results we shall obtain to the more general case where domains of definition matter. Our second simplifying assumption is the standard one we have been making throughout this book of not worrying about smoothness considerations.

Before proceeding let us introduce some better notation for the case of arc length and kinetic energy. If M is a Riemann manifold with Riemann metric g, for any $v_1 \in T_x(M)$ and $v_2 \in T_x(M)$ we shall set

$$(v_1, v_2) = g_x(v_1, v_2), \tag{3}$$

the left-hand side of (3) being undefined if v_1 and v_2 are not tangent vectors at the same point. For any tangent vector v we will set

$$\|v\| = \sqrt{(v, v)} = \sqrt{g_x(v, v)}. \tag{4}$$

For any curve $C:[a, b] \to M$ we put

$$C'(t) = C_*\left(\frac{\partial}{\partial t}\right)_t. \tag{5}$$

Thus the integrals (1) and (2) become

$$L[C] = \int_a^b \|C'(t)\|\, dt \quad \text{and} \quad E[C] = \frac{1}{2}\int_a^b \|C'(t)\|^2\, dt.$$

1. Legendre transformations. Let V be a vector space and ω a linear differential form on V. Then ω determines a map, $\mathscr{L}$, of $V \to V^*$. It is defined as follows: For each $v \in V$, ω_v is a linear form on $T_v(V)$. By the discussion at the end of Section 5 of Chapter II, there is a natural identification, l, of $T_v(V)$ with V (depending only on the vector space structure).

Thus $\omega_v \circ l^{-1}$ is a linear function on V, i.e., an element of V^*. We define this element of V^* to be $\mathscr{L}(v)$. Suppose that the mapping $\mathscr{L}$ is a diffeomorphism of V onto V^*. Then the inverse map, $\mathscr{L}^{-1}$, also comes from a form ω^* on V^*. In fact, at each $v^* \in V^*$, $\mathscr{L}^{-1}(v^*) \in V$ is a linear form on V^* and therefore can be regarded as a linear form on $T_{v^*}(V^*)$ in view of the identification of $T_{v^*}(V^*)$ with V^*. Thus $\mathscr{L}^{-1}(v^*)$ determines an element of $T_{v^*}^*(V^*)$ for each $v^* \in V^*$; i.e., $\mathscr{L}^{-1}$ determines a form ω^* on V^*. It is clear from the definitions that the map from $V^* \to V$ determined by ω^* is $\mathscr{L}^{-1}$.

If $d\omega = 0$, then $d\omega^* = 0$. In fact, let $e_1, \ldots, e_n$ be a basis of V and $e_1^*, \ldots, e_n^*$ the dual basis of V^*. Let $\dot{x}^1, \ldots, \dot{x}^n$ be coordinates on V corresponding to $e_1, \ldots, e_n$ and $y^1, \ldots, y^n$ coordinates on V^* corresponding to the basis $e_1^*, \ldots, e_n^*$. If $\omega = \sum w_i\, dx^i$ then $\mathscr{L}$ is given by $\mathscr{L}(\sum \dot{x}^i e_i) = \sum w_i(\dot{x}_1, \ldots, \dot{x}_n)e_i^*$. Thus if $\omega = dL$, the map $\mathscr{L}$ is given by

$$\mathscr{L}(\dot{x}_1, \ldots, \dot{x}_n) = \left(\frac{\partial L}{\partial \dot{x}_1}(\dot{x}^1, \ldots, \dot{x}^n), \ldots, \frac{\partial L}{\partial \dot{x}_n}(\dot{x}^1, \ldots, \dot{x}^n)\right). \tag{1.1}$$

Now let H be the function defined on V^* by

$$\begin{aligned} H(y) &= \langle \mathscr{L}^{-1}(y), y\rangle - L \circ \mathscr{L}^{-1}(y) \\ \text{or} \quad H(y^1, \ldots, y^n) &= \sum \dot{x}^i y^i - L(\dot{x}^1, \ldots, \dot{x}^n), \end{aligned} \tag{1.2}$$

where, in (1.2), the $\dot{x}^i$'s are regarded as functions of the y's via $\mathscr{L}^{-1}$. Then

$$dH = \sum \dot{x}^i\, dy^i + \sum y^i\, d\dot{x}^i - \sum \frac{\partial L}{\partial \dot{x}^i}\, d\dot{x}^i.$$

By (1.1), $\partial L/\partial \dot{x}^i(\dot{x}^1, \ldots, \dot{x}^n) = y^i(\dot{x}^1, \ldots, \dot{x}^n)$. If we regard the $\dot{x}^i$'s as functions of the y's via $\mathscr{L}^{-1}$, the last two terms in the last equation cancel. Thus $dH = \sum \dot{x}^i\, dy^i$, so that $\omega^* = dH$.

Let us consider the special case (of kinetic energy) where $L(v) = \frac{1}{2}(v, v)$ relative to a scalar product (,). In terms of coordinates $L(\dot{x}^1, \ldots, \dot{x}^n) = \frac{1}{2}\sum a_{ij}\dot{x}^i\dot{x}^j$. Then $\partial L/\partial x^i = \sum a_{ij}\,\dot{x}^j$. Thus, if $v = \sum \dot{x}^i e_i$ and $w = \sum \dot{z}^i e^i$, $\langle w \mid \mathscr{L}(v)\rangle = \sum a_{ij}\,\dot{z}^i\dot{x}^j = (w, v)$. In other words, the map $\mathscr{L}$ is just the standard map of $V \to V^*$ given by the scalar product (,). This is a linear map and is a diffeomorphism of V onto V^* so long as (,) is non-degenerate. Let us compute H in this case. According to (1.2),

$$H(w^*) = \langle \mathscr{L}^{-1}(w^*) \,|\, w^* \rangle - \tfrac{1}{2}(\mathscr{L}^{-1}(w^*), \mathscr{L}^{-1}(w^*)) \tag{1.3}$$

if $w^* = \mathscr{L}(v)$, this reads as $H(\mathscr{L}(v)) = \langle v \,|\, \mathscr{L}(v)\rangle - \frac{1}{2}(v, v) = (v, v) - \frac{1}{2}(v, v) = L(v)$. In other words $H(\mathscr{L}(v)) = L(v)$. We can also describe H by observing that (1.3) is the same as saying $H(w^*) = \frac{1}{2}(w^*, w^*)$ where (,) is the scalar product on V^* induced by (,) on V.

Remarks. 1. In the general case, the functions L and H are determined by $\omega = dL$ and $\omega^* = dH$ up to additive constant. In practice, however, we are given L. We then define H by (1.2), and observe that this definition is independent of the choice of basis $e_1, \ldots, e_n$: The arbitrary additive constant in H is fixed by (1.2) as $H(0, \ldots, 0) = L \circ \mathscr{L}^{-1}(0, \ldots, 0)$.

2. If ω is defined on an open subset, U, of V the same definitions given above provide a map, $\mathscr{L}$, of $U \to V^*$. We wish to apply this remark to $\mathscr{L}^{-1}$. That is, suppose that $\omega = dL$ (defined on all of V, say) gives a map, $\mathscr{L}$, of $V \to V^*$, which is an immersion. That is, suppose that $\mathscr{L}$ is locally one-to-one and bidifferentiable but not necessarily one-to-one or onto. Then for any $\dot{x} \in V$, the form $\omega^* = dH$, defined by (1.2) in a neighborhood of $y = \mathscr{L}(\dot{x})$ gives $\mathscr{L}^{-1}$. Both $\mathscr{L}^{-1}$ and ω^* are not well defined about y—they depend on the choice of $\dot{x}$. However, the form $\sigma = \mathscr{L}^*(\omega^*)$ is well defined at $\dot{x}$ and thus on all of V. Similarly the function $E = H \circ \mathscr{L}$ is well defined on V and we have $\sigma = dE$. Notice that σ is intrinsically defined by ω, whereas E depends on the choice of arbitrary constant fixed by (1.2). The function E is called the *energy* of the (Lagrangian) function, L. Thus for the case $L = \frac{1}{2}(v, v)$ of kinetic energy, (1.3) says $L = E$.

3. If L depends on some auxiliary parameters, i.e., if L is a function on $N \times V$, then L defines a differentiable map of $N \times V \to N \times V^*$ (which is the identity on the N component). If for each $n \in N$ this map

is a diffeomorphism of $n \times V \to n \times V^*$, then the H defined by (1.2) induces a diffeomorphism of $N \times V^*$ onto $N \times V$ inverse to that determined by L.

Let M be a differentiable manifold and L a differentiable function on $T(M) \times R$. For each fixed point $q \in M$ and each fixed $t_0 \in R$, the submanifold $T_q(M) \times t_0$ of $T(M) \times R$ is a vector space. The function L (restricted to $T_q(M) \times t_0$) thus induces a map of $T_q(M) \times t_0 \to T_q^*(M) \times t_0$. We thus have a map, $\mathscr{L}$, of $T(M) \times R \to T^*(M) \times R$. This map is called the *Legendre transformation* (corresponding to L).

If $x^1, \ldots, x^n$ are local coordinates on a neighborhood $U \subset M$, then we can introduce coordinates $(x^1, \ldots, x^n, \dot{x}^1, \ldots, \dot{x}^n, t)$ on $\pi^{-1}(U) \times R \subset T(M) \times R$. The coordinates are defined by $x^i(v) = x^i(\pi(v))$ and the $\dot{x}^i$ are determined by $v = \sum \dot{x}^i(v)\,(\partial/\partial x^i)_{\pi(v)}$. Similarly, on $T^*(M) \times R$ we have local coordinates $(x^1, \ldots, x^n, y^1, \ldots, y^n, t)$ where the y^i are given by $w^* = \sum y^i(w^*)\, dx^i{}_{\pi(w)^*}$ for those w^* with $\pi(w^*) \in U$. In terms of these coordinates the transformation $\mathscr{L}$ is given by

$$x^i \circ \mathscr{L} = x^i$$

$$t \circ \mathscr{L} = t$$

(1.4) and

$$y^i \circ \mathscr{L} = \frac{\partial L}{\partial \dot{x}^i}$$

where $L = L(x^1, \ldots, x^n, \dot{x}^1, \ldots, \dot{x}^n, t)$.

If $\mathscr{L}$ is an immersion, the function L will be called a *regular* Lagrangian (function). In this case, for any $(v, t) \in T(M) \times R$, the map $\mathscr{L}^{-1}$ defined about $\mathscr{L}((v, t))$ comes from a function H. Here H as given by (1.2), is defined locally about $\mathscr{L}((v, t))$ and depends on the choice of (v, t). The function $E = H \circ \mathscr{L}$ *is* well defined on $\sigma(M) \times R$, however. It is called the *energy* of L.

EXERCISE 1.1. In the case of classical mechanics we are given a Riemann metric on M and a function V on M (called the *potential energy*). The function L is taken to be $L = T - V \circ \pi$, where $T(v) = \frac{1}{2}(v, v)$ is the kinetic energy. Show that in this case $E = T + V \circ \pi$.

We can resume the previous discussion in:

THEOREM 1.1. *Let L be a differentiable function on $T(M) \times R$. Then L defines a differentiable map of $T(M) \times R \to T^*(M) \times R$ called the*

Legendre transformation, given in local coordinates by (1.4). *If this map is an immersion of* $T(M) \times R$ *into* $T^*(M) \times R$, *then the function* L *will be called a regular Lagrangian (function). In this case, the map* $\mathscr{L}^{-1}$ *comes locally in a similar way from a function* H *on* $T^*(M) \times R$, *called the Hamiltonian. The function* $E = H \circ \mathscr{L}$ *is globally well defined on* $T(M) \times R$. *If* M *is a Riemann manifold and* $L(v, t) = \frac{1}{2}\|v\|^2$, *then* L *is a regular Lagrangian and* $\mathscr{L}$ *coincides on each tangent space with the map of* $T_q(M) \to T_q^*(M)$ *given by the scalar product. Furthermore, in this case,*

$$E = H \circ \mathscr{L} = L. \tag{1.5}$$

2. Necessary conditions. The fundamental problem in the calculus of variations (in one independent variable) is the following: Let M be an n-dimensional differentiable manifold, and L a differentiable function on $T(M) \times R$. Let $C: [a, b] \to M$ be a curve on M. Then C determines a curve, $\tilde{C}$, on $T(M) \times R$ defined by

$$\tilde{C}(t) = (C'(t), t) \tag{2.1}$$

for each $t \in [a, b]$. Therefore, we can consider the integral

$$I(C) = \int_a^b L(\tilde{C}(t))\, dt.$$

The problem is to find curves C which are "local" minima for I; i.e., that satisfy $I(C) \leqslant I(C')$ for all "nearby" curves C'. In order to make the problem precise we have to define what we mean by "nearby." We first remark that the integral $I(C)$ makes sense for "piecewise differentiable" curves. A continuous map $C: [a, b] \to M$ is called piecewise differentiable if we can subdivide $[a, b]$ into subintervals on each of which C is differentiable. More precisely,

DEFINITION 2.1. *A continuous map of* $[a, b] \to M$ *is called a piecewise differentiable curve if we can find points* $t_0 = a < t_1 < \cdots < t_k = b$ *such that* C *restricted to each interval* $[t_i, t_{i+1}]$ *is the restriction of a differentiable map* C_i *of* $(t_i - \epsilon, t_{i+1} + \epsilon)$ *into* M.

For any piecewise differentiable curve C we define

$$I_a^b(C) = \sum \int_{t_i}^{t_{i+1}} L(\tilde{C}(t))\, dt. \tag{2.2}$$

Definition 2.2. *A curve C is called a strong local minimum if there is a neighborhood* $\mathcal{N}_{(W,Z,\epsilon,0)}$ *of C, such that for all piecewise differentiable* $C^1 \in \mathcal{N}_{(W,Z,\epsilon,0)}$ *with*

$$C^1(a) = C(a) \quad \text{and} \quad C^1(b) = C(b), \tag{2.3}$$

we have

$$I(C) \leqslant I(C^1). \tag{2.4}$$

C is called a strict strong local minimum if for all $C^1 \in \mathcal{N}_{(W,Z,\epsilon,0)}$ *not identical with C we have*

$$I(C) < I(C^1). \tag{2.5}$$

Thus a strong local minimum is a curve C for which (2.4) holds for all curves satisfying (2.3) which are uniformly pointwise close to C, including curves whose tangents may not be close to those of C. In Definition 2.2, we may always take Z to be the distinguished t coordinate system of $[a, b]$ since $[a, b]$ is compact.

Definition 2.3. *A curve C is called a weak local minimum if there are W and* ϵ *such that* (2.4) *holds for all piecewise differentiable curves satisfying* (2.3) *and* $C_j^1 \in \mathcal{N}_{(W,Z,\epsilon,1)}(C_j)$. *Here* C_j^1 *and* C_j *denote the restrictions of* C^1 *and C respectively to the interval* $[s_j, s_{j+1}]$, *where* $a = s_0 < \cdots < s_r = b$ *denote the union of the set of points where* C^1 *and C may not be differentiable. The curve* C^1 *is called a strict weak local minimum if we replace* (2.4) *by* (2.5).

Thus C is a weak local minimum if (2.4) holds for all curves satisfying (2.3) and which are uniformly close to C together with their first derivatives, wherever these derivatives make sense.

Our problem is to determine necessary and sufficient conditions for a curve C to be a (strong or weak) local minimum. To do this it is convenient to transfer considerations from $T(M) \times R$ to $T^*(M) \times R$.

For this purpose it is convenient to introduce some preliminary notation that isn't completely accurate. If L is a regular Lagrangian then it is easy to see that the map $\mathcal{L}$ is a diffeomorphism of some open set, $U \subset T(M) \times R$ containing the image of $\tilde{C}$ onto an open subset of $T^*(M) \times R$. All considerations in this section will center on a neighborhood of $\tilde{C}$. Rather than cluttering up the notation with the set U and its image, etc., we shall talk as if $\mathcal{L}$ is a diffeomorphism of $T(M) \times R$ onto

$T^*(M) \times R$. In all the discussion, however, we will only be using the property that it is locally a diffeomorphism.

Let $\mathscr{L}$ be the Legendre transformation of $T(M) \times R \to T^*(M) \times R$ given by L. [It should be noted that the mapping $\mathscr{L}$ commutes with the obvious projections of $T(M) \times R$ and $T^*(M) \times R$ onto $M \times R$.] The importance of passing to the cotangent bundle is the following: Let ρ denote the projection of $T^*(M) \times R$ onto $T^*(M)$, and let $\bar{\theta} = \rho^*(\theta)$ where θ is the fundamental one form on $T^*(M)$, cf. Section 7 of Chapter III. Then we have

LEMMA 2.1. *Let C be a differentiable map of $[a, b] \to M$. Let $\bar{C} = \mathscr{L} \circ \tilde{C}$ (where $\tilde{C}$ is given by* (2.1)). *Then*

$$\int_a^b L(\tilde{C}(s))\, ds = \int_{[a,b]} \bar{C}^* \omega \tag{2.6}$$

where

$$\omega = \bar{\theta} - H\, dt.$$

To keep the notations straight, it may be convenient to introduce a diagram.

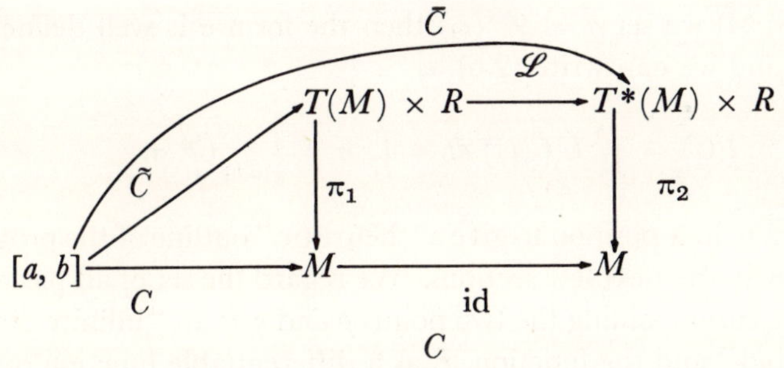

FIGURE 2.1

To prove Lemma 2.1, we note that both sides of (2.6) decompose into the appropriate sums if we subdivide the interval (a, b). We can (by suitable subdivision and change of notation) therefore assume that $C(a, b)$ lies in some coordinate neighborhood, U, of M. Let $x^1, \ldots, x^n$ be a coordinate system on U. On $\pi_1^{-1}(U) \subset T(M) \times R$ we have coordinates $x^1, \ldots, x^n, \dot{x}^1, \ldots, \dot{x}^n, t$. Let $x^1(t), \ldots, x^n(t), \dot{x}^1(t), \ldots, \dot{x}^n(t), t$ denote the curve $\tilde{C}$. Then the left-hand side of (2.6) is just

$$\int_a^b L(x^1(t), \ldots, x^n(t), \dot{x}^1(t), \ldots, \dot{x}^n(t), t)\, dt.$$

On the other hand, the map $\mathscr{L}$ is given, in these coordinates by

$$\mathscr{L}(x^1, \ldots, x^n, \dot{x}^1, \ldots, \dot{x}^n, t) = \left(x^1, \ldots, x^n, \frac{\partial L}{\partial \dot{x}^1}, \ldots, \frac{\partial L}{\partial \dot{x}^n}, t\right)$$

and ω by

$$\omega = \sum y^i \, dx^i - H \, dt.$$

Thus the right-hand side of (2.6) is given by

$$\int_a^b \left(\sum y^i(t) \frac{dx^i}{dt} - H(x^1(t), \ldots, x^n(t), y^1(t), \ldots, y^n(t), t) \right) dt$$
$$= \int_a^b \left(\sum \frac{\partial L}{\partial \dot{x}^i} \frac{dx^i}{dt} - H \right) dt.$$

But by the definition of $\tilde{C}$, $(dx^i/dt)(t) = \dot{x}^i(t)$ and by (1.2) this last integral becomes

$$\int_a^b \left(\sum \frac{\partial L}{\partial \dot{x}^i} \dot{x}^i - \sum \frac{\partial L}{\partial \dot{x}^i} \dot{x}^i + L \right) dt = \int_a^b L \, dt,$$

proving (2.6). If we set $\sigma = \mathscr{L}^*(\omega)$ then the form σ is well defined on $T(M) \times R$ and we can write (2.6) as

$$(2.7) \qquad I(C) = \int_a^b L(\tilde{C}(t)) \, dt = \int_{\tilde{C}} \sigma = \int_{[a,b]} \tilde{C}^* \sigma.$$

[1] We are now in a position to give a "heuristic" outline of the program we will follow in the next few sections. We regard the set of all piecewise differentiable curves joining the two points p and q as an "infinite dimensional manifold" and the function, I, as a differentiable function on this manifold. Let us denote the space of all piecewise differentiable curves joining p and q by Ω_{pq}. A minimizing curve is then a point of the infinite dimensional manifold, Ω_{pq}, where the function I has a minimum. Let us review the facts for the corresponding case of a differentiable function on a (finite dimensional) manifold.

Let N be a differentiable manifold and f a differentiable function on N. If $p \in N$ is a point where f achieves a local minimum then we must

---

[1] The discussion from here until page 160 is merely meant as motivation for what follows. The actual results we use will not depend on them.

have $df_p = 0$. In fact, if we introduce local coordinates this says that all the partial derivatives of f vanish at p, which is exactly the criterion supplied by elementary calculus. [Another way of saying the same thing is that p must be a critical point of f (regarded as a map of $N \to R$) in the sense of Section 3 of Chapter II.] This condition is necessary but not sufficient. To obtain more information we must look at the "second derivatives" of f at p: Let X_p and Y_p be two tangent vectors at p, and let X and Y be vector fields extending them in a neighborhood of p. Let us consider the expression $X_p \langle Y, df \rangle$. It is clearly linear in X_p. On the other hand, we have

$$\begin{aligned} X_p\langle Y, df\rangle &= (\mathscr{L}_X\langle Y, df\rangle)_p \\ &= \langle \mathscr{L}_X Y, df\rangle_p + \langle Y, \mathscr{L}_X\, df\rangle_p. \end{aligned}$$

The first expression vanishes since $df_p = 0$. The second expression can be written as $\langle Y, d\langle \mathscr{L}_X f\rangle\rangle_p = \langle Y, d\langle X, df\rangle\rangle_p = Y_p\langle X, df\rangle$. Thus

$$X_p\langle Y, df\rangle = Y_p\langle X, df\rangle.$$

From this we see that the expression $X_p\langle Y, df\rangle$ depends only on Y_p (not on the extension Y) and is linear in Y_p. We thus obtain a symmetric bilinear form $H_{f,p}$ on the tangent space given by

$$H_{f,p}(X_p, Y_p) = X_p\langle Y, df\rangle = Y_p\langle X, df\rangle.$$

This bilinear form is called the *Hessian* of f at p. In terms of a coordinate system, if $X_p = \sum \xi^i\, (\partial/\partial x^i)_p$ then

$$H_{f,p}(X_p, X_p) = \sum\left(\frac{\partial^2 f}{\partial x^i\, \partial x^j}\right)_p \xi^i\, \xi^j.$$

It is a result of calculus that in order for p to be a minimum of f it is necessary that the Hessian be positive semidefinite and it is sufficient that the Hessian be positive definite. Let us now attempt to formulate these assertions for the "infinite dimensional manifold" of all curves joining p and q.

The first notion we should define is that of "tangent vector" at a point of this infinite dimensional manifold. A curve on Ω_{pq} is a family $C_t(\cdot)$ of curves with $C_t(a) \equiv p$ and $C_t(b) = q$. From this it is clear that a tangent vector of Ω_{pq} at C should be a vector field along the curve C vanishing at p and q; that is, a piecewise differentiable family of vectors

$X_s \in T_{C(s)}(M)$ with $X_a = X_b = 0$. Let us now transcribe the condition $dI_C = 0$. For this purpose let $X_C(= X)$ be a tangent vector at C. Suppose that X can be extended to a vector field, X, defined on some neighborhood of the image of C in M—thus $X_s = X_{C(s)}$. [Notice that this need not be possible in general. For instance, C may cross itself so that $C(s_1) = C(s_2)$ with $X_{s_1} \neq X_{s_2}$. In our heuristic discussion we shall ignore this possibility. A simple way out of this difficulty would be to consider sufficiently small pieces of C. If C minimizes so must any piece.] The vector field X gives rise to a vector field, $\tilde{X}$, in a neighborhood of $\tilde{C}$ in $T(M) \times R$. In fact, let $T(X)$ be the variational vector field of X (defined at the end of Chapter II) and put $\tilde{X} = (T(X), 0)$. The reader should verify that the values of $\tilde{X}$ along the curve $\tilde{C}$ depend only on X_C and not on the extension, X. Let α_s be the local one parameter group generated by X and let $C_s = \alpha_s \circ C$. Then $\tilde{C}_s = \tilde{\alpha}_s \circ \tilde{C}$ where $\tilde{\alpha}_s$ is generated by $\tilde{X}_s$. According to (2.7) we have

$$I[C_s] = \int_{\tilde{C}_s} \sigma = \int_{\tilde{\alpha}_s \circ \tilde{C}} \sigma = \int_{\tilde{C}} \tilde{\alpha}_s^* \sigma.$$

If C is to be a minimum we must have

$$\left.\frac{dI[C_s]}{ds}\right|_{s=0} = 0.$$

But

$$\lim_{s=0} \frac{\tilde{\alpha}_s^* \sigma - \sigma}{s} = \mathscr{L}_{\tilde{X}}\, \sigma$$

so that we must have

$$\int_{\tilde{C}} \mathscr{L}_{\tilde{X}} \sigma = 0.$$

Now $\mathscr{L}_{\tilde{X}}\sigma = \tilde{X} \lrcorner\, d\sigma + d\langle \tilde{X}, \sigma\rangle$ so that we obtain

$$\int_{\tilde{C}} \tilde{X} \lrcorner\, d\sigma + \int_{\tilde{C}} d\langle \tilde{X}, \sigma\rangle = 0.$$

But

$$\int_{\tilde{C}} d\langle \tilde{X}, \sigma\rangle = \langle \tilde{X}, \sigma\rangle_{\tilde{C}(b)} - \langle \tilde{X}, \sigma\rangle_{\tilde{C}(a)}.$$

Let us evaluate $\langle \tilde{X}, \sigma\rangle$. By the definition of σ we get $\langle \tilde{X}, \sigma\rangle = \langle \mathscr{L}_* \tilde{X}, \omega\rangle = \langle \mathscr{L}_* \tilde{X}, \bar{\theta} - H dt\rangle$. Now $\langle \mathscr{L}_* \tilde{X}, dt\rangle = 0$ since $\langle \tilde{X}, dt\rangle$

$= 0$. By the definition of $\bar{\theta}$ we get $\langle \mathscr{L}_* \tilde{X}, \bar{\theta} \rangle = \langle \rho_* \mathscr{L}_* \tilde{X}, \theta \rangle$. Let (v^*, t) be a point of $T^*(M) \times R$. Then by the definition of θ (cf. Section 7 of Chapter III) we get $\langle \mathscr{L}_* \tilde{X}, \bar{\theta} \rangle = \langle X_{\pi(v^*)}, v^* \rangle$. Letting $(v^*, t) = \mathscr{L}(\tilde{C}(a))$ we obtain

$$\langle \tilde{X}, \sigma \rangle_{\tilde{C}(a)} = \langle X_a, v^* \rangle = 0.$$

Similarly for b. Thus $\int d\langle \tilde{X}, \sigma \rangle = 0$ and we obtain that

$$\int_{\tilde{C}} \tilde{X} \rfloor d\sigma = 0$$

for all X. Now

$$\int_{\tilde{C}} \tilde{X} \rfloor d\sigma = \int_a^b \langle \tilde{X}_t \wedge \tilde{C}'(t) \mid d\sigma \rangle \, dt.$$

The integrand can be written as $\langle \tilde{X}_t \wedge \tilde{C}'(t) \mid d\sigma \rangle = -\langle \tilde{X}(t), \tilde{C}'(t) \rfloor d\sigma \rangle$. If $\tilde{X}$ were an arbitrary vector field on $T(M) \times R$, we would be able to conclude from the fact that $\int_a^b \langle \tilde{X}_t, \tilde{C}'(t) \rfloor d\sigma \rangle \, dt \equiv 0$ that $\tilde{C}'(t) \rfloor d\sigma \equiv 0$. The vector fields $\tilde{X}$ are not completely arbitrary, however. (They must come from vector fields X on M vanishing at p and q.) Nevertheless, it turns out that the way σ is defined will allow us to conclude that

$$\tilde{C}'(t) \rfloor d\sigma = 0.$$

The form $d\sigma$ is of rank $2n$ on $T(M) \times R$. There is thus a unique vector field, Y, on $T(M) \times R$ satisfying

$$Y \rfloor d\sigma = 0 \quad \text{and} \quad \langle Y, dt \rangle \equiv 1.$$

The vector field Y is known as the *Euler* vector field. (The corresponding differential equations in terms of local coordinates are known as *Euler's differential equation.*) We thus see that if C is a local minimum then $\tilde{C}$ is an integral curve of the Euler vector field. The details of this argument will be presented in this section.

Next we try to formulate quadratic conditions involving the Hessian. Let X_C and Y_C be two "tangent vectors" at C and let X and Y be vector fields on M extending them. We have already seen that if $C_s = \alpha_s \circ C$ where α_s is generated by X then

$$\frac{dI[C_s]}{ds} = \int_{\tilde{C}_s} \mathscr{L}_{\tilde{X}} \sigma.$$

If we then take the derivative at C in the direction of the "tangent vector" Y_C we obtain the expression

$$\int_{\tilde{C}} \mathscr{L}_{\tilde{Y}} (\mathscr{L}_{\tilde{X}}\sigma).$$

This expression is clearly linear in $\tilde{Y}$ along $\tilde{C}$ and is therefore linear in Y_C. On the other hand,

$$\int_{\tilde{C}} \mathscr{L}_{\tilde{Y}}(\mathscr{L}_{\tilde{X}}\sigma) = \int_{\tilde{C}} \mathscr{L}_{[\tilde{X},\tilde{Y}]}\sigma + \int_{\tilde{C}} \mathscr{L}_{\tilde{X}}(\mathscr{L}_{\tilde{Y}}\sigma).$$

But $[\tilde{X}, \tilde{Y}] = \widetilde{[X, Y]}$ so that $\int_{\tilde{C}} \mathscr{L}_{[\tilde{X},\tilde{Y}]}\,\sigma = \int_{\tilde{C}} \mathscr{L}_{\widetilde{[X,Y]}}\,\sigma = 0$ by Euler equations. Thus

$$\int_{\tilde{C}} \mathscr{L}_{\tilde{Y}}(\mathscr{L}_{\tilde{X}}\sigma) = \int_{\tilde{C}} \mathscr{L}_{\tilde{X}}(\mathscr{L}_{\tilde{Y}}\sigma)$$

is a symmetric bilinear form in X_C and Y_C. It is known as the *second variation.*

Our "second order" conditions are related to the positive definiteness of the second variation. In Section 4 we will make an assumption on L which will guarantee that the second variation is positive definite *if the curve C is sufficiently short.* It turns out that if C surpasses a critical length then the second variation is no longer positive definite and C is no longer a minimum. The details will be presented in Section 5.

Let us[2] proceed to obtain necessary conditions for a weak local minimum. We first observe that if C is a weak local minimum for L on the interval $[a, b]$, the restriction of C to any subinterval $[d, e]$ is also a weak local minimum. In fact, if C^* is a piecewise differentiable curve defined on $[d, e]$ with $C^*(d) = C(d)$ and $C^*(e) = C(e)$, we can extend C^* to $[a, b]$ by setting $C^*(t) = C(t)$ for $a \leqslant t \leqslant d$ and $e \leqslant t \leqslant b$. Then $I_d^e(C^*) < I_d^e(C)$ would imply $I_a^b(C^*) < I_a^b(C)$. Thus if C is a local minimum on $[a, b]$ it must also be a local minimum on $[d, e]$.

In obtaining necessary conditions, let us first restrict attention to a subinterval, $[d, e]$, on which C is differentiable and which is mapped by C into a coordinate neighborhood of M. For C to be a minimum, we would expect that $(d/ds)\, I[C_s]|_{s=0} = 0$ for any "differentiable" one parameter family of curves C_s with $C_0 = C$. Such a one parameter family of curves we regard as a rectangle and will apply Stokes' theorem. Let D_ϵ be a rectangle in the t, s plane defined by $-\epsilon \leqslant s \leqslant \epsilon$ and $d \leqslant t \leqslant e$ and let D_ϵ^+ be the subset of D_ϵ where $s \geqslant 0$ and D_ϵ^- where $s \leqslant 0$.

[2] Our legal discussion begins here.

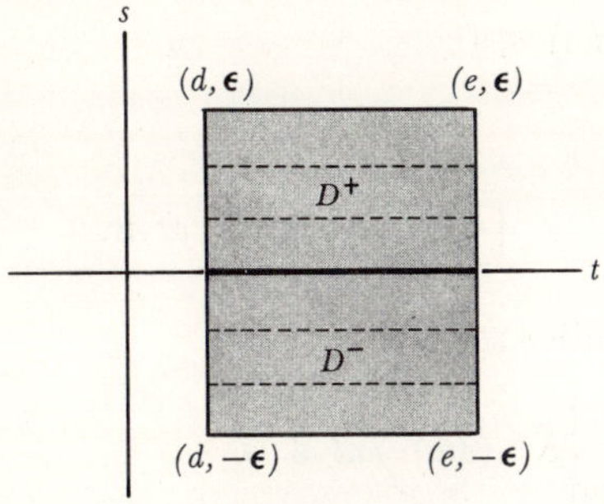

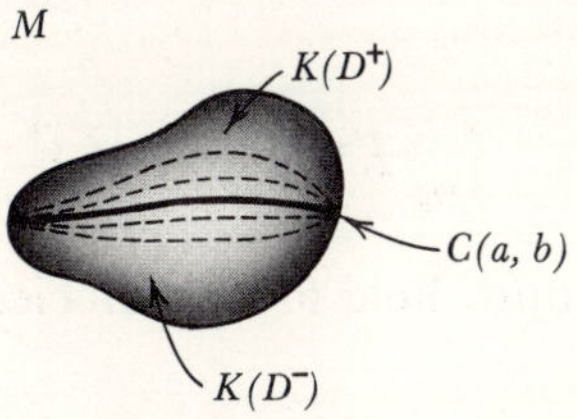

Let K be a differentiable map of (some neighborhood of) $D_\epsilon \to M$ satisfying

$$K(t, 0) = C(t) \qquad d \leqslant t \leqslant e \tag{2.8}$$

and

$$K(d, s) = C(d), K(e, s) = C(e) \qquad -\epsilon \leqslant s \leqslant \epsilon. \tag{2.9}$$

For each fixed value of s, $K(\cdot, s)$ is a differentiable curve on M. It therefore defines curves $\tilde{K}(\cdot, s)$ on $T(M) \times R$ and $\bar{K}(\cdot, s)$ in $T^*(M) \times R$. We thus have maps $\tilde{K}: D \to T(M) \times R$ and $\bar{K}: D \to T^*(M) \times R$. A direct verification in terms of local coordinates shows that these maps are differentiable. If C is a weak local minimum, then for sufficiently small values of $|s|$ we must have

$$I_d^e(C) \leqslant I_d^e(K(\cdot, s)), \quad \text{i.e.,} \quad \int_{[d,e]} \bar{C}^*(\omega) \leqslant \int_{[d,e]} \bar{K}(\cdot, s)^*(\omega). \tag{2.10}$$

In other words,

$$\int_{[d,e]} \bar{K}^*(\cdot, 0)(\omega) - \int_{[d,e]} \bar{K}^*(\cdot, s)(\omega) \leqslant 0.$$

In view of (2.9), $\bar{K}^*\omega$ vanishes along the vertical of D. Therefore, we can write (2.10) as

$$\int_{\partial D_\eta^+} \bar{K}^*(\omega) \leqslant 0.$$

By Stokes' theorem

$$\int_{D_\eta^+} \bar{K}^* (d\omega) \leqslant 0. \tag{2.11}$$

Now

$$\int_{D_\eta^+} \bar{K}^* (d\omega) = \int_0^\eta \int_a^b \left\langle \left(\frac{\partial}{\partial t}\right)_{(t,s)} \wedge \left(\frac{\partial}{\partial s}\right)_{(t,s)} \middle| \bar{K}^* \, d\omega \right\rangle dt \, ds.$$

This must hold for all sufficiently small η so that we have

$$\int_a^e \left\langle \left(\frac{\partial}{\partial t}\right)_{(t,0)} \wedge \left(\frac{\partial}{\partial s}\right)_{(t,0)} \middle| \bar{K}^* (d\omega) \right\rangle dt \leqslant 0.$$

Similarly

$$\int_{\partial(D_\eta^-)} \bar{K}^* (\omega) \geqslant 0$$

so that

$$\int_a^b \left\langle \left(\frac{\partial}{\partial t}\right)_{(t,0)} \wedge \left(\frac{\partial}{\partial s}\right)_{(t,0)} \middle| \bar{K}^* (d\omega) \right\rangle dt \geqslant 0.$$

Putting the two together we obtain

$$\int_a^b \left\langle \left(\frac{\partial}{\partial t}\right)_{(t,0)} \wedge \left(\frac{\partial}{\partial s}\right)_{(t,0)} \middle| \left(\bar{K}^* (d\omega)\right)_{(t,0)} \right\rangle dt = 0.$$

But

$$\left\langle \left(\frac{\partial}{\partial t}\right)_{(t,0)} \wedge \left(\frac{\partial}{\partial s}\right)_{(t,0)} \middle| \left((\bar{K}^* \, d\omega)\right)_{(t,0)} \right\rangle$$

$$= \left\langle \bar{K}_* \left(\frac{\partial}{\partial t}\right)_{(t,0)} \wedge \bar{K}_* \left(\frac{\partial}{\partial s}\right) \middle| d\omega_{K(t,0)} \right\rangle$$

$$= \left\langle \bar{K}_* \left(\frac{\partial}{\partial s}\right)_{(t,0)} \middle| \bar{K}_* \left(\frac{\partial}{\partial t}\right)_{(t,0)} \lrcorner \, d\omega_{K(t,0)} \right\rangle.$$

Thus we have

$$\int_a^b \left\langle \bar{K}_* \left(\frac{\partial}{\partial s}\right)_{(t,0)} \middle| \bar{C}_* \left(\frac{\partial}{\partial t}\right) \lrcorner \, d\omega_{C(t)} \right\rangle dt = 0. \tag{2.12}$$

Equation (2.12) is to hold for all K's satisfying (2.8) and (2.9). We now derive the consequences of this condition.

Let Y_q be a tangent vector at a point q of $T^*(M) \times R$. In terms of the local coordinates we have introduced, we can write

$$Y_q = \sum_{i=1}^n X^i \left(\frac{\partial}{\partial x^i}\right)_q + \sum_{i=1}^n Y^i \left(\frac{\partial}{\partial y^i}\right)_q + T \frac{\partial}{\partial t}.$$

The form $d\omega$ has the expression

$$d\omega = \sum dy^i \wedge dx^i - dH \wedge dt.$$

Therefore

$$\begin{aligned} Y_q \lrcorner\, d\omega_q = \sum_{i=1}^{n} \left(Y^i + T\frac{\partial H}{\partial x^i}\right) dx^i - \sum_{i=1}^{n} \left(X^i - T\frac{\partial H}{\partial y^i}\right) dy^i \\ - \sum_{i=1}^{n} \left(\frac{\partial H}{\partial x^i} X^i + \frac{\partial H}{\partial y^i} Y^i\right) dt. \end{aligned} \tag{2.13}$$

If Y_q is the tangent vector to a curve coming from M, e.g., if $Y_q = \bar{C}'(t)$ then $T = 1$, $X^i = dx^i/dt = \dot{x}^i = \partial H/\partial y^i$. This second sum in (2.13) vanishes, in this case.

By definition, the t component of $\bar{K}(s, t)$ is exactly t. Therefore $\langle K_*(\partial/\partial s)_{(t,0)} \,|\, dt\rangle = 0$. Let $X_q(t)$ denote $\bar{C}_*(\partial/\partial t)_t$ and $\bar{X}_q(t) = \bar{K}_*(\partial/\partial s)_{(0,t)}$. Then (2.13) and (2.12) imply that

$$\sum \int_d^e \left(Y^i + \frac{\partial H}{\partial x^i}\right) \bar{X}^i \, dt = 0. \tag{2.14}$$

Now the only restriction on the differentiable functions $\bar{X}_i$ is that they vanish at $t = d$ and $t = e$. In fact, let $\bar{X}_i$ be n functions vanishing at the end points of $[d, e]$. Define the map $K(s, t)$ in terms of local coordinates by

$$x^i \circ K(s, t) = x^i \circ C(t) + s\bar{X}^i(t). \tag{2.15}$$

Then (2.15) defines $K(s, t)$ as a differentiable map, which obviously satisfies (2.8) and (2.9). Furthermore $K_*(\partial/\partial s)_{(t,0)} = \bar{X}^i(t)$.

Thus (2.14) must be satisfied for all functions X_i of t vanishing at the end points d and e. Taking all but one of the $\bar{X}$'s $= 0$ we have

$$\int_d^e (Y^i + (\partial H/\partial x^i))\bar{X}^i \, dt = 0$$

for each i, for arbitrary functions $\bar{X}^i$ vanishing at the end points. The only way that this could happen is for $Y^i + (\partial H/\partial x^i) = 0$ for all $t \in [d, e]$. In fact, if $(Y^i + (\partial H/\partial x^i))(t_0) > 0$ for some $t_0 \in [d, e]$, this function must be positive in some neighborhood of t_0. If we choose $\bar{X}^i$ to be positive at t_0 and vanish outside this small neighborhood the integral $\int_d^e (Y^i + (\partial H/\partial x^i))X^i \, dt$ would be positive. Thus if $Y_q(t) = \bar{C}'(t)$

$$\begin{aligned} X^i &= \frac{\partial H}{\partial y^i} \\ Y^i &= -\frac{\partial H}{\partial x^i} \end{aligned} \tag{2.16}$$

for all t. In view of (2.13) we can resume the previous discussion in

LEMMA 2.2. *Let C be a weak local minimum of L. Then at every point t where C is differentiable the tangent vector $Y_q = \bar{C}'(t)$ satisfies*

$$(2.17) \qquad Y_q \rfloor d\omega_q = 0.$$

Now let us examine the points where C is possibly not differentiable. Let $\bar{t}$ be a point at which C may not be differentiable and let $[t_1, t_2]$ be a small interval containing $\bar{t}$ and no other such points. Then C on $[t_1, t_2]$ is the restriction of a differentiable map defined on $[t_1, \bar{t} + \epsilon]$ to the interval $[t_1, \bar{t}]$ and of a differentiable map defined on $[\bar{t} - \epsilon, t_2]$ to $[\bar{t}, t_2]$. By choosing t_1 and t_2 close enough to $\bar{t}$ we can assume that $C[t_1, t_2]$ lies in a coordinate neighborhood and that there are differentiable maps C_1 and C_2 of $[t_1, t_2] \to M$ such that $C(t) = C_1(t)$ for $t_1 \leqslant t \leqslant \bar{t}$ and $C(t) = C_2(t)$ for $\bar{t} \leqslant t \leqslant t_2$. Let $X^i(t)$ be differentiable functions defined on $[t_1, t_2]$ vanishing at t_1 and t_2. Corresponding to C, C_1, and C_2 use (2.15) to define three maps K, K_1, and K_2 of the rectangle $-\epsilon \leqslant s \leqslant \epsilon$, $t_1 \leqslant t \leqslant t_2$ into M. The maps K_1 and K_2 are differentiable and for each s, $K(\cdot, s)$ is a piecewise differentiable curve (differentiable except at $\bar{t}$). In fact, $K(t, s) = K_1(t, s)$ for $t \leqslant \bar{t}$ and $K(t, s) = K_2(t, s)$ for $t \geqslant \bar{t}$.

For sufficiently small s we must have $I_{t_1}^{t_2}(C) \leqslant I_{t_1}^{t_2}(K(\cdot, s))$. That is,

$$(2.18) \quad \int_{[t_1,\bar{t}]} \bar{C}^*(\omega) + \int_{[\bar{t},t_2]} \bar{C}^*(\omega) \leqslant \int_{[t_1,\bar{t}]} \bar{K}(\cdot, s)^*\omega + \int_{[\bar{t},t_2]} \bar{K}(\cdot, s)^*\omega.$$

Let us add the terms

$$\int_{[0,\epsilon]} K_1(\bar{t}, \cdot)^*\omega \quad \text{and} \quad - \int_{[0,\epsilon]} K_2(\bar{t}, \cdot)^*\omega.$$

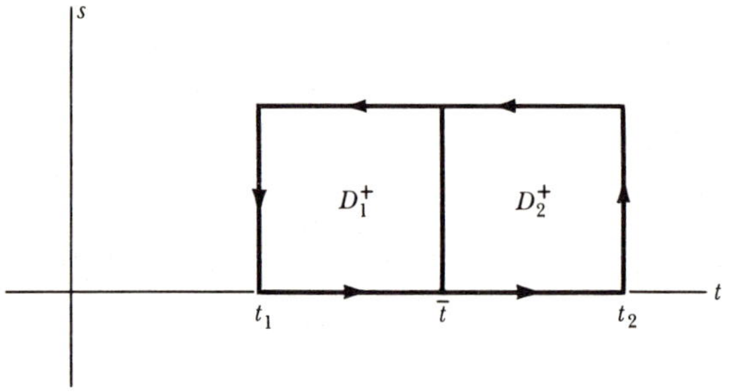

Then

$$\int_{[t_1,\bar{t}]} \bar{C}_1^*\omega + \int_{[0,\epsilon]} K_1(\bar{t}, \cdot)^*\omega - \int_{[t_1,\bar{t}]} \bar{K}^*(\cdot, s)\omega = \int_{D_1^+} K_1^* \, d\omega$$

and

$$\int_{[\bar{t},t_2]} \bar{C}_2^*\omega - \int_{[0,\epsilon]} K_2(\bar{t},\cdot)^*\omega - \int_{[\bar{t},t_2]} \bar{K}^*(0,s)\omega = \int_{D_2^+} K_2^*\, d\omega.$$

By (2.18) we have

$$(2.19)\quad \int_{[0,\epsilon]} K_1(\bar{t},\cdot)^*\omega - \int_{[0,\epsilon]} K_2(\bar{t},\cdot)^*\omega \geqslant \int_{D_1^+} K_1^*\, d\omega + \int_{D_2^+} K_2^*\, d\omega.$$

The right-hand side can be written as

$$\int_0^\epsilon \int_{t_1}^{t_2} \left\langle K_*\left(\frac{\partial}{\partial s}\right)_{t,s} \middle| K_*\left(\frac{\partial}{\partial t}\right)_{t,s} \lrcorner\, d\omega \right\rangle dt\, ds.$$

The inner integral vanishes at $s = 0$ by Lemma 2.2, since the value of an integrand at one point does not change the value of the integral. Thus if we divide (2.19) by ϵ and let $\epsilon \to 0$ we obtain

$$\left\langle \bar{K}_1(t,\cdot)_*\left(\frac{\partial}{\partial s}\right)_{\bar{t},0} \middle| \omega_{\bar{K}_1(t,0)} \right\rangle - \left\langle \bar{K}_2(t,\cdot)_*\left(\frac{\partial}{\partial s}\right)_{\bar{t},0} \middle| \omega_{K_2(t,0)} \right\rangle \geqslant 0.$$

If we argue with negative s instead of positive s this would replace $\geqslant 0$ by $\leqslant 0$. Finally, let $\bar{x}^1, \ldots, \bar{x}^n, \bar{y}_-^1, \ldots, \bar{y}_-^n, t$ be the coordinates of $\bar{K}_1(\bar{t}, 0)$ and $\bar{x}^1, \ldots, \bar{x}^n, \bar{y}_+^1, \ldots, \bar{y}_+^n, \bar{t}$ the coordinates of $\bar{K}_2(t, 0)$. Then this last equation can be written as

$$\sum (\bar{y}_+^i - \bar{y}_-^i)\bar{X}^i(\bar{t}) = 0$$

for all functions $\bar{X}^i$. But this implies that

$$(2.20)\qquad \bar{y}_+^i = \bar{y}_-^i.$$

In other words we have $\bar{C}_1(\bar{t}) = \bar{C}_2(\bar{t})$. Now both $\bar{C}_1$ and $\bar{C}_2$ satisfy the same differential equation (2.16), $\bar{C}_1$ for $t_1 \leqslant t \leqslant \bar{t}$ and $\bar{C}_2$ for $\bar{t} \leqslant t \leqslant t_2$. The uniqueness theorem for differential equations then implies that $\bar{C}_1$ and $\bar{C}_2$ match up at $\bar{t}$ to form a differentiable curve. Thus

LEMMA 2.3. (Hilbert). *If C is a local weak minimum then C is differentiable.*

We now assemble the various facts obtained so far.
Let us examine condition (2.17) where X_q is given by

$$Y_q = \sum X^i\left(\frac{\partial}{\partial x^i}\right)_q + \sum Y^i\left(\frac{\partial}{\partial y^i}\right)_q + T\frac{\partial}{\partial t}\cdot$$

If $T = 0$, then (2.13) shows that (2.17) can only hold if $Y_q = 0$. Thus on any non-zero Y_q satisfying (2.17) we must have $\langle Y_q | dt_q \rangle \neq 0$. (This shows that $d\omega$ is a form of rank $2n$ in the $2n + 1$ dimensional manifold $T^*(M) \times R$.) The form ω is not well defined on $T^*(M) \times R$. However, the form $\sigma = \mathscr{L}^*\omega$ is well defined on $T(M) \times R$. Equation (2.17) is satisfied by the vector $Y_q = \bar{C}' = \mathscr{L}_*\tilde{C}'$. Thus we have

$$\tilde{C}' \rfloor d\sigma = 0 \quad \text{and} \quad \langle \tilde{C}', dt \rangle = 1.$$

Since $d\omega$ is of rank $2n$ on $T^*(M) \times R$, we conclude that $d\sigma$ is of rank $2n$ on $T(M) \times R$. Thus there is a unique vector field, X, called the *Euler vector field* on $T(M) \times R$ defined by

$$X \rfloor d\sigma = 0 \qquad \langle X, dt \rangle = 1. \tag{2.21}$$

If C is a local minimum then C is differentiable and $\tilde{C}$ is an integral curve of X. Let us examine the form of X. Suppose

$$X = \sum X^i \frac{\partial}{\partial x^i} + \sum \dot{X}^i \frac{\partial}{\partial \dot{x}^i} + T \frac{\partial}{\partial t}$$

in terms of local coordinates. Then

$$\mathscr{L}_* X = \sum X^i \frac{\partial}{\partial x^i} + Y^i \frac{\partial}{\partial y^i} + \frac{\partial}{\partial t}$$

where the functions X^i and Y^i are given by (2.16). By the definition of $\mathscr{L}^{-1}$ we have $\dot{x}^i = \partial H / \partial Y^i$ so that we can write

$$X^i = \dot{x}^i \quad \text{and} \quad T = 1. \tag{2.22}$$

We can describe conditions (2.22) in a more intrinsic fashion. By abuse of language, let π denote the projection of $T(M) \times R$ onto M. Then we can thus rewrite (2.22) as

$$\pi_* X_{(v,t)} = v, \qquad \langle X, dt \rangle = 1. \tag{2.23}$$

A vector field, X, on $T(M) \times R$ satisfying (2.23) is called a *second order differential equation* on M. Thus the Euler vector field is a second order differential equation.

EXERCISE 2.1. Show that an integral curve of a second order differential equation is automatically of the form $\tilde{C}$ where C is a curve on M.

DEFINITION 2.4. *Equations* (2.21) *are called Euler's equations. When they are written locally as* (2.17) *they are called "Euler's equations in Hamiltonian form." The vector field, X, given by* (2.21) *is called the Euler vector field. A differentiable curve, C, on M such that $\tilde{C}$ is an integral curve of X is* *called an extremal. In the case of kinetic energy of a Riemann manifold the extremals are called geodesics.*

Thus we have proved

THEOREM 2.1. (Euler). *Let L define a regular problem in the calculus of variation. A necessary condition that a piecewise differentiable curve C be a weak local minimum for L is that C be differentiable and $\tilde{C}$ be an integral curve of X, where X is defined by* (2.21).

If C is an extremal, then $\tilde{C}$ is given, in local coordinates by $x^1(t), \ldots, x^n(t), y^1(t), \ldots, y^n(t), t$. Euler's equations say that $dx^i/dt = \partial H/\partial y^i = \dot{x}^i$ and $dy^i/dt = -\partial H/\partial x^i$. But by (1.2),

$$\frac{\partial H}{\partial x^i} = \frac{\partial}{\partial x^i}\left\{\sum (\dot{x}^i \circ \mathscr{L}^{-1})\, y^i - L \circ \mathscr{L}^{-1}\right\}$$

$$= \sum \frac{\partial(\dot{x}^i \circ \mathscr{L}^{-1})}{\partial x^i}\, y^i - \frac{\partial L \circ \mathscr{L}^{-1}}{\partial x^i} - \frac{\partial L \circ \mathscr{L}^{-1}}{\partial \dot{x}^i}\frac{\partial \dot{x}^i \circ \mathscr{L}}{\partial x^i} = -\frac{\partial L \circ \mathscr{L}^{-1}}{\partial x^i}.$$

Since $y^i = \partial L/\partial \dot{x}^i$ we can thus write Euler's equations, locally, as second order differential equations

$$\frac{dx^i}{dt} = \dot{x}^i$$

(2.24)

$$\frac{d\left(\dfrac{\partial L}{\partial \dot{x}^i}\right)}{dt} = \frac{\partial L}{\partial x^i}.$$

If L is the energy of a Riemann metric, L is given by $L = \frac{1}{2}\sum g_{ij}\dot{x}^i\dot{x}^j$ so that $\partial L/\partial \dot{x}^i = \sum g_{ij}\dot{x}^j$. Then (2.24) become

$$\frac{d}{dt}\left(\sum g_{ij}\dot{x}^j\right) = \frac{1}{2}\sum_{kl}\frac{\partial g_{kl}}{\partial x^i}\,\dot{x}^k\dot{x}^l$$

or

$$\sum g_{ij}\frac{d^2x^j}{dt^2} + \sum \frac{\partial g_{ij}}{\partial x^k}\frac{dx^k}{dt}\frac{dx^j}{dt} = \frac{1}{2}\sum \frac{\partial g_{kl}}{\partial x^i}\frac{dx^k}{dt}\frac{dx^l}{dt}.$$

But

$$\sum \frac{\partial g_{ik}}{\partial x^l}\,\dot{x}^k\dot{x}^l = \sum \frac{\partial g_{il}}{\partial x^k}\,\dot{x}^k\dot{x}^l$$

so we can write

$$\sum g_{ij} \frac{d^2x^j}{dt^2} = \frac{1}{2} \sum_{k,l} \left(\frac{\partial g_{kl}}{\partial x^i} - \frac{\partial g_{ik}}{\partial x^l} - \frac{\partial g_{il}}{\partial x^k} \right) \frac{dx^k}{dt} \frac{dx^l}{dt}.$$

Let (g^{ij}) be the inverse matrix to g_{ij} and set

$$\Gamma^j_{kl} = \frac{1}{2} \sum_{i,k,l} g^{ij} \left(\frac{\partial g_{kl}}{\partial x^i} - \frac{\partial g_{ik}}{\partial x^l} - \frac{\partial g_{il}}{\partial x^k} \right). \tag{2.25}$$

Then the Equations (2.24) for geodesics take the form

$$\frac{d^2x^j}{dt} = \sum \Gamma^j_{kl} \frac{dx^k}{dt} \frac{dx^l}{dt}. \tag{2.26}$$

EXERCISE 2.2. Let Ω be a linear differential form on a manifold N. Suppose that $C: [a, b] \to N$ is a curve such that

$$\int_{[a,b]} C^*\Omega \leqslant \int_{[c,d]} C^{1*}\Omega$$

for all "nearby" curves C^1 defined on a nearby interval $[c, d]$ with $C^1(c) = C(a)$ and $C^1(d) = C(b)$. Show that the tangent vectors to C all satisfy $Y \rfloor d\Omega = 0$. Thus the equations (2.17) or (2.21) are necessary conditions for a "variational problem" of $T(M) \times R$, namely, minimizing $\int \sigma$ among all nearby curves of $T(M) \times R$.

We now consider the case of "variable endpoints." Let N_1 and N_2 be submanifolds of M. A curve C is called a weak (respectively strong) minimum of L relative to N_1 and N_2 if it satisfies the conditions of Definition 2.3 (respectively 2.2) where (2.3) is replaced by

$$C^1(a) \in N_1 \quad \text{and} \quad C^1(b) \in N_2. \tag{2.27}$$

Thus, while in Definitions 2.2 and 2.3 we required that the comparison curves to have the same endpoints as C, we now allow the endpoints to vary on N_1 and N_2.

THEOREM 2.2. *Let C be a weak minimum relative to N_1 and N_2. Then C is an extremal of L and furthermore*

$$\langle v_1, \bar{C}(a) \rangle = \langle v_2, \bar{C}(b) \rangle = 0 \quad \textit{for all} \quad v_1 \in T_{C(a)}(N_1) \quad \textit{and} \quad v_2 \in T_{C(b)}(N_2). \tag{2.28}$$

The first part of Theorem 2.2 is obvious: If C is a weak minimum relative to N_1 and N_2, it certainly is a weak minimum in the sense of Definition 2.3. Thus Theorem 2.1 implies that C is an extremal. It suffices to prove one of the equalities (2.23). Let $h = (x^1, \ldots, x^n)$ be a coordinate system in a neighborhood U of $C(b)$ such that $N_2 \cap U$ is the set of points where $x^{k+1} = \cdots = 0$. Let the vector v_2 be given by $v_2 = a_1(\partial/\partial x^1)_{C(b)} + \cdots + a_k(\partial/\partial x^k)_{C(b)}$. Choose t_0 close enough to b so that $C(t) \in U$ for $t_0 \leqslant t \leqslant b$. Let $\bar{X}_1(t), \ldots, \bar{X}_k(t)$ be differentiable functions with $\bar{X}(t) = 0$ for $t \leqslant t_0$ and $\bar{X}_i(b) = a_i (i = 1, \ldots, k)$ and $\bar{X}_i(b) = 0$ for $i > k$. Choose ϵ so small that $(x^1(t) + s\bar{X}^1(t), \ldots, x^n(t) + s\bar{X}^n(t)) \in h(U)$ for $t_0 \leqslant t \leqslant b$ and $|s| \leqslant \epsilon$ where $h \circ C(t) = (x^1(t), \ldots, x^n(t))$. Let K be the map of the rectangle $D = \{|s| \leqslant \epsilon, a \leqslant t \leqslant b\}$ into M defined by $K(s, t) = C(t)$ for $a \leqslant t \leqslant b$, and $x^i \circ K(s, t) = x^i(t) + s\bar{X}^i(t)$ for $t_0 \leqslant t \leqslant b$. Then as before, we have $\int_{[a,b]} \bar{C}^*(\omega) \leqslant \int \bar{K}(s, \cdot)^*\omega$ for sufficiently small $|s|$. As in the proof of Lemma 2.2 this implies that, for $s > 0$

$$\int_{[0,s]} \bar{K}^*(\cdot, t)\omega \geqslant \int_{D_+} \bar{K}^* \, d\omega;$$

dividing by s and letting $s \to 0$ we obtain, since C is an extremal, $\langle \bar{K}^*(\cdot, t)(\partial/\partial s)_{(0,b)} \,|\, \omega_{\bar{C}(b)} \rangle \geqslant 0$. Doing the same for negative s we obtain $\langle \bar{K}^*(\cdot, t)(\partial/\partial s)_{(0,b)} \,|\, \omega_{\bar{C}(b)} \rangle \leqslant 0$. Thus

$$\left\langle \bar{K}_*(\cdot, t) \left(\frac{\partial}{\partial s}\right)_{(0,b)} \middle| \, \omega_{\bar{C}(b)} \right\rangle = 0.$$

But

$$\bar{K}_*(\cdot, t) \left(\frac{\partial}{\partial s}\right)_{(0,b)} = a_1 \frac{\partial}{\partial x^1} + \cdots + a_k \frac{\partial}{\partial x^k}$$

and

$$\omega_{\bar{C}(b)} = \sum y^i(b)\, dx^i_b - H\, dt, \quad \text{where} \quad \bar{C}(b) = (\sum y^i(b)\, dx^i_b, b).$$

Thus

$$\left\langle \bar{K}_*(\cdot, t) \left(\frac{\partial}{\partial s}\right)_{(0,b)} \middle| \, \omega_{\bar{C}(b)} \right\rangle = \sum_{i=1}^{k} y^i(b) a_i = \langle v_2 | \bar{C}(b) \rangle = 0,$$

proving Theorem 2.2.

EXERCISE 2.3. Show that if L is the kinetic energy associated to the Riemann metric g condition (2.23) says that $C'(a)$ is orthogonal to $T_{C(a)}(N_1)$

relative to the inner product on $T_{C(a)}(M)$ given by g and $C'(b)$ is orthogonal to $T_{C(b)}(N_2)$, i.e.,

$$(v, C'(a)) = 0 \qquad \text{for} \quad v \in T_{C(a)}(N_1)$$

and

$$(v, C'(b)) = 0 \qquad \text{for} \quad v \in T_{C(b)}(N_2).$$

3. Conservation laws. According to Section 2, every regular Lagrangian determines a vector field, X, on $T(M) \times R$, defined by (2.21). For convenience in discussion we will assume in this section that $\mathscr{L}$ is a diffeomorphism of $T(M) \times R$ onto $T^*(M) \times R$. Then ω is well defined on $T^*(M) \times R$. The vector field $Y = \mathscr{L}_* X$ is well defined on $T^*(M) \times R$ and is determined by

$$Y \rfloor d\omega = 0 \quad \text{and} \quad \langle Y, dt \rangle = 1.$$

This vector field determines a local flow, α, on $T^*(M) \times R$. That is, to every point $p \in T^*(M) \times R$, there is a neighborhood U of p and an interval $[-\epsilon, \epsilon]$ on the real line with maps α_s: $U \to T^*(M) \times R$. Equation (2.21) implies that this flow leaves the form $d\omega$ invariant. In fact, $\mathscr{L}_Y \, d\omega = Y \rfloor d(d\omega) + d(Y \rfloor d\omega)$. Thus

$$\mathscr{L}_Y \, d\omega = 0 \tag{3.1}$$

or

$$\alpha_s^* \, d\omega = d\omega. \tag{3.2}$$

Since $\langle Y \mid dt \rangle = 1$, the flow α takes simultaneous points into simultaneous points. More precisely, $t(\alpha_s(p)) = t(p) + s$ for any $p \in T^*(M) \times R$ and s such that $\alpha_s(p)$ is defined. Thus α_s determines (locally) a family of transformations, β_s, of $T^*(M) \to T^*(M)$ defined by

$$\beta_s(q) = (\rho \circ \alpha_s(q, 0)) \tag{3.3}$$

[where ρ is the projection of $T^*(M) \times R$ onto $T^*(M)$].
[Again, the transformations β_s are only defined locally: To every $q \in T^*(M)$ there is a neighborhood U of q and an interval $[-\epsilon, \epsilon]$ of s values for which (3.2) makes sense.] The family β_s is not necessarily a one parameter group: $\beta_{s_1+s_2} \neq \beta_{s_2}\beta_{s_1}$ in general, since Y may depend on t. Now $d\omega$ is of the form $d\omega = \rho^* \, d\theta - dH \wedge dt$ so that $\alpha_s^* \, d\omega = \alpha_s^* \rho^* \, d\theta - \alpha_s^* \, dH \wedge dt$ (because $\alpha_s^* \, dt = dt$). Let ι_t denote the map of $T^*(M) \to T^*(M) \times R$ sending q into (q, t) so that $\iota_t^* \, d\omega = d\theta$. We can then rewrite (3.3) as $\beta_s = \rho \circ \alpha_s \circ \iota_0$. But $\iota_s^* \, d\omega = d\theta$. Thus it follows from (3.1)

that $d\theta = \iota_s^* \alpha_s^* \, d\omega = \iota_0^*(\alpha_s^* \rho^* \, d\theta - \alpha_s^* \, dH \wedge dt) = \iota_0^* \alpha_s^* \rho^* \, d\theta$ since $\iota_s^* \, dt = 0$. In other words, we have

$$\beta_s^* \, d\theta = d\theta, \tag{3.4}$$

a theorem first proved by Poincaré. If we apply (3.4) to $d\theta \wedge \cdots \wedge d\theta = \mathbf{\Omega}$ [the Liouville form on $T^*(M)$] we have

$$\beta_s^* \, \mathbf{\Omega} = \mathbf{\Omega}, \tag{3.5}$$

i.e., the "measure" given by $\mathbf{\Omega}$ is invariant. This was first discovered by Liouville.

Let us now consider the case where L (and hence H) does not depend on t.

THEOREM 3.1. *In the case that H is independent of t, the vector field Y on $T^*(M) \times R$ is of the form $Y = (V, \partial/\partial t)$ where V is the vector field on $T^*(M)$ determined by*

$$V \lrcorner \, d\theta = -dH. \tag{3.6}$$

In fact, in terms of local coordinates, (3.6) is equivalent to (2.21). But if $V = \sum X^i(\partial/\partial x^i) + \sum Y^i(\partial/\partial y^i)$ then $V \lrcorner \, d\theta = \sum Y^i \, dx^i - X^i \, dy^i = -dH$ by (3.6). By Theorem 6.1 of Chapter II, (3.5) uniquely determines V.

Now $\mathscr{L}_V H = V \lrcorner \, dH = -V \lrcorner \, V \lrcorner \, d\theta = 0$, thus

COROLLARY 3.1. (Conservation of energy.) *Under the hypothesis of Theorem* 3.1

$$\mathscr{L}_V H = 0, \quad (\text{and } \mathscr{L}_X H = 0), \tag{3.7}$$

thus the sets defined by $H =$ const on $T^*(M)$ [or on $T^*(M) \times R$] are invariant under the (local) flow generated by V (or by X).

Let h be a constant such that $H = h$ defines a submanifold, W_h, of $T^*(M)$. Corollary 3.1 implies that the vector field V is tangent to W_h (i.e., that $V_x \in T_x(W_h)$ for all $x \in W_h$). Let θ_h denote the form θ restricted to W_h. Then we have

$$\text{(3.8)} \qquad \begin{gathered} V \textit{ induces a vector field } V_h \textit{ on } W_h \textit{ and} \\ V_h \lrcorner \, d\theta_h = 0. \end{gathered}$$

According to Exercise 2.2, the solution curves of V_h can be obtained as extremals of a variational problem on W_h; namely that of minimizing $\int \theta$ among all curves of W_h. This, in itself, is not very interesting, as we are interested in curves of M, not of W_h. In certain cases, however, Equation (3.2) comes from a new variational problem in M. For example, let g be a Riemann metric on M and define L on $T(M) \times R$ by $L(v, t) = \frac{1}{2}g_x(v, v)$ if $v \in T_x(M)$. The map $\mathscr{L}$ is, on each tangent space, $T_x(M)$, the map of $T_x(M) \to T_x^*(M)$ induced by the non-singular metric g. Furthermore, $H \circ \mathscr{L} = L$. If C is an extremal of L, the conservation of energy thus implies that the $\|C'(t)\|^2$ and therefore $\|C'(t)\|$ is constant. Furthermore, the submanifold W_1 can be described by observing that $\mathscr{L}^{-1}(W_1)$ is the set of vectors of unit length of $T(M)$. If C is a curve of M and $\rho(\bar{C})$ lies in W_1, then

$$\int (\rho\bar{C})^*\theta = \int (\rho\bar{C})^* \sum y^i\, dx^i = \frac{1}{2}\int_a^b \left(\sum \frac{\partial L}{\partial \dot{x}^i}\frac{dx^i}{dt}\right) dt = \int_a^b \|C'(t)\|^2\, dt.$$

Since $\|C'(t)\|^2 = 1$ is a constant, this integral is equal to $\int_a^b \|C'(t)\|\, dt$ which is just the arc length of C. Thus if C minimizes arc length among all curves parametrized proportional to arc length then Exercise 2.2 shows that C is an extremal of the kinetic energy L—i.e., geodesic.

It turns out that if a curve, C, minimizes arc length, then under suitable "reparametrization" proportional to arc length it becomes an extremal of L.

EXERCISE 3.1 (Principle of least action). Prove this last assertion in the case that C is a differentiable curve whose tangent vector never vanishes.

Since there are some technical difficulties involved if we allow curves with a vanishing tangent vector, we shall postpone the proof of this assertion until Section 7.

We close this section by discussing the problem of deciding when the flow α_s is globally defined for the case of the energy of a Riemann manifold. According to the existence theorem on differential equations, if $\alpha_s(p)$ is defined for $0 \leqslant s < s_0$ and lies within some compact subset K of $T^*(M) \times R$, then $\alpha_s(p)$ is defined for $0 \leqslant s \leqslant s + \epsilon$ for some small ϵ. (The similar statement for $s_0 < 0$.) Thus, if α_s can not be globally defined, there must be some point $p \in T^*(M) \times R$ and an s_0 such that $\alpha_s(p)$ is defined for $0 \leqslant s < s_0$ (or $0 \geqslant s > s_0$) and the limit as $s \to s_0$ of $\alpha_s(p)$

does not exist. This can happen. To see this, we just observe that if $\alpha_s(p)$ is to tend to a limit, certainly $\pi\alpha_s(p)$ must. But $\pi\alpha_{s_0}(p)$ is a geodesic of M. If we let M' be the submanifold of M obtained by throwing away a point of this geodesic, i.e., $M' = M - q$ with $q = \pi\alpha_{s_0}(p)$ then $\pi\alpha_s(p)$ does not tend to any limit in M' as $s \to s_0$. In other words, if M is a manifold for which α_s is globally defined, $M - q$ for any q no longer has this property. It seems intuitively clear, that the reason α_s is not globally defined on $T^*(M - q) \times R$, is that $M - q$ is "not complete." We will now indeed show that every Riemann manifold, M, is a metric space relative to a natural metric and if this metric space is complete, then α_s is globally defined. (It should be remarked that the question of global definability of α depends on the Riemann metric and not merely on the differentiable structure of M. For instance, if M is the sphere, with its spherical metric, α_s is not globally defined for $M - q$. But $M - q$ is diffeomorphic to the Euclidean plane for which α_s is globally defined.)

THEOREM 3.2. *Let M be a Riemann manifold. Define $d(x_1, x_2)$ by*

$$d(x_1, x_2) = \text{g.l.b.} \int_a^b \|C'(t)\|\, dt$$

for all piecewise differentiable curves C with $C(a) = x_1$ and $C(b) = x_2$. Then d is a metric on M. If M is a complete metric space relative to this metric then α_s is globally defined on $T^*(M) \times R$.

We first check that d is a metric. It is immediate that $d(x_1, x_2) = d(x_2, x_1)$. Also, if $C(a) = x_1$, $C(b) = x_2$, and $D(b) = x_2$, $D(c) = x_3$ then C and D fit together to form a piecewise differentiable curve joining x_1 to x_3. Thus

$$d(x_1, x_3) \leqslant \int_a^b \|C'(t)\|\, dt + \int_b^c \|D'(t)\|\, dt$$

or

$$d(x_1, x_3) \leqslant d(x_1, x_2) + d(x_2, x_3).$$

What is left to show is that

$$d(x_1, x_2) = 0 \quad \text{implies} \quad x_1 = x_2.$$

Let U be a coordinate neighborhood of x_1 with coordinates $x^1, \ldots, x^n$ such that $U = h^{-1}(B_2^n)$, and $x_1 = h^{-1}(0)$. For every point $y \in h^{-1}(B_1^n)$, g_y is positive definite. Since $h^{-1}(B_1^n)$ is compact, we can find a sufficiently small ϵ such that $\|v_y\|^2 > \epsilon\,(\langle v_y \,|\, dx_y^1\rangle^2 + \cdots + \langle v_y \,|\, dx_y^n\rangle^2)$ for all

$y \in h^{-1}(B_1^n)$. In particular, let C be any curve joining x_1 to x_2, i.e., $C(a) = x_1$ and $C(b) = x_2$. Let t_1 be the least upper bound of those t_0 satisfying $a \leqslant t_0 \leqslant b$ and $C(t) \in h^{-1}(U)$ for all t with $a \leqslant t < t_0$. Then

$$\int_a^b \|C'(t)\| > \int_a^{t_2} \|C'(t)\|\, dt \qquad \text{for any } t_2 \leqslant t_1.$$

But

$$\int_a^{t_2} \|C'(t)\|\, dt > \epsilon \int_a^{t_2} |\langle C'(t), dx^i_{C(t)}\rangle|\, dt > \epsilon \left|\int_a^{t_2} \langle C'(t), dx^i_{C(t)}\rangle\, dt\right|$$

$$= \epsilon\, |x^i(t_2)| \qquad \text{for all } i.$$

If $x_1 \neq x_2$ any curve joining x_1 to x_2 must either leave $h^{-1}(B_1^n)$ [in which case $|x^i(t_2)|$ becomes at least 1/2 for some i] or x_2 lies in $h^{-1}(B_1^n)$ and $|x^i(t_2)| \geqslant |x^i(x_2)|$ for some t_2. In any event, this implies that $d(x_1, x_2) > 0$.

Now that we know that d is a metric, we proceed to the proof of the second half of Theorem 3.2. Assume that M is complete (i.e., complete in the d metric). Let $p \in T^*(M) \times R$ be such that $\alpha_s(p)$ is defined for $0 \leqslant s < s_0$. We assert that the curve $\pi(\alpha_s(p))$ converges as $s \to s_0$. In fact, s is a parameter on $\pi(\alpha_s(p))$ proportional to arc length. Thus $d(\pi(\alpha_s(p))), \pi(\alpha_{s'}(p))) \leqslant k|s - s'|$ for $0 \leqslant s, s' < s_0$. By completeness, $\pi(\alpha_s(p))$ tends to some point $q \in M$. If $s_0 - s$ is small enough, $\pi\alpha_s(p)$ will lie in some coordinate neighborhood about q (with compact closure). If $dx^1, \ldots, dx^n$ are coordinates for sufficiently large k we have

$$k\|v\|^2 > (\langle v\,|\,dx_y^1\rangle^2 + \cdots + \langle v\,|\,dx_y^n\rangle^2),\ \pi(v) = y.$$

Now if v is the tangent vector to $\pi\alpha_s(p)$ then v has constant length and we have $|\langle v\,|\,dx^i\rangle| < K$ for some large K. Thus $\alpha_s(p)$ lies in a compact neighborhood in $T^*(M) \times R$ for $s < s_0$ (and s close to s_0). Thus α_s can be continued past s_0, proving Theorem 3.2.

We will see in Section 7, that the converse is also true: If α_s is globally defined, then M is complete.

4. Sufficient conditions. In Section 2 we obtained necessary conditions for a curve to be a minimum (Theorem 2.1). In this section we will give sufficient criteria for an extremal to be a minimum. For this purpose we make some elementary observations. Let U be an open subset of M, let I be a subinterval of R and let φ be a map of $U \times I \to T(M) \times R$ such that $(\pi \times \text{id}) \circ \varphi(x, t) = (x, t)$ for $x \in U$ and $t \in I$. For each fixed t, $\varphi(\cdot, t)$ gives a vector field on U; thus φ represents a "system of differential equations" on U. Suppose, furthermore that

$$(4.1)\quad L \circ \varphi = 0 \quad \text{and} \quad L > 0 \quad \text{in} \quad (\pi \times \text{id})^{-1}(U \times I) - \varphi(U \times I).$$

LEMMA 4.1. *Let U, I, L, and φ be as above, $[a, b] \subset I$ and $C: [a, b] \to M$ be a solution curve of φ lying in U, i.e., let*

a) $C([a, b]) \subset U$

and

b) $\tilde{C}(t) = \varphi(C(t), t)$.

Then C is a minimum curve for L. In fact, if D is any piecewise differentiable curve $D: [a, b] \to U$ with $D(a) = C(a)$ then

$$I(C) < I(D) \qquad \textit{unless } C = D. \tag{4.2}$$

The proof is completely trivial. In fact, according to a) and (4.1), $I(C) = 0$. If $D \neq C$, then for at least one $t \in [a, b]$, $\tilde{D}(t) \neq \tilde{C}(t)$. Now $\tilde{D}(t) \in (\pi \times \mathrm{id})^{-1}(U \times I)$ so $L(\tilde{D}(t)) > 0$ and therefore $I(D) > 0$.

LEMMA 4.2. *Suppose we replace, in Lemma 4.1, the second part of condition (4.1) by $L > 0$ in $W - \varphi(U \times I)$ where W is some neighborhood of $\varphi(U \times I)$ in $(\pi \times \mathrm{id})^{-1}(U \times I)$. Then we can conclude (4.2) for all $D: [a, b] \to U$ with $D(a) = C(a)$ such that $\tilde{D}(t) \in W$ (when $\tilde{D}$ is defined).*

In fact, the argument is the same.

The conditions imposed on L in the last two lemmas are not, in general, fulfilled in a non-trivial way. For instance, if L is the energy of a Riemann manifold, the only map φ giving (4.1) assigns to each $x \in U$ the zero tangent vector at x. The integral curves C then become the constant maps which surely satisfy (4.2) but this isn't very exciting. Nevertheless, Lemmas 1 and 2 will give us useful information. To see this, we make the following remark: Let S be a function defined on $U \times I$. Define the function $\tilde{S}$ on $\pi^{-1}(U \times I)$ by

$$\tilde{S}(v, t) = v(S) + \frac{\partial S}{\partial t} \tag{4.3}$$

or in local coordinates

$$\tilde{S}(x^1, \ldots, x^n, \dot{x}^1, \ldots, \dot{x}^n, t) = \sum \frac{\partial S}{\partial x^i} \dot{x}^i + \frac{\partial S}{\partial t}. \tag{4.4}$$

Then

$$\int_a^b \tilde{S}(\tilde{D}(t)) = S(D(b), b) - S(D(a), a) \tag{4.5}$$

for any differentiable (and hence, for any piecewise differentiable) curve $D:[a, b] \to U$. Therefore, we can assert

LEMMA 4.3. *Let* $\varphi: U \times I \to T(M) \times R$ *be such that* $(\pi \times \mathrm{id}) \circ \varphi = \mathrm{id}$, *and* C *satisfies* (a) *and* (b) *of Lemma* 4.1. *Suppose that there is a function* S *defined on* $U \times I$ *such that*

$$(4.6) \qquad (L - \tilde{S}) \circ \varphi = 0$$

and

$$(4.7) \quad (L - \tilde{S})(p) > 0 \qquad \text{for } p \in (\pi \times \mathrm{id})^{-1}(U \times I) - \varphi(U \times I).$$

Then for any curve $D:[a, b] \to U$ (*where* $[a, b] = I$), *with* $D(a) = C(a)$ *we have*

$$(4.8) \qquad I(C) < I(D) - S(D(b), b) + S(C(b), b) \ \text{unless } D = C.$$

If (4.7) *holds only for* $p \in W - \varphi(U \times I)$, *where* W *is a neighborhood of* $\varphi(U \times I)$ *in* $(\pi \times \mathrm{id})^{-1}(U \times I)$, *then* (4.8) *is true for those* D *satisfying* $\tilde{D}(t) \in W$ *for* $t \in [a, b]$.

The proof is the same as that of Lemma 4.1 using (4.5).

If $D(b) = C(b)$, and $D(a) = C(a)$, then (4.8) reduces to (4.2). This means that C is a strict local minima to the fixed end point problem. [It is a strong or weak minimum according to whether (4.7) holds for all $p \in (\pi \times \mathrm{id})^{-1}(U \times I) - \varphi(U \times I)$ or only for those in W.] Thus, in order to prove that a given extremal, C, is a minimum, it is sufficient to find a φ such that C is an integral curve of φ and an S such that (4.6) and (4.7) hold. Before discussing when such φ and S exist we make one further observation.

We will say that L is positive definite if, for all points, p, of $T(M) \times R$ the matrix

$$(L_{\dot{x}^i\dot{x}^j})(p) = \frac{\partial^2 L}{\partial \dot{x}^i \, \partial \dot{x}^j}(p)$$

is positive definite. The matrix $(L_{\dot{x}^i\dot{x}^j})(p)$ depends on the local coordinate system $x, \ldots, x^n$ about $\pi(p)$. However, the condition of positive definiteness is independent of coordinates.

This can be proved directly by looking at the formula for changes of coordinates. We will prove it by giving the quadratic form $\sum L_{\dot{x}^i\dot{x}^j} j\dot{x}^i\dot{x}^j$ an invariant definition. We do this by observing that $\dot{x}^1, \ldots, \dot{x}^n$ are *linear* coordinates on the vector space $T_{\pi(p)}(M)$. Let f be function on a vector space V and $\dot{x}^1, \ldots, \dot{x}^n$ coordinates of V relative to some basis, $e_1, \ldots, e_n$. We want to attach an invariant meaning to the quadratic form whose matrix, relative to this basis is $(\partial^2 f/\partial \dot{x}^i\, \partial \dot{x}^j)$. Notice that any vector v_1 induces a vector field $l^{-1}(v_1)$ on V, (assigning to each $v \in V$ the tangent vector $l_v^{-1}(v_1)$). If $v_1 = \sum \xi_1^i e_i$ then $l^{-1}(v_1) = \sum \xi_1^i(\partial/\partial \dot{x}^i)$. For $p \in V$, define the bilinear forms $B_p(f)$ by $B_p(f)(v_1, v_2) = l_p^{-1}(v_2)\, \langle l^{-1}(v_1)\,|\, df\rangle$. In terms of coordinates $\langle l^{-1}(v_1)\,|\,df\rangle = \sum (\partial f/\partial \dot{x}^i)\xi_1^i$. If $v_2 = \sum \xi_2^i e_i$ then

$$l_p^{-1}(v_2)\, \langle l^{-1}(v_1)\,|\,df\rangle = \sum \frac{\partial^2 f}{\partial \dot{x}^i \partial \dot{x}^j}\, \xi_1^i \xi_2^j.$$

Thus the bilinear forms $B_p(f)$ are symmetric and their matrix is given by $(\partial^2 f(p)/\partial \dot{x}^i\, \partial \dot{x}^j)$. This proves that the condition that $(L_{\dot{x}^i\dot{x}^j})(p)$ be positive definite is independent of local coordinates.

LEMMA 4.4. *If L is positive definite, then condition* (4.6) *and*

$$\frac{\partial}{\partial \dot{x}^i}(L - \tilde{S})\circ \varphi = 0 \tag{4.9}$$

implies that (4.7) *hold for all $p \in W$ for some neighborhood W of $\varphi(U \times I)$. If L is the energy of a Riemann metric*[3] *then* (4.9) *implies* (4.7). [*Condition* (4.9), *although expressed in terms of local coordinates is easily seen to be independent of choice of coordinates. It says that if v is a vector tangent to $T_x(M)$ (more precisely to $(\pi \times \mathrm{id})^{-1}(x, t)$) at $\varphi(x, t)$, then $v(L - \tilde{S}) = 0$.*]

In fact, the function $\tilde{S}$ depends linearly on the $\dot{x}^i$ so that $\partial^2\tilde{S}/\partial\dot{x}^i\partial\dot{x}^j = 0$. Thus if $p = \varphi(x, t)$ the function $(L - \tilde{S})$ has the property that $(\partial(L - \tilde{S})/\partial \dot{x}^i)(p) = 0$, and that

$$\frac{\partial^2}{\partial \dot{x}^i\, \partial \dot{x}^j}(L - \tilde{S})(p) = (L_{\dot{x}^i\dot{x}^j})(p)$$

is a positive definite matrix. This means that $L - \tilde{S}$, as a function of the $\dot{x}^i$ achieves a strict minimum at p. But this minimum value is zero by (4.6). In other words, for values of the $\dot{x}^i$ close enough to $\dot{x}^i(p)$, $(L - \tilde{S}) > 0$. Applying this to all $p \in \varphi(U \times I)$ we get the first half of Lemma 4.4. If L is the energy of a Riemann metric the L is clearly positive definite. Furthermore, $L - \tilde{S}$ is a quadratic function on $T_x(M)$ for fixed $x \in U$, and $t \in I$. Using the Taylor expansion for $L - \tilde{S}$ about $\varphi(x, t)$, it is immediate that (4.6) and (4.9) imply (4.7).

---

[3] More generally, if L is quadratic (not necessarily homogeneous) in the $\dot{x}^i$

In the case of positive definite Lagrangians we are looking for maps φ for which there is an S satisfying (4.6) and (4.9). It is now convenient to apply the Legendre transformation to the whole setup and pass to $T^*(M) \times R$.[4] If φ is a map of $U \times I \to T(M) \times R$ then $\psi = \mathscr{L} \circ \varphi$ maps $U \times I$ into $T^*(M) \times R$. That is, we have the diagram:

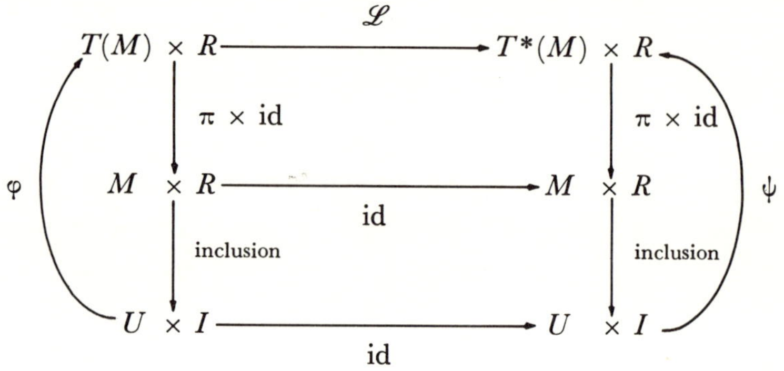

Given φ we know ψ and conversely. How do the conditions (4.6) and (4.9) read in terms of ψ? Let $x^1, \ldots, x^n, \dot{x}^1, \ldots, \dot{x}^n, t$ be local coordinates on $T(M) \times R$ and $x^1, \ldots, x^n, y^1, \ldots, y^n, t$ be the corresponding local coordinates on $T^*(M) \times R$. Let

$$\dot{x}^i(x, t) = \dot{x}^i \circ \varphi(x, t)$$

and

$$y^i(x, t) = y \circ \psi(x, t)$$

so that

$$y^i(x, t) = \frac{\partial L}{\partial \dot{x}^i} (\varphi(x, t)).$$

Then

$$\frac{\partial(L - \tilde{S})}{\partial \dot{x}^i} \circ \varphi = \frac{\partial L}{\partial x^i} \circ \varphi - \frac{\partial S}{\partial x^i}$$

by (4.4). Thus (4.9) becomes

(4.10) $$y^i(x, t) = \frac{\partial S}{\partial x^i}.$$

[4] Since we are going to be interested only in the neighborhood of a curve $\tilde{C}$ we may assume that $\mathscr{L}$ is a diffeomorphism.

Also

$$L - \tilde{S} = L - \sum \frac{\partial S}{\partial x^i} \dot{x}^i - \frac{\partial S}{\partial t} = L - \sum y^i \dot{x}^i - \frac{\partial S}{\partial t}.$$

But $L - \sum y^i \dot{x}^i = -H$ so that (4.6) becomes

$$\frac{\partial S}{\partial t} + H \circ \psi = 0. \tag{4.11}$$

We can regard (4.10) and (4.11) as a single partial differential equation

$$\frac{\partial S}{\partial t} + H\left(x^1, \ldots, x^n, \frac{\partial S}{\partial x^1}, \ldots, \frac{\partial S}{\partial x^n}, t\right) = 0. \tag{4.12}$$

This equation is known as the Hamilton-Jacobi equation. Rather than to look directly for a solution, S, of this equation and define $y^i(x, t)$ by (4.10), it is more convenient to proceed somewhat differently. We observe that (4.10) and (4.11) can be written more succinctly as

$$\psi^* \omega = dS. \tag{4.13}$$

In fact, $\psi^* \omega = \sum y^i(x, t)\, dx^i - (H \circ \psi)\, dt$ and

$$dS = \sum \frac{\partial S}{\partial x^i} dx^i + \frac{\partial S}{\partial t} dt.$$

Comparing the coefficients of dx^i and dt gives (4.10) and (4.11). We are therefore looking for maps ψ for which there is a function S such that $\psi^* \omega = dS$. There will certainly be such an S if $\psi^*\, d\omega = 0$ and $H^1(U \times I) = 0$. In other words

LEMMA 4.5. *Let L be positive definite. Let $H^1(U \times I) = 0$ and $\psi: U \times I \to (\pi \times \mathrm{id})^{-1} U \times I$ satisfy $(\pi \times \mathrm{id})\psi = \mathrm{id}$ and*

$$\psi^*\, d\omega = 0. \tag{4.14}$$

Then if S is defined by (4.13) *and $\varphi = \mathscr{L}^{-1}\psi$, then φ and S satisfy* (4.6) *and* (4.7) *for all p in some neighborhood of $\varphi(U \times I)$. If L is the energy of a Riemann metric, then* (4.7) *holds for all $p \in (\pi \times \mathrm{id})^{-1}(U \times I)$.*

In our quest for a map ψ satisfying (4.13), we will split the problem up into two parts.

i) to find a submanifold N of $T^*(M) \times R$ satisfying $\iota^*(\omega) = dS^*$ where ι is the injection of $N \to T^*(M) \times R$ and S^* a suitable function on N;

ii) to find a submanifold N satisfying i) such that $\pi \times \text{id}$ restricted to N is a diffeomorphism of N onto $U \times I$, (for suitable $U \subset M$ and $I \subset R$).

Let us first examine condition i). In view of ii) we are interested in $(n + 1)$-dimensional submanifolds N.

LEMMA 4.6. *Let N_0 be an n-dimensional submanifold of $T^*(M) \times R$. Suppose that the Euler field, $Y = \mathscr{L}_* X$, is nowhere tangent to N_0 and $\iota_0^*(d\omega) = 0$ where ι_0 is the injection of $N_0 \to T^*(M) \times R$. Let N be the submanifold swept out, locally, by N under the flow, α, generated by Y. Then $\iota^*(d\omega) = 0$ if ι is the injection of $N \to T^*(M) \times R$.*

What we have to show is that for every pair of vectors Z_1 and Z_2 tangent to N at $p \in N$, $\langle Z_1 \wedge Z_2 \mid d\omega_p \rangle = 0$.

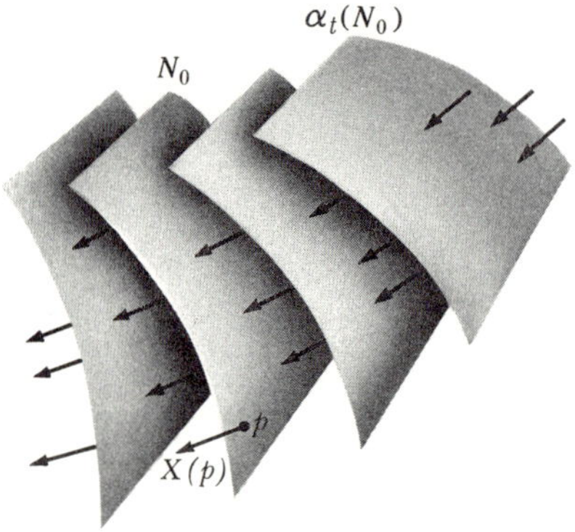

Now the tangent space $T_p(N)$ is spanned by $\alpha_{t^*}(T_{\alpha_{-t}(p)}(N_0))$ and $Y(p)$ where $\alpha_{-t}(p) \in N_0$. If Z_1 and Z_2 both lie in $\alpha_{t^*}(T_{\alpha_{-t}(p)}(N_0))$ then, since $\alpha_{t^*}\, d\omega = d\omega$,

$$\begin{aligned} \langle Z_1 \wedge Z_2 \mid d\omega_p \rangle &= \langle \alpha_{t^*} Z_1 \wedge \alpha_{t^*} Z_2 \mid d\omega_p \rangle \\ &= \langle Z_1 \wedge Z_2 \mid d\omega_{\alpha_{-t}(p)} \rangle \\ &= 0 \qquad\qquad \text{where } Z_1, Z_2 \in T_{\alpha_{-t}(p)}(N_0). \end{aligned}$$

By linearity, the only remaining case is $\langle Y \wedge Z_1 | d\omega_p \rangle = \langle Z_1 | Y(p) \lrcorner\, d\omega \rangle = 0$, by (2.21). This proves Lemma 4.6.

We should remark that since $\langle Y | dt \rangle = 1$, *any simultaneous submanifold, i.e., any manifold on which* $t =$ const, *is nowhere tangent to* Y.

As an example of an N_0 satisfying the hypothesis of Lemma 4.6, let $x^1, \ldots, x^n$ be a coordinate system in a neighborhood U about a point p and let $v^* = \sum y_0^i \, dx_p^i \in T_p^*(M)$. Let N_0 be the submanifold of $T_p^*(M) \times R$, consisting, in local coordinates, of all points of the form

$$(x^1, \ldots, x^n, y_0^1, \ldots, y_0^n, t_0),$$

i.e., defined by the equation $y^i = y_0^i$ and $t = t_0$. Then $\iota_0^* (d\omega) = \iota_0^*(\sum dy^i \wedge dx^i - dH \wedge dt) = 0$ since $\iota_0^* (dy^i) = \iota_0^* (dt) = 0$. If we choose U to be diffeomorphic to E^n, then $N = N_0 \times I$ is diffeomorphic to E^{n+1}, therefore $H^1(N) = 0$. By Lemma 4.6, $\iota^* (d\omega) = 0$ or $\iota^* (\omega) = dS^*$ where S^* is some function on N. Furthermore, $\pi \times \mathrm{id}$ is a diffeomorphism of some neighborhood of $(q, t_0) = \iota^{-1} \circ (\pi \times \mathrm{id})^{-1} (p, t_0)$ onto some neighborhood of (p, t_0) in M. To prove this, it suffices to check $(\pi \times \mathrm{id})_* \circ \iota_*$ is a non-singular map of $T_{(q,t_0)} (N) = T_q(N_0) + T_{t_0}(R)$ onto $T_{(p,t_0)}(M \times R)$. But π is a diffeomorphism of N_0 onto U and $\iota_*(\partial/\partial t)_{(q,t_0)} = X(\iota(q, t_0))$ thus $(\pi \times \mathrm{id})_* \, \iota_*(\partial/\partial t)_{(q,t_0)}$ has a non-zero component relative to $\partial/\partial t$. This implies that $(\pi \times \mathrm{id})_* \circ \iota_*$ is non-singular. Thus N satisfies conditions i) and ii). Therefore, if we define $\psi = \iota \circ \eta$, where η is the inverse of $(\pi \times \mathrm{id}) \circ \iota : N \to M$, and $S = \psi^* S^*$, then ψ and S satisfy the conditions of Lemma 4.5. Since v^* is arbitrary, we have

THEOREM 4.1. *Let L be a positive definite Lagrangian and C an extremal of L with* $C(t_0) = p$. *Then there exists a sufficiently small neighborhood U about* p_0, *an interval I about* t_0 *and a function S defined on* $U \times I$ *such that any curve B in a sufficiently small* C^1 *neighborhood of C, with* $B(t) \in U$ *for all* $t \in I$, *and*

$$B(t_0) = p \tag{4.15}$$

satisfies

$$I(C) < I(B) + S(C(t), t) - S(B(t), t). \tag{4.16}$$

In particular, C is a strict weak local minimum on any interval $[t_0, t]$ *where* $t \in I$. *If L is the energy of a Riemann manifold,*[3] *then* (4.15) *holds for all B such that* $B(t) \in U$ *for all* $t \in I$, *so that C is a strict strong local minimum for* ψ *on* $[t_0, t]$.

In fact, since the v^* in the discussion preceding the theorem is arbitrary, we can take $v^* = \bar{C}(t_0)$. If we then define φ to be $\varphi = \mathscr{L}^{-1}\psi$, it follows from the previous discussion that the hypotheses of Lemma 4.3 are satisfied, implying Theorem 4.1. A little more care in our choice of coordinates gives

THEOREM 4.2. *Let L be a positive definite Lagrangian and C an extremal of L with $C(t_0) = p \in K$, and $\langle v, \bar{C}(t_0)\rangle = 0$ for all $v \in T(K)$, where K is a submanifold of M. There exists an interval I about t_0, such that C is a local minimum relative to (N_1, N_2) where $N_1 = K$ and $N_2 = C(t)$ for any $t \in I$.*

In fact, choose coordinates $x^1, \ldots, x^n$ such that K is given by $x^{k+1} = \cdots = x^n = 0$. The coordinates of $v^* = \bar{C}(t_0)$ will then be of the form $(x_0^1, \ldots, x_0^n, 0, \ldots, 0, y_0^{k+1}, \ldots, y_0^n, t_0)$. In particular, $\psi((K \cap U) \times t_0)$ is the set of all points of the form $(x^1, \ldots, x^k, 0, \ldots, 0, \ldots, 0, y_0^{k+1}, \ldots, y_0^n, t_0)$ so that if j is the injection of $K \cap U \to M$, $j^*\psi^*(\omega) = 0$. Thus S *is constant on* $j(K \cap U)$. Thus, as before, Theorem 4.2 follows from Lemma 4.3, taking $a = t$ and $b = t_0$ in the lemma.

5. Conjugate and focal points, Jacobi's condition. In the last section we showed that for positive definite L an extremal minimizes "locally." That is, if K is a submanifold and C is an extremal with $C(t_0) \in K$ and $\langle v, \bar{C}(t_0)\rangle = 0$ for all $v \in K$, then $C(t)$ provides a minimum for t close to t_0. The proof we gave, however, does not tell on how large an interval C actually minimizes. It can indeed happen that C does minimize for $t_0 \leqslant t \leqslant t_1$ but not for any larger interval. The purpose of this section is to describe the point t_1.

For this purpose, let us introduce some notation. If K is a submanifold of M, we will denote by $T^0(K)$ the submanifold of $T^*(M)$ consisting of those points $v^* \in T^*(M)$ for which $\pi(v^*) \in K$ and $\langle v | v^*\rangle = 0$ for all $v \in T_{\pi(v^*)}(K)$. It is easy to see that $T^0(K)$ is indeed a submanifold and has dimension n. If K is a point, p, then $T^0(K)$ is just $T_p^*(M)$. If K is given, in local coordinates by $x^{k+1} = \cdots = x^n = 0$, then $T^0(K)$ is given by $x^{k+1} = \cdots = x^n = y^1 = \cdots = y^k = 0$. Therefore, if $\iota_0 : T^0(K) \to T^*(M) \times R$ by sending $T^0(K) \to T^0(K) \times t_0$, then $\iota_0^*\omega = 0$. We might therefore take $T^0(K)$ to be our N_0 in Lemma 4.6. The one trouble is, of course, that π is quite singular on $T^0(K)$, mapping it into K, the difference in dimension being $n - k$. Nevertheless, what *is* true is

LEMMA 5.1. *Let K be a submanifold of M and let $\gamma : T^0(K) \times R \to T^*(M) \times R$ be the map defined by*

$$\gamma(v^*, t) = \alpha_t(v^*, 0).$$

For each $v^* \in T^0(K)$ *there is an* $\epsilon > 0$ *such that* $(\pi \times \mathrm{id})_* \circ \gamma_*$ *is nonsingular for* $0 < |t| < \epsilon$.

Definition 5.1. *The points* $t \neq 0$ *for which* $(\pi \times \mathrm{id})_* \circ \gamma_*$ *is singular at* (v^*, t) *are called focal points of K along C, where C is the extremal with* $\bar{C}(0) = v^*$. *If K is a point, focal points are called conjugate points.*

Lemma 5.1 says that t is not a focal point of C if $|t|$ is small. A focal point is, roughly speaking, a point where "two infinitesimally near extremals starting transversal to K, meet." For instance, if K is a curve in the plane and L is energy of the Euclidean metric, the extremals under consideration are just straight lines perpendicular to K. Two "infinitesimally close" such lines will "meet" at the so-called center of curvature of K.

Thus q is the focal point of K along C. The picture seems to indicate (and we shall give a proof soon enough) that for x on C, C locally minimizes the distance from x to K provided that x lies on the interval (p, q), and does not minimize locally if x lies beyond q. We shall indeed prove that geodesics minimize up to the first focal point.

There are no conjugate points in the plane since the extremals emanating from a fixed point spread out. However, let us look at the surface of the (unit) sphere. Here the extremals are great circles. If p is the north pole, all extremals leaving p meet at the south pole. Again, these great circles minimize distance until they have gone through the south pole. Once they have passed the south pole they *don't even locally* minimize distance. The reader should give a proof of this fact following the figure on page 184.

Before proceeding to the proof of Lemma 5.1, we first discuss the so-called "equations of variation" of a system of differential equations, introduced at the end of Chapter III.

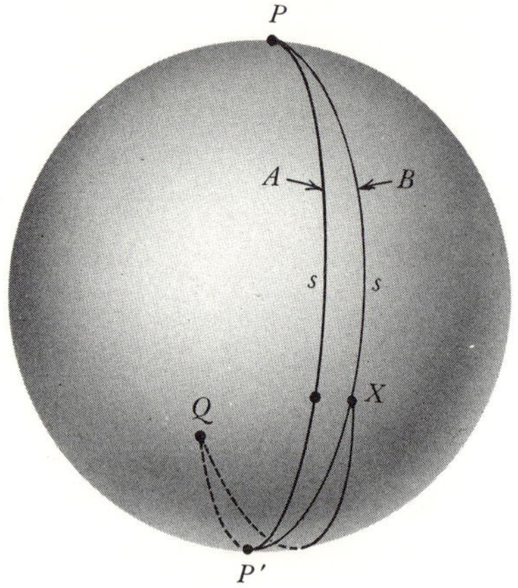

Let N be a differentiable manifold and Z a vector field on N. Then Z induces a vector field $T(Z)$ on $T(M)$. It is defined as follows: Let α_t be the local one parameter group of diffeomorphisms of N generated by Z. That is, for each point p of N there is a neighborhood U and an interval I about 0 in R where a family of diffeomorphisms $\alpha_t : U \to N$ (for $t \in I$) is defined, (cf. Section 8 of Chapter II). These α_t induce maps α_{t*} of $T(U) \to T(N)$. It is easy to check that these maps α_{t*} define a local one parameter group of transformations of $T(N)$ and therefore determine a vector field which we call $T(Z)$. By definition, $\pi\alpha_{t*} = \alpha_t$ so that $\pi_*(T(Z)_v) = Z_{\pi(v)}$ for any $v \in T(N)$ and the integral curves of $T(Z)$ project onto the integral curves of Z.

EXERCISE 5.1. The vector field $T(Z)$ measures the behavior of "infinitesimally near" integral curves of Z. More precisely, let K map the rectangle $0 < \alpha < \epsilon$, $a < t < b$ into N such that $K(\alpha, \cdot)$ is an integral curve of Z for each α. Show that the curves $C_\alpha : [a, b] \to T(N)$ defined by $C_\alpha(t) = K_* \, (\partial/\partial\alpha)_{(\alpha,t)}$ are integral curves of $T(Z)$.

Let $z^1, \ldots, z^n$ be a coordinate system on N, and let the flow α_t be given by

$$z^i \circ \alpha_t = \varphi^i(z^1, \ldots, z^n, t).$$

Then α_{t^*} is given by

$$\alpha_{t^*}\left(\frac{\partial}{\partial z^j}\right)_p = \sum_j \frac{\partial \varphi^i}{\partial z^j}\left(\frac{\partial}{\partial z^i}\right)_{\alpha_t(p)}.$$

Thus if $z^1, \ldots, z^n, \dot{z}^1, \ldots, \dot{z}^n$ are the corresponding local coordinates on $T(N)$, the map α_{t^*} is given by

$$z^i \circ \alpha_{t^*} = \varphi^i(z^1, \ldots, z^n, t)$$

$$\dot{z}^i \circ \alpha_{t*} = \sum_j \dot{z}^j \frac{\partial \varphi^i}{\partial z^j}.$$

Taking the derivative of these equations with respect to t at $t = 0$ gives

$$\text{(5.1)} \qquad \frac{dz^i}{dt} = Z^i, \qquad \frac{d\dot{z}^i}{dt} = \sum \dot{z}^j \frac{\partial Zi}{\partial z^j}$$

if $Z = \sum Z^i\,(\partial/\partial z^i)$. In other words, the vector field $T(Z)$ is given by

$$\text{(5.2)} \qquad T(Z) = \sum Z^i \frac{\partial}{\partial z^i} + \sum\left(\sum \dot{z}^j \frac{\partial Z^i}{\partial z^j}\right)\frac{\partial}{\partial \dot{z}^i}.$$

Our interest in the previous discussion is the case where $N = T^*(M) \times R$ and Z is the Euler vector field, $Y = \mathscr{L}_* X$. Thus, we obtain a vector field, $T(Y)$, on $T(T^*(M) \times R)$ which we shall call the *Jacobi vector field*. Let $x^1, \ldots, x^n$ be local coordinates on M. Let $x^1, \ldots, x^n$, $y^1, \ldots, y^n, t$ be the corresponding coordinates on $T^*(M) \times R$ and $x^1, \ldots$, $x^n, y^1, \ldots, y^n, t, \dot{x}^1, \ldots, \dot{x}^n, \dot{y}^1, \ldots, \dot{y}^n, \dot{t}$ coordinates on $T(T^*(M) \times R)$. Then $T(Y)$ is of the form

$$\begin{aligned} T(Y) = {} & \sum \frac{\partial H}{\partial y^i}\frac{\partial}{\partial x^i} - \sum \frac{\partial H}{\partial x^i}\frac{\partial}{\partial y^i} + \frac{\partial}{\partial t} \\ & + \sum\left(\sum \frac{\partial^2 H}{\partial y^i \partial x^j}\dot{x}^j + \frac{\partial^2 H}{\partial y^i \partial y^j}\dot{y}^j + \frac{\partial^2 H}{\partial y^i \partial t}\right)\frac{\partial}{\partial \dot{x}^i} \\ & - \sum\left(\sum \frac{\partial^2 H}{\partial x^i \partial x^j}\dot{x}^j + \frac{\partial^2 H}{\partial y^j \partial x^i}\dot{y}^j + \frac{\partial^2 H}{\partial x^i \partial t}\right)\frac{\partial}{\partial \dot{y}^i}\ . \end{aligned}$$

In particular, let $V \in T_u(T^*(M) \times R)$, given by

$$V = \sum \xi_0^i \left(\frac{\partial}{\partial x^i}\right)_{(u,0)} + \sum \eta_0^i \left(\frac{\partial}{\partial y^i}\right)_{(u,0)} + \tau_0 \left(\frac{\partial}{\partial t}\right)_{(u,0)}$$

be a tangent vector at some point $(u, 0) \in T^*(M) \times R$. Then the components of

$$\alpha_{s*}(V) = \sum \xi^i(s) \left(\frac{\partial}{\partial x^i}\right)_{\alpha_s(u,0)} + \eta^i(s) \left(\frac{\partial}{\partial y^i}\right)_{\alpha_s(u,0)} + \tau(s) \left(\frac{\partial}{\partial t}\right)_{\alpha_s(u,0)}$$

satisfy the differential equation

$$\begin{aligned} \frac{d\xi^i}{ds} &= \sum \left(H_{y^i x^j}\xi^j + H_{y^i y^j}\eta^j + H_{y^i t}\tau\right) \\ \frac{d\eta^i}{ds} &= -\left(\sum H_{x^i x^j}\xi^j + H_{x^i y^j}\eta^j + H_{x^i t}\tau\right) \qquad (5.4) \\ \frac{d\tau}{ds} &= 0, \end{aligned}$$

where $H_{y^i x^j}$, for instance, is the *function of s* defined by

$$\begin{aligned} H_{y^i x^j}(s) &= \frac{\partial^2 H}{\partial y^i \, \partial x^j}(x^1(s), \ldots, x^n(s), y^1(s), \ldots, y^n(s), t(s)) \\ &= \frac{\partial^2 H}{\partial y^i \, \partial x^j} \circ \bar{C}(s) \end{aligned}$$

where C is the extremal with $(\bar{C}(0)u) = (u, 0)$; and the same for the other functions $H_{x^i x^j}$, $H_{y^i y^j}$, $H_{y^i t}$ and $H_{x^i t}$ occurring on the right of (5.4). Now (5.4) implies that $\tau = 0$, if $\tau_0 = 0$. Thus if V is tangent to a simultaneous manifold (*i.e.*, $\tau_0 = 0$) so is $\alpha_{s*}(V)$, and the components of $\alpha_{t*}(V)$ satisfy the differential equations

$$\begin{aligned} \frac{d\xi^i}{dt} &= \sum \left(H_{y^i x^j}\xi^j + H_{y^i y^j}\eta^j\right) \\ \frac{d\eta^i}{dt} &= -\sum \left(H_{x^i x^j}\xi^j + H_{x^i y^j}\eta^j\right). \qquad (5.5) \end{aligned}$$

Notice that Equations (5.5) are *linear* differential equations which can be written as

$$\frac{d}{dt}\begin{pmatrix} \xi^1 \\ \vdots \\ \xi^n \\ \eta^1 \\ \vdots \\ \eta^n \end{pmatrix} = \left(\begin{array}{c|c} A & B \\ \hline C & A^t \end{array}\right) \begin{pmatrix} \xi^1 \\ \vdots \\ \xi^n \\ \eta^1 \\ \vdots \\ \eta^n \end{pmatrix} \qquad (5.6)$$

where A, B, and C are $n \times n$ matrices whose entries are given by (5.5). The Equations (5.5) or (5.6) are called the Jacobi differential equations. What is important for us is that the matrix B is a positive definite symmetric matrix. This is so since $(\partial^2 H/\partial y^i \partial y^j)$ is just the Jacobian matrix (along a fixed $T_p^*(M)$) of the map $\mathscr{L}^{-1}$. Thus it is the inverse of $(L_{\dot{x}^i\dot{x}^j})$, and the inverse of a positive definite matrix is positive definite.

We are now in a position to prove Lemma 5.1. In order to show that $(\pi \times \mathrm{id})_* \gamma_{t*}$ is non-singular, it certainly suffices to show $\pi_* \circ \gamma_{t*}$ sends no non-zero tangent vector $V_0 \in T_u(T^0(K))$ into zero. If $\xi^1, \ldots, \xi^n$, $\eta^1, \ldots, \eta^n$ are the components of a vector V, then $\pi_*(V) = \sum \xi^i\,(\partial/\partial x^i)$. What we want to show, therefore, is the following:

There is an $\epsilon > 0$ such that for all $0 < |t| < \epsilon$, and any $V_0 \in T_u(T_0(K))$ with $V_0 = \sum \xi_0^i\,(\partial/\partial x^i) + \eta_0^i\,(\partial/\partial y^i)$ we have

$$\pi_* \gamma_{t*}(V_0) = \sum \xi^i(t) \left(\frac{\partial}{\partial x^i}\right) \neq 0$$

where the functions $\xi^i(t)$ are (part of) a solution to the system of differential equations (5.5), *with initial conditions $\xi^i(0), \eta^i(0) = \xi_0^i, \eta_0^i$.*

First let us examine the case where K is a point. Then $T^0(K)$ is $T_p^*(M)$ and the initial conditions require that V_0 be tangent to $T_p^*(M)$, i.e., $\pi_* V_0 = 0$ or $\xi_0^1 = \cdots = \xi_0^n = 0$. Thus we have a family of linear transformations, Q_t assigning to each column vector $(\eta_0^1, \ldots, \eta_0^n)$ the vector $(\xi^1(t), \ldots, \xi^n(t))$ whose components are the solution of (5.4) satisfying $\xi^i(0) = 0, \eta^i(0) = \eta_0$. We would like to show that the Q_t are non-singular for small t. By the mean value theorem

$$\xi^i(t) = t\frac{d\xi^i}{dt}(0) + O(t^2)$$

so that

$$\xi^i(t) = t \sum H_{y^i y^j} \eta_0^j + O(t^2).$$

In other words, the transformation Q_t differs from tP by terms of order t^2, where P is the linear transformation defined by

$$\xi^i = \sum H_{y^i y^j} \eta_0^j.$$

Now Q_t is non-singular if and only if $(1/t)Q_t$ is (for non-zero t). But P is non-singular and $(1/t)\,Q_t$ is close to P which is non-singular. Since the set of non-singular transformations is open, this implies that Q_t is non-singular for small, non-zero, $|t|$.

In the more general case of a submanifold K, let us assume that our coordinates have been chosen so that K is given by $x^{k+1} = \cdots = x^n = 0$. Then the conditions that $V_0 \in T^0(K)$ are that

$$\xi_0^{k+1} = \cdots = \xi_0^n = \eta_0^1 = \cdots = \eta_0^k = 0.$$

Again, we have transformations, Q_t, assigning to each vector $(\xi_0^1, \ldots, \xi_0^k, \eta_0^{k+1}, \ldots, \eta_0^n)$ the vector $(\xi^1(t), \ldots, \xi^n(t))$ where the ξ's are solutions of (5.4) with

$$\xi^i(0) = \xi_0^i \quad \text{for } i \leqslant k, \qquad \xi^i(0) = 0 \quad \text{for } i > k,$$
$$(d\xi^i/dt)(0) = 0 \quad \text{for } i \leqslant k \quad \text{and} \quad (d\xi^i/dt)(0) = \eta_0^i \quad \text{for } i > k.$$

Again, by the mean value theorem we have

$$(5.6)\qquad \begin{aligned} \xi^i(t) &= \xi_0^i + t\sum_{j=1}^{k} H_{y^i x^j}\xi_0^j + t\sum_{j=k+1}^{n} H_{y^i y^j}\eta_0^j + O(t^2), && i \leqslant k \\ \xi^i(t) &= \phantom{\xi_0^i +{}} t\sum_{j=1}^{k} H_{y^i x^j}\xi_0^j + t\sum_{j=k+1}^{n} H_{y^i y^j}\eta_0^j + O(t^2) && i > k. \end{aligned}$$

Let P_t denote the linear transformation given by $(\xi_0^1, \ldots, \xi_0^k, \eta_0^{k+1}, \ldots, \eta_0^n) \to (\xi^1(t), \ldots, \xi^n(t))$ given by

$$\xi^i(t) = \xi_0^i + t\sum_{j=1}^{k} H_{y^i x^j}\xi_0^j + t\sum_{j=k+1}^{n} H_{y^i y^j}\eta_0^j, \qquad i \leqslant k$$

$$\xi^i(t) = \sum_{j=1}^{k} H_{y^i x^j}\xi_0^j + \sum_{j=k+1}^{n} H_{y^i y^j}\eta_0^j, \qquad i > k \text{ (notice the lack of } t\text{)}.$$

Then (5.6) says that $L_t Q_t = P_t + O(t)$ where the matrix of L_t is

$$\begin{pmatrix} \begin{matrix} 1 & & 0 \\ & \ddots & \\ 0 & & 1 \end{matrix} & 0 \\ 0 & \begin{matrix} t^{-1} & & \\ & \ddots & \\ & & t^{-1} \end{matrix} \end{pmatrix}.$$

(the upper left block is $k \times k$)

Since L_t is non-singular, it suffices to prove that $P = \lim_{t \to 0} P_t$ is non-singular; since $L_t Q_t \to P$, and the set of non-singular transformations is open. But the matrix of P has the form

$$\begin{array}{c} \\ \\ \\ k \\ \\ \\ \end{array}\left(\begin{array}{ccc|ccc} 1 & & & & & \\ & \ddots & & & 0 & \\ & & 1 & & & \\ \hline \multicolumn{3}{c|}{\text{———}} & H_{y^{k+1}y^{k+1}} & \cdots & H_{y^{k+1}y^{n}} \\ \multicolumn{3}{c|}{\text{———}} & & & \\ \multicolumn{3}{c|}{\text{———}} & H_{y^{n}y^{k+1}} & \cdots & H_{y^{n}y^{n}} \end{array}\right).$$

(with k marking the column and row of the partition)

Thus the determinant of P is just

$$\begin{vmatrix} H_{y^{k+1}y^{k+1}} & H_{y^{k+1}y^{n}} \\ H_{y^{n}y^{k+1}} & H_{y^{n}y^{n}} \end{vmatrix}$$

which is positive, since the matrix $(H_{y^i y^j})$ is a symmetric *positive definite* matrix.[5]

Thus we have proved Lemma 5.1.

We are now in a position to prove that extremals minimize up to the first focal point. More precisely,

THEOREM 5.1. *Let L be a positive definite Lagrangian defined on $T(M) \times R$. Let K be a submanifold of M and C an extremal with $C(t_0) \in K$, and $\bar{C}(t_0) \in T^0(K)$. Let t_1 be the first focal point of t_0 along C and $t_0 < \bar{t} < t_1$. Then C is a weak local minimum on the interval $[t_0, \bar{t}]$ relative to the variable end point problem $\{K, C(\bar{t})\}$. If L is quadratic (for instance the kinetic energy of a Riemannian metric) then C is a strong minimum on this interval.*

---

[5] Recall that if (A_{ij}) $(i, j = 1, \ldots, n)$ is a positive definite matrix, so is the $(n - k) \times (n - k)$ submatrix where $(i, j = k + 1, \ldots, n)$. In fact, the matrix gives the restriction of the quadratic form associated to (A_{ij}) to the subspace of vectors whose first k-coordinates vanish; and the restriction of a positive definite form is positive definite.

We wish to choose the "initial" manifold N_0 of Lemma 4.6 so that $\bar{C}(t_0) \in N_0$ and the projection $\pi_* \circ \alpha_s{}^*(T_{\bar{C}(t_0)}(N_0))$ is non-singular for $0 \leqslant s \leqslant \bar{t} - t_0$. To simplify notations let us take $t_0 = 0$ and $s = t$. We already know, that if we take $N_0 = T^0(K)$, the projection is non-singular for $0 < t < t_1$. Unfortunately, it is highly singular for $t = 0$. Our approach is to take N_0 close to $T^0(K)$ so that the projection remains non-singular for $0 < t \leqslant \bar{t}$ but is no longer singular at the origin.

In fact, for any compact subinterval, say $[\delta, t]$ *of* $0 < t < t_1$, *the projection will be non-singular for* N_0 *close enough to* $T^0(K)$.

Indeed, since we are really studying the solutions of the Jacobi equations, this is a consequence of the continuous dependence of solutions of a differential equation on initial conditions.

Let us introduce coordinates $x^1, \ldots, x^n$ at $C(0)$ so that K is given by $x^{k+1} = \cdots = x^n = 0$. Let (G_{ij}) be the $(n - k) \times (n - k)$ matrix given by

$$G_{ij} = H_{y^i y^j}(\bar{C}(0)) \qquad i, j = k + 1, \ldots, n.$$

If $\bar{C}(0)$ has the coordinates $(x_0^1, \ldots, x_0^n, 0, \ldots, 0, y_0^{k+1}, \ldots, y_0^n)$ introduce the submanifolds, N_0^ϵ, of $T^*(M)$ given by

$$\begin{aligned} x^i &= x_0^i + u^i & i &= 1, \ldots, k \\ x^i &= \quad \epsilon \sum_{j=k+1}^{n} G_{ij} u^j & i &= k + 1, \ldots, n \\ y^i &= 0 & i &= 1, \ldots, k \\ y^i &= u^i & i &= k + 1, \ldots, n. \end{aligned}$$

For $\epsilon = 0$, N_0^ϵ reduces to $T^0(K)$. For any ϵ, the point $u^i = 0$ is just $\bar{C}(0)$. On the tangent space $T_{\bar{C}(0)}(N_0^\epsilon)$ let us introduce the coordinates $\upsilon^1, \ldots, \upsilon^n$, corresponding to the basis $\partial/\partial u^1, \ldots, \partial/\partial u^n$. The passage to the υ, coordinates is given by

$$\begin{aligned} \xi_0^i &= \upsilon^i & i &= 1, \ldots, k \\ \xi_0^i &= \epsilon \sum_{k+1}^{n} G_{ij} \upsilon^j & i &= k + 1, \ldots, n \\ \eta_0^i &= 0 & i &= 1, \ldots, k \\ \eta_0^i &= \upsilon^i & i &= k + 1, \ldots, n. \end{aligned}$$

Let $\xi^i(t)$, $\eta^i(t)$ be the solution of (5.5) with these initial conditions. We then get a linear map, Q_t^ϵ from $(\upsilon^1, \ldots, \upsilon^n)$ to $(\xi^1(t), \ldots, \xi^n(t))$. We wish to choose ϵ so that Q_t^ϵ is non-singular for $0 \leqslant t \leqslant \bar{t}$. By the mean value theorem, Q_t^ϵ is given by

$$\xi^i(t) = \upsilon^i + t \sum_{i=1}^{k} H_{y^i x^j}(0)\upsilon^i + \epsilon t \sum_{j,l=k+1}^{n} H_{y^i x^j} G_{jl} \upsilon^l + t \sum_{j=k+1}^{n} H_{y^i y^j}(0)\upsilon^j + O(t^2), \qquad (i \leqslant k)$$

$$\xi^i(t) = \epsilon \sum_{j=k+1}^{n} G_{ij}\upsilon^j + t \sum_{j=k+1}^{n} G_{ij}\upsilon^j + t \sum_{j=1}^{k} H_{y^i x^j}\upsilon^j + \epsilon t \sum_{j,l=k+1}^{n} H_{y^i x^j} G_{jl} \upsilon^l + O(t^2) \qquad \text{for } (i = k+1, \ldots, n).$$

Thus the matrix Q_t^ϵ has the form

$$\overset{k}{\left(\begin{array}{c|c} I + tA & \epsilon t B + tC \\ \hline tD + \epsilon t E & (\epsilon + t)G \end{array}\right)} + 0(t^2).$$

Let $P_{\epsilon+t}$ be the matrix

$$\left(\begin{array}{ccc|ccc} 1 & & & & & \\ & \ddots & & & 0 & \\ & & 1 & & & \\ \hline & & & \epsilon + t & & \\ & 0 & & & \ddots & \\ & & & & & \epsilon + t \end{array}\right).$$

For any ϵ in the compact interval $0 \leqslant \epsilon \leqslant \epsilon_0$, the limit as $t \to 0^+$ of $P_{\epsilon+t}^{-1}\, Q_t^\epsilon$ is the non-singular matrix

$$\left(\begin{array}{c|c} I & 0 \\ \hline 0 & G \end{array}\right).$$

Since the set of non-singular matrices is open, we therefore find a $\delta > 0$ so that $P_{\epsilon+t}^{-1}\, Q_t$ is non-singular for $0 \leqslant \epsilon \leqslant \epsilon_0$ and $0 < t < 2\delta$, and therefore Q_t^ϵ is non-singular for this range. Now choose $\epsilon > 0$ so small that Q_t^ϵ is non-singular for $\delta \leqslant t \leqslant \bar{t}$. Thus Q_t^ϵ is non-singular for $0 < t < \bar{t}$. On the other hand Q_0^ϵ is the non-singular matrix

$$\left(\begin{array}{c|c} I & 0 \\ \hline 0 & \epsilon G \end{array}\right).$$

Thus Q_t^ϵ is non-singular for $0 \leqslant t \leqslant \bar{t}$, proving Theorem 5.1.

We have shown that extremals minimize up to the first focal point. We will now prove the following theorem due to Jacobi.

THEOREM 5.2. *Extremals do not minimize past the first focal point. More precisely, let K be a submanifold of M and C an extremal with* $C(t_0) \in K$, $\bar{C}(t_0) \in T^0(K) \times t_0$. *If* $\bar{t}$ *is a focal point of K along C then C is not a weak minimum on any interval* (t_0, t_1) *where* $t_0 < \bar{t} < t_1$ *(or* $t_0 > \bar{t} > t_1$*).*

A rough idea of the proof is the following: At the focal point $\bar{t}$ two "infinitesimally close" extremals meet. If actually two arbitrarily close *minimizing* extremals meet at $\bar{t}$, they would have to give the same value to the integral I (as they both minimize and are both in neighborhoods of each other). Thus we would have two minimizing extremals C and D meeting at $\bar{t}$. They could not be tangent to each other, for then they would coincide.

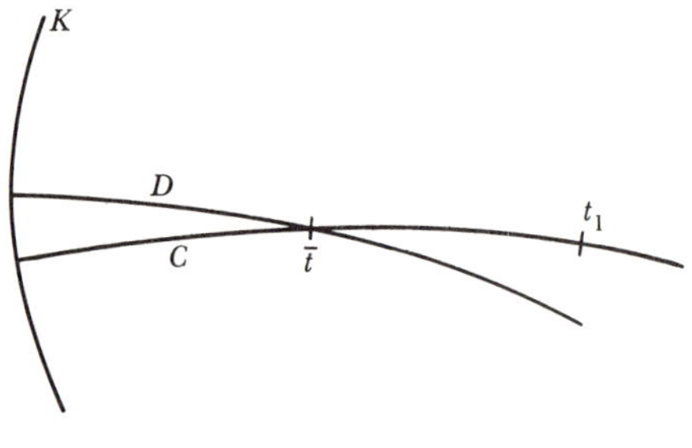

But then, the curve, B, equal to D on $(t_0, \bar{t})$ and to C on $(\bar{t}, t_1)$ would be a piecewise differentiable curve with $I_{t_0}^{t_1}(B) = I_{t_0}^{t_1}(C)$. If C is a minimum, so is B, contradicting Theorem 2.1 as B is not differentiable at $\bar{t}$.

The proof above doesn't quite work, mainly because at a conjugate point there do not have to be arbitrarily near extremals that meet, only "infinitesimally near." However, we can make the proof work by giving an "infinitesimal version" of the preceding argument. For this purpose it is convenient to have a single coordinate system along all of C,[6] we achieve this by

LEMMA 5.2. *Let* $C: (t_0, t_1) \to M$ *be a differentiable curve. We can write* $C = F \circ B$ *where* F *is a differentiable map of an open subset* ,U, *of* E^n *into* M *which is locally a diffeomorphism and* $B: (t_0, t_1) \to U$ *is a differential curve in* U.

Thus Lemma 5.2 says that every curve can be "thickened."

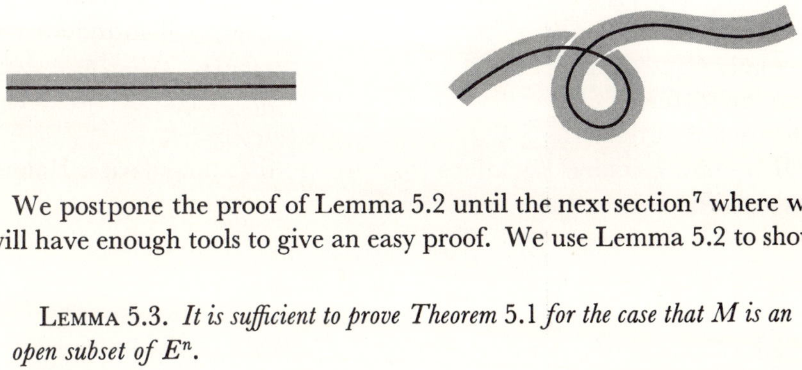

We postpone the proof of Lemma 5.2 until the next section[7] where we will have enough tools to give an easy proof. We use Lemma 5.2 to show

LEMMA 5.3. *It is sufficient to prove Theorem* 5.1 *for the case that* M *is an open subset of* E^n.

---

[6] The idea of the proof is to regard the Jacobian equations [Equation (5.5)] as the "Euler equations" of the second variation (cf. the discussion in Section 2). Now both the Jacobi equations and the second variation are intrinsically defined. However, the second variation is defined for vector fields X_t along the curve C. Now $X_t \in T_{C(t)}(M)$ so that for each t, the vector X_t lies in a different vector space. We cannot regard X as a curve in some fixed vector space unless we have some way of identifying all of the tangent spaces along C. One way of doing this is to introduce what is known as a linear connection on M, cf. Chapter VII. Our approach will be to move everything over to Euclidean space where the identification is the standard one. In the Riemann case there is a canonically associated connection so that the proof could be presented in a more intrinsic fashion.

[7] Cf. Exercise 6.3.

In fact we apply Lemma 5.2 to write $C = F \circ B$. Since F is locally a diffeomorphism of $U \to M$, induces a differentiable map, F_*, of $T(U)$ into $T(M)$ [and hence of $T(U) \times R \to T(M) \times R$ which is a diffeomorphism locally]. Then $L \circ (F_* \times \text{id})$ is a regular, positive definite, Lagrangian on U and for any curve D in U, the integral $I[D]$ relative to $L \circ (F_* \times \text{id})$ is the same as $I[F \circ D]$ relative to L. If $D(0) \in F^{-1}(K \cap F(U))$ then $F(D(0)) \in K$. Therefore, if B is not a minimum for $L \circ (F_* \times \text{id})$ then $C = F \circ B$ is not a minimum for L. On the other hand, $(F_* \times \text{id}) \circ \mathscr{L}_U$ is locally a diffeomorphism of $T_*(U) \times R \to T_*(M) \times R$ where $\mathscr{L}_U$ is the Legendre transformation corresponding to $L \circ (F_* \times \text{id})$. Furthermore, $\alpha_t \circ (F_* \times \text{id}) \circ \mathscr{L}_U = (F_* \times \text{id}) \circ \mathscr{L}_U \alpha_{Ut}$ where α_{Ut} is the flow on $T_*(U) \times R$. This implies that if $\bar{t}$ is not a focal point of $F^{-1}(K)$ along B, then $\bar{t}$ is not a focal point of K along C; or, the other way around, if $\bar{t}$ is a focal point of K along C then it is also a focal point of $F^{-1}(K)$ along B. Thus if B does not minimize past its focal points, neither does C, proving Lemma 5.3. We will assume, therefore, for the rest of this section, that M is an open subset of E^n.

Since we are now in Euclidean space, we have global coordinates $x^1, \ldots, x^n$, $y^1, \ldots, y^n$, etc. We can now introduce global coordinates consistent with Equations (5.5)–(5.6). That is, we will introduce coordinates $\xi^1, \ldots, \xi^n, \eta^1, \ldots, \eta^n, t$ on the submanifold of $T(T^*(M) \times R)$ consisting of the simultaneous vectors lying in the spaces $T_{\bar{C}(t)}(T^*(M) \times R)$: The vectors of the form $\sum \xi^i(\partial/\partial x^i)_{\bar{C}(t)} + \sum \eta^i(\partial/\partial y^i)_{\bar{C}(t)}$.

If we now examine Equations (5.5) we see that they have a Hamiltonian form, i.e., we can write them as

$$\frac{d\xi^i}{dt} = \frac{\partial \mathscr{H}}{\partial \eta^i}, \quad \frac{d\eta^i}{dt} = -\frac{\partial \mathscr{H}}{\partial \xi^i} \tag{5.7}$$

where

$$\mathscr{H} = \frac{1}{2} \sum_{i,j=1}^{n} \left(H_{x^i x^j} \xi^i \xi^j + 2H_{x^i y^j} \xi^i \eta^j + H_{y^i y^j} \eta^i \eta^j \right). \tag{5.8}$$

Thus, if we consider $\xi^1, \ldots, \xi^n$ as coordinates on a vector space V corresponding to some choice of basis and $\eta^1, \ldots, \eta^n$ the associated coordinates on $T^*(V)$ we would like to regard $\mathscr{H}$ as a Hamiltonian of a regular variational problem and (5.5) [or (5.7)] as its Euler equations. By Theorem 1.1, we know that we can do this if $(\partial^2 \mathscr{H}/\partial \eta^i \, \partial \eta^j)$ is non-singular. But $(\partial^2 \mathscr{H}/\partial \eta^i \, \partial \eta^j) = (H_{y^i y^j})$ and this is non-singular since L is regular. We have, therefore, a regular Lagrangian, Λ, on $T(V) \times R$

and a Legendre transformation $\mathcal{X}: T(V) \times R \to T^*(V) \times R$ such that

$$\Lambda = -\mathscr{H} \circ \mathcal{X} + \sum \dot{\xi}^i \eta^i \tag{5.9}$$

and

$$\dot{\xi}^i = \frac{\partial \mathscr{H}}{\partial \eta^i} = \sum_k (H_{x^k y^i} \xi^k + H_{y^i y^k} \eta^k). \tag{5.10}$$

Substituting (5.10) into (5.9) we obtain

$$\Lambda \circ \mathcal{X}^{-1} = \frac{1}{2} \sum (-H_{x^i x^j} \xi^i \xi^j + H_{y^i y^j} \eta^i \eta^j). \tag{5.11}$$

We also have

LEMMA 5.4. *If* $\Gamma(t)$ *is an extremal of* Λ *then*

$$I_{\Lambda t_1}^{t_2}[\Gamma] = \frac{1}{2} \sum_i (\xi^i(t_2)\eta^i(t_2) - \xi^i(t_1)\eta^i(t_1)) \tag{5.12}$$

where

$$\Gamma(t) = (\xi^1(t), \ldots, \xi^n(t)).$$

In fact, since $\mathscr{H}$ is a homogeneous function of degree 2 in ξ^i, and η^i,

$$\mathscr{H} = \frac{1}{2} \sum \left(\xi^i \frac{\partial \mathscr{H}}{\partial \xi^i} + \eta^i \frac{\partial \mathscr{H}}{\partial \eta^i} \right).$$

If $\xi^i(t)$, $\eta^i(t)$ are solutions of (5.7), this implies that

$$\mathscr{H} \circ \Gamma = \frac{1}{2} \sum (-\xi^i \dot{\eta}^i + \eta^i \dot{\xi}^i).$$

By (5.9) this implies

$$\Lambda \circ \tilde{\Gamma} = \Lambda \circ \mathcal{X}^{-1} \tilde{\Gamma} = \frac{1}{2} \sum \left(\xi^i \dot{\eta}^i + \eta^i \dot{\xi}^i \right) = \frac{1}{2} \frac{d}{dt} \sum \left(\xi^i(t) \eta^i(t) \right)$$

giving (5.12).

We would still like to know more explicitly the relationship between Λ and L. It is given by

LEMMA 5.5. *Let* $\Gamma(t) = (\xi^1(t), \ldots, \xi^n(t))$ *be a differentiable curve,* $\Gamma: (t_0, t_1) \to V$. *Let* $K_\Gamma: (-\epsilon, \epsilon)(t_0, t_1) \to M$ *be defined by*

$$x^i \circ K_\Gamma(s, t) = x^i(t) + s\xi^i(t)$$

(*where the extremal* C *is given by* $(C(t) = (x^1(t), \ldots, x^n(t))$. *Then*

$$(5.13)\quad \frac{\partial}{\partial s}\int_{t_0}^{t_1} L \circ \tilde{K}_\Gamma(0, t)\, dt = \sum_{i=1}^{n} [y^i(t_1)\xi^i(t_1) - y^i(t_0)\xi^i(t_0)],$$

and

$$(5.14)\quad \frac{\partial^2}{\partial s^2}\int_{t_0}^{t_1} L \circ \tilde{K}_\Gamma(0, t)\, dt = 2\int_{t_0}^{t_1} \Lambda \circ \Gamma(t)\, dt.$$

Thus Λ measures the second derivative or "second variation" of L in the "direction" Γ.

Proof. Let us denote $x^i \circ K(s, t)$ by $x^i(s, t)$. Then $\mathscr{L}$ defines $y^i(s, t)$ by

$$(5.15)\quad y^i(s, t) = \frac{\partial L}{\partial \dot{x}^i}\left(x^i(s, t), \frac{\partial x^i}{\partial t}(s, t), t\right)$$

and

$$(5.16)\quad \begin{aligned} L\left(x^1(s, t), \ldots, x^n(s, t), \frac{\partial x^i}{\partial t}(s, t), \ldots, \frac{\partial x^n}{\partial t}(s, t), t\right) \\ = -H(x^1(s, t), \ldots, x^n(s, t), y^i(s, t), \ldots, y^n(s, t)) \\ + \sum y^i(s, t)\frac{\partial x^i}{\partial t}(s, t). \end{aligned}$$

Differentiating (5.16) with respect to s yields

$$\frac{\partial L}{\partial s} = -\sum \frac{\partial H}{\partial x^i}\frac{\partial x^i}{\partial s} - \sum \frac{\partial H}{\partial y^i}\frac{\partial y^i}{\partial s} + \sum \frac{\partial y^i}{\partial s}\frac{\partial x^i}{\partial t} + \sum y^i \frac{\partial^2 x^i}{\partial s \partial t}.$$

The two middle terms cancel by (5.15) since $\partial H/\partial y^i = \partial x^i/\partial t$; furthermore, $\partial x^i/\partial s = \xi^i$. Therefore, we can rewrite the equation as

$$(5.17)\quad \frac{\partial L}{\partial s} = -\sum \frac{\partial H}{\partial x^i}\xi^i + \sum y^i\dot{\xi}^i.$$

Adding and subtracting $\sum (\partial y^i/\partial t)\xi^i$ we obtain

$$(5.18)\quad \frac{\partial L}{\partial s} = -\sum\left(\frac{\partial H}{\partial x^i} + \frac{\partial y^i}{\partial t}\right)\xi^i + \frac{\partial}{\partial t}\left(\sum y^i\xi^i\right).$$

If we set $s = 0$, then $\partial H/\partial x^i + \partial y^i/\partial t = H_{x^i} + \dot{y}^i = 0$ since C is an extremal. Thus

$$\left.\frac{\partial L}{\partial s}\right|_{s=0} = \frac{d}{dt}\left(\sum y^i\xi^i\right)$$

and this proves (5.13).

If we differentiate (5.17) with respect to s we obtain

$$\frac{\partial^2 L}{\partial s^2} = -\sum_{i,j} \frac{\partial^2 H}{\partial x^i \partial x^j} \xi^i \xi^j - \sum_{i,j} \frac{\partial^2 H}{\partial x^i \partial y^j} \xi^i \frac{\partial y^j}{\partial s} + \sum_j \frac{\partial y^j}{\partial s} \dot{\xi}^j. \tag{5.19}$$

Now, $\partial x^i/\partial t = \partial H/\partial y^i$ so differentiating with respect to s yield

$$\dot{\xi}^i = \sum \frac{\partial^2 H}{\partial x^j \partial y^i} \xi^j + \sum \frac{\partial^2 H}{\partial y^i \partial y^j} \frac{\partial y^j}{\partial s}. \tag{5.20}$$

If we multiply (5.20) by $\partial y^i/\partial s$ and sum over i we obtain

$$\sum \dot{\xi}^i \frac{\partial y^i}{\partial s} = \sum \frac{\partial^2 H}{\partial x^i \partial y^j} \xi^i \frac{\partial y^j}{\partial s} + \sum \frac{\partial^2 H}{\partial y^i \partial y^j} \frac{\partial y^i}{\partial s} \frac{\partial y^j}{\partial s}.$$

Substituting this into (5.19) yields

$$\frac{\partial^2 L}{\partial s^2} = -\sum \frac{\partial^2 H}{\partial x^i \partial x^j} \xi^i \xi^j + \sum \frac{\partial^2 H}{\partial y^i \partial y^j} \frac{\partial y^i}{\partial s} \frac{\partial y^j}{\partial s}. \tag{5.21}$$

On the other hand, if we set $s = 0$ in (5.20) we obtain

$$\dot{\xi}^i = \sum_j \left(H_{x^j y^i} \xi^j + H_{y^i y^j} \left.\frac{\partial y^j}{\partial s}\right|_{s=0} \right).$$

Comparing this with (5.10) and the fact that $\mathcal{A}^{-1}$ is one-to-one shows that

$$\left.\frac{\partial y^i}{\partial s}\right|_{s=0} = \eta^i.$$

Substituting this into (5.21) gives

$$\left.\frac{\partial^2 L}{\partial s^2}\right|_{s=0} = -\sum H_{x^i x^j} \xi^i \xi^j + \sum H_{y^i y^j} \eta^i \eta^j. \tag{5.22}$$

Comparing this with (5.11) completes the proof of Lemma 5.5.

We are now in a position to prove Theorem 5.1. Let us introduce coordinates on E^n such that K is given [in some neighborhood of $C(t_0)$] by $x^{k+1} = \cdots = x^n = 0$. If C is to be a weak minimum relative to the end point problem $B(t_0) \in K$, $B(t_1) = C(t_1)$, then we must have

$$I_{t_0}^{t_1} [K_\Gamma(s, \cdot)] \geqslant I_{t_0}^{t_1} [C]$$

for sufficiently small s if Γ is a curve given $\Gamma(t) = (\xi^1(t), \ldots, \xi^n(t))$ and satisfying $\xi^{k+1}(t_0) = \cdots = \xi^n(t_0) = 0$ and $\xi^1(t_1) = \cdots = \xi^n(t_1) = 0$.

Indeed $K_\Gamma(s, \cdot)$ is a nearby curve satisfying the same end conditions. Since $C = K_\Gamma(0, \cdot)$, we must have

$$\frac{\partial}{\partial s} I_{t_0}^{t_1} [K_\Gamma(0, \cdot)] = 0 \tag{5.23}$$

and

$$\frac{\partial^2}{\partial s^2} I_{t_0}^{t_1} [K_\Gamma(0, \cdot)] \geqslant 0, \tag{5.24}$$

for all such Γ.

Now Equation (5.13) shows that (5.23) holds for all Γ subject to our conditions. In fact, the conditions on $\bar{C}(t_0)$ are $y^i(t_0) = 0$ for $i \leqslant k$ so that $\sum y^i(t_0)\xi^i(t_0) = 0 = \sum y^i(t_1)\xi^i(t_1)$. [In reality (5.23) is equivalent to Euler's equations and (2.23) giving another proof of Theorem 2.1.]

Condition (5.24) implies, by Lemma 5.5, that

$$\int_{t_0}^{t_1} \Lambda \circ \tilde{\Gamma}(t)\, dt \geqslant 0 \tag{5.25}$$

for all Γ subject to our conditions. Now $\Gamma(t) \equiv 0$ certainly satisfies our conditions and gives a value of zero to the integral. In other words, (5.25) asserts that 0 *is a minimum value for* (5.25) *among all curves* Γ *satisfying our end conditions.*

Now assume that there is a conjugate point $\bar{t}$, with $t_0 < \bar{t} < t$. By the definition of conjugate point, there is a non-zero vector $V = (\xi_0^1, \ldots, \xi_0^k, 0, \ldots, 0; 0, \ldots, 0, \eta_0^{k+1}, \ldots, \eta_0^n)$ such that the solution Γ given by $\xi^i(t), \eta^i(t)$ of (5.5) with $\xi^i(t_0) = \xi_0^i, \ldots, \eta^i(t_0) = \eta_0$ has $\xi^i(\bar{t}) = 0$ for all i. Since Γ is a solution of (5.5) or (5.7) we have, by Lemma 5.4

$$\int_{t_0}^{t} L \circ \tilde{\Gamma}(t)\, dt = \sum (\xi^i(\bar{t})\eta^i(\bar{t}) - \xi^i(t_0)\eta^i(t_0)) = 0. \tag{5.26}$$

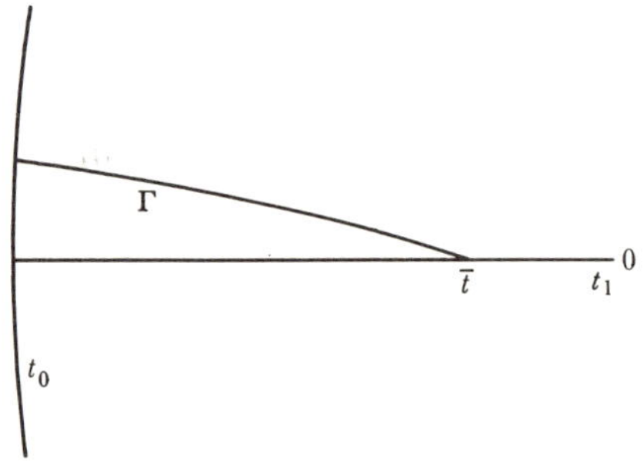

Consider the curve $\underline{\Gamma}$ which coincides with Γ for $t_0 \leqslant t \leqslant \bar{t}$ and with 0 for $\bar{t} \leqslant t \leqslant t_1$. It gives the value 0 to $\int_{t_0}^{t_1} \Lambda \circ \tilde{\underline{\Gamma}}\, dt$, therefore is a minimum and hence an extremal of Λ. On the other hand, $\underline{\Gamma}$ is *not* differentiable at $\bar{t}$, for otherwise, by uniqueness of solutions to (5.5) it would have to vanish identically. This contradicts Theorem 2.1 (applied to Λ) and proves Theorem 5.1.

6. The Riemannian case. According to Theorem 3.1, since $L = \frac{1}{2}\| \;\|^2$ is independent of t, there is vector field, V, on $T^*(M)$ such that $Y = (V, (\partial/\partial t))$. As $\mathscr{L}$ induces a diffeomorphism of $T(M)$ onto $T^*(M)$, V determines a vector field Z on $T(M)$. We have, in fact, already obtained an expression for Z in local coordinates (cf. the end of Section 2). The local flow, δ_t, generated by Z is called the *geodesic flow on* $T(M)$. As we have seen in Section 3, δ_t is globally defined whenever M is complete. We now wish to study flow δ_t and the vector field Z in some detail. First of all, if $v \in T(M)$ then $\pi_*(Z_v) = v \in T_{\pi(v)}(M)$. This is indeed true for all vector fields on $T(M)$ which correspond to "second order differential equations" on M. Next, we take account of the behavior of Z under "dilitation." That is, the multiplicative group of reals, R^*, (i.e., all $t \neq 0$) acts as a group of transformations of $T(M)$ for any manifold M: If $v \in T(M)$ define $\mu_t(v) = tv$. It is easy to see that μ defines a differentiable action of R^* on $T(M)$; $\mu_{t_1} \circ \mu_{t_2} = \mu_{t_1 t_2}$ and $\pi \circ \mu_t = \pi$. In terms of coordinates $x^i \circ \mu_t = x^i$ and $\dot{x}^i \circ \mu_t = t\dot{x}^i$. Therefore, $\mu_{t*}(\partial/\partial \dot{x}^i)_v = t(\partial/\partial \dot{x}^i)_{tv}$. Therefore, it follows from (2.26) (or directly from the definitions!) that $\mu_t^* Z = (1/t)Z$.[8]

Definition 6.1. *Let M be a differentiable manifold. If W is a vector field on $T(M)$ such that*

$$\pi_*(W_v) = v \qquad \text{for all } v \in T(M) \tag{6.1}$$

and

$$\mu_{t*}W = \frac{1}{t} W \tag{6.2}$$

then W is called a spray. A spray, W, is called complete if the flow generated by W is globally defined.

[8] Recall that for any vector field, Z, we have defined $\mu_t^* Z$ by $(\mu_t^* Z)_v = \mu_{-t*}(Z_{\mu t(v)})$.

Thus we can resume the previous discussion by saying that if M is a Riemann manifold, then the vector field Z is a spray. We shall call Z the *geodesic* spray of M.

Suppose W is a vector field on the tangent bundle of a manifold M and let p be a point of M. Then for a sufficiently small neighborhood V of 0 in $T_p(M)$ and interval I about 0 in R, W induces a map $V \times I \to M$. In fact, if δ_t is the flow generated by W choose V and I so that $\delta_t(v)$ is defined for $v \in V$ and $t \in I$. Then the map σ defined by $\sigma(v, t) = \pi \circ \delta_t(v)$ is a differentiable map of $V \times I \to M$. Suppose W satisfies (6.1). For any vector $v \in V$ define the curve C_v by $C_v(t) = \pi \circ \delta_t(v)$; (6.1) implies that $C_v'(t) = \delta_t(v)$. Suppose W also satisfies (6.2). Then we have

$$\mu_t \circ \delta_s = \delta_{s/t} \circ \mu_t. \tag{6.3}$$

In fact, from Exercise 8.4 of Chapter II, we know that $\mu_t \circ \lambda_s = \lambda_s \circ \mu_t$ if λ is the flow generated by $\mu_{t*}W$. But $\mu_{t*}W = (1/t)W$ so $\lambda_s = \delta_{s/t}$ establishing (6.3). Therefore, in the case of a spray, the map σ satisfies

$$\sigma(tv, s) = \sigma(v, ts). \tag{6.4}$$

In particular, for a sufficiently small neighborhood, U, of the origin in $T_p(M)$, the map $\sigma(v, t)$ is defined for all $|t| \leqslant 2$. We define the map $\exp: U \to M$ by

$$\exp(v) = \sigma(v, 1). \tag{6.5}$$

(If W is a complete spray, then exp and σ are defined for all $v \in T_p(M)$). Let us examine the map $\exp_*: T_0(T_p(M)) \to T_p(M)$. The line tv is mapped by exp into the curve $C_v(t)$ according to (6.4). Thus $\exp_*$ takes the tangent to the line tv into the tangent to $C_v(t)$ at $t = 0$. But the tangent to $C_v(t)$ at $t = 0$ is just v. Thus $\exp_*$ sends the tangent to tv into v. But this is just the identification map, l_0, of $T_0(T_p(M))$ with $T_p(M)$ given in Section 5 of Chapter II. In other words, $\exp_*: T_0(T_p(M)) \to T_p(M)$ coincides with l_0. To resume, we have proved

THEOREM 6.1. *Let W be a spray and δ the flow it generates. For any $p \in M$, and any interval I of real numbers, there is a neighborhood V of* 0 *in $T_p(M)$ such that the map $\sigma: V \times I \to M$ given by $\sigma(v, t) = \pi\, \delta_t(v)$ is defined and satisfies* (6.4). *If we define* exp *by* (6.5) *then* exp *takes lines through the origin into the solution curves of W, and $\exp_*$ coincides at the origin with l_0.*

Since $\exp_*$ is non-singular at the origin it defines a diffeomorphism of some neighborhood of the origin of $T_p(M)$ onto a neighborhood, U, of

p of M. We shall call such U together with the map $\exp^{-1}$ a *normal* chart about p. We call U a normal neighborhood of p. If we choose a basis of $T_p(M)$, then the corresponding (linear) coordinates on $T_p(M)$ induce a local coordinate system on U, which we call a *normal coordinate system at* p. If $(x^1, \ldots, x^n)$ is a normal coordinate system at p then $x^i(p) = 0$ and a solution curve of W passing through p is given by $x^i(t) = tx^i_0$.

We can apply the preceding to the case where M is a Riemann manifold and $W = Z$ is the geodesic spray. In this case we can relate the geodesic flow δ to the flow α on $T^*(M) \times R$. In fact, it follows immediately from the definitions that $\alpha_s(v^*, t) = (\mathscr{L} \circ \delta_s \circ \mathscr{L}^{-1}v^*, t + s)$. In particular, $(\pi \times \text{id}) \circ \alpha_s(v^*, 0) = \pi\sigma(v, s) \times s$. In particular, if U is a normal neighborhood about p, then the mapping $(\pi \times \text{id}) \circ \alpha_s$ is a diffeomorphism of $\mathscr{L} \circ \mu_{1/s}(\exp^{-1}U)$ onto $U \times s$.

We also know, from Theorem 4.1, that if C is a geodesic on a Riemann manifold M with $C(0) = p$, then C is a strong local minimum of the energy up to the first conjugate point of p (relative to the fixed end point problem).

We observe next, that for any compact set, U, in $T_p(M)$, there is a positive $\rho(U)$ such that for any geodesic C with $C(0) = p$ and $C'(0) \in U$, all conjugate points, t, of 0 along C satisfy $|t| > \rho$. (This indeed follows from an obvious continuity argument.)

LEMMA 6.1. (Gauss' lemma). *Let S_r denote the sphere of radius r in $T_p(M)$, i.e., $S_r = \{v \mid \|v\| = r\}$. Let C be a geodesic parametrized by arc length with $C(0) = p$. (Note that for small enough r, $\exp S_r$ is a submanifold of M; we shall denote it by Σ_r. For any r, $C(r) \in \Sigma_r$). Then for sufficiently small r, $C'(r)$ is orthogonal to Σ_r.*

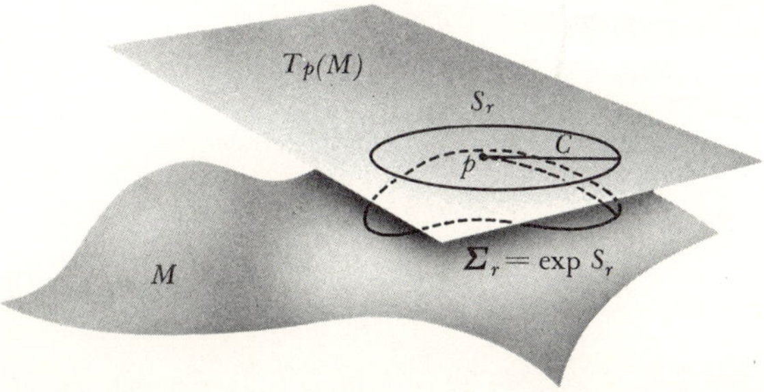

(Rays through the origin of $T_p(M)$ are obviously orthogonal to the spheres about the origin in the Euclidean metric of $T_p(M)$. What Lemma 6.1 says is their images under exp are still orthogonal in the metric of M.)

Proof. Choose r less than the minimal distance to the conjugate points, i.e., $r < \rho(S_1)$. Then all geodesics C with $C'(0) \in S_1$ minimize the energy on the interval $(0, r)$, relative to the fixed end point problem. On the other hand, the value of this energy is exactly r. That is, r is the minimum value of the energy for all curves B on the interval $(0, r)$ with $B(0) = p$, $B(r) \in \Sigma_r$. Thus, by Theorem 2.2, C must be orthogonal to Σ_r.

Let us rephrase this last result. The function r is well defined on $T_p(M)$ and differentiable except at the origin. On any normal neighborhood, U, the function $\tau = r \circ \exp^{-1}$ is defined, and differentiable except at p. Let C be any geodesic starting at p and parametrized by arc length with $C(t) = q$. Then $\langle C'(t), d\tau_q \rangle = \langle l(t), dr \rangle = 1$ [where $l(t)$ is a line through the origin in $T_p(M)$ whose tangent vector has length 1]. Lemma 6.1 asserts that if t is sufficiently small, the hyperplane in $T_q(M)$ defined by $d\tau_q = 0$ is orthogonal to the unit vector $C'(t)$. If we extend $C'(t)$ to an orthonormal basis, $C'(t), e_2, \ldots, e_n$, of $T_q(M)$, we have $\langle C'(t), d\tau_q \rangle = 1$, $\langle e_j, d\tau_q \rangle = 0$ for $j \geqslant 2$. Thus the dual basis to $C'(t), e_2, \ldots, e_n$, is $d\tau_q, e_2^*, \ldots, e_n^*$. Thus, if g_q is the metric at q, $g_q = d\tau_q^2 + (e_2^*)^2 + \cdots + (e_n^*)^2$. In other words, we have proved

LEMMA 6.2. *Let g be a Riemann metric on M and $p \in M$. There is a sufficiently small normal neighborhood W of p such that*

$$g = d\tau^2 + h \tag{6.6}$$

at all points of $W - p$, where h is a positive semidefinite symmetric differential form. Furthermore, $\langle v_q \odot v_q, h_q \rangle = 0$ implies that v_q is tangent to the (unique) geodesic C joining p to q lying in W.

In fact, we just take $h_q = (e_2^*)^2 + \cdots + (e_n^*)^2$ and Lemma 6.2 is immediate. Lemma 6.2 implies

LEMMA 6.3. *Let p and W be as in Lemma 6.2. For any $q \in W$, the geodesic, C, joining p to q provides an absolute minimum for arc length, i.e.,*

$\int \|C'(t)\| \, dt = d(p, q)$. Any curve B joining p to q whose length is $d(p, q)$ must satisfy $B(t) = C(\alpha(t))$ where α is some monotone function.

In fact, let $q \in \Sigma_r$. Then $\int_0^r \|C'(t)\| \, dt = r$. If B is any curve with $B(0) = p$, $B(a) = q$, let t_1 be the first point where $B(t_1) \in \Sigma_r$ [i.e., $t_1 = \text{g.l.b. } t \in (0, a)$ such that $B(t) \in \Sigma_r$]. Then

$$\int_0^a \|B'(t)\| \, dt$$

$$\geqslant \int_0^{t_1} \|B'(t)\| \, dt = \int_0^{t_1} \sqrt{\langle B'(t) \circledcirc B'(t), d\tau^2\rangle + \langle B'(t) \circledcirc B'(t), h\rangle} \, dt$$

$$\geqslant \int_0^{t_1} (\langle B'(t) \circledcirc B'(t), d\tau^2\rangle)^{1/2} \, dt = \int_0^{t_1} (\langle B'(t), d\tau\rangle^2)^{1/2}$$

$$\geqslant \int_0^{t_1} \langle B'(t), d\tau\rangle \, dt = (\tau(t_1) - \tau(0)) = r.$$

Thus $r = d(p, q)$. Furthermore, if B has length r all of the above inequalities must be equalities. The first implies

$$\int_{t_1}^a \|B'(t)\| \, dt = 0,$$

or B never goes outside of Σ_r. The second implies that $\langle B'(t) \circ B'(t), h\rangle \equiv 0$, so that $B'(t)$ is always tangent to some geodesic through p; this means that $B(t)$ must be on one fixed geodesic, the only choice is the geodesic joining p to q. If the third inequality is an equality, then $\langle B'(t), d\tau\rangle \geqslant 0$ so that $B(t) = C(\alpha(t))$ for some monotone non-decreasing function α.

From Lemma 6.3 we see that

the topology on M given by the metric, d, coincides with the manifold topology.

In fact, if $d(p_i, p) \to 0$ then eventually p_i lies in W and d coincides with the radial distance in $\exp^{-1}(W)$. Thus $d(p_i, p) \to 0$ implies $p_i \to p$. The converse is trivial.

LEMMA 6.4. *Let p and q be any two points of M. Let C be a piecewise differentiable curve joining p to q such that $L[C] \equiv d(p, q)$. Then C is, up to monotone reparametrization, a geodesic joining p to q. More succinctly, if C is a curve, $C(0) = p$, $C(1) = q$, with $E[C] = d^2(p, q)$ then C is a geodesic.*

In fact, if C were not a geodesic, there would be a point $r = C(t_0)$ and normal neighborhood, W, of r so that C lies in W for t_1 close to t_0, and is not a geodesic in this interval. Join t_0 to t_1 by a geodesic, γ, and replace C by the new curve, B, coinciding with C outside the interval (t_0, t_1) and with γ on (t_0, t_1). Lemma 6.3 now implies $L[B] < L[C]$ contradicting $L[C] = d(p, q)$.

The second part of the Lemma follows from the observation that

$$\left(\int_0^1 \|C'(t)\|\, dt\right)^2 \leqslant \int_0^1 \|C'(t)\|^2\, dt$$

(Cauchy's inequality) with equality only if $\|C'(t)\|$ is constant. Thus if $E[C] \leqslant d^2(p, q)$ then $L[C] \leqslant d(p, q)$ and C is parametrized proportionally to arc length.

For $p \in M$ let $\mathsf{B}_r(p)$ denote exp B_r; i.e., for small r $\mathsf{B}_r(p) = \{q \mid d(p, q) < r\}$. We now show that for sufficiently small r, B_r is "convex."

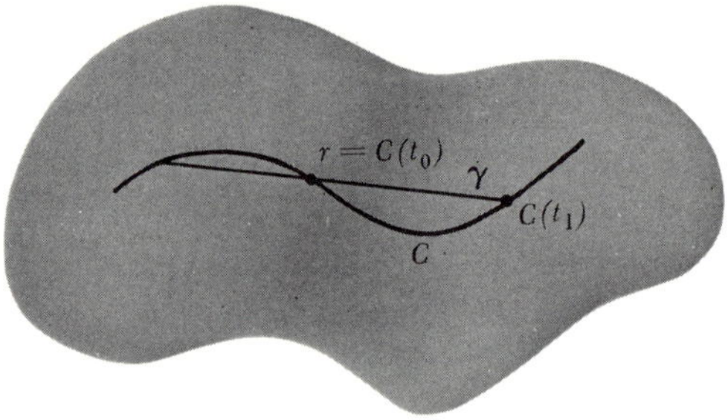

Lemma 6.5. *For every $p \in M$ we can find an $\epsilon > 0$ such that for $r \leqslant \epsilon$ the spheres Σ_r are convex. More precisely, if C is a geodesic with $C(0) \in \Sigma_r$ and $C'(0)$ tangent to Σ_r, then for sufficiently small $|t| > 0$, $C(t)$ lies outside $\overline{\mathsf{B}}_r$.*

Proof. First choose ϵ so small that we lie in a normal neighborhood of p and let $x^1, \ldots, x^n$ be normal coordinates, so that Σ_r is given by $(x^1)^2 + \cdots + (x^n)^2 = r^2$. Let $F(x^1, \ldots, x^n) = (x^1)^2 + \cdots + (x^n)^2$, and put $x^i \circ C(t) = x^i(t)$. Then since $C(0) \in S_r$, we have $F(x^1(0), \ldots, x^n(0)) = r$ since C is tangent to Σ_r at 0, $\langle C'(0), dF\rangle = 0$ or

$$\left.\frac{dF \circ C}{dt}\right|_{t=0} = 2 \sum x^i \left.\frac{dx^i}{dt}\right|_{t=0} = 0.$$

Now

$$\frac{d^2F\circ C}{dt^2} = 2\left(\sum \frac{dx^i}{dt}\frac{dx^i}{dt} + x^i\frac{d^2x^i}{dt^2}\right).$$

But C is a geodesic so that by (2.26)

$$\frac{d^2x^i}{dt^2} = -\sum_{j,k}\Gamma^i_{jk}\frac{dx^j}{dt}\frac{dx^k}{dt}.$$

Thus we can write

$$\frac{d^2F\circ C}{dt^2} = 2\left(\sum_i \frac{dx^i}{dt}\frac{dx^i}{dt} - \sum_{j,k} x^i\,\Gamma^i_{jk}\frac{dx^j}{dt}\frac{dx^k}{dt}\right)$$

$$= 2\sum\left(\delta^j_i - \sum_k x^k\,\Gamma^k_{ij}\right)\frac{dx^i}{dt}\frac{dx^j}{dt}.$$

Since Γ^k_{ij} are bounded, for sufficiently small x^k, the matrix $\delta^j_i - \sum x^k\Gamma^k_{ij}$ is positive definite. Thus, for sufficiently small ϵ,

$$\left.\frac{dF\circ C}{dt}\right|_{t=0} = 0, \qquad \frac{d^2F\circ C}{dt^2} > 0.$$

In other words, 0 is a strict minimum point for $F\circ C$ or $F\circ C(t) > r$ for small $|t| > 0$, proving Lemma 6.5.

THEOREM 6.2. *For every $p \in M$ there is a $\delta > 0$ such that $\mathsf{B}_\delta(p)$ is geodesically convex. That is, for any points $q_1, q_2 \in \mathsf{B}_\delta$ there is a unique geodesic, C, (up to parametrization) joining q_1 to q_2 lying in B_δ. Furthermore, $L[C] = d(p, q)$ and C is uniquely determined (up to parametrization) by this property.*

We first observe that to every $q \in M$ there is $\rho(q)$ such that $\mathsf{B}_{\rho(q)}$ is a normal neighborhood of q and satisfies the conclusion of Lemma 6.3 relative to q. Furthermore, ρ depends nicely on q. (In fact, it is the minimum distance to a conjugate point.) Thus there is a $K > 0$ such that $\rho(q) > K$ for $q \in \mathsf{B}_\epsilon(p)$. Let $\kappa = K/2$. Then $\mathsf{B}_\kappa(p)$ has the property that for any $q \in \mathsf{B}_\kappa(p)$, $\mathsf{B}_\kappa(p) \subset \mathsf{B}_K(q)$. In particular, for any $q_1, q_2 \in \mathsf{B}_\kappa(p)$, $q_1 \in \mathsf{B}_K(q_2)$. In other words, by Lemma 6.3

for any $q_1, q_2 \in \mathsf{B}_\kappa(p)$ there is a unique geodesic, joining q_1 to q_2 of length $d(q_1, q_2)$.

We wish to choose δ so small that when q_1, q_2 lie in $\mathsf{B}_\delta(p)$, so does this geodesic joining them. To do this, we take $\delta < \kappa$ and $< \frac{1}{4}\epsilon$ (where ϵ is given in Lemma 6.5).

Suppose $q_1, q_2 \in \mathsf{B}_\delta(p)$, and let C be the geodesic joining them. Then

$$d(q_1, q_2) \leqslant d(p, q_1) + d(p, q_2) \leqslant 2\delta < \tfrac{1}{2}\epsilon.$$

If q is any point on C

$$d(p, q) \leqslant d(q, q_1) + d(q_1, p) \leqslant d(q_1, q_2) + d(q_1, p) < \tfrac{3}{4}\epsilon,$$

by the above. Let $r = \max d(p, q)$ for q on C. It clearly suffices to show that $r \leqslant \max\,(d(p, q_1), d(p, q_2))$. If $r > \max\,(d(p, q_1), d(p, q_2))$ then r must be achieved at some interior point t_0 of C. But then $C'(t_0)$ must be tangent to Σ_r. Since $r < \frac{3}{4}\epsilon$, Lemma 6.5 implies that $C(t)$ lies outside B_r for small $|t - t_0| > 0$, contradicting the assertion that $r = \max d(p, q)$. This proves Theorem 6.2.

The exponential map that we have been studying so far has been a mapping from $T_p(M) \to M$. As we have seen, for any submanifold, K, the space $T^0(K) \subset T^*(M)$ is a reasonable generalization of $T_p^*(M)$. In the Riemannian situation it is more convenient to consider $T^\perp(K)$: The submanifold of $T(M)$ consisting of those tangent vectors v such that $\pi(v) \in K$ and $(w, v) = 0$ for all $w \in T_{\pi(v)}(K)$. ($T^\perp(K)$ is, of course, just the image of $T^0(K)$ under $\mathscr{L}^{-1}$.)

More precisely, let f be an immersion of $K \to M$. We define $T_f^\perp(K)$ to be the manifold consisting of all pairs (p, v) where $p \in K$ and $v \in T(M)$ with $\pi(v) \in T_{f(p)}(M)$ and $(w, v) = 0$ for $w \in f_* T_p(K)$. It is convenient to regard K as the submanifold of $T^\perp(K)$ of the form $(p, 0)$. In this set-up many of the previous results of this section generalize. We will state some of the important ones as exercises.

Exercise 6.1. Define exp: $T_f^\perp(K) \to M$ as follows: Put $\exp\,(p, v) = C(1)$ where C is the geodesic with $C(0) = p, C'(0) = v$. Show that exp is well defined and differentiable in some neighborhood of K in $T_f^\perp(M)$. If M is complete exp is defined and differentiable on all of $T_f^\perp(K)$.

Exercise 6.2. Show that there is a neighborhood W of K on which exp is an immersion.

Exercise 6.3. Show that if $K = R$ [i.e., (K, f) is a curve] then $T_f^\perp(K)$ is diffeomorphic to R^n. (This together with Exercise 6.2 gives a proof of Lemma 5.2.)

EXERCISE 6.4. Let f be a proper embedding of $K \to M$. Show that there is a neighborhood W of K in $T_f^{\perp}(K)$ such that the distance from $\exp(p, v)$ to $f(K)$ is $\|v\|$.

We know from Theorem 5.1, that geodesics minimize energy up to the first focal (conjugate) point and not beyond. We can make the same assertions about length.

THEOREM 6.3. *Let K be a submanifold of M and C a geodesic parametrized by arc length with $C(0) \in K$, $C'(0) \in T^{\perp}(K)$. Let $t_1 > 0$ be the first focal point of K along C. If $t_2 > t_1$, there is a curve B close to C with $B'(0) \in K$, $B(t_2) = C(t_2)$ such that $L[B] < L[C]$. In particular a geodesic does not minimize length past a conjugate point. If $t_2 < t_1$ then C provides a local minimum for arc length.*

In fact, $E[C] = L[C] = t_2$. There is a curve, B, for which $E[B] < E[C]$. By Cauchy's inequality $(L[B])^2 \leqslant E[B]\, t_2 < t_2$ so $L[B] < t_2 = L[C]$. This establishes the first part. We leave the second part of the theorem as an exercise to the reader.

7. Completeness. In this section we continue to study geodesics on Riemann manifolds. In particular, we wish to study the geometric properties of the metric d. Our main goal is to prove the two following assertions:

1) that completeness of the geodesic flow is equivalent to the completeness of M under d (the converse to the last assertion in Theorem 3.2);
2) that on a complete Riemann manifold every two points can be joined by a geodesic.

First we make some general observations about metric spaces.

DEFINITION 7.1. *Let M be a metric space. By an arc is meant a continuous map of a closed interval $a \leqslant t \leqslant b$ into M. A path is a continuous map of $a \leqslant t < b$ into M. An arc, C, is called a segment if*

$$d(C(t_1), C(t_2)) + d(C(t_2), C(t_3)) = d(C(t_1), C(t_3)) \tag{7.1}$$

for any $a \leqslant t_1 \leqslant t_2 \leqslant t_3 \leqslant b$.

Thus the segments play the role of "straight lines." It is clear that if $a \leqslant c \leqslant d \leqslant b$ and p is a segment on $[a, b]$, the restriction of p to $[c, d]$ is also a segment.

We first observe that if M is a Riemann manifold with associated metric, d, and C is a segment joining p to q then C is (up to reparametrization) a minimizing geodesic joining p to q. In fact, we can restrict attention to a small piece of C; i.e., we may assume that the image of C lies in a neighborhood with the properties given in Theorem 6.2. Then $d(r_1, r) + d(r, r_2) = d(r_1, r_2)$ implies that r lies on the unique geodesic in N joining r_1 to r_2. Thus C is locally, and therefore, globally, a geodesic. If $C(0) = p$ and $C(a) = q$, choose $0 = r_1 < \cdots < r_k = a$ such that the restriction of C is the minimizing geodesic joining r_i to r_{i+1}. Then

$$L[C] = \sum d(C(r_i), C(r_{i+1})) = d(p, q)$$

since C is a segment.

Conversely, if C is a minimizing geodesic, then every restriction of C to a subinterval is and this implies that C is a segment.

DEFINITION 7.2. *A path, $C: a \leqslant t < b \rightarrow M$ is called a ray if the image of C is closed and the restriction of C to $a \leqslant t \leqslant c$ is a segment for all $a \leqslant c < b$.*

Thus in the Euclidean plane, the rays are the straight lines extending to infinity in one direction. If we consider the plane with some points missing we can have rays of finite length. On a complete Riemann manifold all rays have infinite length. (Exercise!)

LEMMA 7.1. *Let $t_1 < t_2 < \cdots < t_{m+1}$ be real numbers and C_i a segment defined on the interval $t_i \leqslant t \leqslant t_{i+1}$, with $C_i(t_i) = C_{i-1}(t_i) = a_i$, $(i = 2, \cdots, m)$. Also put $a_1 = C(t_1)$, $a_m = C_m(t_{m+1})$. Then if*

$$d(a_1, a_2) + \cdots + d(a_m, a_{m+1}) = d(a_1, a_{m+1}), \tag{7.2}$$

the map $C: t_1 \leqslant t \leqslant t_{m+1} \rightarrow M$, given by $C(t) = C_i(t)$ for $t_i \leqslant t \leqslant t_{i+1}$, is a segment.

Proof. By induction, it is clearly sufficient to prove this for $m = 2$. For any $t_1 \leqslant x \leqslant y \leqslant t_3$ let us set $d(x, y) = d(C(x), C(y))$. We wish to show that for any three numbers x, y, z with $t_1 \leqslant x \leqslant y \leqslant z \leqslant t_3$,

$$d(x, y) + d(y, z) = d(x, z). \tag{7.3}$$

If $t_1 \leqslant x \leqslant y \leqslant z \leqslant t_2$, or $t_2 \leqslant x \leqslant y \leqslant z \leqslant t_3$ there is nothing to prove, since p_1 and p_2 are segments. Suppose $t_1 \leqslant x \leqslant t_2 \leqslant y \leqslant z \leqslant t_3$. Since C_1 and C_2 are segments

$$d(a_1, x) + d(x, a_2) = d(a_1, a_2)$$
$$d(a_2, y) + d(y, z) = d(a_2, z)$$
$$d(a_2, z) + d(z, a_3) = d(a_2, a_3).$$

Adding, and using (7.2), we obtain,

$$d(a_1, x) + d(x, a_2) + d(a_2, y) + d(y, z) + d(z, a_3) = d(a_1, a_3).$$

By the triangle inequality, $d(x, a_2) + d(a_2, y) \geqslant d(x, y)$, so that

$$d(a_1, x) + d(x, y) + d(y, z) + d(z, a_3) \leqslant d(a_1, a_3).$$

The triangle inequality also implies

$$d(a_1, x) + d(x, z) + d(z, a_3) \geqslant d(a_1, a_3).$$

Comparing the last two inequalities gives

$$d(x, z) \geqslant d(x, y) + d(y, z)$$

or

$$d(x, y) + d(y, z) = d(x, z)$$

proving the lemma.

LEMMA 7.2. *Let M be a connected Riemann manifold and d the associated metric. For every $p \in M$ there is a number $\rho > 0$ such that for any $q \in M$,*

1) *If $d(p, q) \leqslant \rho$ there is a segment, C, defined for $0 \leqslant t \leqslant d(p, q)$ with $C(0) = p$ and $C(d(p, q)) = q$.*
2) *If $d(p, q) > \rho$ there is an $x \in M$ such that $d(p, x) + d(x, q) = d(p, q)$ and $d(x, p) = \rho$.*

Proof. Choose ρ less than the radius of a normal neighborhood of p. Then 1) is clearly satisfied (cf. Exercise 7.1). Let us prove 2). Choose a sequence, C_i, of curves joining p to q such that $\lim L[C_i] = d(p, q)$. Let x_i be a point of $\Sigma_\rho \cap \operatorname{Im} C_i$. Then $d(x_i, q) \leqslant L[C_i] - \rho$. Since Σ_ρ is compact, we can choose a convergent subsequence. Let x be a limit point. Then $d(x, q) \leqslant d(p, q) - \rho$ and $d(p, x) + d(x, q) = d(p, q)$.

THEOREM 7.1. *Let M be a connected Riemann manifold with associated metric d. For any two points p and q, there is either a segment joining p to q or a ray C with $C(0) = p$, satisfying $d(p, x) + d(x, q) = d(p, q)$ for all x on $\operatorname{Im} C$.*

The two alternative possibilities are illustrated in the figure on page 210.

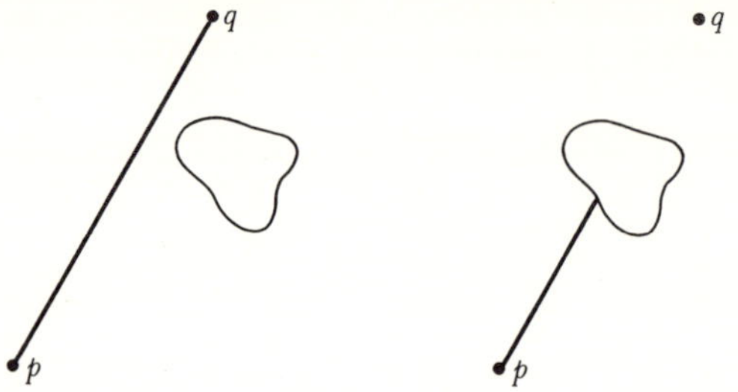

Proof. Suppose that there is no segment joining p and q. We will construct a sequence of points a_i, numbers t_i, and segments C_i such that $a_1 = p$ and

i) $d(a_1, a_2) + \cdots + d(a_{k-1}, a_k) + d(a_k, q) = d(a_1, q)$ for all k.

ii) $t_{i+1} - t_i = d(a_i, a_{i+1})$ $i = 1, \ldots, k - 1$.

iii) C_i is defined on the interval (t_i, t_{i+1}) with $C_i(t_i) = a_i, C_i(t_{i+1}) = a_{i+1}$ $(i = 1, \ldots, k - 1)$.

We construct the sequence by induction. If $k = 1$, the assertions are vacuous. Assume we have defined $a_1, \ldots, a_k, t_1, \ldots, t_k$. Now $d(a_1, a_k) + d(a_k, q) \leqslant d(a_1, a_2) + \cdots + d(a_{k-1}, a_k) + d(a_k, q) = d(a_1, q)$ so that

$$d(a_1, a_k) + d(a_k, q) = d(a_1, q) \tag{7.4}$$

and

$$d(a_1, a_2) + \cdots + d(a_{k-1}, a_k) = d(a_1, a_k). \tag{7.5}$$

There is no segment joining a_k to q; otherwise, by Lemma 7.1, there would be a segment joining p to q. According to Lemma 7.2, there is thus an x with $d(a_k, x) > 0$, $d(a_k, x) + d(x, q) = d(a_k, q)$ and a segment joining a_k to x. Consider the set, S_k, of all such x and let $d_k = \sup_{x \in S_k} d(a_k, x)$. Choose $a_{k+1} \in S_k$ so that

$$d(a_k, a_{k+1}) > \tfrac{1}{2} d_k. \tag{7.6}$$

By (7.4) and (7.5)

$$d(a_1, a_2) + \cdots + d(a_k, a_{k+1}) + d(a_{k+1}, q) = d(a_1, q).$$

Let $t_{k+1} = t_k + d(a_k, a_{k+1})$ and define C_k on (t_k, t_{k+1}) to be the geodesic parametrized by arc length joining a_k to a_{k+1}. Then conditions i), ii), and iii) are satisfied.

Let $\bar{t} = \lim\ t_i = \sum t_k - t_{k-1} < d(p, q)$. Define the path C on $0 \leqslant t \leqslant \bar{t}$ by $C(t) = C_i(t)$ if $t_i \leqslant t \leqslant t_{i+1}$. By Lemma 7.1 every subarc of C is a segment. We wish to show that the image of C is closed. Indeed, what we wish to prove is that there is no point $z \in M$ such that $z = \lim_{s_i \to \bar{t}} C(s_i)$ for some sequence $s_i \to \bar{t}$. But given any $\epsilon > 0$ we can find a $\delta > 0$ such that $d(C(s), C(t)) < \epsilon$ for $s, t > \bar{t} - \delta$. In fact, by construction $d(C(s), C(t)) = |s - t|$. Thus if $\lim C(s_i) = z$, we would have $\lim_{t \to \bar{t}} C(t) = z$ and we could extend C to a segment on $0 \leqslant t \leqslant \bar{t}$ by putting $C(\bar{t}) = z$. But then $\lim_{k \to a} d(a_k, z) = 0$ and (7.4) gives

$$d(a_1, z) + d(z, q) = d(a_1, q).$$

Now $z \neq q$ since there is no segment joining p to q. Therefore, we can find, by Lemma 7.2, an x such that

$$d(z, x) + d(x, q) = d(z, q), \qquad d(z, x) > 0$$

and a segment joining z to x.

For this x, we have

$$d(a_k, z) + d(z, x) + d(x, q) = d(a_k, z) + d(z, q) = d(a_k, q).$$

Thus

$$d(a_k, z) + d(z, x) = d(a_k, x).$$

By Lemma 7.2, we can find a segment joining a_k to x. Thus, for all k,

$$d_k > d(z, x) > 0. \tag{7.7}$$

On the other hand, by (7.6) $d_k < 2d(a_k, a_{k+1})$ so $d_k \to 0$, a contradiction. Thus there can be no limit point z; i.e., C is a ray. Q.E.D.

Notice, in the previous proof, $\{a_i\}$ was a Cauchy sequence without a limit point. This can't happen if M is complete. Thus

THEOREM 7.2. *In a complete connected Riemann manifold, every two points p and q can be joined by a geodesic of length $d(p, q)$.*

Another way of saying the same thing is that for any $p \in M$, exp *maps T_p onto M.*

We have seen that if M is complete, the geodesic flow is complete. We can now prove the converse. Assume that the geodesic flow is complete. This means that every geodesic can be extended to $(-\infty, \infty)$, i.e., there is no ray of finite length. We wish to conclude from this that M is complete. It suffices to show that every Cauchy sequence has a limit point.

Let x_i be a Cauchy sequence. Theorem 7.1 (and the fact that there are no rays of finite length) implies that there are minimizing geodesics, C_i, joining x_1 to x_i. Since x_i is a Cauchy sequence, $d(x_1, x_i)$ converges to some number, d. On the other hand, the vectors $C_i'(0) \in T_{x_1}(M)$ lie on the unit sphere. Choose a convergent subsequence, $C_{i_j}'(0)$ tending to a vector v. Let $x = C(d)$ where C is the geodesic with $C'(0) = v$. Then the completeness of the geodesic flow implies $x_{i_j} \to x$. Thus

THEOREM 7.3. *Let M be a connected Riemann manifold. Then M is complete (as a metric space with metric d) if and only if the geodesic flow is complete.*

We close this section by showing that every manifold has a Riemann metric which is complete. First,

LEMMA 7.3. *Let M_1 and M_2 be connected Riemann manifolds, and f an isometry from M_1 to M_2, i.e., $\|f_*(v)\| = \|v\|$ for any $v \in T(M_1)$. Then for any $p, q \in M_1$, $d(f(p), f(q)) \leqslant d(p, q)$.*

Proof. If C is a curve in M_1 joining p to q, $f \circ C$ joins $f(p)$ to $f(q)$. But $L[C] = L[f \circ C]$ so $d(p, q) = \inf L[C] = \inf L[f \circ C] \geqslant \inf_B L[B]$ where B ranges over all curves joining $f(p)$ to $f(q)$.

THEOREM 7.4. *Let M be a differentiable manifold. There exists a Riemann metric, g, on M which is complete.*

Proof. Let f be a proper embedding of M into Euclidean space, E^N (cf. Theorem 4.4 of Chapter II). Let $g = f^*(ds^2)$ where $ds^2 = dx_1^2 + \cdots + dx_N^2$ is the Euclidean metric on E^N. By Lemma 7.4, a Cauchy sequence, $\{x_i\}$ in M is mapped into a Cauchy sequence in E^N. Since E^N is complete and f is proper, $\{x_i\}$ converges.

8. Isometries. Let M be a Riemann manifold and $\tau : M \to M$ an isometry: i.e., τ is a diffeomorphism of M onto itself preserving the Riemann metric. The τ must also preserve the various geometric objects derived from the metric. For instance, let $\tau_* : T(M) \to T(M)$ be the induced mapping on $T(M)$. Then τ_* must leave the geodesic spray invariant, i.e., $(\tau_*)^*X = X$, and $\tau_* \delta_t = \delta_t \tau_*$, where X is the geodesic spray and δ the geodesic flow.

THEOREM 8.1. *Let* T *be a family of isometries of a Riemann manifold* M *and* S *the set of (simultaneous) fixed points of* τ (*i.e.*, $x \in S$ *if and only if* $\tau x = x$ *for all* $\tau \in \mathrm{T}$). *Then* $S = \bigcup N_i$ *where each* N_i *is a submanifold of* M. *Furthermore, every* $p \in S$ *has a neighborhood,* U, *and coordinates,* $x^1, \ldots, x^n$, *such that every* $\tau \in \mathrm{T}$ *is a linear transformation in terms of these coordinates.*

Let $p \in S$ and U a normal neighborhood of p. For any $\tau \in \mathrm{T}$, $\tau \circ \exp = \exp \circ \tau_*$ for all $v \in \exp^{-1} U$, since τ is an isometry. As $\tau_* : T_p(M) \to T_p(M)$ is linear, this says that τ is linear in terms of the normal coordinate system. The set of simultaneous fixed points of a family of linear transformations of a vector space is a linear subspace. Thus $S \cap U$ is a submanifold of U proving that S is a union of submanifolds.

It should also be remarked that if S is connected it actually is a submanifold. Also, if S contains a piece of a geodesic it must contain the whole geodesic.

THEOREM 8.2. *Let* T *be a finite*[9] *group of diffeomorphisms of a manifold* M, *and let* S *be the set of simultaneous fixed points. Then* S *is a union of submanifolds and every* $p \in S$ *has a neighborhood,* U, *and coordinates* $x_1, \ldots, x_n$ *in terms of which every* $\tau \in \mathrm{T}$ *is linear.*

It suffices to show that T is a set of isometries for some Riemann metric. Let g be any Riemann metric on M. (There exists one by Theorem 7.4.) Then $a = \sum_{\tau \in \mathrm{T}} \tau^* g$ is a Riemann metric (as the sum of positive definite bilinear forms is positive definite). Furthermore,

$$\beta^* a = \sum_{\tau \in \mathrm{T}} (\beta\tau)^* g = \sum_{\tau \in \mathrm{T}} \tau^* g = a \qquad \text{for } \beta \in \mathrm{T}$$

so the τ's are isometries.

[9] For those readers who know about Haar measure, we can replace finite by compact, and sum by integral.

CHAPTER V

Lie Groups

In geometry, a central role is played by the groups of automorphisms of the various structures that arise. Frequently, these groups possess differentiable structures. In this chapter, we study the relationship between the algebraic and differential properties of such groups.

1. Definitions. We now give the basic definition

DEFINITION 1.1. *A Lie group is a differentiable manifold, G, with a differentiable map* $\varphi: G \times G \rightarrow$ *which makes G into a group. That is,* φ *satisfies*

1) $\varphi(a, \varphi(b, c)) = \varphi(\varphi(a, b), c)$
2) *there is an* $e \in b$ *such that* $\varphi(e, b) = b$ *for all* b

and

3) *there is a differentiable map:* $b \rightarrow G$, *sending* $a \rightarrow a^{-1}$ *satisfying* $\varphi(a, a^{-1}) = e$.

We will denote $\varphi(a, b)$ *by ab.*

More succinctly, a Lie group is a group which has the structure of a differentiable manifold and for which the mapping of $G \times G \rightarrow G$ given by $(x, y) \to xy^{-1}$ is differentiable. In fact, the map $x \to x^{-1}$ can be written as $(e, x) \to ex^{-1}$ so 3) holds, and $(x, y) \to xy$ is the composite of the two differentiable maps $(x, y) \to (x, y^{-1})$ and $(x, y^{-1}) \to xy$. Something a bit stronger is also true.

LEMMA 1.1. *Let G be a group and a manifold so that the map of $G \times G \to G$ given by $(x, y) \to xy$ is differentiable. Then G is a Lie group.*

What we must show is that if the map $(x, y) \to xy$ is differentiable, then the map $x \to x^{-1}$ is differentiable. We first observe that it is sufficient to show that the map $x \to x^{-1}$ is differentiable in some neighborhood, N, of the identity, e. In fact, for any $a \in G$, the mappings $x \to xa$ (which we shall call R_a) and $x \to ax$ (which we shall call L_a) are differentiable. Now $x^{-1} = a(xa)^{-1}$ so that $x \to x^{-1}$ can be factored through $x \to R_a x \to (R_a x)^{-1} \to L_a(R_a x)^{-1}$. If we choose a so that $xa \in N$, we see that $x \to x^{-1}$ is differentiable, at x. Let us now show that there is such a neighborhood, N, of e: The maps L_x are all diffeomorphisms of G onto G. If we choose a coordinate neighborhood of the identity, the map L_x depends differentiably on x (in terms of the local coordinates). By the implicit function theorem, therefore, there is some neighborhood, N, of e such that $(L_x)^{-1}y$ depends differentiably on x for all $x, y \in N$. Thus $(L_x)^{-1}e = x^{-1}$ depends differentiably on x; i.e., the map $x \to x^{-1}$ is differentiable.

THEOREM 1.1. *Let G be a Lie group, M a differentiable manifold, and f a one-to-one immersion of $M \to G$ such that*

$$(1.1) \quad f(x)f(y) \in f(M) \quad \text{and} \quad (f(x))^{-1} \in f(M) \quad \text{for all } x, y \in M.$$

Define multiplication on M by $xy = f^{-1}(f(x)f(y))$. Suppose this map is continuous. Then M is a Lie group under this multiplication.

In fact, condition (1.1) says that $f(M)$ is a subgroup of G. Thus, the map, $\tilde{\varphi}$, of $M \times M \to G$ given by $(x, y) \to f(x)f(y)$ is differentiable. Let $z(x, y) = f^{-1}(f(x)f(y))$. We wish to show that the map, φ, of $M \times M \to M$ given by $(x, y) \to z(x, y)$ is differentiable. Since φ is continuous it suffices to show that there is a neighborhood, U, of $z(x, y)$ such that for every differentiable function, g, defined in U, $g \circ \varphi$ is differentiable at (x, y). Since f is an immersion, we can find a neighborhood V of $f(z)$, and U of z with coordinates $v^1, \ldots, v^n$ such that $f(U)$ is given by those points of V where $v^{p+1} = \cdots = v^n = 0$, and, $v^1, \ldots, v^p$

are coordinates on U. Thus any function g on U can be extended to a function $\tilde{g}$ by setting $\tilde{g}\ (v^1, \ldots, v^n) = g(v^1, \ldots, v^p, 0, \ldots, 0)$. But $g \circ \varphi = \tilde{g} \circ \varphi$ which is differentiable.

It turns out that the assumption that φ be continuous is superfluous, cf. Section 3 below. In view of this we make the following definition:

DEFINITION 1.2. *Let G_1 and G_2 be Lie groups, and f an isomorphism of $G_1 \to G_2$ which is a one-to-one immersion. We shall call the pair (G_1, f) a Lie subgroup of G_2. In general, a differentiable map of $G_1 \to G_2$ which is a group homomorphism is called a Lie homomorphism.*

Thus a Lie subgroup of a Lie group is a subgroup in the group theoretical sense which is also an immersed submanifold. Notice that it need not be an embedded submanifold as we will see by the following examples.

Examples.

(1) An n-dimensional vector space, V, regarded as a group under addition is clearly a Lie group.

(2) The product of any number of Lie groups is a Lie group when provided with the product structures of group and manifold.

(3) The torus T^n being the product of n circles, S^1, is a Lie group. On T^n we can introduce angular "coordinates" $(\theta_1, \ldots, \theta_n)$ (where θ_i is defined mod 1) so that $\theta_i(p + q) = \theta_i(p) + \theta_i(q)$. (Since multiplication is commutative on T^n it is convenient to write it as $+$.) For any (λ_1, λ_2) the map, f, of $R \to T^2$ given by $f(t) = (\lambda_1 t, \lambda_2 t)$ is clearly an immersion and a homomorphism. If $(\lambda_1 t, \lambda_2 t) = (\lambda_1 s, \lambda_2 s)$, then $\lambda_1(s - t) = m$, $\lambda_2(s - t) = n$ for some integers m and n, i.e., $\lambda_1/\lambda_2 = m/n$ is a rational number. Thus, if λ_1/λ_2 is irrational (R, f) defines a Lie subgroup of T^2. It is easy to see, that in this case $f(R)$ is dense in T^2, *a fortiori* it is not an embedding.

(4) The group $GL(n)$ is a Lie group.

(5) The group $O(n)$ of orthogonal $n \times n$ matrices is a Lie group. In fact, $A \in O(n)$ if and only if $A \in GL(n)$ and $AA^t = I$ (where A^t is the transpose of A), i.e., if $A = (A^t)^{-1}$. Consider the transformation $\tau: GL(n) \to GL(n)$ given by $\tau(B) = (B^t)^{-1}$. Now τ is clearly differentiable and $\tau^2(B) = (((B^t)^{-1})^t)^{-1} = B$. Thus the set of diffeomorphisms of $GL(n)$ given by {identity, τ} is a finite group. According to Theorem 8.2 of Chapter 4, the fixed points of τ, i.e., $O(n)$, form a union of submanifolds of $GL(n)$. An elementary argument shows that $O(n)$ has two connected components, $O(n)^+$ consisting of those elements with determinant 1 and $O(n)^-$ consisting of those whose determinants is -1. Since

$O_n^+ = O_n^- \cdot A$ for a suitable A, $O(n)^+$ and $O(n)^-$ have the same dimension and thus $O(n)$ is an embedded submanifold of $GL(n)$.

2. The invariant forms and the Lie algebra. Let G be a Lie group. We shall say that a differential form, Ω, defined on G is left invariant if $L_a^*\Omega = \Omega$ for all $a \in G$;[1] similarly, Ω is right invariant if $R_a^*\Omega = \Omega$ for all $a \in G$. If ω_1 and ω_2 are left (or right) invariant forms, so are $b\omega_1 + c\omega_2$ and $\omega_1 \wedge \omega_2$ for any constants b and c. Also, $L_a^*\, d\omega = dL_a^*\omega$; thus if ω is left invariant so is $d\omega$. In other words, the space of left (or right) invariant forms is a subalgebra of the algebra of all forms of b and is closed under d. We shall call these invariant algebra $\mathscr{I}_l(G)$ (left invariant) and $\mathscr{I}_r(G)$ (right invariant).

THEOREM 2.1. *Let G be a Lie group and e the identity element. The mapping sending $\omega \to \omega_e$ is an isomorphism of $\mathscr{I}_l(G)$ onto $\wedge(T_e^*(G))$.*

The restriction map is obviously a homomorphism. We wish to show that it is one to one and onto. We first show that it is one to one. We wish to show that if ω, ω' are in $\mathscr{I}_l(G)$ and $\omega_e = \omega'_e$, then $\omega_a = \omega'_a$ for all $a \in G$.

Now $(L_a^*\omega)_e = \omega_a \circ \wedge(L_{a*})$ and $(L_a^*\omega')_e = \omega'_a \circ \wedge(L_{a*})$.

Since ω and ω' are left invariant $(L_a^*\omega)_e = \omega_e = \omega'_e = (L_a^*\omega')e$ so $\omega_a \circ \wedge(L_{a*}) = \omega'_a \circ \wedge(L_{a*})$. As L_{a*} is a one to one map of $T_e \to T_a(M)$, this implies that $\omega_a = \omega'_a$.

We now show that the map is onto. Let $\alpha \in \wedge\ T_e(G)$. Define ω by $\omega_a = \alpha \circ \wedge(L_{a*^{-1}})$.

We let the reader verify that the ω so defined is differentiable. Then $\omega_e = \alpha$ and

$$\begin{aligned}(L_b^*\omega)_a &= \omega_{ba} \circ (\wedge L_{b*}) = \alpha \circ \wedge(L_{(ba)*^{-1}}) \circ \wedge(L_{b*})\\ &= \alpha \circ \wedge(L_{a*^{-1}}) = \omega_a\end{aligned}$$

so that ω is left invariant.

In particular, the space of left invariant *linear* differential forms is n-dimensional (if n is the dimension of G). Let $\omega^1, \ldots, \omega^n$ be a basis for this space. Then

$$d\omega^i = \sum_{1 \leqslant j < k \leqslant n} C_{jk}^i \omega^j \wedge \omega^k \tag{2.1}$$

[1] Recall that $L_a(x) = ax$ and $R_a x = xa$.

since the $d\omega^i$ are left invariant. If we apply L_a^* to both sides of (2.1) we obtain that $C^{jk} \circ L_a = C_i^{jk}$; i.e., *the* C_i^{jk} *are constants.*

These constants are known as the constants of structure of G. Equation (2.1) is known as the *Maurer-Cartan equations.*

Let X be a vector field on G. We say that X is left (right) invariant if $L_a^*X = X$ for all $a \in G(R_a^*X = X)$. If X and Y are left invariant, so are $X + Y$ and $[X, Y]$. Thus *the left invariant vector fields form a Lie algebra.* Let X be a left invariant vector field, and T_t the local one parameter group it generates. $T_t x$ is, in particular, defined for all x lying in some neighborhood U of e, and all t in some interval I about 0. For y and x sufficiently close to e, and sufficiently small t, $T_t L_y x$ and $L_y T_t x$ are defined and are equal. $T_t(yx) = y(T_t x)$. Taking $x = e$, we have

$$T_t y = y T_t e. \tag{2.2}$$

Equation (2.2) is valid for y close to e and t close to zero. We can extend its validity to all values of t and y. In fact, for any $t \in R$, choose an n so that (2.2) is valid for t/n. Then define $T_t y$ to be $y(T_{t/n}e)^n$. By the uniqueness of the solution to differential equations, this definition is consistent. Also, for any $z \in G$, define $T_t z = z(T_t e)$. Since X is left invariant, the one parameter group of transformations so defined has X as its infinitesimal generator. The mapping of $R \to G$ given by $t \to T_t e$ defines a one parameter subgroup. In fact $T_{(t+s)}e = T_t(T_s e) = (T_s e) T_t(e)$. In resumé,

THEOREM 2.2. *The space of left invariant vector fields on G form an n-dimensional Lie algebra, which we shall denote by g. Each $X \in g$ generates a globally defined one parameter group of transformations, T_t, of G. The curve $T_t e$ is a Lie subgroup of G and for any $y \in G$, (2.2) holds.*

The only statement we have not yet verified is that dim $g = n$. This follows from the fact that the map of $g \to T_e(G)$ sending $X \to X_e$ is one to one. (The proof is the same as for invariant forms.) We shall sometimes identify g with $T_e(G)$. Let $X_1, \ldots, X_n$ be a basis of G. Let $\alpha^1, \ldots, \alpha^n$ be the basis of $T_e(G)$ dual to $X_{1e}, \ldots, X_{ne}$, and let $\omega^1, \ldots, \omega_i^n$ be the left invariant forms corresponding to $\alpha_1, \ldots, \alpha_n$. Then

$$\langle X_i | \omega^j \rangle = \delta_i^j .$$

By [Chapter III, Equation (1.10)]

$$\langle X_j \wedge X_k \,|\, d\omega^i \rangle = X_j\langle X_k \,|\, \omega^i \rangle - X_k \langle X_j \,|\, \omega^i \rangle - \langle [X_j, X_k], \omega^i \rangle.$$

As the first two terms on the right vanish (they are Lie derivatives of constants) we have

$$\langle X_j \wedge X_k \mid d\omega^i \rangle = -\langle [X_j, X_k], \omega^i \rangle.$$

On the other hand, by (2.1)

$$\langle X_j \wedge X_k \mid d\omega^i \rangle = \langle X_j \wedge X_k \mid \sum C^i_{lm} \omega^l \wedge \omega^m \rangle = C^i_{jk}.$$

Thus, for $j < k$, $\langle [X_k, X_j], \omega^i \rangle = C^i_{jk}$, or

$$[X_k, X_j] = \sum C^i_{jk} X_i. \tag{2.3}$$

[It will be convenient to define C^i_{jk} for all j and k to be antisymmetric. Then (2.3) is valid for all k and j.]

Let G be a Lie group and f a Lie homomorphism of $H \to G$. The map f_{*e} of $T_e(H)$ into $T_e(G)$ can be regarded as a linear map, f_*, of the Lie algebra, h, of H into the Lie algebra, g, of G. For any $x \in H$, $L_{f(x)} \circ f = f \circ L_x$. Thus,

$$L_{f(x)*} f_{*e}(T_e(H)) = f_{*x}(T_x(H)). \tag{2.4}$$

Equation (2.4) says that any $X \in h$ is f related to $f_*(X)$ (cf. Chapter II, Section 8). Thus, by Theorem 8.4 of Chapter II, for any $X, Y \in h$, $f_*([X, Y]) = [f_*(X), f_*(Y)]$. Thus *$f_*$ is a Lie algebra homomorphism of h into g.* In particular $f_*(h)$ is a subalgebra of g.

If f is an isomorphism, then (2.2) says that any left invariant vector field tangent to $f(H)$ at e is tangent to $f(H)$ at all points. Thus $f_*(h) \subset g$ spans a differential system having $f(H)$ as an integral manifold.

We are now in a position to show how to remove the assumption about continuity of φ in Theorem 1.1. In fact, we wish to show that the map, φ, of $M \times M \to M$ is differentiable. Let g_1 be the tangent space to $f(M)$ at e. This generates a subspace of left invariant vector fields which we shall also denote by g_1. If we could show that g_1 is a subalgebra then the differential system generated by g_1 is completely integrable. Condition (1.1) implies that $f(M)$ is a maximal integral manifold of this differential system. By (1.1) the differential map of $M \times M \to G$ actually maps into $f(M)$. Applying Exercise 5.2 of Chapter III yields the desired result. That g_1 is a subalgebra also follows from (1.1) as will be seen from the discussion in Section 3.

Let g_1 be a subalgebra of g. It defines a differential system which is completely integrable. Let G_1 be the maximal integral manifold of g_1 containing e. For any $y \in G$, L_y leaves the differential system, g_1, invariant and therefore maps any maximal integral manifold of g_1 onto a maximal integral manifold. If $y \in G_1$ then $L_{y^{-1}}y = e$, so that $L_{y^{-1}}$ takes G_1 into an integral manifold containing e, i.e., into itself. Thus if $x, y \in G_1$ so is $y^{-1}x$. By Theorem 1.1, this implies that G_1 is a Lie subgroup of G.

Any maximal integral manifold, K, of this completely integrable system is a coset of G_1. In fact, since the system is invariant under left translation, any left translate of K is also a maximal integral manifold. If we left translate by the inverse of an element of K we obtain a maximal integral manifold containing e. Thus K is a coset of G_1.

THEOREM 2.3. *Let f_1 and f_2 be two differentiable maps of a connected manifold M into a Lie group G such that*

$$f_1^*(\omega^i) = f_2^*(\omega^i) \tag{2.5}$$

for all left invariant forms ω^i. Then f_1 and f_2 differ by a left translation. That is, there is an $x \in G$ such that $f_2 = L_x \circ f_1$.

Proof. Consider the direct product group $G \times G$. Let π_1 and π_2 denote the projections onto the first and second components. The forms $\pi_1^*(\omega^i)$ and $\pi_2^*(\omega^i)$ are linearly independent left invariant forms on $G \times G$. The forms $\pi_2^*(\omega^i) - \pi_2^*(\omega^i)$ clearly define the completely integrable system, $\mathscr{D}$, corresponding to the diagonal subgroup [of all elements of the form (x, x)]. Let h be the map of M into $G \times G$ given by $h(y) = (f_1(y), f_2(y))$ for all $y \in M$. Then (2.5) asserts that $h^*(\pi_1^*(\omega^i) - \pi_2^*(\omega^i)) = 0$. Since M is connected, h must map M into a maximal integral leaf of $\mathscr{D}$. Thus h maps M into a coset of the diagonal, i.e., there exist x_1 and $x_2 \in G$ so that $x_1 f_1(y) = x_2 f_2(y)$ for all $y \in M$. Thus f_1 and f_2 differ by a translation.

COROLLARY. *Let G be a connected Lie group and f a diffeomorphism of G onto itself preserving the left invariant forms. Then f is a left translation.*

Just take $M = G$.

THEOREM 2.4. *Let G be a Lie group with left invariant forms $\omega^1, \ldots, \omega^n$, and constants of structure c^i_{jk}. Let M be a differentiable manifold and $\theta_1, \ldots, \theta_n$ forms on M satisfying*

$$d\theta^i = \sum c^i_{jk}\theta^j \wedge \theta^k. \tag{2.6}$$

Then every $y \in M$ has a neighborhood U and a diffeomorphism f of $U \to G$ so that

$$\omega^i = f^*(\theta^i). \tag{2.7}$$

Any two such maps differ by left translation. If M and G are simply connected then M is diffeomorphic to G.

The second part of the theorem is an immediate consequence of the first and Theorem 2.3. The last part would follow from standard facts about simply connected spaces. Thus, all that we have to prove is the first part.

Instead of looking for the map (from M to G) directly, we look for its graph. The graph of any map is a submanifold of $M \times G$ whose projection onto the first factor is a diffeomorphism. Let π_1 and π_2 denote the projection of $M \times G$ onto the M and G. Put $\bar{\theta}_i = \pi_1^* \theta^i$ and $\bar{\omega}^i = \pi_2^* \omega^i$. The projection π_1 is locally a diffeomorphism on an n-dimensional submanifold, K, of $M \times G$ if $\bar{\theta}_1, \ldots, \bar{\theta}_n$ are linearly independent on K. Then K is the graph (locally) of a map of (an open set of) m into G. This map is a diffeomorphism if the projection, π_2, of K onto G is a diffeomorphism, which will happen if the forms $\bar{\omega}_1, \ldots, \bar{\omega}_n$ are independent on K. Condition (2.7) becomes

$$\bar{\theta}^i - \bar{\omega}^i = 0, \qquad i = 1, \ldots, n \tag{2.8}$$

on K. Now (2.8) is a differential system and we are looking for integral manifolds. But (2.8) is completely integrable. In fact,

$$\begin{aligned} d(\bar{\theta}^i - \bar{\omega}^i) &= \sum c^i_{jk}(\bar{\theta}^j \wedge \bar{\theta}^k - \bar{\omega}^j \wedge \bar{\omega}^k) \qquad \text{[by (2.6)]} \\ &= \sum c^i_{jk}(\bar{\theta}^j \wedge (\bar{\theta}^k - \bar{\omega}^k) + (\bar{\theta}^j - \bar{\omega}^j) \wedge \bar{\omega}^k). \end{aligned}$$

Thus through any point (y, x) of $M \times G$ there passes an n-dimensional integral manifold of (2.8). Since the forms $\bar{\theta}^1, \ldots, \bar{\theta}^n$, $\bar{\omega}^1, \ldots, \bar{\omega}^n$ are independent, at every $z \in M \times G$ the forms $\bar{\theta}^1_z, \ldots, \bar{\theta}^n_z$ and $\bar{\omega}^1_z, \ldots, \bar{\omega}^n_z$ are independent on the subspace of $T_z(M \times G)$ defined by $\bar{\theta}^i_z - \bar{\omega}^i_z = 0$. Thus K defines a diffeomorphism of a neighborhood of y in M onto a neighborhood of x in G satisfying (2.7).

Thus we see that the constants c^i_{jk} do indeed determine the structure of the group G. Theorems 2.3 and 2.4 can be regarded as uniqueness theorems: The group G (if it is simply connected) is determined by its constants of structure. We can ask for the corresponding existence theorem: Given any constants c^i_{jk}, does there exist a Lie group having c^i_{jk}

as its constants of structure? Note that the c^i_{jk} must satisfy certain conditions. They must be antisymmetric in j and k. Furthermore, if we take the exterior derivative of (2.1) we obtain

$$\begin{aligned} 0 &= \sum c^i_{jk}(d\omega^j \wedge \omega^k - \omega^j \wedge d\omega^k) \\ &= \sum_{j<k} c^i_{jk}\Big(\sum_{l<m} c^j_{lm}\omega^l \wedge \omega^m \wedge \omega^k - \sum_{l<m} c^k_{lm}\omega^j \wedge \omega^l \wedge \omega^m\Big). \end{aligned}$$

This implies that

$$\sum_r (c^r_{jk}c^s_{rl} + c^r_{kl}c^s_{rj} + c^r_{lj}c^s_{rk}) = 0. \tag{2.9}$$

We let the reader verify that (2.9) can also be regarded as the necessary and sufficient condition that the multiplication defined by (2.3) be a Lie algebra; it is just Jacobi's identity.

Therefore we must ask, given constants c^i_{jk} which are antisymmetric in j and k and satisfy (2.9), does there exist a Lie group having the c^i_{jk} as constants of structure? The answer to this question is yes. We shall not prove this fact since it depends on some delicate algebraic considerations that are out of the scope of this book.

3. Normal coordinates, exponential map. Let G be a Lie group. We will now construct a spray on G. We must define a vector field, Z, on $T(G)$. For each $v \in T(G)$, there is a unique $X \in g$ such that $X_{\pi(v)} = v$ (in other words, for each tangent vector at any point of G extends to a unique left invariant vector field on G). Then $T(X)$ (the variational field) is a vector field on $T(G)$. We set $Z_v = T(X)_v$. We leave to the reader the task of verifying that the rule $v \to Z_v$ is differentiable, i.e., defines a vector field. It follows immediately from the definition of the variation, $T(X)$, that $\pi_* Z_v = v$ and $\mu_{t*}(Z_v) = (1/t)\, Z_{\mu_t v}$ for any $t \neq 0 \in R$. Thus Z is a spray, we shall call it the left invariant spray. Normal coordinates of this spray at e (given by the exponential map) will be called *normal coordinates of* G.

Let X be a left invariant vector field and C a solution curve of X. The curve C' in $T(G)$ is then a solution curve of $T(X)$. For each t, $C''(t) = T(X)_{C'(t)}$ and $\pi(C'(t)) = X_{C(t)}$ so that $C''(t) = Z_{C'(t)}$. In other words, C' is a solution curve of Z, i.e., C is a geodesic of Z.

Let us consider the exponential map of Z at e. It is a differentiable map of $T_e(G) \to G$. We can regard it as a map of $g \to G$. If $X \in g$, the curve $\exp tX$ is a geodesic of Z passing through e and tangent to X_e at e.

On the other hand, by the previous remarks, T_t is also a geodesic of the spray tangent to X_e at e (where T_t is the one parameter family of transformations generated by X). Thus, by uniqueness, $\exp tX = T_t e$. In view of Theorem 6.1 of Chapter IV, we have proved

THEOREM 3.1. *The differentiable map* exp *of g into G is a diffeomorphism of some neighborhood of O in g onto a neighborhood of e. For any $X \in g$, the map $t \to \exp tX$ defines a one-dimensional Lie subgroup of G and all Lie homomorphisms of $R \to G$ are of this form.*

Only the last assertion of the theorem remains to be proved. Let $f: R \to G$ be a Lie homomorphism. Define the one parameter family, T_t, of diffeomorphisms of $G \to G$ by $T_t x = xf(t)$. Then $T_{t+s}x = xf(t+s) = xf(t)f(s) = T_s(T_t x)$ so that T_t is a one parameter group. Furthermore $yT_t x = y(xf(t)) = yxf(t)$ so $L_y T_t = T_t L_y$. Thus, if X is the infinitesimal generator of T_t, $X \in g$. Thus $f(t) = T_t e = \exp tX$.

Let $f: G_1 \to G_2$ be a Lie homomorphism. Any $X \in g_1$ is f related to $f_*(X) \in g_2$. Thus, if T_{1t} is the one parameter family of transformations of G_1 generated by X and T_{2t} the one parameter group of transformations of G_2 generated by $f_*(X)$, then $f \circ T_{1t} = T_{2t} \circ f$. If we apply this to e, we obtain

(3.1) $$\exp \circ f_* = f \circ \exp.$$

Various facts follow immediately from (3.1):

COROLLARY 3.1. *Every 1–1 Lie homomorphism of $G_1 \to G_2$ is an immersion.*

In fact, if $f_{*p}(v) = 0$ for some $v \neq 0 \in T_p(G_1)$, then $f_*(X) = 0$ where $X_e = L_{p^{-1}*}(v) \neq 0$. But then $\exp \circ f_*(tX) = e = f(\exp tX)$. Since $X \neq 0$, $\exp tX \not\equiv e$, and f cannot be 1–1.

COROLLARY 3.2. *Every continuous homomorphism, φ, of $R \to G$ is differentiable.*

It clearly suffices to show that φ is differentiable near zero. Let U be a normal neighborhood of e and $x^1, \ldots, x^n$ a normal coordinate system. If $q = \exp X \in U$ and $q^m = \exp mX \in U$ for $1 \leqslant m \leqslant k$, then $x^i(q^m) = mx^i(q)$ [since $x^i(\exp tX) = tx^i(\exp X)$]. Choose $\epsilon > 0$ so small that

$\varphi(t) \in U$ for all $|t| < \epsilon$. For any $0 < t_0 < \epsilon$ and $0 < |m| \leqslant n$ $x^i \circ \varphi\,(mt_0/n) = mx^i \circ \varphi(t_0/n)$, and $x^i \circ \varphi(t_0/n) = (1/n)\, x^i \circ \varphi(t_0)$. Thus $x^i \circ \varphi(mt_0/n) = (m/n)\, x^i \circ \varphi(t_0)$. By continuity,

$$x^i \circ \varphi(rt_0) = rx^i \circ (\varphi(t_0))$$

for all $|r| \leqslant 1$. Thus φ is linear in terms of normal coordinates, and therefore differentiable.

Now let us see what multiplication looks like in terms of normal coordinates. Let U be a normal neighborhood of the identity, $X_1, \ldots, X_n$ a basis of g and $x^1, \ldots, x^n$ the corresponding normal coordinates. Let $V \subset U$ be a neighborhood of e such that $ab \in U$ for all $a, b \in V$. In other words, under the multiplication map of $G \times G \to G$, $V \times V$ is mapped into U. Let y^i and z^i be the functions on $V \times V$ given by $y^i = x \circ \pi_1$, $z^i = x^i \circ \pi_2$ where π_1 and π_2 are the projections of $V \times V$ onto the first and second components, respectively. Then $y^1, \ldots, y^n, z^1, \ldots, z^n$ form a coordinate system on $V \times V$ and in terms of these coordinates, multiplication is given by

$$x^i = f^i(y^1, \ldots, y^n;\, z^1, \ldots, z^n),$$

that is, for any $a, b \in V$, $x^i(ab) = f^i(x^1(a), \ldots, x^n(a);\, x^1(b), \ldots, x^n(b))$. We wish to find the first few terms in the Taylor expansion of the f^i Since $ea = ae = e$,

$$f^i(0, \ldots, 0;\, z^1, \ldots, z^n) = z^i$$

and

$$f^i(y^1, \ldots, y^n;\, 0, \ldots, 0) = y^i.$$

Thus

(3.2) $\quad f^i(y^1, \ldots, y^n;\, z^1, \ldots, z^n) = y^i + z^i + \text{second-order terms.}$

Put another way, for $Y, Z \in g$

(3.3) $\quad \exp t(Y + Z) = \exp tY \exp tZ + O(t^2).$

Now let us compute the second-order terms of the f^i. Let $Y = \sum y^i X_i$ and $Z = \sum z^i X_i$. For any function, ψ on G, we have, for every fixed s,

$$\frac{d}{dt}\psi(\exp sY \exp tZ) = \lim_{h=0} \frac{1}{h}[\psi(\exp sY \exp tZ \exp hZ) - \psi(\exp sY \exp tZ)]$$

$$= \lim_{h=0} \frac{1}{h}[\psi(T_h \exp sY \exp tZ) - \psi(\exp sY \exp tZ)]$$

where T_h is the one parameter group generated by Z. Thus

$$\frac{d}{dt}\psi(\exp sY \exp tZ) = Z\psi(\exp sY \exp tZ)$$

since Z is left invariant. Iterating this equation, we obtain

$$\frac{d^r}{dt^r}\psi(\exp sY \exp tZ) = Z^r\psi(\exp sY \exp tZ).$$

In particular, by Taylor's formula,

$$x^i(\exp sY \exp tZ) = x^i(\exp sY) + t(Zx^i)(\exp sY) + t^2(Z^2x^i)\exp sY + O(t^3).$$

Similarly, for any ψ, $\frac{d}{ds}\psi(\exp sY) = (Y\psi)(\exp sY)$. Applying this to the members of the right-hand side of the above equation, we get

$$x^i(\exp sY \exp tZ) = s(Yx^i)(e) + t(Zx^i)e + s^2(Y^2x^i)(e) + st(YZx^i)e + t^2(Z^2x^i)(e) + \text{third-order terms}. \tag{3.4}$$

Thus,

$$f^i(y^1, \ldots, y^n; z^1, \ldots, z^n) = y^i + z^i + Y^2x^i(e) + YZx^i(e) + Z^2x^i(e) + \text{third-order terms}. \tag{3.5}$$

In particular,

$$f^i(y^1, \ldots, y^n; z^1, \ldots, z^n) - f^i(z^1, \ldots, z^n; y^1, \ldots, y^n) = [Y, Z]x^i(e) + \text{third-order terms}. \tag{3.6}$$

Now (3.3) asserts that $x^i(\exp Y \exp Z) = x^i(\exp Y) + x^i(\exp Z)$ + second-order terms. Thus, for $(a, b) \in V \times V$,

$$x^i(ab) - x^i(a) - x^i(b) = \sum g_{ijk}(a, b)x^j(a)x^k(b) \tag{3.7}$$

where the g_{ijk} are differentiable functions on $V \times V$. Take $a = cd$ and $b = d^{-1}c^{-1}dc$ in (3.7), obtaining

$$x^i(dc) - x^i(cd) - x^i(d^{-1}c^{-1}dc) = \sum g_{ijk}(cd, d^{-1}c^{-1}dc)x^j(cd)x^k(d^{-1}c^{-1}dc).$$

Now $x^i(d^{-1}c^{-1}dc)$ vanishes when either d or c is the identity and therefore is a term of second order. Thus

$$x^i(d^{-1}c^{-1}dc) = x^i(cd) - x^i(dc) + \text{third-order terms}.$$

Taking $c = \exp Y$ and $d = \exp Z$, (3.6) implies that[2]

$$x^i((\exp Z)^{-1}(\exp Y)^{-1}(\exp Z)(\exp Y)) = x^i(\exp[Y, Z]) + \text{third-order terms.} \tag{3.8}$$

Now $d(d^{-1}c^{-1}de) = c^{-1}dc$. Taking $a = d, b = d^{-1}c^{-1}dc$ in (3.7), remarking that $x^i(d^{-1}c^{-1}de)$ is of second order, we obtain

$$x^i(c^{-1}dc) = x^i(d) + x^i(d^{-1}c^{-1}dc) + \text{third-order terms.}$$

Taking $c = \exp Y$ and $d = \exp Z$ and using (3.6), we obtain

$$x^i((\exp Y)^{-1}\exp Z \exp Y) = x^i(Z) + x^i(\exp[Y, Z]) + \text{third-order terms.} \tag{3.9}$$

For any $x \in G$, let $A(x)$ denote the inner automorphism corresponding to x. That is, $A(x)$ is the Lie isomorphism of $G \to G$ given by

$$A(x)y = x^{-1}yx. \tag{3.10}$$

Then Equation (3.9) can be written as

$$x^i[A(\exp Y) \exp Z] = x^i(Z) + x^i(\exp[Y, Z]) + \text{third-order terms.} \tag{3.11}$$

Let us now derive some consequences of Equations (3.2)–(3.11).

LEMMA 3.1. *Let* $g = V_1 + \cdots + V_k$ *be a direct sum decomposition of* g *(as a vector space). For any* $X \in g$, $X = X_1 + \cdots + X_k$, $(X_i \in V_i)$, *put*

$$\varphi(X) = \exp X_1 \exp X_2 \ldots \exp X_k. \tag{3.12}$$

Then this map $\varphi : g \to G$ *is a diffeomorphism of some neighborhood of* 0 *in* g *onto a neighborhood of* e *in* G.

It is obvious that φ is a differentiable map. According to (3.2), $\varphi(X) = \exp(X_1 + \cdots + X_k) +$ second-order terms. Thus $\varphi_* : T_0(g) \to T_e(G)$ is just l_0 (Theorem 6.1 of Chapter IV); i.e., φ is a diffeomorphism, locally.

---

[2] Recall that by definition, $x^i(\exp W) = (Wx^i)(e)$ for any $W \in g$.

THEOREM 3.2. *Every continuous homomorphism of a Lie group G_1, into a Lie group G_2 is differentiable.*

By translation, it suffices to prove that f is differentiable in some neighborhood of e. Let $X_1, \ldots, X_n$ be a basis for the Lie algebra g_1 of G_1. Let f be the homomorphism of $G_1 \to G_2$. Then the map $t \to f(\exp tX_i)$ is a continuous homomorphism of $R \to G_2$ and therefore is differentiable by Corollary 3.2. Thus, there exist Y_i in the Lie algebra g_2 of G_2 such that $f(\exp tX_i) = \exp tY_i$. Since f is a homomorphism,

$$f((\exp t_1X_1)(\exp t_2X_2) \ldots (\exp t_nX_n)) = (\exp t_1Y_1) \ldots (\exp t_nY_n). \tag{3.13}$$

Now the map, F, of $g_1 \to G_2$ sending $X = t_1X_1 + \cdots + t_nX_n$ into $(\exp t_1Y_1) \cdots (\exp t_nY_n)$ is clearly differentiable. Equation (3.13) says that $F = f \circ \varphi$ (where φ is the map given by (3.12) corresponding to the decomposition $g = \{X_1\} + \cdots + \{X_n\}$. By Lemma 3.1, φ is a diffeomorphism at 0. Thus $f = F \circ \varphi^{-1}$ at some neighborhood of e, and is thus differentiable there.

Theorem 3.2 implies that the topology of a Lie group determines its differentiable structure: If G_1 and G_2 are Lie groups and f is a homeomorphism and a homomorphism of G_1 onto G_2, then f is a diffeomorphism.

Let us examine Equations (3.10) and (3.11) some more. For each $x \in G$, the transformation $A(x)$ leaves e fixed. It therefore induces a transformation $A(x)_*$ of $T_e(G) = g$ into itself. For any $x, y \in G$, $A(xy) = A(x)A(y)$ so that $A(xy)_* = A(x)_*A(y)_*$. Thus $x \to A(x)_*$ is a representation of G by linear transformations on g.

DEFINITION 3.1. *The linear transformation $x \to A(x)_*$ is called the adjoint representation. The linear transformation $A(x)_*$ will be denoted by $Ad\, x$.*

Let X and Y be elements of g. According to (3.11) the curve $\exp(sX)$ is mapped by $A(\exp tY)$ into $\exp(sX + st[Y, X])$ up to terms of third order in s and t. Thus

$$\frac{d}{dt} Ad(tY)X|_{t=0} = [Y, X]. \tag{3.14}$$

DEFINITION 3.2. *The representation of g as a Lie algebra of linear transformations of g by left bracket is called the adjoint representation of g. Thus*

$$(Ad\ Y)X = [Y, X].$$

By what we have just remarked we have,

$$Ad(\exp Y)X = \exp(Ad\ Y)X. \tag{3.15}$$

4. Closed subgroups. Using the results of the previous section we will prove

THEOREM 4.1. *Let G be a Lie group and H a closed subset of G which is (algebraically) a subgroup. Then H is a Lie subgroup of G. (More succinctly, every closed subgroup of a Lie group is a Lie subgroup.)*

According to Theorem 1.1, to show that H is a Lie group, it suffices to show that it is a submanifold of G. To do this, it is sufficient to show (according to Exercise 2.4 of Chapter II) that every $p \in G$ has a neighborhood U such that $U \cap H$ is a submanifold of U. If $p \notin H$, then since H is closed there is a U such that $U \cap H = \varnothing$. If $p \in H$, since H is a subgroup, we can left translate the whole story by p^{-1} and are reduced to the case where $p = e$. What we must show is:

There is a neighborhood, U, of e such that $U \cap H$ is a submanifold of U.

We will eventually get U as a sufficiently small normal neighborhood. If H actually were a closed Lie subgroup, we would expect $U \cap H$ to be $\exp(V \cap h)$ where V is a neighborhood of 0 and h the Lie algebra of H. Our first step is to construct the missing Lie algebra of H.

LEMMA 4.1. *Let H be a closed subgroup of G and h the set of all $X \in g$ such that* $\exp tX \in H$ *for all t. Then h is a subalgebra of g.*

It is clear that if $X \in h$ then $sX \in h$ since $\exp t(sX) = \exp tsX$. In order to show that h is closed under addition and Lie multiplication, it is convenient to have the following lemma:

LEMMA 4.2. *Let X_i be a sequence of elements of g with $X_i \to X$ and $t_i \to 0$ a sequence of non-zero real numbers. If* $\exp t_iX_i \in H$ *for all i then* $\exp tX \in H$ *for all t, i.e., $X \in H$.*

Notice that since $\exp - t_iX_i = (\exp t_iX_i)^{-1} \in H$, we can take the $t_i > 0$. Similarly, we need only prove $\exp tX \in H$ for $t > 0$. For each $t > 0$ define $k_i(t)$ by

$$k_i(t) = \left[\frac{t}{t_i}\right] \left(= \text{largest integer} \leqslant \frac{t}{t_i}\right).$$

Then

$$\frac{t}{t_i} - 1 < k_i(t) < \frac{t}{t_i}$$

so that

$$\lim_{i=\infty} t_ik_i(t) = t.$$

Since $\exp t_iX_i \in H$, $\exp k_i(t)t_iX_i = (\exp t_iX_i)^{k_i(t)} \in H$. But $\lim_{i=\infty} k_i(t)t_iX_i = tX$, and since H is closed and exp is continuous, $\exp tX \in H$, proving Lemma 4.2.

Now to prove Lemma 4.1. Let $X, Y \in h$. Then $\exp tX \exp tY \in H$. On the other hand,

$$\exp tX \exp tY = \exp(t(X + Y) + tZ_t)$$

where $Z_t \to 0$ as $t \to 0$. Taking t_i a sequence of positive numbers approaching zero and $X_i = X + Y + Z_{t_i}$ we can apply Lemma 4.2 and conclude that $X + Y \in h$. Similarly,

$$(\exp tX)(\exp tY)(\exp tX)^{-1}(\exp tY)^{-1} = \exp(t^2[X, Y] + t^2Z_t)$$

so that $[X, Y] \in h$. This proves Lemma 4.1.

We now wish to show that for some neighborhood U of e, $U \cap H = \exp h$. For this purpose, choose a complimentary subspace, say h', to h in g. Thus $g = h + h'$.

Lemma 4.3. *There is a neighborhood, V', of 0 in h' such that* $0 \neq X' \in V$ *implies* $\exp X' \notin H$.

Let us put a Euclidean metric on h' and let K be the set of all $X' \in h'$ with $1 \leqslant \|X'\| \leqslant 2$. Suppose Lemma 4.3 were false, then there would exist a sequence $Y_i \to 0$ with $\exp Y_i \in H$. We can clearly choose integers n_i such that $X_i = n_iY_i \in K$. Since K is compact, we can assume (passing to a subsequence if necessary) that $X_i \to X \neq 0$. Furthermore, $(1/n_i) \to 0$

and $\exp(1/n_i)X_i \in H$. By Lemma 4.2, this implies that $X \in h$, contradicting the definition of h'.

Now let W be a neighborhood of 0 in g for which the map $\varphi: g \to G$ defined by $\varphi(X + X') = \exp X' \exp X$ is a diffeomorphism (cf. Lemma 2.1). Choose W of the form $W' \times W''$ where $W'' \subset V'$. Let $U = \varphi(W)$. If $x \in U$, then $x = \exp X' \exp X$ for $X \in h$, $X' \in V'$. If $x \in U \cap H$, then $\exp X' \exp X \in H$ implies (since $\exp X \in H$) that $\exp X' \in H$ as H is a group. Thus $X' = 0$. In other words, $U \cap H = U \cap \exp h$, proving Theorem 4.1.

Notice, that in the course of the proof we have not only established that H carries a differentiable structure, but also that G/H does. More precisely,

THEOREM 4.2. *Let G be a Lie group and H a closed subgroup. Then G/H carries a unique differentiable structure such that*

i) *the projection map, π, of $G \to G/H$ is differentiable;*

ii) *every $p \in G/H$ has a neighborhood W and a differentiable map $\varphi: W \to G$ such that $\pi \circ \varphi = \text{id}$.*

Left translation by any element of G is a diffeomorphism of G/H.

Let us first observe, that the last statement of the theorem is true for any differentiable structure on G/H satisfying i) and ii). In fact, left translation by z^{-1} is the inverse of left translation by z so it is sufficient to prove that left translation is differentiable. Let $p \in G/H$ and W, ψ as in ii). For any $q \in W$, $zq = \pi(z\psi(q))$ by definition. That is, left translation by z on W can be written as $\pi \circ L_z \circ \psi$, which is differentiable.

This implies that it is sufficient to prove existence and uniqueness of the differentiable structure at one point of G/H. In fact, if a coordinate system exists on a neighborhood, U, of p, one obtains a coordinate system about every point by translation. It follows from i) and ii) (on U) that this defines a (unique) differentiable structure on G/H.

Therefore we will show the existence and uniqueness at the coset of the identity. We take as our neighborhood, W, the image of the neighborhood U of e constructed in the proof of Lemma 4.3 (Theorem 4.1). Every $x \in U$ can be uniquely written as $x = \exp X' \exp X$. In other words, there is a well defined map σ of $W \to h'$ sending $xH \to X'$. Choose a coordinate system on h' thus introducing (via σ) coordinates on W. Furthermore, the map $\psi = \exp \circ \sigma$ of $W \to U$ is clearly differentiable and satisfies ii). The map, φ, of $g \to G$ sending $X + X'$ into $(\exp X')(\exp X)$ is a diffeomorphism of some neighborhood of zero onto U. Thus the map, ρ, of $U \to h'$ sending $x = (\exp X')(\exp X)$ into X' is differentiable. But $\pi = \sigma^{-1} \circ \rho$ and since σ defines the differentiable structure on W, this shows that π is differentiable, proving the existence of a differentiable structure on G/H satisfying i) and ii).

To prove uniqueness, let us assume that W has two differentiable structures with two coordinate maps σ_1 and σ_2 and maps ψ_1 and ψ_2 into G. We would like to show that $\sigma_1\sigma_2^{-1}$ is differentiable. But $\sigma_1\sigma_2^{-1} = \sigma_1\pi\, \psi_2\sigma_2^{-1}$. But $\sigma_1\pi$ is a differentiable map of G into E^k by i) and $\psi_2\sigma_2^{-1}$ is a differentiable by ii); thus $\sigma_1\sigma_2^{-1}$ is differentiable, proving Theorem 4.2.

For later use, it is convenient to have the following lemma:

Lemma 4.4. *Let G be a Lie group, H a topological group*[3] *and f a continuous homomorphism of $H \to G$. There exists a unique smallest Lie algebra h such that if V is any neighborhood of* 0 *in h, for all sufficiently small neighborhoods U of e in H, $f(U) \subset \exp V$.*

It is clear that we can take h as the intersection of all subalgebras with the property above (since any intersection of subspaces of a finite-dimensional space reduces to a finite intersection).

We can also characterize h more geometrically:

Exercise 4.1. Show that $X \in h$ if and only if there is a sequence $h_i \to e$ in H and numbers $n_i \to \infty$ such that $X = \lim n_i \exp^{-1} X_i$. Thus h is the space of "limiting directions" of $f(H)$. (The proof is similar to that of Lemma 4.1.)

5. Invariant metrics. In Euclidean space, E^n, the straight lines have two characterizations. On the one hand, they are geodesics—they minimize length. On the other hand, they are the cosets of the one parameter subgroups of E^n, if we regard E^n as a Lie group. It turns out that this is not just an accident. It is a consequence of the fact that right and left translations are isometries of the Euclidean metric. We shall now establish the corresponding result in general.

---

[3] By a topological group is meant a group H which is a topological space and on which the group operations are continuous; i.e., the map of $H \times H \to H$ sending $(x, y) \to xy^{-1}$ is continuous.

Definition 5.1. *Let G be a Lie group. A Riemann metric,* $\mathbf{a}$, *on G is said to be bi-invariant if L_x and R_x are isometries of* $\mathbf{a}$ *for all $x \in G$.*

Notice that if $\mathbf{a}$ is a bi-invariant metric on G then A_x is an isometry for any $x \in G$; in fact $A_x = R_{x^{-1}}L_x$. Also the mapping of $G \to G$ sends x into x^{-1} is also an isometry. In fact, at e, this map sends every tangent into its negative and so preserves length on $T_e(G)$. Let us now show that it is an isometry at any $y \in G$. For this purpose, write the map $x \to x^{-1}$ as $x \to xy^{-1} \to (xy^{-1})^{-1} = yx^{-1} \to y^{-1}yx^{-1} = x^{-1}$. Since $L_{y^{-1}}$ and $R_{y^{-1}}$ are isometries, it is sufficient to check that the map $xy^{-1} \to (xy^{-1})^{-1}$ is an isometry at $x = y$, which we have already verified.

Theorem 5.1. *Let* $\mathbf{a}$ *be a bi-invariant Riemann metric on the Lie group G. Then every geodesic is a left translate of a one-parameter group and conversely.*

In the proof of Theorem 5.1, it is sufficient (by left translation) to restrict attention to a neighborhood of e. We wish to show that every geodesic passing through e is a one-parameter group and conversely. For this purpose we state the following lemma.

Lemma 5.1. *Let σ be an isometry of a Riemann manifold M with $\sigma^2 = \mathrm{id}$. Let x be an isolated fix point of σ. Let U be a neighborhood of x with unique geodesics containing no other fix points. If $y \in U$ and $\sigma(y) \in U$, the geodesic joining y to $\sigma(y)$ passes through x.*

Let γ be the geodesic joining y to $\sigma(y)$. Since $\sigma^2 = 1$, $\sigma(\gamma)$ is the geodesic joining $\sigma(y)$ to y. Thus σ maps γ into itself reversing directions. Therefore it must contain a fix point which must be x.

Let us apply Lemma 5.1 to the following situation: Take $z \in G$ to be close to e and $z(t)$ the (unique) one-parameter group joining e to z with $z(1) = z$. Let $x = z(1/2)$ and define σ by $\sigma(y) = xy^{-1}x$. Then $\sigma^2(y) = x(xy^{-1}x)^{-1}x = y$ so $\sigma^2 = 1$ and σ is an isometry. If $\sigma(y) = y$, then $xy^{-1}x = y$, or $(y^{-1}x)^2 = 1$. If y and x are close enough (in particular if both are close to e) this implies that $y = x$. Thus x is an isolated fix point of σ.

Now $\sigma(z) = xz^{-1} = e$. Thus the geodesic joining e to z passes through x. Repeating this argument for both halves of the one-parameter group

$z(t)$, we obtain that the geodesic joining e to z contains $z(r)$ for every $r = p/2^n$. By continuity, this implies that $z(t)$ is the geodesic.

Thus every one-parameter group is a geodesic. But this implies that the exponential map of the geodesic spray of **a** at e coincides with Exp, proving Theorem 5.1.

Let us now examine the problem of existence of an invariant metric on a Lie group. If we define an arbitrary scalar product on $g = T_e(G)$, we can extend it to a Riemann metric on G by left translation, i.e., define (v, w) for $v, w \in T_x(a)$ by $(v, w) = (L_{x^{-1}*} v, L_{x^{-1}*} w)$. It is clear that this metric is invariant under left translation. How about right translation? For $v, w \in g$ we must have $(R_{x*} v, R_{x*} w) = (v, w)$. Now $(R_{x*} v, R_{x*} w) = (L_{x^{-1}*} R_{x*} v, L_{x^{-1}*} R_{x*} w)$ and $R_{x*} L_{x^{-1}*} = Ad\,x$. Thus we must have

$$(Ad\,x(v)\; Ad\,x(w)) = (v, w). \tag{5.1}$$

Conversely, if (5.1) holds for all $x \in G$, then (v, w) extends to a bi-invariant metric on G. In other words,

Lemma 5.2. *Let* (,) *be a scalar product on g. In order that* (,) *extend (by left translation) to a bi-invariant metric on G it is necessary and sufficient that* (,) *be invariant under the adjoint representation, i.e., that* (5.1) *hold.*

Thus we are faced with the following problem: Given a representation, ρ, of a Lie group G (in particular the adjoint representation) on a vector space V, when does there exist an invariant scalar product on V? In other words, when will there be a (positive definite) scalar product (,) on V satisfying

$$(\rho(x)v, \rho(x)w) = (v, w), \tag{5.2}$$

for all $x \in G$, $v, w \in V$?

Notice that the orthogonal group is compact. (Exercise !). If $\rho(G)$ preserves (,), it is a subgroup of the orthogonal group of (,) and therefore $\overline{\rho(G)}$ must be compact.

Conversely, let us suppose that $\overline{\rho(G)} = H$ is a compact subgroup of all non-singular linear transformations of V. According to Theorem 4.1, H is a Lie group. We would like to find a scalar product (,) on V so that $(hv, hw) = (v, w)$ for all $v, w \in V$, $h \in H$. To do this, let $\omega_1, \ldots, \omega_k$ be a basis of left invariant forms of H, ($k = \dim H$) and let $\Omega = \omega_1 \wedge \cdots \wedge \omega_k$. Since Ω vanishes nowhere, it defines an orientation on H. For any continuous function f on H, we can thus form the integral $\int_H f\,\Omega$. Now let $(\ ,\ )_0$ be any positive definite scalar product on V. Define $(\ ,\ )$ by

$$(5.3) \qquad (v, w) = \int_{x \in G} (xv, xw)_0\, \Omega.$$

We leave to the reader the verification that $(\ ,\ )$ is an invariant, positive definite, scalar product on V. Thus we have proved

THEOREM 5.2. *Let ρ be a linear representation of a Lie group G on a vector space V. In order that V possesses a scalar product invariant under G, it is necessary and sufficient that $\overline{\rho(G)}$ be compact. In particular, if G is compact, there exists an invariant scalar product for any ρ.*

EXERCISE 5.1. Show that any representation of a compact Lie group is fully reducible.

EXERCISE 5.2. Let G be a compact Lie group acting differentiably on a manifold M. Show that there is a Riemann metric on M so that G becomes a group of isometries. (Thus we can apply Theorem 8.2 of Chapter IV.)

By Lemma 5.2 we now have

THEOREM 5.3. *Let G be a Lie group. In order that G possesses a bi-invariant Riemann metric, it is necessary and sufficient that $\overline{Ad(G)}$ be compact. In particular, on every compact Lie group there is an invariant metric.*

As a corollary we have

EXERCISE 5.3. Let G be a compact connected Lie group. Every $x \in G$ lies on a one-parameter subgroup. (This is *not* true, in general, for non-compact Lie groups.)

6. Forms with values in a vector space. It is convenient to extend the formalism of differential forms to forms taking values in a vector space. Up until now, a linear differential form on M was a real linear functional on every tangent space of M (varying differentiably from point to point). We would like to consider linear maps from every tangent space to a fixed vector space W. For instance, if G is a Lie group, we defined the left invariant forms $\omega_1, \ldots, \omega_n$ corresponding to a choice of basis of the

Lie algebra g. These forms are somewhat unelegant in that they depend on the choice of basis. On the other hand, there is a well defined vector valued form, ω, which assigns to each tangent vector $v \in T_p(G)$ the element of g (regarded as a left invariant vector field) which takes the value v at p. We would expect that the equations of structure of G take a simple form in terms of ω and indeed they do. First we must set up a bit of formalism.

Let M be a manifold and W a finite-dimensional vector space. By a W valued differential form of degree q we shall mean an element of $W \otimes \wedge^q(M)$. In other words, if Ω is a W valued differential form, it is a linear map of $\wedge^q T_x(M) \to W$ for each $x \in M$ (and varies in a differentiable manner); i.e., an element of $W \otimes \wedge^q T_x^*(M)$ for each $x \in M$. The set of all W valued differential q-forms is a vector space in the obvious way; and, in fact it is a module over the ring of differentiable functions on M. We will denote this space by $\wedge_W^q(M)$, and the space of all W valued differential forms by $\wedge_W(M)$; thus $\wedge_W(M) = \wedge_W^0(M) + \cdots + \wedge_W^n(M)$.

Given a vector field, X, on M, we define, in the obvious way, the linear maps $\mathscr{L}_X : \wedge_W^q(M) \to \wedge_W^q(M)$ and interior product by X mapping $\wedge_W^q(M) \to \wedge_W^{q-1}(M)$. Furthermore we have the operator $d: \wedge_W^q(M) \to \wedge_W^{q+1}(M)$, with the usual relations of Section 1 of Chapter III holding. Given a map φ of $M_1 \to M_2$ there is induced a linear map $\varphi^* : \wedge_W(M_2) \to \wedge_W(M_1)$, etc. All of these definitions and assertions can be made by choosing a basis, $e_1, \ldots, e_k$ of W. Then every $\Omega \in \wedge_W(M)$ can be written as $\Omega = \sum \Omega_i \otimes e_i$ where Ω_i are (real valued) exterior differential forms. We then have $d\Omega = \sum d\Omega_i \otimes e_i$ and an immediate verification shows that this definition is independent of the choice of $e_1, \ldots, e_k$. The same goes for all the previous assertions.

Let W_1 and W_2 be vector spaces and let l be a linear map of $W_1 \to W_2$. Then there is the obvious linear map, $l_\#$, of $\wedge_{W_1}(M) \to \wedge_{W_2}(M)$. The map $l_\#$ clearly commutes with all the operations just mentioned. Thus, $l_\# \, d\Omega = dl_\#\Omega$, $l_\# \mathscr{L}_X\Omega = \mathscr{L}_X l_\#\Omega$, etc.

Let W_1, W_2, and W_3 be three vector spaces and ρ a bilinear map of $W_1 \times W_2$ into W_3. We will define a bilinear map $\rho_\#$ of $\wedge_{W_1}(M) \times \wedge_{W_2}(M)$ into $\wedge_{W_3}(M)$ as follows:

Let $e_1, \ldots, e_k$ be a basis for W_1 and $f_1, \ldots, f_l$ be a basis for W_2. Let $\omega = \sum \omega_i e_i \in \wedge_{W_1}(M)$ and $\Omega = \sum \Omega_j f_j \in \wedge_{W_2}(M)$. Put

$$\rho(\omega, \Omega) = \sum \omega_i \wedge \Omega_j \otimes \rho(e_i, f_j). \tag{6.1}$$

An easy check shows that (6.1) is independent of change of coordinates.

It follows from (6.1) that

$$d\rho_{\#}(\omega, \Omega) = \rho_{\#}(d\omega, \Omega) + (-1)^q \rho_{\#}(\omega, d\Omega) \tag{6.2}$$

if ω is of degree q. Also

$$\mathscr{L}_X \rho_{\#}(\omega, \Omega) = \rho_{\#}(\mathscr{L}_X \omega, \Omega) + \rho_{\#}(\omega, \mathscr{L}_X \Omega) \tag{6.3}$$

for any vector field X on M. If φ is a map of a manifold $V \to M$ we have

$$\varphi^* \rho_{\#}(\omega, \Omega) = \rho_{\#}(\varphi^* \omega, \varphi^* \Omega). \tag{6.4}$$

These are two special cases which will be of importance to us and we wish to establish notation for each. Let $W = W_1 = W_2 = W_3$ be a Lie algebra and ρ be the Lie bracket. In this case we will denote $\rho(\omega_1, \omega_2)$ by $[\omega_1 \wedge \omega_2]$. If ω_1 is of degree p and ω_2 of degree q, we have

$$[\omega_1 \wedge \omega_2] = (-1)^{pq+1}[\omega_2 \wedge \omega_1]. \tag{6.5}$$

[The $(-1)^{pq}$ comes from the law of multiplication of $\bigwedge(M)$ and the $+1$ from the fact that the Lie bracket is anti-commutative.] If ω_1, ω_2 and ω_3 are forms of degree p, q, and r, Jacobi's identity becomes

$$(-1)^{pr}[\omega_1 \wedge [\omega_2 \wedge \omega_3]] + (-1)^{qp}[\omega_2 \wedge [\omega_3 \wedge \omega_1]] + (-1)^{rq}[\omega_3 \wedge [\omega_2 \wedge \omega_1]] = 0.$$

Another case of importance to us is where $W_2 = W_3 = V$ is a vector space and W_1 is a linear space of linear transformation of V. Thus W_1 is a subspace of V. In this case we shall write $\rho(\omega_1, \omega_2)$ simply by $\omega_1 \wedge \omega_2$.

Now let us return to the case of a Lie group G and its fundamental g valued form ω. Let $e_1, \ldots, e_n$ be a basis of g so that $\omega = \omega^1 \otimes e_1 + \cdots + \omega^n \otimes e_n$ where the ω_i are a basis of left invariant forms. By (2.1) we have

$$d\omega^i = \sum_{j<k} c^i_{jk} \omega^j \wedge \omega^k$$

so that $d\omega = \sum c^i_{jk} \omega^j \wedge \omega^k \otimes e_i$. By (2.3) $[e_k, e_j] = \sum c^i_{jk} e_i$. Thus by (6.1),

$$[\omega \wedge \omega] = \sum_{\text{all } j,k,i} \omega^k \wedge \omega^j \otimes c^i_{jk} e.$$

Comparing the two expressions (note the interchange of j and k and the fact that the first sum is over $j < k$ and the second over all j, k) we obtain the equations of structure of G as

$$d\omega = -\tfrac{1}{2}[\omega \wedge \omega]. \tag{6.6}$$

CHAPTER VI

Differential Geometry of Euclidean Space

In this chapter we will study the differential geometry of submanifolds of Euclidean space; in particular two-dimensional surfaces in E^3. It is natural to expect that a lot of information can be obtained from the study of the group of Euclidean motions, $\mathscr{E}(n)$. In fact, most of the equations of classical differential geometry are fairly direct consequences of the structure equations of $\mathscr{E}(n)$. We therefore begin with a study of this group.

1. The equations of structure of Euclidean space. Every Euclidean motion is the product of a translation and a Euclidean motion keeping a given point fixed. If we take as our fixed point the origin of E^n, then every Euclidean motion is of the form rt where t is a translation and r is an orthogonal transformation, i.e., $r \in O(n)$. Let $\mathscr{T}$ denote the subgroup of translations. It is a commutative group which we can identify with E^n—identify a vector with translation through that vector. The group $\mathscr{T}$ is a normal subgroup of $\mathscr{E}(n)$. In fact, let $r \in O(n)$ and t_v be translation through the vector v. Then t_v is the transformation of E^n sending $x \to x + v$. Let us look at $r^{-1}t_vr$. It sends $x \to rx \to rx + v \to x + r^{-1}v$. Thus $r^{-1}t_vr = t_{r^{-1}v}$. In other words, if we identify $\mathscr{T}$ with E^n, *then* $A(r)$ *acting on* $\mathscr{T}$ *becomes identified with the usual action of* r^{-1} *on* E^n *for any* $r \in O(n)$.

Now the Lie algebra of $\mathscr{E}(n)$ splits, *as a vector space*, into a direct sum of the Lie algebra of $\mathscr{T}$ and $O(n)$. Under our identification of $\mathscr{T}$ with E^n, we can identify the Lie algebra of $\mathscr{T}$ with E^n also. Thus the Lie algebra of $\mathscr{E}(n)$ can be regarded as the direct sum $E^n + O(n)$. Let us examine the bracket operation in terms of these identifications. Since $\mathscr{T}$ is commutative,

$$[v_1, v_2] = 0 \qquad \text{for} \quad v_1, v_2 \in E^n.$$

Since $O(n)$ is a subgroup, $[X_1, X_2]$ is the bracket operation of X_1, $X_2 \in o(n)$. Now let us examine $[X, v]$ for $X \in o(n)$, $v \in E^n$. According to the above

$$A(\exp tX)v = \exp(-tX)v \tag{1.1}$$

where the right side is the action of the orthogonal transformation $\exp(-tX)$ on the vector v. Thus we can write, by (1.1),

$$[X, v] = -Xv. \tag{1.2}$$

Now let us consider the equations of structure of $\mathscr{E}(n)$. For this purpose, we will write the fundamental linear differential form of $\mathscr{E}(n)$ as $\omega + \Omega$ corresponding to the decomposition of the Lie algebra of $\mathscr{E}(n)$ into $E^n + O(n)$. Thus ω is a vector valued form with values in E^n and Ω has values in $O(n)$. The equations of structure now take the form

$$d\omega + d\Omega = -\tfrac{1}{2}[(\omega + \Omega) \wedge (\omega + \Omega)].$$

Expanding the right-hand side, we have

$$-\tfrac{1}{2}[(\omega + \Omega) \wedge (\omega + \Omega)] = -\tfrac{1}{2}[\omega \wedge \omega] - [\Omega \wedge \omega] - \tfrac{1}{2}[\Omega \wedge \Omega].$$

But $[\omega \wedge \omega] = 0$ as $\mathscr{T}$ is commutative so that

$$d\omega + d\Omega = -[\Omega \wedge \omega] - \tfrac{1}{2}[\Omega \wedge \Omega].$$

On both sides of this equation we have a sum of two forms, one with values in E^n and one with values in $O(n)$. They both must be separately equal. If we use (1.2) we obtain

$$d\omega = \Omega \wedge \omega \tag{1.3}$$

and

$$d\Omega = -\tfrac{1}{2}[\Omega \wedge \Omega]. \tag{1.4}$$

These equations are known as the *equations of structure of Euclidean space.* Let us rewrite (1.3) and (1.4) in terms of a coordinate system for E^n and

$O(n)$. In terms of the standard basis $e_1, \ldots, e_n$ of E^n, the form ω can be written as

$$\omega = \sum_{i=1}^{n} \omega^i \otimes e_i. \tag{1.5}$$

Let $\{E_{ij}\}_{i<j}$ be a basis for $O(n)$ where E_{ij} is the linear transformation defined by

$$\begin{aligned} E_{ij}\, e_j &= e_i \\ E_{ij}\, e_i &= -e_j \\ E_{ij}\, e_k &= 0 \qquad k \neq i \text{ or } j. \end{aligned} \tag{1.6}$$

The formula for the bracket in this basis is then given by

$$[E_{ij}, E_{pq}] = -\delta_j^p E_{iq} + \delta_i^p E_{jq} + \delta_i^q E_{pj} - \delta_j^q E_{pi}, \tag{1.7}$$

as can be checked by an easy computation. (Here $E_{lm} = -E_{ml}$ for $l > m$.) In terms of this basis we can write

$$\Omega = \sum_{i<j} \Omega^{ij} E_{ij}. \tag{1.8}$$

Then

$$\begin{aligned} \Omega \wedge \omega &= \left(\sum_{i<j} \Omega^{ij} \otimes E_{ij}\right) \wedge \left(\sum_{k=1}^{n} \omega^k \otimes e_k\right) \\ &= \sum_{i<j,k=1,\ldots,n} (\Omega^{ij} \wedge \omega^k) \otimes (E_{ij} e_k) \\ &= \sum_{i<j} (\Omega^{ij} \wedge \omega^j) \otimes e_i - (\Omega^{ij} \wedge \omega^i) \otimes e_j \\ &= \left(\sum_{l<j} \Omega^{lj} \wedge \omega^j - \sum_{i<l} \Omega^{il} \wedge \omega^i\right) \otimes e_j. \end{aligned}$$

We can simplify this last expression if we define $\Omega^{ij} = -\Omega^{ji}$ for $i > j$. Then we have

$$\Omega \wedge \omega = \sum_{j=1}^{n} (\Omega^{lj} \wedge \omega^j) \otimes e_l.$$

If we now compare the coefficients of e_l on both sides of (1.3) we obtain

$$d\omega^l = \sum \Omega^{jl} \wedge \omega^j. \tag{1.9}$$

Now let us compute $[\Omega \wedge \Omega]$. We have,

$$[\Omega \wedge \Omega] = \left[\left(\sum_{i<j} \Omega^{ij} \otimes E_{ij}\right) \wedge \left(\sum_{p<q} \Omega^{pq} \otimes E_{pq}\right)\right]$$
$$= \sum_{\substack{i<j\\ p<q}} (\Omega^{ij} \wedge \Omega^{pq}) \otimes [E_{ij}, E_{pq}]$$
$$= \sum_{\substack{i<j\\ p<q}} (\Omega^{ij} \wedge \Omega^{pq}) \otimes (-\delta_j^p E_{iq} + \delta_i^p E_{jq} + \delta_i^q E_{pj} - \delta_j^q E_{pi}).$$

Collecting the terms, we get

$$\sum_{l,m}\left(-\sum_{\substack{l<p\\ p<m}} \Omega^{lp} \wedge \Omega^{pm} + \sum_{\substack{p<l\\ p<m}} \Omega^{pl} \wedge \Omega^{pm} + \sum_{\substack{p<m\\ l<p}} \Omega^{pm} \wedge \Omega^{lp} - \sum_{\substack{m<p\\ l<p}} \Omega^{mp} \wedge \Omega^{lp}\right) \otimes E_{lm}$$
$$= -\sum_{l,m,p} (\Omega^{lp} \wedge \Omega^{pm}) \otimes E_{lm}$$
$$= -2\sum_{\substack{l<m\\ p}} (\Omega^{lp} \wedge \Omega^{pm}) \otimes E_{lm}.$$

Comparing the coefficient of E_{lm} in (1.4) yields

$$d\Omega^{lm} = \sum \Omega^{lp} \wedge \Omega^{pm}. \tag{1.10}$$

We now give another interpretation to the equations of structure (1.3) and (1.4) [or (1.9) and (1.10)] which will be fundamental for what follows. We first observe that we can identify the group $\mathscr{E}(n)$ with the set of all orthogonal frames of E^n: By an orthogonal frame we mean a pair consisting of an origin and a set of n orthonormal vectors. Thus a frame is given by $(v; f_1, f_2, \ldots, f_n)$ where $v \in E^n$ and $f_1, \ldots, f_n$ are an orthonormal basis. Let $\mathcal{O}(n)$ denote the set of all orthogonal frames of E^n. There is a map φ of $\mathscr{E}(n) \to \mathcal{O}(n)$ defined as follows: $\varphi(t_v r) = (v;\ re_1, \ldots, re_n)$ (where $e_1, \ldots, e_n$ is the standard basis of E^n). Thus φ assigns to every Euclidean motion, the frame obtained by applying that motion to the standard frame. It is easy to see that φ is a one-to-one map of $\mathscr{E}(n)$ onto $\mathcal{O}(n)$. We make $\mathcal{O}(n)$ into a differentiable manifold so that φ is a diffeomorphism. Thus, we can transfer the forms ω and Ω to $\mathcal{O}(n)$, and we wish to give them a geometric interpretation. In order to do this, we first study the left invariant vector fields.

LEMMA 1.1. *Let π be the projection map of $\mathcal{O}(n)$ onto E^n given by $\pi(v; f_1, \ldots, f_n) = v$. Then π is differentiable. Let $\underset{\sim}{e}_i$ denote the left invariant* *vector field on $\mathscr{E}(n)$ corresponding to the standard basis element e_i of E^n (under the identification of E^n with the Lie algebra of translations). Then*

$$l_v(\pi_*\varphi_*(\underline{e}_i)_{(v,f_1,\ldots,f_n)}) = f_i. \tag{1.11}$$

(Recall that l_v is just the standard identification of the tangent space to a vector space with the vector space.)

The first part of the lemma is trivial. Let us verify (1.11). The vector field e_i is the infinitesimal generator of the one parameter group of transformations of $\mathscr{E}(n)$ given by *right* translations by $\exp(se_i)$ for all real s. These transformations send $t_v r \to t_v r t_{se_i}$. But $rt_{se_i} = t_{sre_i} r$, so $t_v \to t_{v+sre_i} r$. Thus $\varphi_*(\underline{e}_i)$ is the infinitesimal generator of the one parameter group on $\mathcal{O}(n)$ which sends $(v; f_1, \ldots, f_n) \to (v + sf_i; f_1, \ldots, f_n)$. This implies (1.11). The same argument shows that $\pi_*\varphi_*(E_{ij}) = 0$ for $E_{ij} \in \mathcal{O}(n)$.

Now let X be a tangent vector at the point $z = (v; f_1, \ldots, f_n)$ of $\mathcal{O}(n)$. We can write $l_v\pi_*(X) = \sum X_i f_i$.

The left invariant vector field $\sum X_i \underline{e}_i$ has the property that $\varphi_*(\sum X_i \underline{e}_i)_z$ differs from X by a linear combination of $\varphi_*(E_{ij})$. Thus

$$\langle X, ((\varphi^{-1})_*\omega)_z\rangle = \sum X_i e_i. \tag{1.12}$$

For the sake of simplicity, let us, from now on, drop the φ in the identification of $\mathcal{O}(n)$ with $\mathscr{E}(n)$ so that we can write (1.12) as

$$\langle X, \omega_z\rangle = \sum X_i e_i.$$

Let us also leave out the symbol l_v wherever it should be clear that we are making such an identification. Then we can describe the form ω as follows:

If X is tangent to $\mathcal{O}(n)$ at $z = (v; f_1, \ldots, f_n)$, write $\pi_*(X) = \sum X_i f_i$. Then $\langle X, \omega_z\rangle = \sum X_i e_i$. In particular

$$\langle X, \omega_z^i\rangle = (\pi_*(X), f_i). \tag{1.13}$$

Thus the forms ω^i have the following simple geometrical interpretation: If we make an "infinitesimal displacement" of the frame $z = (v; f_1, \ldots, f_n)$ by the vector X, then $\langle X, \omega_z^i\rangle$ tells us how much we have moved the origin, v, of z in the f_i direction.

Let us recast (1.13) in a somewhat more convenient form. The mapping π can be regarded as a vector valued function on $\mathcal{O}(n)$. What is $d\pi$? An immediate verification shows that $\langle X_z, d\pi_z\rangle = l_v\pi_*(X)$ which we have decided to write simply as $\pi_*(X)$. The map which assigns to

$(v;f_1,\ldots,f_n)$ the vector f_i is also a vector valued function on $\mathcal{O}(n)$ which we denote by just f_i. On the other hand, the scalar product (,) is a bilinear map of $E^n \times E^n \to R$ so it makes sense to talk of the scalar product of forms with values in E^n. In view of these remarks, we can write (1.13) as

$$\omega^i = (d\pi, f_i). \tag{1.14}$$

What is (df_i, f_j)? It is a linear differential form measuring the amount of rotation involved in our infinitesimal displacement on $\mathcal{O}(n)$. It is an instructive exercise to the reader to verify directly that $(df_i, f_j) = \Omega^{ij}$. We will prove this soon. For the moment let us denote (df_i, f_j) by Ω'^{ij} and deduce some of its properties. Since $(f_i, f_j) = \delta_i^j =$ constant, we obtain, by exterior differentiation

$$(df_i, f_j) + (f_i, df_j) = 0,$$

or

$$\Omega'^{ij} = -\Omega'^{ji}.$$

Now apply d to (1.14). We obtain

$$\begin{aligned} d\omega^i &= (dd\pi, f_i) - (d\pi, df_i) = -(d\pi, df_i) \\ &= -(d\pi, \textstyle\sum \Omega'^{ij} \otimes f_j) \qquad \text{(since } df_i = \textstyle\sum \Omega'^{ij} \otimes f_j) \\ &= \textstyle\sum \Omega'^{ij} \wedge (d\pi, f_j) \\ &= \textstyle\sum \Omega'^{ij} \wedge \omega^j. \end{aligned}$$

If we compare this last equation with (1.9) we can conclude that $\Omega'^{ij} = \Omega^{ij}$. In fact, we have,

$$\sum \Omega'^{ij} \wedge \omega^j = \sum \Omega^{ij} \wedge \omega^j$$

or

$$\sum (\Omega'^{ij} - \Omega^{ij}) \wedge \omega^j = 0.$$

By Cartan's lemma (Theorem 4.4 of Chapter I), we can conclude that for each j,

$$(\Omega'^{ij} - \Omega^{ij}) = \sum a_k^{ji}\, \omega^k$$

where the a_k^{ji} are symmetric in i and k. On the other hand, both Ω'^{ij} and Ω^{ij} are anti-symmetric in i and j. Therefore, we can apply the following lemma which we will have occasion to use several times.

LEMMA 1.2. *Let a_k^{ij} be symmetric in j and k and anti-symmetric in i and j. Then $a_k^{ij} = 0$.*

In fact,

$$a^i_{jk} = -a^j_{ik} = -a^j_{ki} = a^k_{ji}$$

and

$$a^i_{jk} = a^i_{kj} = -a^k_{ij} = -a^k_{ji}.$$

Thus

$$a^i_{jk} = a^k_{ji} = -a^k_{ji} = 0.$$

Thus, we have proved that $\Omega'^{ij} = \Omega^{ij}$ or

$$\Omega^{ij} = (df_i, f_j). \tag{1.15}$$

Now we can give another proof of (1.10).

$$\begin{aligned} d\Omega^{ij} = (df_i, df_j) &= \left(\sum_k \Omega^{ik} \otimes f_k, \sum_l \Omega^{jl} \otimes f_l\right) \\ &= \sum_{k,l} \Omega^{ik} \wedge \Omega^{jl}(f_k, f_l) \\ &= \sum_k \Omega^{ik} \wedge \Omega^{jk}. \end{aligned}$$

Resuming, we have

$$\omega^i = (d\pi, f_i) \tag{1.14}$$

$$\Omega^{ij} = (df_i, f_j) \tag{1.15}$$

$$d\omega^i = \sum \Omega^{ji} \wedge \omega^j \tag{1.9}$$

$$d\Omega^{ij} = \sum \Omega^{ik} \wedge \Omega^{kj}. \tag{1.10}$$

We conclude this section with the observation that since the ω^i and Ω^{ij} correspond to left invariant forms on $\mathscr{E}(n)$, we can apply Theorems 2.3 and 2.4 of Chapter V. In particular, *any transformation of $\mathscr{O}(n)$ preserving ω and Ω is a Euclidean motion* (Corollary to Theorem 2.3) *and*

Proposition 1.1. *Let M be a manifold of* dim $n + n(n-1)/2$ *and θ^i, Θ^{ij} be forms on M satisfying*

$$d\theta^i = \sum \Theta^{ji} \wedge \theta^j$$
$$d\Theta^{ij} = \sum \Theta^{ik} \wedge \Theta^{jk},$$

then there exists (locally) a diffeomorphism, f, of M onto $\mathscr{O}(n)$ so that $f^*(\omega^i) = \theta^i$ and $f^*(\Omega^{ij}) = \Theta^{ij}$.

2. The equations of structure of a submanifold. One of our principal objects of study in this chapter is the geometry of submanifolds of Euclidean space. The method that we shall use is the so-called "method of moving frames" first invented by Darboux and exploited to the fullest by Cartan. The idea is to choose frames "tangent" to the submanifold and see what the equations of structure mean for these frames.

DEFINITION 2.1. *Let φ be an immersion of a k-dimensional manifold M into E^n. By $\mathcal{O}_\varphi(M)$ is meant the set of all pairs $(x; f_1, \ldots, f_n)$ where $x \in M$ and $f_1, \ldots, f_n$ is an orthonormal basis of E^n such that*

$$f_1, \ldots, f_k \in l_{\varphi(x)}(\varphi_* T_x(M)). \tag{2.1}$$

Roughly speaking, $\mathcal{O}_\varphi(M)$ is the set of all frames of E^n whose first k vectors are tangent to $\varphi(M)$ [and whose last $(n - k)$ are orthogonal to $\varphi(M)$]. Thus if $n = 3$ and $k = 1$ or 2, a point of $\mathcal{O}_\varphi(M)$ is as is illustrated in Figures 6.1 and 6.2.

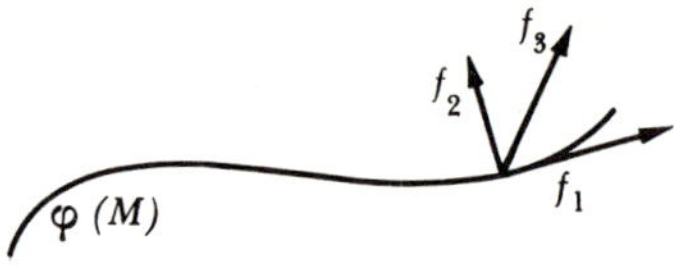

FIGURE 6.1

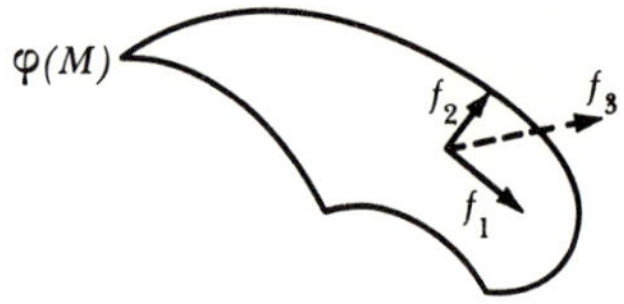

FIGURE 6.2

It is an easy exercise to show that $\mathcal{O}_\varphi(M)$ carries a natural differentiable structure and there is a differentiable projection, which we shall call π_φ, of $\mathcal{O}_\varphi(M)$ onto M.

DEFINITION 2.2. *Let φ and M be as in Definition 2.1. Denote **by** $\hat{\varphi}$ the mapping of $\mathcal{O}_\varphi(M) \to \mathcal{O}(n)$ given by*

$$\hat{\varphi}(x; f_1, \ldots, f_n) = (\varphi(x); f_1, \ldots, f_n). \tag{2.2}$$

It is easy to see that $\hat{\varphi}$ is an immersion of $\mathcal{O}_\varphi(M) \to \mathcal{O}(n)$ and that $\pi \circ \hat{\varphi} = \varphi \circ \pi_\varphi$. Thus we have commutativity of the diagram

$$\begin{array}{ccc} \mathcal{O}_\varphi(M) & \xrightarrow{\hat{\varphi}} & \mathcal{O}(n) \\ \downarrow{\scriptstyle \pi_\varphi} & & \downarrow{\scriptstyle \pi} \\ M & \xrightarrow{\varphi} & E^n. \end{array}$$

We can consider the forms $\hat{\varphi}^*(\omega)$ and $\hat{\varphi}^*(\Omega)$ on $\mathcal{O}_\varphi(M)$.

DEFINITION 2.3. *We define the forms,* θ, Θ, θ^i *and* Θ^{ij} *by*

$$\theta = \hat{\varphi}^*(\omega), \tag{2.3}$$

$$\Theta = \hat{\varphi}^*(\Omega), \tag{2.4}$$

$$\theta^i = \hat{\varphi}^*(\omega^i), \tag{2.5}$$

and

$$\Theta^{ij} = \hat{\varphi}^*(\Omega^{ij}). \tag{2.6}$$

In case we wish to make the dependence on φ *more explicit we shall call these forms* θ_φ, Θ_φ, *etc.*

We then have, pulling back the equations of structure of Euclidean space via φ^*,

$$d\theta = \Theta \wedge \theta, \tag{2.7}$$

$$d\Theta = -\tfrac{1}{2}[\Theta \wedge \Theta], \tag{2.8}$$

or

$$d\theta^i = \sum_j \Theta^{ji} \wedge \theta^j \tag{2.9}$$

$$d\Theta^{ij} = \sum \Theta^{ip} \wedge \Theta^{pj}. \tag{2.10}$$

These are known as the *equations of structure of* (M, φ).

Before we study these equations in greater detail, we wish to make an important observation. In the study of Euclidean geometry, we regard two objects as equivalent if they differ from one another by a Euclidean motion. In particular, we would regard two maps, φ_1 and φ_2, of $M \to E^n$ as equivalent if $\varphi_1 = \alpha \circ \varphi_2$ where α is a Euclidean motion. The importance of the space $\mathcal{O}_\varphi(M)$ and the forms θ and Θ lies in

THEOREM 2.1. *Let* φ_1 *and* φ_2 *be two maps of* $M \to E^n$. *Then* $\varphi_1 = \alpha \circ \varphi_2$ *where* α *is a Euclidean motion of* E^n *if and only if there is a diffeomorphism,* ψ, *of* $\mathcal{O}_{\varphi_1}(M)$ *onto* $\mathcal{O}_{\varphi_2}(M)$ *with* $\theta_{\varphi_1} = \psi^*(\theta_{\varphi_2})$, $\Theta_{\varphi_1} = \psi^*(\Theta_{\varphi_2})$. *In other words,* $\mathcal{O}_\varphi(M)$ *and the forms* θ_φ *and* Θ_φ *determine* φ *up to a Euclidean motion.*

It follows from the definitions that $\varphi_1 = \alpha \circ \varphi_2$ if and only if $\mathcal{O}_{\varphi_1}(M) = \alpha(\mathcal{O}_{\varphi_2}(M)$ and $\hat{\varphi}_1 = \alpha \circ \hat{\varphi}_2 \circ \psi$ [where α is also to be regarded as acting on $\mathcal{O}(n)$]. If we regard $\mathcal{O}(n)$ as the group of Euclidean motions, then the forms ω^i and Ω^{ij} are just the left invariant forms and Theorem 2.1 is an immediate consequence of Theorem 2.3 of Chapter V.

Now let us come back to the study of the forms θ^i and Θ^{ij}. Recall that the forms ω^i on $\mathcal{O}(n)$ measure the "infinitesimal changes" of the base point relative to the i-th basis vector of the frame. When we pull back to $\mathcal{O}_\varphi(M)$ by φ^*, we are considering only those frames whose base point lies on $\varphi(M)$. The "infinitesimal change" of base point should therefore have no component orthogonal to the tangent space of $\varphi(M)$. On the other hand, [by (2.1)] in $\mathcal{O}(n)$ we are restricting attention to those frames whose first k vectors are tangent and whose last $n - k$ vectors are normal to $\varphi(M)$. Thus the last $n - k$ components of θ should vanish, i.e.,

$$\theta^i = 0 \qquad i = k+1, \ldots, n. \tag{2.11}$$

We give a formal proof of this. In fact, the forms θ^i and Θ^{ij} can be given a simple expression exhibiting this fact: Let E^i be the function on $\mathscr{F}$ assigning to each frame its i-th component, that is, $E^i(p, f^1, \ldots, f^n) = f^i$. Then the functions $\varphi \circ \pi_\varphi$ and $E^1, \ldots, E^n$ are differentiable vector valued functions on $\mathscr{F}(M)$. Then it follows immediately from the definitions that

$$\theta^i = (d(\varphi \circ \pi_\varphi), E^i) \tag{2.12}$$

and

$$\Theta^{ij} = (dE^i, E^j). \tag{2.13}$$

Thus for any tangent vector $X \in T_u(\mathscr{F}(M))$ we have $\langle X, \theta^i \rangle = (\langle X, d(\varphi \circ \pi_\varphi) \rangle, E^i(u)) = (\varphi_* \circ \pi_* X, E^i(u)) = 0$ if $i > k$.[1] In particular (2.11) follows from (2.12) and the definitions. We give a more detailed discussion of this in the next section.

3. The equations of structure of a Riemann manifold. In the course of proving (2.11), we can prove somewhat more: That the θ^i and Θ^{ij} for $i, j \leqslant k$ depend only on the Riemann metric induced on M by φ and not on φ. Let us make this distinction a little clearer.

The map φ, induces a Riemann metric on M (by applying φ^* to the Euclidean metric on E^n). Two different maps (that is, maps which do not differ by merely a Euclidean motion) may induce the same metric on M. For instance, any two Riemann manifolds of dimension one are

[1] As usual we are suppressing the symbol l_v. Strictly speaking we should write $\langle X, d(\varphi \circ \pi_\varphi) \rangle = l_{\varphi \circ \pi(u)} \circ \pi_* \circ \varphi_* X$, instead of $\langle X, d(\varphi \circ \pi_*) \rangle = \varphi_* \circ \pi_* X$.

locally isometric. In fact, arc length can always be introduced, locally, as a parameter—put $s(t) = \int_{t_0}^{t} \|\partial/\partial t\|^2 \, dt$. On the other hand, it is geometrically obvious that not every pair of curves differ from one another by a Euclidean motion. One must take into account the way the curve is "bent" in space, something that is extrinsic, not describable purely in terms of the Riemannian geometry (which in this case is trivial). This "bending" (we shall see) is measured by Θ^{ij}, with $j > k$.

As a second example, consider the two maps of a portion of the plane into E^3 given by

$$\varphi_1\colon \begin{aligned} x \circ \varphi_1(u, v) &= u \\ y \circ \varphi_1(u, v) &= v \\ z \circ \varphi_1(u, v) &= 0 \end{aligned} \qquad \varphi_2\colon \begin{aligned} x \circ \varphi_1(u, v) &= u \\ y \circ \varphi_1(u, v) &= \sin v \\ z \circ \varphi_1(u, v) &= \cos v. \end{aligned}$$

Thus φ_1 maps into a plane and φ_2 maps into a cylinder, so that they differ by more than a Euclidean motion. Under both maps the Euclidean metric of E^3 pulls back to $du^2 + dv^2$. Here again, the fact that the cylinder is "bent" and the plane is not, is measured by Θ^{13} and Θ^{23}.

In order to establish (2.11) and that θ depends only on the Riemannian structure, we will construct the analogue of θ for any Riemann manifold.

Let M be a k-dimensional Riemann manifold. By $\mathcal{O}(M)$ we shall mean the set of pairs $(p, X_1, \ldots, X_k)$ where $X_1, \ldots, X_k$ is an orthonormal basis of $T_p(M)$. We can make $\mathcal{O}(M)$ into a differentiable manifold: Let π denote the projection map of $\mathcal{O}(M)$ onto M, sending $(p, X_1, \ldots, X_k)$ to p. Let U be a coordinate neighborhood of M. Let $Y_1, \ldots, Y_k$ be a set of vector fields on U which is an orthonormal basis at each point. Then for each $(p, X_1, \ldots, X_k)$ in (U) there is a unique $A \in \mathcal{O}(k) = (a_{ij})$ so that $X_i = \sum a_{ij} Y_j$. This defines a one to one map of $\pi^{-1}(U)$ onto $U \times O(k)$, once we have chosen $Y_1, \ldots, Y_k$. We can choose $Y_1, \ldots, Y_k$ as follows: Let $x^1, \ldots, x^k$, be a coordinate system on U. The vector fields $\partial/\partial x^1, \ldots, \partial/\partial x^k$ are independent everywhere on U. If we apply the Gram-Schmidt orthonormalization procedure to $\partial/\partial x^1, \ldots, \partial/\partial x^n$ then we obtain orthonormal vector fields $Y_1, \ldots, Y_k$. Thus we have constructed maps for each coordinate system U. The reader can check that the conditions of Theorem 2.1 of Chapter II are satisfied and so $\mathcal{O}(M)$ becomes a differentiable manifold, and π is a differentiable map. [A comparison with Section 6 of Chapter II shows that we can regard $\mathcal{O}(M)$ as a submanifold of bundle of all frames $\mathcal{F}(M)$. It consists of those frames of $\mathcal{F}(M)$ which are orthonormal.] The manifold $\mathcal{O}(M)$ is called the *bundle of orthonormal frames of M*.

Now let us define the fundamental form ω_M on $\mathcal{O}(M)$. Let Z be a tangent vector at the point $z = (p, X_1, \ldots, X_k)$ of $\mathcal{O}(M)$. Then $\pi_*(Z)$ is a tangent vector to M at p. Denote by $\langle Z, \omega_M^i \rangle$, the component of $\pi_*(Z)$ with respect to X_i, that is,

$$\langle Z, \omega_M^i \rangle = (\pi_*(Z), X_i) \qquad \text{(scalar product)}. \tag{3.1}$$

This defines ω_M^i as linear differential forms on M. We set $\omega_M = (\omega_M^1, \ldots, \omega_M^k)$ so that ω_M is a linear differential form with values in E^k. [Observe that if M is E^k with the Euclidean metric and we make the standard identifications, $\mathcal{O}(M) = \mathcal{O}(k)$ and $\omega_M = \omega$.]

The forms ω_M^i allow a simple expression for the Riemann metric. In fact, if we denote the Riemann metric by ds^2, we have

$$\pi^*(ds^2) = (\omega_M, \omega_M) = (\omega_M^1)^2 + \cdots + (\omega_M^k)^2 \tag{3.2}$$

[where by $(\omega_M^i)^2$ we mean the symmetric product $\omega_M^i \circ \omega_M^i$]. Let us verify (3.2). Let Z be a tangent vector at the point $Z = (p, X_1, \ldots, X_k)$. Then

$$\begin{aligned} \langle Z, \pi^*(ds^2) \rangle &= \langle \pi_*(Z), ds^2 \rangle \\ &= (\pi_*(Z), \pi_*(Z)) \\ &= \sum (\pi_*(Z), X_i)^2 \qquad \text{(since } X_i \text{ is an orthonormal basis)} \\ &= \sum \langle Z, \omega_M^i \rangle^2 \qquad \text{by (2.1).} \end{aligned}$$

Another useful expression is the formula for the volume. Recall, that the volume form is a scalar product on the one-dimensional space $\wedge^k(T_p(M))$ for each $p \in M$. Let us denote the associated quadratic form by dV^2. Then

$$\pi^*(dV^2) = (\omega_M^1 \wedge \cdots \wedge \omega_M^k) \circledcirc (\omega_M^1 \wedge \cdots \wedge \omega_M^k). \tag{3.3}$$

If M is an oriented Riemann manifold, there is a unique k-form, Ω, such that $\Omega > 0$ and so that $\Omega \circledcirc \Omega = dV^2$. By abuse of language we shall denote this form by dV and call it the volume element of M. Let us denote by $\mathcal{O}^+(M)$ the submanifold of $\mathcal{O}(M)$ consisting of those frames which induce the positive orientation, i.e., such that $\langle X_1 \wedge \cdots \wedge X_k, dV \rangle = +1$. Let us continue to denote by π and ω_M^i their restrictions to $\mathcal{O}^+(M)$. Then on $\mathcal{O}^+(M)$ we have

$$\pi^*(dV) = \omega_M^1 \wedge \cdots \wedge \omega_M^k. \tag{3.4}$$

We leave (3.3) and (3.4) as exercises to the reader.

Now let φ be an immersion of a k-dimensional manifold M into E^n and put the induced Riemann metric on M. Then for every $p \in M$, φ_* is an isometry of $T_p(M) \to T_{\varphi(p)}(E^n)$ (which we identified with E^n). Thus φ^* takes the orthonormal basis $X_1, \ldots, X_k$ into a set of orthonormal vectors $e_1, \ldots, e_k$, tangent to $\varphi_*(T_p(M))$. Conversely, given any orthonormal set $e_1, \ldots, e_k$ tangent to $\varphi_*(T_p(M))$ there is a unique orthonormal basis $X_1, \ldots, X_k$ of $T_p(M)$ so that $\varphi_* X_i = e_i$. Thus we have a map, ρ, of $\mathcal{O}_\varphi(M) \to \mathcal{O}(M)$ sending $(p, e_1, \ldots, e_n) \to (p, X_1, \ldots, X_k)$ where

$$\varphi_*(X_i) = e_i.$$

It is easy to see that ρ is differentiable and that $\pi_\varphi = \pi \circ \rho$, thus we have the diagram

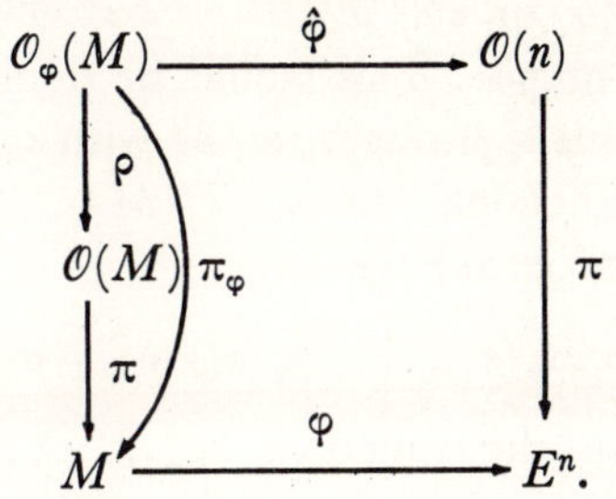

It is now easy to verify that

$$\rho^*(\theta) = \omega_M, \tag{3.5}$$

or

$$\begin{aligned} \rho^*(\theta^i) &= \omega_M^i \qquad \text{for } i \leqslant k \\ &= 0 \qquad\qquad i \geqslant k. \end{aligned} \tag{3.6}$$

We leave this verification as an exercise to the reader.

The forms Θ^{ij} for i and $j \leqslant k$ also come from $\mathcal{O}(M)$. That is, *there exist unique forms* Ω_M^{ij}, $(i, j = 1, \ldots, k)$ *on* $\mathcal{O}(M)$ *(depending only on the Riemann structure of* M*) such that*

$$\rho^*(\Omega_M^{ij}) = \Theta^{ij}. \tag{3.7}$$

The forms Ω_M^{ij} are anti-symmetric and satisfy

$$d\omega_M^i = \sum_{j=1}^{k} \Omega_M^{ji} \wedge \omega_M^j. \tag{3.8}$$

Observe that (3.8) together with anti-symmetry uniquely determine the Ω_M^{ij}. In fact, the forms $\omega_M^1, \ldots, \omega_M^k$ are linearly independent so we can argue exactly as in Section 1. This implies that the Ω_M^{ij}, if constructed so as to satisfy (3.7) for some φ, does not depend on φ but rather on the Riemann structure of M.

Notice also, that the forms Ω_M^{ij} are uniquely determined by (3.7): If $\rho^*(\Omega_M^{ij}) = \rho^*(\Omega_M'^{ij})$ then $\rho^*(\Omega_M^{ij} - \Omega_M'^{ij}) = 0$. But ρ is surjective, implying that $\Omega_M^{ij} - \Omega_M'^{ij} = 0$. This fact allows us to construct the Ω_M^{ij} from the Θ^{ij} locally on small open sets of $\mathcal{O}(M)$. The forms thus constructed then must agree on the overlap by uniqueness.

To construct the forms Ω_M^{ij} on an open set U of $\mathcal{O}(M)$, it is sufficient to find a map, σ, of $U \to \mathcal{O}_\varphi(M)$ with $\rho \circ \sigma = \text{ident}$. Then take $\Omega_M^{ij} = \sigma^*(\Theta^{ij})$. Now $\Theta^{ij} = \hat{\varphi}^*(de_i, e_j)$. If two points of $\mathcal{O}_\varphi$ lie over the same point of $\mathcal{O}(M)$, the first k elements of their frame coincide. Thus $\rho^* \circ \sigma^* \circ \hat{\varphi}^*(e_i) = \hat{\varphi}^* e_i$ or $\rho^*(\Omega_M^{ij}) = \rho^* \circ \sigma^* \Theta^{ij} = \Theta^{ij}$. We would like to construct such a map σ. This would be done if we can find $n - k$ vector valued functions $e_{k+1}, \ldots, e_n$ on M with $e_{k+1}(p), \ldots, e_n(p)$ orthonormal and all orthogonal to $\varphi_*(T_p(M))$. Then for each frame $z = (p; x_1, \ldots, x_k)$ of $\mathcal{O}(M)$ we set

$$\sigma(z) = (p, \varphi_*(x_1), \ldots, \varphi_*(x_k), e_{k+1}(p), \ldots, e_n(p)).$$

We wish to define the functions $e_{k+1}, \ldots, e_n$ in a neighborhood of a point $q \in M$. We proceed as follows. Let W be a subspace of E^n of dimension $n - k$ which is complementary to $\varphi_*(T_q(M))$. Then W is complementary to $\varphi_*(T_p(M))$ for all p in some neighborhood, U, of q. For such a p, the orthogonal projection of W onto $[\varphi_*(T_q(M))]^\perp$ is nonsingular. If we choose a fixed basis $v_{k+1}, \ldots, v_n$ of W, we then obtain a basis $v_{k+1}(p), \ldots, v_n(p)$ for $[\varphi_*(T_p(M))]^\perp$ at each $p \in U$. Applying the Gram-Schmidt orthonormalization process gives us the desired functions $e_{k+1}, \ldots, e_n$.

Thus we have established the existence of the Ω_M^{ij} on any $\mathcal{O}(M)$ provided that the Riemann metric can be obtained from an immersion in E^n. An extremely deep theorem due to Nash asserts that every Riemann metric can be obtained this way. This would imply the existence of the Ω_M^{ij} for any Riemann manifold. Actually, however, we can give a direct proof of the existence of the Ω_M^{ij} by a very simple local construction. We prefer to postpone this until Chapter VII, where we can place the problem in a more general context. In the next sections we study some of the geometrical information contained in these forms.

Observe that the antisymmetry of the Ω_M^{ij} and (3.8) uniquely determine them. In fact, the forms $\omega_M^1, \ldots, \omega_M^k$ are linearly independent so we can argue exactly as in Section 1. Notice also that the forms Ω_M^{ij} are uniquely determined by (3.7): If $\rho^*(\Omega_M^{ij}) = \rho^*(\Omega_M'^{ij})$ then $\rho^*(\Omega_M^{ji} - \Omega_M'^{ij})$

$= 0$. But ρ_* is onto implying that $\Omega_M^{ij} - \Omega_M'^{ij} = 0$. The first uniqueness shows that the forms Ω_M^{ij} depend only on the Riemann structure of M.

Let us resume, for the convenience of later reference, the various formulae we have obtained for immersed submanifolds of Euclidean space:

THEOREM 3.1. *Let φ be an immersion of a k-dimensional manifold into E^n. The manifold $\mathcal{O}_\varphi(M)$ consists of all pairs $(p; f_1, \ldots, f_n)$ where $p \in M$ and $f_1, \ldots, f_n$ is an orthonormal frame with $f_1, \ldots, f_k$ tangent to $\varphi(M)$ at p and $f_{k+1}, \ldots, f_n$ orthogonal to $\varphi(M)$ at p. The functions f_i can be regarded as vector valued functions on $\mathcal{O}_\varphi(M)$ as can the map $\varphi \circ \pi_\varphi$. Then the form θ^i and Θ^{ij} on $\mathcal{O}_\varphi(M)$ defined by*

$$\theta^i = (dp \circ \pi_\varphi, f_i) \tag{3.9}$$

and

$$\Theta^{ij} = (df_i, f_j) \tag{3.10}$$

satisfy

$$\theta^i = 0 \qquad i = k + 1, \ldots, n \tag{2.11}$$

$$d\theta^i = \sum_{j=1}^{n} \Theta^{ij} \wedge \theta^j \tag{2.9}$$

and

$$d\Theta^{ij} = \sum \Theta^{ip} \wedge \Theta^{pj}. \tag{2.10}$$

Furthermore, the forms Θ^{ij}, $i < j \leqslant k$ depend only on the Riemann structure on M induced by φ.

Let us return, for a moment, to the forms Ω_M^{ij} of (3.7) and (3.8). Suppose that the forms Ω_M^{ij} satisfied the equation

$$d\Omega_M^{ij} = \sum \Omega_M^{ik} \wedge \Omega_M^{kj}. \tag{3.11}$$

Then by Proposition 1.1, there exists (locally) a diffeomorphism, φ, of $\mathcal{O}_M$ onto $\mathcal{O}(k)$ such that $\varphi^* (\omega^i) = \omega_M^i$. In particular,

$$\varphi^*((\omega^1)^2 + \cdots + (\omega^k)^2) = (\omega_M^1)^2 + \cdots + (\omega_M^k)^2$$

or φ is a local isometry of M with E^k. Thus

PROPOSITION 3.1. *Let M be a Riemann manifold with forms Ω_M^{ij} satisfying* (3.8) *and* (3.11). *Then M is locally isometric to Euclidean space.*

4. Curves in Euclidean space. In this section we use the results of Section 2 to obtain some information about the differential geometry of curves in Euclidean space. Everything that we will do here works, modulo easy generalizations, in n-dimensional space. We will restrict our attention, for simplicity, to curves in three space. Let C be a curve; i.e., C is an immersion of an interval, I, in E^3. Then $\mathcal{O}_C(I)$ is two-dimensional —the two degrees of freedom correspond to motion along the curve and rotation of f_2 and f_3 about f_1, cf. Figure 6.1 of Section 2. On $\mathcal{O}_C(I)$ the forms θ^2 and θ^3 vanish and the forms θ^1 and Θ^{ij} are given by

$$(4.1)\qquad \begin{aligned} d\hat{C} &= \theta^1 f_1 \\ de_1 &= \Theta^{12}e_2 + \Theta^{13}e_3 \\ de_2 &= \Theta^{21}e_1 + \Theta^{23}e_3 \\ de_3 &= \Theta^{31}e_1 + \Theta^{32}e_2. \end{aligned}$$

The quadratic form $(\theta^1)^2 = \pi^*(ds^2)$ is just the element of arc length of the curve, lifted to $\mathcal{O}_C(I)$. Notice that at each point of the curve there are exactly two choices for f_1 which can be described by an orientation of the curve. In particular the quadratic form (df_1, df_1) depends only on the point of the curve; i.e., we can write

$$(df_1, df_1) = \pi^*(K^2\, ds^2)$$

where $K \geqslant 0$ is a function on I. The function K will be called *the absolute curvature* of C. It is clearly an invariant of C under Euclidean motions.

We shall now assume that a choice of orientation has been made so that there is a definite choice of f_1; i.e., we pass to a submanifold of $\mathcal{O}(I)$ [which by abuse of language we shall continue to denote by $\mathcal{O}(I)$]. Since now dC and f_1 are both vector valued forms on I we can write the first equation of (4.1) as $\theta^1 = (d\hat{C}, f_1) = \pi^*(dC, f_1)$. We define the form ds by $ds = (dC, f_1)$ so that we have $\theta^1 = \pi^*(ds)$. [This is consistent as $ds^2 = (ds)^2$]. Very often, in classical differential geometry, one is given more than a curve—one is given also a definite choice of f_2 (and hence f_3) at every point of the curve. For instance, if the curve lies on a certain surface, S, under study, it is natural to require that f_2 be tangent to S. Therefore, we make the following definition:

Definition 4.1. *Let C be a curve, i.e., $C: I \to E^3$. By a ribbon over C is meant a differentiable map, n, of $I \to \mathcal{O}_C(I)$ with $\pi(n) = C$. Thus a ribbon amounts to a definite (differentiable) choice of normal vector at every point of C.*

Via the map n, the forms Θ^{ij} pull back to I. In particular we can write

$$
\begin{aligned}
n^*(\Theta^{12}) &= k\,ds \\
n^*(\Theta^{23}) &= \tau\,ds \qquad (4.2)\\
n^*(\Theta^{13}) &= w\,ds.
\end{aligned}
$$

The function k is called the *curvature* of the ribbon, the function τ is called the *torsion* and the function w is called the *twist*. These three functions are obviously invariants of the ribbon under Euclidean motion. Conversely, let R_1 and R_2 be two ribbons over the curves C_1 and C_2 defined on the interval I. There is an obvious diffeomorphism, ψ, of $\mathcal{O}_{C_1}(I)$ onto $\mathcal{O}_{C_2}(I)$; in fact every frame of $\mathcal{O}_{C_1}(I)$ differs from the frame of R_1 at the same point of I by a rotation through a definite angle, and the same for $\mathcal{O}_{C_2}(I)$. Now just let ψ take R_1 into R_2, it then extends to a diffeomorphism of $\mathcal{O}_{C_1}(I)$ onto $\mathcal{O}_{C_2}(I)$. An easy computation (which we leave as an exercise to the reader) shows that if k, τ, and w, *are the same for* R_1 *and* R_2, *then* $\psi^*\Theta_2^{ij} = \Theta_1^{ij}$. *Thus, by Theorem* 2.1, ψ *is a congruence.*

Let C be a curve whose absolute curvature is strictly positive everywhere. Then there is a natural ribbon, called the *Frenet ribbon*, associated to C defined as follows: Since the curvature does not vanish, we have that $de_1/dt \neq 0$. We define e_2 to be the unique unit vector such that

$$\frac{de_1}{dt} = Ke_2. \qquad (4.3)$$

We then choose e_3 so that $e_1 \wedge e_2 \wedge e_3 > 0$ (relative to the natural orientation of E^3). The vector e_2 is called the *principal normal* and the vector e_3 is called the *binormal* of the curve. The frame e_1, e_2, e_3 is called the *Frenet trihedral* of the curve. If we compare (4.3) with (4.1) we see that $\theta^{13} \equiv 0$, i.e., the Frenet ribbon has no twist Equation (4.1) can be written as

$$
\begin{aligned}
\frac{de_1}{ds} &= Ke_2 \\
\frac{de_2}{ds} &= Ke_1 - \mathsf{T}e_3 \qquad (4.4)\\
\frac{de_1}{ds} &= \mathsf{T}e_3.
\end{aligned}
$$

The function T in this case is called the *torsion* of the curve. The curvature measures how much the curve differs from being a straight line.

(Thus a curve with $K = 0$, is a straight line.) The torsion measures how far the curve departs from lying in a plane. (If the torsion vanishes identically the curve is planar.) Applying *the previous remarks we see that two curves with non-vanishing curvature are congruent if and only if they have the same curvature and torsion.*

There is a simple but important geometrical operation that can be performed with a ribbon. It consists, intuitively speaking, of flattening the ribbon out on a plane so as to obtain a plane curve. In precise language let R be a ribbon. Thus we have functions $e_1(s)$, $e_2(s)$, $e_3(s)$ with

$$\frac{de_1}{ds} = Ke_2 - we_3$$

$$\frac{de_2}{ds} = -Ke_1 - \tau e_3$$

$$\frac{de_3}{ds} = -we_1 + \tau e_2.$$

Construct the curve $D(R)$ in E^2 by setting

$$\frac{dD(R)(s)}{ds} = f_1(s) \tag{4.5}$$

where $f_1(s)$ is a unit tangent vector satisfying

$$\frac{df_1(s)}{ds} = k(s)\, f_2(s) \tag{4.6}$$

with $f_2(s)$ the unit vector orthogonal to $f_1(s)$ and satisfying $f_1 \wedge f_2 > 0$. Equation (4.6) is just a system of homogeneous linear equations of degree two. Therefore, it has a solution unique up to an initial choice of $f_1(s_0)$. Thus $D(R)$ is determined up to initial position and direction. We fix, once and for all, such a set of initial conditions: Say $D(R(0)) = 0$, $D(R)'(0) = \partial/\partial x'$. The map $R \to D(R)$ which assigns the curve $D(R)$ to the ribbon R is called "developing the ribbon on the plane". Equation (4.6) expresses the fact that we are untwisting the ribbon and flattening out the curve along the ribbon: $k(s)$ is, after all, the component of change of e_1 tangent to the ribbon.

EXERCISE 4.1. Show that developing any ribbon over a straight line yields a straight line.

EXERCISE 4.2. Let C be a plane curve; i.e., $C(s)$ lies in a fixed plane, P. Let R be the ribbon over C defined by choosing e_2 to lie in P. If we identify P

with E^2, show that $D(C)$ differs from C by a plane Euclidean motion. Conclude: A curve C lies in a plane if and only if there is a ribbon R over C whose torsion vanishes.

EXERCISE 4.3. Let C be a curve lying in the plane P. Let R be the ribbon over C obtained by choosing e_2 to be always normal to P. Show that $D(R)$ is a straight line.

EXERCISE 4.4. Let C be any curve. Show that there exists a ribbon R over C with $D(R)$ a straight line. Furthermore, R is unique if the curvature of C is nowhere zero.

EXERCISE 4.5. Let C be a curve with nowhere vanishing curvature and let R be its Frenet ribbon. Show that $D(R)$ and C have the same curvature.

As we have already seen, for a plane curve there is a natural ribbon associated: We simply define e_2 to be the unique normal vector lying in the plane so that $e_1 \wedge e_2$ gives the positive orientation of the plane, and e_3 is the normal to the plane. Then Equation (4.4) becomes

$$\frac{de}{ds} = e_1$$

$$\frac{de_1}{ds} = K(s)e_2 \tag{4.7}$$

$$\frac{de_2}{ds} = -K(s)e_1.$$

An immediate consequence of Theorem 2.1 is that two curves with the same curvature differ from each other by a Euclidean motion. The curvature measures "which way" and how much the curve is turning in the plane. The sign of the curvature depends on the direction in which we traverse the curve.

If $K(s) > 0$ then the curve C stays on one side of the tangent line near s. To see this let us assume (by change of parameter on C and a Euclidean motion) that $s = 0$, $C(0) = (0, 0)$ and $e_1 = (1, 0)$ so the tangent line is just the x_1 axis. Then (4.7) says that

$$C(0) = (0, 0), \qquad C'(0) = (1, 0) \quad \text{and} \quad C''(0) = (0, K(0)).$$

Thus if we write $C(t) = (x_1(t), x_2(t))$ we have $x_2(t) = \frac{1}{2}K(0)t^2 +$ higher-order terms. Thus if $K(0) > 0$, $x_2(t)$ is positive for small enough nonzero values of t.

Equation (4.7) suggests another way of looking at the curvature, K. Let us regard the function e_1 as a map into the unit circle. Then $e_2(s)$ is the unit tangent vector at the point to the unit circle at the point $e_1(s)$. Then (4.7) says that K is the Jacobian of this map when we parametrize both C and the unit circle by arc length. But arc length on the unit circle is the way we measure angles. Thus if $\tau(s)$ denotes the angle (defined mod 2π) that $e_1(s)$ makes with the x_1-axis, say, we have

$$K(s) = \frac{d\tau}{ds}. \tag{4.8}$$

Now let us examine closed curves in the plane, that is, immersions, C, of the circle, S^1, in E^2. We can still introduce arc length as a local parameter and study such things as curvature, etc. In particular we can look at the tangential map, sending each point $x \in S^1$ into the unit tangent vector of C at x. This can be viewed as a map of S^1 into the unit circle of E^2, or, if we like, as a map of S^1 into itself. As such, we can look at the degree of this map. This integer is of importance because of the following considerations: Let us call a family, C_u, of curves, depending on a real parameter u, a *smooth family* if C_u is continuous on the C^1 topology; i.e., if for any $\mathcal{N}_{(W,Z,\epsilon,1)}C_{u_0}$ there is a $\delta > 0$ so that $C_u \in \mathcal{N}_{(W,Z,\epsilon,1)}$ when $|u - u_0| < \delta$. Thus C_u is close to C_{u_0} in the C^1 sense when u is close to u_0. We call two curves, C_0 and C_1 *smoothly homotopic* if there is a smooth family of curves C_u defined for $0 \leqslant u \leqslant 1$ coinciding with C_0 at $u = 0$ and C_1 at $u = 1$.[2] Then we have

THEOREM 4.1. (Whitney). *Let C_0 and C_1 be two curves in the plane. Let n_0 be the degree of the tangential map of C_0 and n_1 the degree of the tangential map of C_1. Then C_0 is smoothly homotopic to C_1 if and only if $n_0 = n_1$.*

Before proceeding to the proof of Theorem 4.1 we study the degree of maps of S^1 into itself a little more closely. On a circle, S^1, we can introduce angular coordinates, τ, defined mod 2π. Although the function τ is not globally defined, the form $\omega = d\tau$ is. [In fact, S^1 is a Lie group and choosing the appropriate basis for the (one-dimensional) Lie algebra of S^1, the form ω is just the fundamental one form on S^1.] Integrating ω over S^1 gives $\int_{S^1} \omega = 2\pi$. If $S^{1\prime}$ is a second copy of the circle with

---

[2] Remember that for each u the map C_u is an *immersion*, i.e., $C_u'(t) \neq 0$ for all t.

angular coordinate τ' and form ω', and if f is a differential map of $S^1 \to S^{1'}$ then deg (f) is given by

$$\deg(f) = \frac{1}{2\pi}\int_{S^1} f^*(\omega'). \tag{4.9}$$

We might be tempted to write (4.9) as

$$\deg(f) = \frac{1}{2\pi}\int_0^{2\pi}\left(\frac{d\tau'}{d\tau}\right) d\tau. \tag{4.10}$$

We must be careful, of course, since τ' is only defined mod 2π. What (4.10) suggests, however, is that the degree measures the number of times the function f wraps S^1 around $S^{1'}$. This is so, and in fact we can define the degree this way for any continuous (not necessarily differentiable) map of $S^1 \to S^{1'}$.

To this end, we first remark that angular coordinates mean that the circle can be viewed as the reals mod 2π. That is, there is a natural projection, p, of R^1 onto S^1 which is locally a homeomorphism. For any space X and any map $f: S^1 \to X$ we can associate the map $\hat{f}: R \to X$ given by $\hat{f} = f \circ p$. The function $\hat{f}$ satisfies the equation $\hat{f}(x + 1) = \hat{f}(x)$. Conversely, given a map $\hat{f}: R^1 \to X$ satisfying $\hat{f}(x + 1) = \hat{f}(x)$ there is a map $f: S^1 \to X$ such that $\hat{f} = f \circ p$. The corresponding statements are true about the strip $R^1 \times I$ and the cylinder $S^1 \times I$, where I is the unit interval. Suppose now that $X = S^1$ and that we are given a map $F: S^1 \times I \to S^1$. We lift F to a map $\hat{F}$ of $R^1 \times I \to S^1$ satisfying $\hat{F}(x + 1, s) = \hat{F}(x, s)$. Since $\hat{F}$ is uniformly continuous, we can subdivide $R^1 \times I$

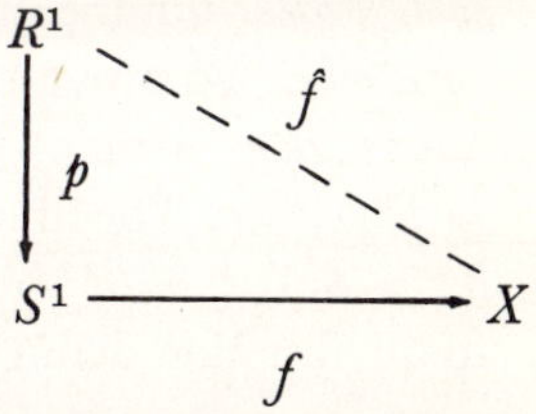

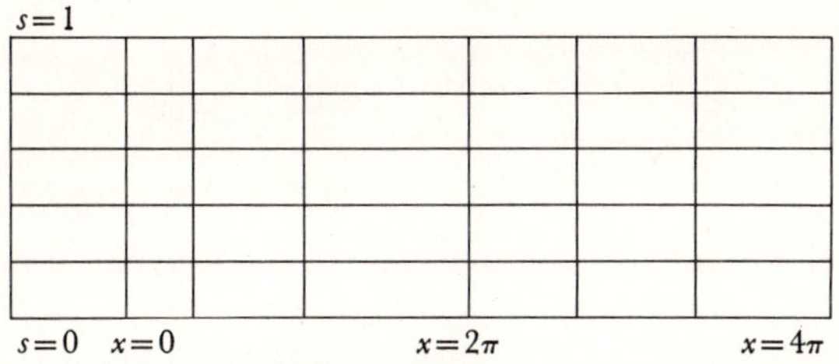

FIGURE 6.3

into a rectangular mesh such that the image of each rectangle is contained in an interval of S^1 of length $< \pi/4$. Now choose a point y_0 of R^1 such that $p(y_0) = \hat{F}(0, 0)$. Since p is locally a homeomorphism and we have chosen our mesh sufficiently refined, there is a unique map F^* defined in the rectangle into R^1 such that $pF^* = \hat{F}$ and $F^*(0, 0) = y_0$. Then we can uniquely extend the map F^* to the rectangle directly above this one so as to satisfy $pF^* = \hat{F}$ and agree with F^* along the common boundary. Continuing in this way, we define F^* on the first vertical strip. Then we can start at the bottom rectangle of the next step and continue until the function F^* is defined on the square $0 \leqslant s \leqslant 1, 0 \leqslant x \leqslant 2\pi$. Notice that F^* is characterized by the properties $pF^* = \hat{F}, F^*(0, 0) = y_0$. (This implies that if we would have started with a $y_0^1 = y_0 + k$ then the new $F^{*\prime}$ would satisfy $F^{*\prime} \rightarrow F^* = k$.) Since $\hat{F}(x + 1, s) = \hat{F}(x, s)$, $F^*(1, s) - F^*(0, s) = 2\pi n$ for some integer n. By the parenthetical remark made above, this integer does not depend on the choice of y_0. If we are given a map $f: S^1 \rightarrow S^1$ we define $F(x, s) = f(x)$. The integer n we get in this way is called the *degree of f*. It is clear that if f is differentiable this coincides with the previous definition of degree and gives meaning to (4.10). The degree measures how many times f wraps the circle around itself. From the previous discussion we see that *two homotopic maps have the same degree*. On the other hand, *two maps of $S^1 \rightarrow S^1$ with the same degree are homotopic*. To prove this we observe that if f has degree n, the procedure above gives a map, f^*, of $R^1 \rightarrow R^1$ such that $pf^* = fp$ and $f^*(x + 1) = f^*(x) + n$. The function $F^*(x, s) = (1 - s)f^* + sg(x)$, where $g(x) = nx$, $F^*(x + 1, s) = F^*(x, s) + n$ and thus $pF^*(x, s)$ provided homotopy of f to the function defined by g. Thus all maps of degree n are homotopic to g.

If we compare (4.8) with (4.10), we see that the degree of the tangential map of a curve C is $(\frac{1}{2}\pi) \int_0^L K(s)\, ds$ (where L is the arc length of C). Instead of having to worry about arc length it is more convenient to introduce the linear differential form $K\, ds$ and write

$$\text{deg of tangential map} = \frac{1}{2\pi} \int_C K\, ds. \tag{4.11}$$

Now let us prove Theorem 4.1. If C_0 is smoothly homotopic to C_1 then the tangential map of C_0 is homotopic to the tangential map of C_1 so that $n_0 = n_1$. To prove the converse let us observe first that if C is a curve, the family C_u given by $C_u(s) = uC(s)$ [multiplication of the vector $C(s)$ by the real number u] is a smooth family of curves for $u \neq 0$. The length of C_u is uL if L is the length of C. Thus we may assume (after making smooth homotopies if necessary) that C_0 and C_1 have total length 2π.

Let h be an E^2 valued function on S^1, which doesn't vanish anywhere.

Suppose that h is given by $h(\tau)$ in terms of angular coordinates. Then the curve $C(\tau) = \int_0^\tau h(t)\,dt$ is a differentiable curve provided that $C(2\pi) = C(0)$, i.e., that $\int_0^{2\pi} h(t)\,dt = 0$. Let $h(0, 0)$ be the tangential map of C_0 and $h(\cdot, 1)$ the tangential map of C_1. Since they have the same degree they are homotopic. Thus there is a continuous map $(\tau, u) \to h(\tau, u)$ mapping $S^1 \times I \to S^1$. [Here we are viewing S^1 as the unit circle so $h(\tau, u) \in E^2$ is a unit vector.]

Let

$$C_u(\tau) = \int_0^\tau h(t, u)\,dt - \frac{\tau}{2\pi}\int_0^{2\pi} h(t, u)\,du.$$

This would provide a smooth homotopy if we knew that the tangent vector, $C_u'(\tau)$, did not vanish anywhere. But

$$C_u'(\tau) = h(\tau, u) - \frac{1}{2\pi}\int_0^{2\pi} h(t, u)\,dt.$$

If $h(t, u)$ is never a constant in t for a fixed u, then

$$\left\| \frac{1}{2\pi}\int_0^{2\pi} h(t, u)\,du \right\| < 1 = \|h(t, u)\|.$$

The map $h(\cdot, u)$ is certainly not constant if $n_0 = n_1 \neq 0$. All that remains to be shown is that two non-constant maps h_0 and h_1 of $S^1 \to S^1$, both of degree zero are homotopic via non-constant maps. To this end, let h_0^* and h_1^* be the lifted maps of h_0 and h_1. By suitable rotations, we can assume that $h_0^*(0) = h_1^*(0)$ and $\sup h_1^*(x) > 0$, $\sup h_2^*(x) > 0$. Then $sh_1^* + (1 - s)h_0^*$ provides the desired homotopy.

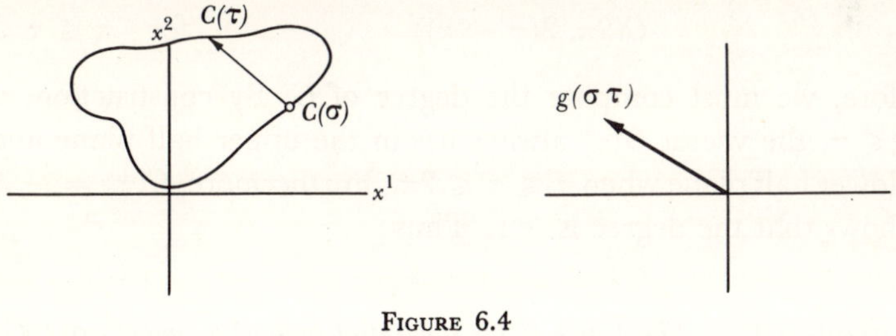

FIGURE 6.4

We are still left with the problem of computing the degree of the tangential map for a given curve. We will compute it for a *simple curve*, i.e., a curve without self-intersections.

Let C be a simple curve. Then introducing arc length $0 \leqslant \tau \leqslant 2\pi$ as a parameter on S^1, we have $C(\tau) = C(\tau')$ implies that $\tau = \tau'$ or $\tau = 0$ and $\tau' = 2\pi$. By a suitable translation and rotation we may assume that

$C(S^1)$ lies in the upper half plane, $C(0) = (0, 0)$ and $C'(0)$ lies on the x_1-axis.

Now define the map, g, of the triangle $0 \leqslant \sigma \leqslant \tau \leqslant 2\pi$ into S^1 as follows:

$$g(\sigma, \tau) = \frac{C(\tau) - C(\sigma)}{\|C(\tau) - C(\sigma)\|} \qquad \begin{array}{l} \text{if } 0 \leqslant \sigma < \tau < 2\pi \\ \text{and } 0 < \sigma < \tau \leqslant 2\pi \end{array}$$

$$g(\sigma, \tau) = C'(\tau) \qquad \text{if } \sigma = \tau$$

$$g(\sigma, 2\pi) = -C'(0).$$

Then g is well defined since $C(\sigma) = C(\tau)$ implies $\sigma = \tau$ or $\sigma = 0$ and $\tau = 2\pi$. It is clear that g is continuous. Now the map $\tau \rightarrow C(\tau, \tau)$

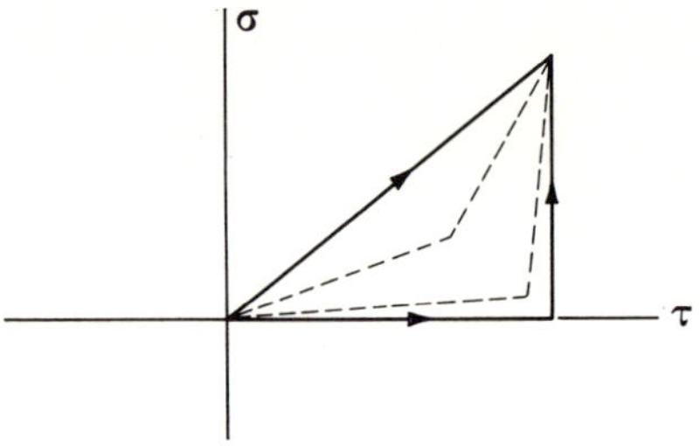

FIGURE 6.5

is just the tangential map. On the other hand, it is homotopic to the map G given by

$$G(\tau) = C(2\tau, 0) \qquad \text{for } 0 \leqslant \tau \leqslant \pi$$
$$C(2\pi, 2(\tau - \pi)) \qquad \text{for } \pi \leqslant \tau \leqslant 2\pi.$$

Therefore, we must compute the degree of G. By construction, when $0 \leqslant \tau \leqslant \pi$, the vector $G(\tau)$ always lies in the upper half plane and lies in the lower half plane when $\pi \leqslant \tau \leqslant 2\pi$. Furthermore, $G(\pi) = - G(0)$. This shows that the degree is ± 1. Thus

THEOREM 4.2. *The degree of the tangential map of a simple curve, C, is* ± 1. *Thus*

$$\int_C K \, ds = \pm 2\pi. \tag{4.12}$$

EXERCISE 4.6. *S*how how to compute this index when there are a (finite number of) transverse regular self-intersections.

It is useful to have a generalization of (4.12) to the case of a curve with corners, that is, for a piecewise differentiable map, C, of S^1 into E^2. Recall that C is piecewise differentiable if it is continuous and if there are a finite number of points $x_1, \ldots, x_n$ on S^1 so that C restricted to each arc $x_1x_2, x_2x_3, \ldots, x_{n-1}x_n, x_nx_1$, is a differentiable map and, in fact, the restriction of a differentiable map defined on a slightly larger arc. We write C_i for the restriction of C to x_ix_{i+1} $(i < n)$ and C_n for the restriction of C to x_nx_1.

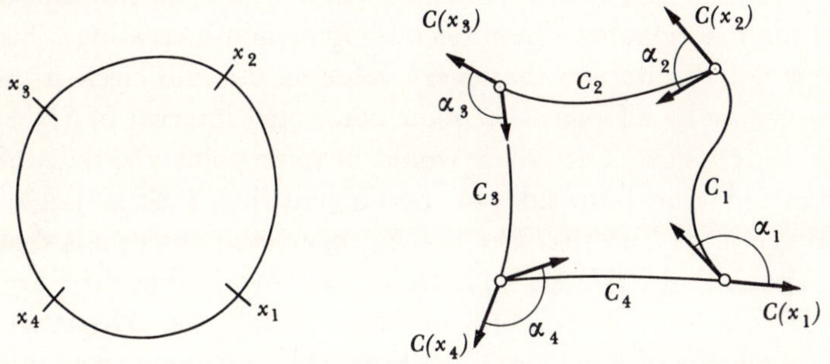

FIGURE 6.6

It makes sense to talk of the tangent vector of C_i at x_i and x_{i+1}. We denote by α_i the exterior angle at x_i, that is, α_i is the angle from the tangent vector to C_{i-1} at x_i to the tangent vector to C_i at x_i. [For $i = 1$ the angle from $C_n'(x_1)$ to $C_1'(x_1)$.] As before, we say that C is simple if it has no self-intersections. Then Equation (4.12) generalizes to

$$\sum_i \int_{C_i} K\,ds + \sum \alpha_i = \pm 2\pi. \tag{4.13}$$

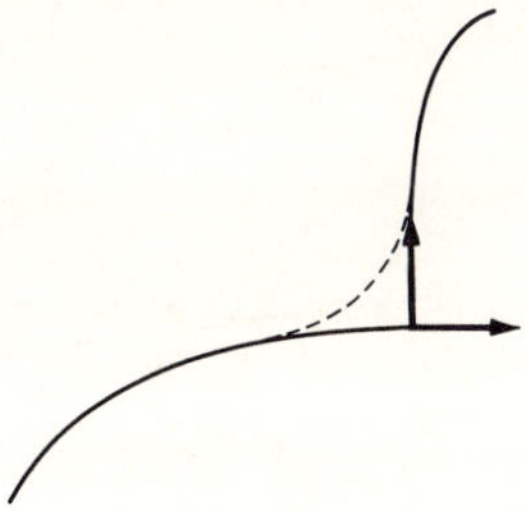

FIGURE 6.7

Equation (4.13) can be obtained from (4.12) by a simple limiting process. It suffices to examine the corners and see the effect of smoothing them out (see Figure 6.7). We leave the details as an exercise to the reader.

A curve C is called convex if $C(S^1)$ lies entirely on one side of its tangent line at every point. An immediate consequence of Theorem 4.2 is

THEOREM 4.3. *Let C be a simple closed curve. Then C is convex if and only if its curvature does not change sign, i.e., if $K \geqslant 0$ everywhere or $K \leqslant 0$ everywhere.*

Proof: To say that K does not change sign means (by 4.8) that the tangential map is monotone (non-decreasing or non-increasing). Since the degree is ± 1 this implies that every value on the unit circle is taken on either once or by all points on some connected interval of S^1. Suppose C were not convex. Then there would be some point y so that there are points of $C(S^1)$ on both sides of the tangent line, l, at y. Let n be the normal vector to l, so that $l = \{v \in E^2 | (n, v) = a\}$ where a is some constant. To say that $C(S^1)$ lies on both sides of l means that there are points $x \in S^1$ so that $(C(x), n) > a$ and so that $(C(x), n) < a$. The tangent line at the points x_1 and x_2 where $(C(x), n)$ takes on maximum and minimum values is parallel to l (see Figure 6.8). Thus two of the tangent vectors at the points y, x_1, x_2 must be equal. Thus one of the arcs joining points must give a constant value to the tangent vector. This is clearly impossible, since it would imply that the two points lie on the same line parallel to l.

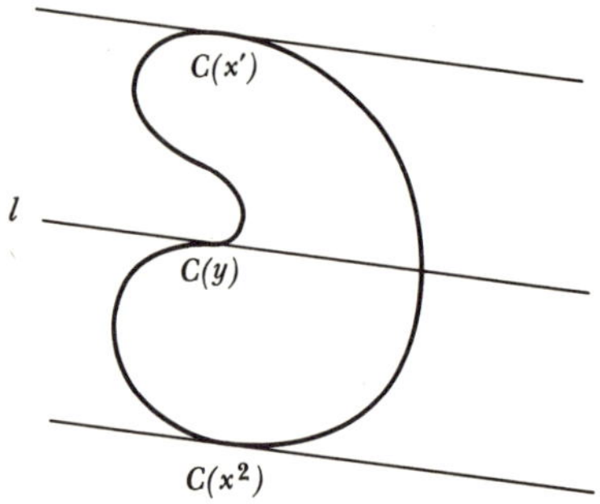

FIGURE 6.8

We give one further application of the degree of a map of $S^1 \to S^1$—the concept of the index of a vector field. Let X be a vector field on E^2 vanishing at the origin. Thus $X = f(\partial/\partial x_1) + g(\partial/\partial x_2)$ where x_1 and x_2 are the Euclidean coordinates on E^2, and $f(0, 0) = g(0, 0) = 0$. Suppose that X does not vanish at any other point near the origin—that is, suppose there is an $\epsilon > 0$ such that $f(x_1, x_2) = g(x_1, x_2) = 0$ and $x_1^2 + x_2^2 < \epsilon^2$ implies that $x_1 = x_2 = 0$. For any circle, C_r, of radius $r < \epsilon$ and centered at the origin we get a map of C_r into the unit circle defined as follows:

$$\text{map each } p \in C_r \to \frac{X_p}{\|X_p\|} \cdot$$

[Here $X_p = T_p(E^2)$ is the value of the vector field X at p. As usual, we identify $T_p(E^2)$ with E^2 and so regard X_p as an element of E^2.]

It is easy to see that this is a differentiable map.

EXERCISE 4.7. Show that the degree of this map is independent of r for $r < \epsilon$. This degree is called *the index* of X at the origin.

EXERCISE 4.8. What is the index of the vector field

$$x_1(\partial/\partial x_1) + x_2(\partial/\partial x_2)?$$

EXERCISE 4.9. What is the index of $x_1(\partial/\partial x_1) - x_2(\partial/\partial x_2)$?

EXERCISE 4.10. What is the index of $x_2(\partial/\partial x_1) - x_1(\partial/\partial x_2)$?

EXERCISE 4.11. Show that the index of $-X$ is the same as the index of X.

EXERCISE 4.12. The index of a vector field at an isolated zero can be defined on any oriented two-dimensional manifold by choosing an appropriate chart in the neighborhood at the zero. Show that this definition is independent of choice of chart.

We make one final observation concerning the curvature of a plane curve. We obtained it as a special case of the curvature of a space curve. That is, we were taking advantage of the fact that the submanifold we are discussing has dimension one. In the case of curves in the plane, however, we also have the special information that the submanifold has co-dimension one. This also allows us to define the curvature and generalize to higher dimension. In fact, if C is an oriented plane curve, the unit normal vector $e_2(s)$ is well defined. We can regard e_2 as a map into the unit circle. According to (4.7) the Jacobian of this map (relative to the arc length parametrization of both the unit circle and the curve C) is just $-K(s)$. We now proceed to generalize this construction to an n-dimensional manifold in $n + 1$-dimensional Euclidean space.

5. The second fundamental form. Let φ be an immersion of an n-dimensional manifold in E^N. We can regard φ as a vector valued function and consider the vector valued form, $d\varphi$. As we have already seen the quadratic differential form $(d\varphi, d\varphi)$ is nothing other than the Riemann metric induced on M by φ. In the classical literature, $(d\varphi, d\varphi)$ is called the *first fundamental form.* Let us now suppose that M is an *oriented* manifold and $N = n + 1$. For each $p \in M$ $\varphi_*(T_p(M))$ can be regarded as an n-dimensional subspace of E^{n+1}. Since M is oriented, $\varphi_*(T_p(M))$ has a preferred orientation. Therefore, there is a unique unit vector, $e_{n+1}(p)$, orthogonal to $\varphi_*(T_p(M))$ which, together with the orientation of $\varphi_*(T_p(M))$ give a positive orientation to E^{n+1}. That is, there is a unique $e_{n+1}(p) \in E^{n+1}$ such that $\varphi_*(X_1 \wedge \cdots \wedge X_n) \wedge e_{n+1}(p) > 0$ where $X_i \in T_p(M)$ and $X_1 \wedge \cdots \wedge X_n > 0$. (Here E^{n+1} is taken with the standard orientation.) We thus have defined a map, e_{n+1}, of $M \to E^{n+1}$. It actually maps M into the unit sphere of E^{n+1} and therefore is called the *spherical map of* φ.

We can, however, regard e_{n+1} as a vector valued function on M. As such, we can construct the quadratic differential form $-(d\varphi, de_{n+1})$;[3] it is called the *second fundamental form of* φ. We will denote the value of this form on $X, Y \in T_p(M)$ by $(X, Y)_{II}$. Its importance lies in the fact that the first and second fundamental forms together determine the map φ. More precisely

THEOREM 5.1. *Let φ_1 and φ_2 be two immersions of an oriented n-dimensional manifold, M, into E^{n+1}. If they induce the same Riemann metric and the same second fundamental form, they differ by a Euclidean motion.*

Proof: We wish to apply Theorem 2.1. To this effect we first observe that there is an obvious diffeomorphism between $\mathcal{O}_{\varphi_1}(M)$ and $\mathcal{O}_{\varphi_2}(M)$. In fact, for the case at hand (M oriented and φ maps M into a Euclidean space of one higher dimension) $\mathcal{O}_\varphi(M)$ is just a two sheeted covering of $\mathcal{O}(M)$. That is, each point of $\mathcal{O}(M)$ determines two points of $\mathcal{O}_\varphi(M)$ and the orientation picks out the preferred one. (This is how we defined the map e_{n+1}, in fact.) In other words, $\mathcal{O}_\varphi(M)$ is given once $\mathcal{O}(M)$ is given and $\mathcal{O}(M)$ is determined by the Riemann metric. Since we are assuming

---

[3] Recall that $(d\varphi, de)$ is a symmetric product of forms, i.e.,

$$\langle X, Y, (d\varphi, de)\rangle = \tfrac{1}{2}[(\langle X, d\varphi\rangle, \langle Y, de\rangle) + (\langle Y, d\varphi\rangle, \langle X, de\rangle)].$$

that φ_1 and φ_2 induce the same Riemann metric we can identify $\mathcal{O}_{\varphi_1}(M)$ and $\mathcal{O}_{\varphi_2}(M)$. To apply Theorem 2.1 we must show that

$$\theta_1^i = \theta_2^i \quad \text{and} \quad \Theta_1^{ij} = \Theta_2^{ij} \quad \text{for all } i \text{ and } j. \tag{5.1}$$

Now by Theorem 3.1, the forms θ^i and Θ^{ij} $(i < j \leqslant n)$ are determined by the Riemann metric. Thus what we must show is

$$\Theta_1^{i,n+1} = \Theta_2^{i,n+1}. \tag{5.2}$$

For this, it will suffice to show that the second fundamental form determines the linear differential forms $\Theta^{n+1,i}$ on $\mathcal{O}_\varphi(M)$. Now

$$\pi_\varphi^*(d\varphi, de_{n+1}) = (d\varphi \circ \pi_\varphi, df_{n+1}).$$

By (3.9) and (3.10) we have

$$d(\varphi \circ \pi_\varphi) = \sum \theta^i f_i$$

and

$$d(f_{n+1}) = \sum_{n=1}^{i} \Theta^{n+1,i} f_i. \tag{5.3}$$

Thus

$$\pi_\varphi^*(d\varphi, de_{n+1}) = \sum \theta^i \circledcirc \Theta^{n+1,i}. \tag{5.4}$$

Since the θ^i are linearly independent at all points, this shows that the forms $\Theta^{n+1,i}$ are determined by the second fundamental form. Thus (5.2) holds, proving Theorem 5.1.

Since the first and second fundamental forms determine the immersion up to Euclidean motion, we now study various properties of these forms and their geometrical interpretation. The first fundamental form is nothing but the Riemann metric. The second fundamental form tells us how the manifold is immersed in E^{n+1}, it gives the "second order terms" of the immersion at each point. More precisely, if $C(t)$ is a curve on M, then [writing $X(t)$ for $\varphi \circ C(t)$]

$$(e_n \circ C(t), X'(t)) \equiv 0$$

so, differentiating, we get

$$\left(\frac{de_{n+1} \circ C(t)}{dt}, X'(t)\right) + \left(e_{n+1}, \frac{d^2X}{dt^2}\right) = 0.$$

The first term is just the second fundamental form evaluated on C', C'. Thus, we can choose an orthonormal coordinate $x^1, \ldots, x^{n+1}$ of E^n so that e_{n+1} is the vector $(0, \ldots, 0, 1)$ and coordinates $u^1, \ldots, u^n$ about p so that the immersion φ is given by

$$x^i = u^i + \sum a^i_{jk}\, u^j u^k + \cdots \qquad i = 1, \ldots, n$$

$$x^{n+1} = \sum A_{jk}\, u^j u^k + \cdots$$

where

$$A_{jk} = \left(\frac{\partial}{\partial u^j}\bigg|_p, \frac{\partial}{\partial u_k}\bigg|_p \right)_{II}.$$

Thus, if $(\ ,\ )_{II}$ is (positive or negative) definite at p, the submanifold $\varphi(M)$ lies, locally, on one side of its tangent plane.

We can also define the curvature of an oriented n-dimensional manifold, M, in E^{n+1}. Let us consider the map e_{n+1} of $M \to S^n$. It induces a linear map e_{n+1*} of $T_p(M) \to T_{e_{n+1}(p)}(S^n)$. This induces a map $\wedge^n (e_{n+1*})$ of $\wedge^n (T_p(M)) \to \wedge^n (T_{e_{n+1}(p)}(S^n))$. Now both of these (one-dimensional) vector spaces have a distinguished basis, since both S^n and M are oriented Riemann manifolds. With respect to this basis, the transformation $\wedge^n (e_{n+1*})$ is given by a scalar which we call the *Gaussian curvature*, $K(p)$ of φ at p. Roughly speaking, K measures the quotient of the volume of the spherical image (under e_{n+1}) of a small piece of M to the volume of M.

We can actually compute the Gaussian curvature from the first and second fundamental forms. In fact, consider the equation

$$0 = d\theta^{n+1} = \sum \Theta^{n+1} \wedge \theta^i$$

which is the case $n + 1$ of (2.9). By Cartan's lemma (Theorem 4.4 of Chapter I), this implies that there are unique A_{ij} with $A_{ij} = A_{ji}$ such that

(5.5) $$\Theta^{n+1,i} = \sum A_{ij}\, \theta^j.$$

Thus, by (5.4)

(5.6) $$\pi^*(d\varphi, de_{n+1}) = \sum A_{ij}\, \theta^i \circledcirc \theta^j.$$

Now let $X_1, \ldots, X_n$ be an orthonormal basis of $T_p(M)$ and $(p; f_1, \ldots, f_{n+1})$ the corresponding point of $\mathcal{O}_\varphi(M)$. Then

$$de_{n+1*}(X_i) = \langle X_i, de_{n+1}\rangle$$

$$= \langle \overline{X}_i, df_{n+1}\rangle$$

where $\overline{X}_i$ is a tangent vector at $(p; f_1, \ldots, f_{n+1})$ with $\pi_* \overline{X}_i = X_i$. But $\langle \overline{X}_i, df_{n+1} \rangle = \sum \langle \overline{X}_i, \Theta^{n+1,j} \rangle f_j$. Using (5.5) and the fact that $\langle \overline{X}_i, \theta^j \rangle = 0$ for $i \neq j$ and $\langle \overline{X}_i, \theta^i \rangle = 1$, we obtain

$$e_{n+1}{}^*(X_i) = \sum A_{ij} f_j.$$

In particular, the Gaussian curvature is given by

$$K = \det(A_{ij}). \tag{5.7}$$

Equation (5.6) tells us what the matrix (A_{ij}) represents. In fact, since the second fundamental form is symmetric, and the first fundamental form is positive definite, there is a unique linear transformation, A_p, such that

$$(A_p X, Y)_I = (X, A_p Y)_I = (X, Y)_{II} \quad \text{for any } X, Y \in T_p(M). \tag{5.8}$$

Equation (5.6) says that $(-A_{ij})$ is a matrix representing this linear transformation, relative to the basis given by the point of $\mathcal{O}_\varphi(M)$. Thus we can write (5.7) as

$$K(p) = (-1)^n \det A_p. \tag{5.9}$$

Since the linear transformation A_p determines the second fundamental form uniquely, we can say that the Riemann metric induced by φ and the linear transformation A_p (for each $p \in M$) together uniquely determine φ up to Euclidean motion.

Let us now insert Equation (5.5) into (2.10). We get

$$\begin{aligned} d\Theta^{ij} &= \sum_{p=1}^{n+1} \Theta^{ip} \wedge \Theta^{pj} = \sum_{p=1}^{n} \Theta^{ip} \wedge \Theta^{pj} + \Theta^{in+1} \wedge \Theta^{n+1,j} \\ &= \sum_{p=1}^{n} \Theta^{ip} \wedge \Theta^{pj} - \sum_{k,l} A_{ik}\theta^k \wedge A_{jl}\,\theta^l. \end{aligned} \tag{5.10}$$

For i and $j \leqslant n$, the terms $d\Theta^{ij}$ and $\Theta^{ip} \wedge \Theta^{pj}$ depend only on the Riemann metric. Therefore *the expression* $\sum A_{ik}\theta^k \wedge A_{jl}\theta^l$ *also depends only on the Riemann metric.* Now $\sum A_{ik}\theta^k \wedge A_{jl}\theta^l$ just represents the transformation $\wedge^2 (A_p)$. Thus we have

THEOREM 5.2. (Gauss' theorem egregium). *Let φ be an immersion of an n-dimensional manifold, M, into E^{n+1}. For each $p \in M$ let A_p be the* *linear transformation of $T_p(M)$ defined by* (5.8). *Then* $\wedge^2 (A_p)$ *depends only on the Riemann metric induced by* φ.

We obtain as an immediate consequence of Theorem 5.2.

COROLLARY 5.1. *If n is even, the Gaussian curvature of the immersion φ depends only on the Riemann metric induced by φ.*

In fact, the Gaussian curvature is given by det A which is given by $\wedge^n (A)$. If $n = 2m$ we can write $\wedge^n (A) (e^1 \wedge \cdots \wedge e^n) = (Ae^1 \wedge Ae^2) \wedge (Ae^3 \wedge Ae^4) \wedge \cdots \wedge (Ae^{n-1} \wedge Ae^n)$ and so $\wedge^n (A)$ is determined by $\wedge^2 (A)$ which depends only on the Riemann metric.

For the case $n = 1$ (i.e., for curves in the plane) all curves are isometric and the curvature can take on all values. Thus Corollary 5.1 is false for $n = 1$. In fact, it is false for any odd n. This follows from

COROLLARY 5.2. *Let φ_1 and φ_2 be two immersions of an n-dimensional manifold, M, into E^{n+1} which induce the same Riemann metric. If n is odd and φ_1 and φ_2 have the same non-zero Gaussian curvature at all points then φ_1 differs from φ_2 by a Euclidean motion.*

We can paraphrase Corollary 5.2 more succinctly by saying that if n is odd the Riemann metric together with the Gaussian curvature (if it is not zero) determines the immersion up to Euclidean motion. Since it is possible to construct immersions which induce the same Riemann metric, have non-zero Gaussian curvature, and do not differ by a Euclidean motion we conclude that Corollary 5.1 is false for n odd.

To prove Corollary 5.2 we just observe that it suffices to show that φ_1 and φ_2 induce the same A. Since they induce the same Riemann metric they induce the same $\wedge^2 (A)$ and therefore the same $\wedge^{n-1} (A)$ since n is odd. If they induce the same $\wedge^n (A)$ (i.e., the same Gaussian curvature) and $\wedge^n (A) \neq 0$ anywhere they induce the same A since (by Cramer's rule) $\wedge^n (A)$ and $\wedge^{n-1} (A)$ determines A when $\wedge^n (A) \neq 0$.

We can give an immediate interpretation to the condition $\wedge^2 (A) = 0$. In fact Equations (5.10) then reduce to the equations of structure of Euclidean space. According to the last assertion of Section 1, we can conclude that there is a local diffeomorphism from $\mathcal{O}(M)$ onto $\mathcal{O}(n)$ carrying the forms θ^i and Θ^{ij} $(i, j \leqslant n)$ onto ω^i and Ω^{ij}. Since

$\pi^* ds^2 = (\theta^1)^2 + \cdots + (\theta^n)^2$ this means that the Riemann metric induced on M by φ is locally equivalent (isometric) to the Riemann metric on E^n, i.e., that M is locally Euclidean. This does not mean that $\varphi(M)$ is an n-dimensional subspace of E^{n+1}. For instance, if φ embeds M as a cylinder, the induced metric is also locally Euclidean. We shall now see that this is essentially the only case where $\wedge^2 (A) \equiv 0$.

The rest of this section will be devoted to the proof of the following two theorems:

THEOREM 5.3. (Hartman-Nirenberg). *Let φ be an immersion of a simply connected*[4] *n-dimensional manifold M into E^{n+1}. Suppose that M is complete with respect to the Riemann metric induced by φ and that $\wedge^2 (A) = 0$. Then M is isometric to E^n and φ immerses M as a cylinder, i.e., there is a Euclidean coordinate system $y^1, \ldots, y^n$ on M so that φ is given by*

$$\varphi(y^1, \ldots, y^n) = \sum_{j=1}^{n-1} a_j y^j + b(y^n)$$

where $a_1, \ldots, a_{n+1}$ are constant vectors in E^{n+1} and b is a vector valued function of one variable.

The theorem is intuitively reasonable: If $\wedge^2 (A) = 0$, then A is of rank at most one. This means, roughly speaking, that the hypersurface, $\varphi(M)$, is bending in at most one direction at each point. What Theorem 5.3 says is that we can piece these directions together.

At the other extreme we have the following generalization of Theorem 4.3.

THEOREM 5.4. (Chern-Lashof). *Let φ be an immersion of a compact oriented n-dimensional manifold, M, into E^{n+1}. Then the following two statements are equivalent:*

1) *The degree of the spherical map is ± 1 and the Gaussian curvature does not change sign (i.e., is always $\geqslant 0$ or $\leqslant 0$), and*
2) *$\varphi(M)$ is a convex hypersurface (i.e., $\varphi(M)$ lies entirely on one side of its tangent hyperplane at each point).*

[4] For the sake of completeness we assume the standard facts about simply connectedness, which are available in any elementary topology book. The reader who is unfamiliar with these facts can modify the arguments to obtain the corresponding local theorems.

As a corollary of Theorem 5.4 we obtain

THEOREM 5.5. (Hadmard). *Let* φ *be an immersion of a compact oriented n-manifold into* E^{n+1} $(n \geqslant 2)$ *whose Gaussian curvature is positive everywhere. Then* $\varphi(M)$ *is a convex hypersurface.*

In fact, if the Jacobian is everywhere positive, the spherical map gives a diffeomorphism of M into the unit sphere, S^n. According to Theorem 4.3 of Chapter III, it maps M onto S^n. Since M is compact, this is a covering map.[5] Since the sphere is simply connected for $n \geqslant 2$ we can conclude that the degree is $+1$. Thus condition 1) of Theorem 5.4 is fulfilled, proving Theorem 5.5.

We now turn to the proofs of Theorem 5.3 and 5.4. For this purpose we return to the study of Legendre transformations introduced in Section 1 of Chapter IV. There we were interested in non-singular Legendre transformations. Here we are particularly interested in the study of degenerate Legendre transformations. More precisely, we are given a connected open subset, D, of an n-dimensional vector space, V, and a differentiable function, f, defined on D. As in Section 1 of Chapter IV, the form df defines a map, $\mathscr{L}$, of D into V^*. If $x^1, \ldots, x^n$ is a coordinate system of V corresponding to a basis of V and $y^1, \ldots, y^n$ is a coordinate system on V^* corresponding to the dual basis then the map is given by

$$y^i = \frac{\partial f}{\partial x^i}.$$

Since we are interested in the case where $\mathscr{L}$ is degenerate, we will want to know the rank of $\mathscr{L}$. We define the function r to be this rank. Thus $r(x) =$ rank of $\mathscr{L}$ at x. We introduce the function r^* by

$$r^*(x) = \min_{U \ni x} (\sup r(z), z \in U).$$

That is, $r^*(x)$ is the maximum rank of $\mathscr{L}$ at all nearby points of D. Clearly, $r(x) \leqslant r^*(x)$. For any integer k we let D_k denote the set of all $x \in D$ such that $r^*(x) \leqslant k$, so D_k is open. We need one further piece of notation. Let H be a k-dimensional hyperplane of V (that is H is a coset of a k-dimensional subspace of V). Then $H \cap D$ is an open subset of H

---

[5] See footnote on page 269. Aside from the proof of Theorem 5.3, this is the only place where we will use facts about covering spaces.

(in the k-dimensional topology of H). Each connected component of this set $H \cap D$ is called a *k-plane section.* Thus a k-plane section through a point $x \in D$ is the connected component containing x of $H \cap D$ where H is a k-dimensional hyperplane passing through x. We now state the main lemma.

LEMMA 5.1. (Chern-Lashof-Hartman-Nirenberg). *Let f be a function defined on an open connected set, D, of an n-dimensional vector space, V, such that its Legendre transformation, $\mathscr{L}$, has rank $< n$ at all points of D. Let $x_0 \in D$ be a point with $r(x_0) = r^*(x_0) = k$. Then $\mathscr{L}$ is constant on an $(n - k)$-plane section, π_{n-k}, of D_k, through x_0 and $r(x) = r^*(x) = k$ for all $x \in \pi_{n-k}$, and $r(x) \geq k$ for $x \in \overline{\pi_{n-k}}$. Furthermore there is a neighborhood, U, of x_0 such that $x \in U$ and $\mathscr{L}(x) = \mathscr{L}(x_0)$ implies that $x \in \pi_{n-k}$.*

We notice several facts before proceeding to the proof of Lemma 5.1. If $r^*(x_0) = k$ [but $r(x_0)$ is not necessarily k] we can find a sequence of points $x_i \to x_0$ with $r^*(x_i) = r(x_i) = k$. Through each of these points Lemma 5.1 provides an $(n - k)$ plane section. By passing to a subsequence, if necessary, we can assume that these sections converge to an $(n - k)$ section through x_0. Since $\mathscr{L}$ is constant on the sections through x_i we can conclude that $\mathscr{L}$ is constant on the section through x_0. Thus we have proved (assuming Lemma 5.1)

LEMMA 5.2. *If $r^*(x_0) = k$, then $\mathscr{L}$ is constant on an $(n - k)$ section of D_k through x_0, which is the limit of the $n - k$ sections given by Lemma* 5.1.

In the case $k = 1$ we can say somewhat more:

LEMMA 5.3. *If $r^*(x_0) = 1$ then there is a unique $(n - 1)$ plane section passing through x_0 on which $\mathscr{L}$ is constant, and $r^* \equiv 1$ on this $(n - 1)$ plane section.*

In fact, since we are in n-space, any two non-parallel $n - 1$ hyperplanes must intersect. Let $\pi_{n-1}(x_0)$ be the section given by Lemma 5.2 and assume that there were another $n - 1$ section π'_{n-1} passing through x_0 on which $\mathscr{L}$ is constant. Let $\{x_i\}$ be the sequence used in the proof of Lemma 5.2. If x_i is close enough to x_0, then $\pi_{n-1}(x_i)$ and π'_{n-1} intersect (and intersect at a point, $\bar{x}_i$, arbitrarily close to x_0). However, by Lemma

5.1, since $\bar{x}_i \in \pi_{n-1}(x_i)$, $r(\bar{x}_i) = r^*(\bar{x}_i) = 1$ and there is a unique plane section passing through $\bar{x}_i$ on which $\mathscr{L}$ is constant. Thus $\pi_n(\bar{x}_i) = \pi_n(x_i) = \pi'_{n-1}$ which is a contradiction. The rest of Lemma 5.3 is immediate.

We now prove Lemma 5.1. We first show that *there is a neighborhood U of x_0 and an $(n - k)$-dimensional hyperplane H such that $U \cap H$ is equal to the set of points, x, of U for which $\mathscr{L}(x) = \mathscr{L}(x_0)$*. Since $r(x_0) = r^*(x_0) = k$, the rank of $\mathscr{L}$ is k in some neighborhood of x_0. Now the Jacobian matrix of $\mathscr{L}$ is $(\partial^2 f/\partial x^i \partial x^j)$. If we replace the basis $e_1, \ldots, e^n$ of V by $f_1, \ldots, f_n$, and make the corresponding change for the dual basis V^*, the Jacobian matrix of $\mathscr{L}$ in terms of the new linear coordinate system is $T^*(\partial^2 f/\partial x^i \partial x^j)T$ where T is the matrix representing the linear change of coordinates. Since the matrix $(\partial^2 f/\partial x^i \partial x^j)$ is symmetric, we can choose T appropriately so that the upper left-hand $k \times k$ submatrix of $T^*(\partial^2 f/\partial x^i \partial x^j)T$ is non-singular. Passing through these new linear coordinates (and calling them $x^1, \ldots, x^n$) we may assume that $(\partial^2 f/\partial x^i \partial x^j)$ $i, j = 1, \ldots, k$ is non-singular at x_0 and therefore in a neighborhood of x_0. Let us introduce functions $y^1, \ldots, y^k$ by putting

$$(5.11) \qquad y^i = \frac{\partial f}{\partial x^i} \quad i = 1, \ldots, k, \quad \text{and} \quad y^i = x^i \qquad \text{for } i > k.$$

Then the functions y^i can be used as coordinates in some neighborhood, U, of x_0. Let us put $g_i = \partial f/\partial x^i$, $i = 1, \ldots, n$. Then the matrix $(\partial g_i/\partial y^j)$ has rank k. Since $g_i = y^i$ for $i = 1, \ldots, k$, we can conclude that $\partial g_i/\partial y^j \equiv 0$ for i and j both $> k$. Thus the g_i for $i > k$ are functions of $y^1, \ldots, y^k$. Now

$$df = \sum \frac{\partial f}{\partial x^i} dx^i = \sum g_i \, dx^i$$

so

$$\sum_{i=1}^{n} dg_i \wedge dx^i = 0.$$

Let us write this as

$$\sum_{i=1}^{k} dy^i \wedge dx^i + \sum_{j=k+1}^{n} dg_j \wedge dy^j = 0.$$

Writing

$$dx^i = \sum_{\alpha=1}^{n} \frac{\partial x^i}{\partial y^\alpha} dy^\alpha \quad \text{and} \quad dg_j = \sum_{\beta=1}^{n} \frac{\partial g_j}{\partial y^\beta} dy^\beta$$

and substituting we get

$$\sum_{i=1}^{k}\sum_{\alpha=1}^{n}\frac{\partial x^i}{\partial y^\alpha}\,dy^i\wedge dy^\alpha+\sum_{j=k+1}^{n}\sum_{\beta=1}^{n}\frac{\partial g_j}{\partial y^\beta}\,dy^\beta\wedge dy^j=0.$$

If we compare the coefficient of $dy^i \wedge dy^j$ where $i \leqslant k < j$ we get

$$\frac{\partial x^i}{\partial y^j}+\frac{\partial g_j}{\partial y^i}=0. \tag{5.12}$$

Since the g_j depend only on $y^1, \ldots, y^k$, (5.12) implies that

$$\frac{\partial^2 x^i}{\partial y^j \partial y^l}=0 \qquad \text{for } j, l > k.$$

Thus

$$x^i=\sum_{j=k+1}^{n}\frac{\partial g_j}{\partial y^i}\,y^j+c^i \qquad i=1,\ldots,k \tag{5.13}$$

where the c^i are functions of $y^1, \ldots, y^k$.

Now the set $\mathscr{L}(x) = \mathscr{L}(x_0)$ is the set of all x where $y^i(x) = y^i(x^0)$, $(i = 1, \ldots, k)$. Thus in (5.13) the $\partial g_j/\partial y^i$ and the c^j are constants and the set $\mathscr{L}(x) = \mathscr{L}(x_0)$ is the intersection of a hyperplane, H, with U.

Now Equation (5.13) gives us the formula for the change of coordinates from the y's to the x's. We can write it as

$$x^i=\sum_{j=k+1}^{n}\frac{\partial g_j}{\partial y^i}\,y^j+c^i \qquad i=1,\ldots,k$$

(5.14) and

$$x^j=y^j \qquad j=k+1,\ldots,n.$$

If we multiply the Jacobian matrix of (5.14) with the Jacobian matrix of (5.11) we get the identity matrix. Looking at the term corresponding to r, s with $r, s \leqslant k$

$$\sum_{i=1}^{k}\frac{\partial g_i}{\partial x^s}\left(\sum_{j=k+1}^{n}x^j\frac{\partial^2 g_j}{\partial y^i\partial y^s}+\frac{\partial c^r}{\partial y^i}\right)=\begin{matrix}1 & \text{if } r=s.\\ 0 & \text{if } r\neq s.\end{matrix}$$

In particular

$$1=\det\left(\frac{\partial^2 f}{\partial x^r\partial x^s}\right)_{r,s=1,\ldots,k}\times\det\left(\sum_{j=k+1}^{n}x^j\frac{\partial^2 g_j}{\partial y^r\partial y^s}+\frac{\partial c^r}{\partial y^s}\right). \tag{5.15}$$

Let us now consider the intersection of H with D_k. Let π_{n-k} be the component of x_0 on $H \cap D_k$. To finish Lemma 5.1 we must show that

$\mathscr{L}$ is constant on π_{n-k} and that $r(x) = r^*(x) = k$ for $x \in \pi_{n-k}$. Since $\pi_{n-k} \subset D_k$, we know that $r^*(x) \leqslant k$ on π_{n-k}. To show that $r(x) = r^*(x) = k$ it thus suffices to show that $r(x) = k$, i.e., the determinant

$$J_k = \det\left(\frac{\partial^2 f}{\partial x^r \partial x^s}\right)_{r,s=1,\ldots,k} \neq 0 \quad \text{for any } x \in \pi_{n-k}.$$

Let $x(t)$ be any curve on π_{n-k} defined for $0 \leqslant t \leqslant 1$, with $x(0) = x_0$. We wish to show that for all t the point $x(t)$ satisfies the conditions above. Let t_0 denote the least upper bound of the set of all t such that

$$\mathscr{L}(x(s)) = \mathscr{L}(x_0) \quad \text{and} \quad J_k(x(s)) \neq 0, \qquad \text{for all } 0 \leqslant s \leqslant t_0.$$

Then it is clear that $\mathscr{L}(x(t)) = \mathscr{L}(x(0))$. To show that $J_k(x(t)) \neq 0$ it is sufficient to show that $J_k(x(s))$ is bounded away from zero as s approaches t. However, as long as $J_k(x(s)) \neq 0$, (5.15) gives us an explicit formula for $J_k(x(s))$. Since the $\partial^2 g^j/\partial y^r \partial y^s$ and $\partial c^r/\partial y^s$ depend only on $y^1, \ldots, y^k$, i.e., on $\mathscr{L}(x(s))$ (which is constant) and since the x^i are bounded, we conclude that

$$\det\left(\sum_{j=k+1}^{n} x^j \frac{\partial^2 g_j}{\partial y^r \partial j^s} + \frac{\partial c^r}{\partial y^s}\right)$$

is bounded and therefore J_k is bounded away from zero. This proves Lemma 5.1.

Let us now prove Theorem 5.3. Since $\wedge^2(A) \equiv 0$, the Riemann manifold M is locally isometric to E^n. That is, to every $p \in E^n$ and $q \in M$ there is a diffeomorphism, φ_p, of some neighborhood of p onto some neighborhood of q which is an isometry. Along any differentiable curve, C, with finite length and $C(0) = p$ we can continue the map φ. In fact if $\varphi(C(t_0))$ has been defined for $0 \leqslant t < t_0$, then the curve $\varphi \circ C$ on M over the interval $0 < t < t_0$ has finite length. Since M is complete, there is a unique point q_{t_0} such that setting $\varphi(C(t_0)) = q_{t_0}$ gives us a continuous curve on $0 \leqslant t \leqslant t_0$. Then the local isometry allows us to continue the map further. In this way every differentiable curve of finite length on E^n gives a map to M. Since E^n is simply connected, this gives us a diffeomorphism, φ, of E^n into M. Since M is connected and the map is an isometry, $\varphi(E^n) = M$. Since the map is an isometry, it is a covering map. Since M is simply connected we conclude that φ is a diffeomorphism of E^n with M. Thus the first part of Theorem 5.3 is proved.

To prove the second part we use Lemmas 5.1 and 5.2. Assume that $\varphi(M)$ is not a hyperplane of E^{n+1}. Then there is at least one point, x_0, where the rank of the normal map is not zero. Since the rank is $\leqslant 1$ (because $\wedge^2(A) \equiv 0$) the rank is exactly 1 at x_0. Let us choose Euclidean coordinates $x^1, \ldots, x^{n+1}$ and a neighborhood, U, of x_0 so that the projection, π, of $\varphi(U)$ onto the hyperplane $x^{n+1} = 0$ is non-singular on U. Then we can introduce $y^1, \ldots, y^n$ as local coordinates on U and the map φ is given by $\varphi(x^1, \ldots, x^n) = (x^1, \ldots, x^n, f(x^1, \ldots, x^n))$ where f is some differentiable function. Now the vector $\nu_f = (\partial f/\partial x^1, \ldots, \partial f/\partial x^n, -1)$ is normal to the surface $\varphi(U)$ at every point of U. The *unit* normal vector is thus

$$\frac{1}{\|\nu_f\|}\nu_f = \frac{1}{\sqrt{1 + \left(\frac{\partial f_1}{\partial x^1}\right)^2 + \cdots + \left(\frac{\partial f}{\partial x^n}\right)^2}}\left(\frac{\partial f}{\partial x^1}, \ldots, \frac{\partial f}{\partial x^n}, -1\right),$$

In terms of these coordinates, the matrix of the map n_* (regarded as a map into E^{n+1}) is

$$\begin{pmatrix} \frac{1}{\|\nu_f\|}\frac{\partial^2 f}{\partial x^1 \partial x^n} - \frac{1}{\|\nu_f\|^2}\frac{\partial f}{\partial x^1}\frac{\partial\|\nu_f\|}{\partial x^1} & \cdots & \frac{1}{\|\nu_f\|}\frac{\partial^2 f}{\partial x^1 \partial x^n} - \frac{1}{\|\nu_f\|^2}\frac{\partial f}{\partial x^n}\frac{\partial\|\nu_f\|}{\partial x^1} & \frac{-1}{\|\nu_f\|^2}\frac{\partial\|\nu_f\|}{\partial x^1} \\ \vdots & & \vdots & \vdots \\ \frac{1}{\|\nu_f\|}\frac{\partial^2 f}{\partial x^n \partial x^1} - \frac{1}{\|\nu_f\|^2}\frac{\partial f}{\partial x^n}\frac{\partial\|\nu_f\|}{\partial x^1} & \cdots & \frac{1}{\|\nu_f\|}\frac{\partial^2 f}{\partial x^n \partial x^n} - \frac{1}{\|\nu_f\|^2}\frac{\partial f}{\partial x^n}\frac{\partial\|\nu_f\|}{\partial x^n} & \frac{-1}{\|\nu_f\|^2}\frac{\partial\|\nu_f\|}{\partial x^n} \end{pmatrix}$$

which has the same rank as the matrix

$$\frac{1}{\|\nu\|_f}\begin{pmatrix} \frac{\partial^2 f}{\partial x^1 \partial x^1} & \cdots & \frac{\partial^2 f}{\partial x^1 \partial x^n} & \frac{-1}{\|\nu_f\|} & \frac{\partial\|\nu_f\|}{\partial x^1} \\ \vdots & & \vdots & \vdots & \vdots \\ \frac{\partial^2 f}{\partial x^n \partial x^1} & \cdots & \frac{\partial^2 f}{\partial x^n \partial x^n} & \frac{-1}{\|\nu_f\|} & \frac{\|\partial\nu_f\|}{\partial x^n} \end{pmatrix}$$

which in turn has the same rank as the $n \times n$ matrix

$$\left(\frac{\partial^2 f}{\partial x^i \partial x^j}\right)$$

since

$$\frac{\partial\|\nu_f\|}{\partial x^i} = \frac{1}{\|\nu_f\|}\sum\left(\frac{\partial f}{\partial x^k}\right)\frac{\partial^2 f}{\partial x^k \partial x^i}.$$

Thus *the rank of the normal map is the same as the rank of the Legendre transformation given by the function* f *of* $x^1, \ldots, x^n$. In particular, by Lemma 5.3, there is a unique $n - 1$ plane section, π_{n-1}, through $\pi\varphi(x_0)$ in $\pi\varphi(U)$ on which $\mathscr{L}$ is constant. But

$$\mathscr{L} = \left(\frac{\partial f}{\partial x^1}, \ldots, \frac{\partial f}{\partial x^n}\right)$$

so to say that $\mathscr{L}$ is constant means that the normal vector to $\varphi(U)$ is constant along σ_{n-1} where $\sigma_{n-1} = (\pi \circ \varphi)^{-1}(\pi_{n-1})$. Thus there is a unique $n - 1$ dimensional hyperplane, P, passing through $\varphi(x_0)$, whose projection on $\pi(\varphi(U))$ is just π_{n-1} and such that $\varphi(\sigma_{n-1}) \subset P$. Since φ is an *isometry* (and M can be identified with E^n) $\varphi^{-1}(\varphi(\pi_{n-1})) = \sigma_{n-1}$ is part of a hyperplane, H, (in the Euclidean geometry of M) through x_0, and φ is linear on σ_{n-1}. Now σ_{n-1} is an open subset of H, and on σ_{n-1}, $r^*(x) = 1$, by Lemma 5.3. Therefore, $r^*(x) = 1$ if x is any boundary point of σ_{n-1} (in H). (Here r and r^* are the rank of the normal map P.) Applying the argument above to x says that there is unique $n - 1$ plane section, σ'_{n-1} through x which gets mapped into a hyperplane of E^{n+1} by φ. The uniqueness implies that the hyperplane giving σ'_{n-1} must be H again. Let S be the set of points, x, of H which have the property that φ maps a neighborhood of x (in H) into a hyperplane. Then we have just seen that S is both open and closed in H. We conclude that φ restricted to H is a linear map of H onto an $n - 1$ dimensional hyperplane of E^{n+1}. Now let z be any other point not on H where $r^*(z) = 1$. Then there is a hyperplane, H', passing through z with the same properties. Since H and H' are unique, they cannot intersect. Thus H and H' are parallel. Thus we have shown that *through every point* $x \in M$ $(=E^n)$ *where* $r^*(x) = 1$ *there is a unique hyperplane,* H_x, *on which* φ *is linear. All of these hyperplanes are parallel.*

Now consider those z for which $r^*(z) = 0$. Let H_z be the hyperplane through z parallel to the above hyperplanes, H_x. Then $r^* \equiv 0$ on H_z. Otherwise there would be a y on H_z with $r^*(y) = 1$ implying $r^* \equiv 1$ on H_y. But H_y must coincide with H_z yielding a contradiction. Thus $r^* \equiv 0$ on H_z. This means that $r \equiv 0$ in some neighborhood of H_z, which means that the normal map is constant in some neighborhood of H_z. Thus, φ is again linear on H_z. Thus we have shown the existence of a family of parallel hyperplanes, H_q, passing through every point, q, of M so that φ is linear on each H_q. If we choose Euclidean coordinates $y^1, \ldots, y^n$ so that these hyperplanes are given by $y^n =$ constant we see that the map φ has the desired form, proving Theorem 5.3.

We now turn to the proof of Theorem 5.4. For this purpose it is convenient to introduce the absolute curvature, $|K|$. Roughly speaking, for any set $D \subset M$ the integral $\int_D |K| \, dA$ measures the total volume on S^n of $n(D)$ where points of S^n are counted k times if they are covered k times (here dA is the volume element given by the Riemann metric on A). A point $x \in M$ is called an extreme point (under φ) if $\varphi(M)$ lies entirely on one side of the hyperplane tangent of M at x. We let $\mathscr{E}$ denote the set of extreme points of M. It is clear that $\mathscr{E}$ is a set with content so that we can form $\int_{\mathscr{E}} |K| \, dA$.

LEMMA 5.4. *Let φ be an immersion of a compact n-dimensional manifold M into E^{n+1} and let $\mathscr{E}$ denote the set of extreme points of M. Then*

$$\int_{\mathscr{E}} |K| \, dA \geqslant \kappa_n \tag{5.15}$$

where κ_n is the volume of the unit sphere.

Proof: For the purpose of proving Lemma 5.4 it is convenient to introduce the projective space P^n instead of S^n. Recall that P^n is the set of all one-dimensional subspaces of E^{n+1}. There is an obvious map, ρ, of S^n onto P^n taking every unit vector, u, into the one-dimensional subspace it generates. For any sufficiently small set $D \subset P^n$, $\rho^{-1}(D)$ splits into two components, $D_1 \cup D_2$ in S^n with $\rho(D_1) = \rho(D_2) = D$. If we define the volume of D to be equal to the volume of D_1 = volume of D_2 we then get a volume density of P^n. The total volume of P^n with this density is $\kappa_n/2$. Now we observe that the map $\rho \circ n$ (where n is the normal map) restricted to E covers each point of P^n at least twice. In fact, let u be any unit vector. Consider the function (u, φ) on M. Since M is compact, it achieves its maximum and minimum at two points x_1 and x_2. These two points must be distinct, for otherwise $\varphi(M)$ would lie in an n-dimensional hyperplane (φ, u) = const which is impossible since M is compact and φ is an immersion. At each of these points $d(u, \varphi) = (u, d\varphi) = 0$ so u is normal to the M at x_1 and x_2. Thus $\rho(u) = \rho \circ n(x_1) = \rho \circ n(x_2)$, that is, $\rho(u)$ is covered (at least) twice. Now $\int_{\mathscr{E}} |K| \, dA = \int_{\mathscr{E}'} |K| \, dA$ where $\mathscr{E}'$ is the set of points of $\mathscr{E}$ where $|K| > 0$. On the other hand, $\int_{\mathscr{E}'} |K| \, dA$ = total volume $(\rho \circ n(\mathscr{E}'))$ (counted with multiplicity) and $\mathrm{vol}((\rho \circ n(\mathscr{E}')) \geqslant \mathrm{vol}(\rho \circ n(\mathscr{E})) - \mathrm{vol}(\rho \circ n(\mathscr{E}''))$ where $\mathscr{E}''$ is the set of points where $|K| = 0$. But $\mathrm{vol}(\rho \circ n(\mathscr{E}'')) = 0$ by Sard's theorem since K is the Jacobian determinant of n. Thus

$$\int_{\mathscr{E}} |K| \, dA \geqslant 2\left(\frac{1}{2}\kappa_n\right) = \kappa_n,$$

proving Lemma 5.4. The reader can verify, as an exercise, that the inequality in (5.5) is actually an equality.

An immediate implication of (5.15) is that $\int_M |K|\, dA \geqslant \kappa_n$.

THEOREM 5.6. (Chern-Lashof). *Let φ be an immersion of a compact n-dimensional manifold M in E^{n+1}. If*

$$\int_M |K|\, dA = \kappa_n \tag{5.16}$$

then $\varphi(M)$ is a convex hypersurface, i.e., $\varphi(M)$ lies on one side of its tangent hyperplane at every point.

Proof. Another way of putting the conclusion of Theorem 5.5 is to say that every point of M is an extreme point, i.e., that $\mathscr{E} = M$. In view of (5.15), Equation (5.16) implies that $\int_{M-\mathscr{E}} |K|\, dA = 0$. Now $\mathscr{E}$ is a closed subset of M and therefore $M - \mathscr{E}$ is open, and therefore $|K|$ must vanish identically on $M - \mathscr{E}$ in order for $\int_{M-\mathscr{E}} |K|\, dA$ to vanish. The condition $|K|(x) \neq 0$ is equivalent to $r(x) = n$ where r is the rank of the normal map. Thus we would prove Theorem 5.6 if we show that $M - \mathscr{E} \neq \varnothing$ implies the existence of an $x \in M - \mathscr{E}$ with $r(x) = n$. What we shall prove is that

LEMMA 5.5. *If there exists an $x \in M - \mathscr{E}$ with $r(x) = k$, $(< n)$ then there exists an $x'' \in M - \mathscr{E}$ with $r(x'') > k$.*

Then proceeding step by step we conclude that $M - \mathscr{E} \neq \varnothing$ implies the existence of an $x \in M - \mathscr{E}$ with $r(x) = n$.

To prove Lemma 5.5 observe that if $r^*(x) > k$ we are done. In fact, since $M - \mathscr{E}$ is open, if $r^*(x) > k$ there is some $x' \in M - \mathscr{E}$ with $r(x') > k$. Thus we need only consider the case where $r(x) = r^*(x) = k$. We now apply Lemma 5.1. As in the proof of Theorem 5.3, we can choose a small neighborhood of x and appropriate Euclidean coordinates on E^{n+1} so that the rank of the normal map is the same as the rank of the Legendre transformation $(\partial f/\partial x^1, \ldots, \partial f/\partial x^n)$. Applying Lemma 5.1 we conclude that there is an $n - k$-dimensional hyperplane H_{n-k} in E^n with the following properties:

1) $\varphi(x) \in H_{n-k} \subset H$ where H is the tangent hyperplane to M at x,
2) $\varphi(M) \cap H_{n-k}$ contains an open neighborhood, U, of $\varphi(x)$ (in H_{n-k}) on which $r = r^* \equiv k$,

and

3) $n(x) = n(x')$ for all x^* with $\varphi(x) \in U$, thus the tangent hyperplane at x^* is still H.

Let W be a maximal open subset of H_{n-k} satisfying conditions 2) and 3). Since M is compact, we know that $W \neq H_{n-k}$. Let y' be a point on the boundary of W. Thus we can choose a sequence of points x_i with $\varphi(x_i) \in W$ such that $x_i \to x'$ and $\varphi(x') = y'$. By continuity, $n(x') = n(x_i)$ so that H is the tangent hyperplane at x'. In particular, $x' \in M - \mathscr{E}$. We now show that $r^*(x') > k$. In fact, since $x' = \lim x_i$ we know that $r^*(x') \geqslant k$. If $r^*(x') = k$ we could choose a neighborhood of x' and appropriate coordinates in E^{n+1} and apply Lemma 5.1 again. Then x' would be on $\overline{\pi_{n-k}}$ and $r(x') = k$. Thus Lemma 5.1 now applied to x' (remembering the uniqueness) implies that $y' \in W$, contradiction. Thus we must have $r^*(x') > k$. This means there is an x'' sufficiently near to x' (and thus in $M - \mathscr{E}$) with $r(x'') > k$. This proves Lemma 5.5 and Theorem 5.5. We leave the converse of Theorem 5.6 as an exercise to the reader.

Now we can prove Theorem 5.4 as a corollary of Theorem 5.6. That 2) implies 1) is obvious. Let us see that 1) implies 2). We have

$$\int |K| \, dA = \pm \int K \, dA = 1 \cdot \kappa_n$$

by Equation (3.8) of Chapter III. Thus Theorem 5.6 implies 5.4.

6. Surfaces. Now let us apply the preceding results to the study of two-dimensional oriented Riemann manifolds and immersions of two-dimensional manifolds in E^3. First we resume the various formulae obtained in Sections 2 and 3 for the case of two-dimensional manifolds. The group $O^+(2)$ is just the circle, so the manifold $\mathcal{O}(M)$ is three-dimensional. Local coordinates on $\mathcal{O}(M)$ are obtained by choosing a coordinate chart, (U, h) on M and differentiable choice of frame at each point of U, that is, a map, ψ, of $U \to \pi^{-1}(U)$ with $\pi \circ \psi =$ identity. Let us write $\psi(x) = (x, f_1(x), f_2(x))$ (where f_1, f_2 is an oriented orthonormal base of $T_x(M)$. Any other orthonormal basis, e_1, e_2 of $T_x(M)$ is given by

$$\text{(6.1)} \qquad \begin{aligned} e_1 &= (\cos \tau) f_1(x) + (\sin \tau) f_2(x) \\ e_2 &= -(\sin \tau) f_1(x) + (\cos \tau) f_2(x). \end{aligned}$$

Here τ is determined mod 2π by these equations. If x^1, x^2 are local coordinates on U then x^1, x^2, τ (where τ is not quite legal) are local coordinates on $\pi^{-1}(U)$. The forms θ^1, θ^2 are defined by

$$\begin{aligned}\langle X, \theta^1\rangle &= (\pi_* X, e_1)\\ \langle X, \theta^2\rangle &= (\pi_* X, e_2)\end{aligned} \tag{6.2}$$

if X is a tangent vector at (x, e_1, e_2) [cf. (3.1) and (3.5) where ρ is the identity].

We have (cf. 3.4)

$$\theta^1 \wedge \theta^2 = \pi^* \, dA \tag{6.3}$$

where dA is the oriented element of area on M. It is convenient to have an expression for θ^1 and θ^2 in terms of local coordinates. Let U, x^1, x^2, and ψ be as above. Then $\overline{\theta^1} = \psi^*(\theta^1)$ and $\overline{\theta^2} = \psi^*(\theta^2)$ are forms on U. For each $x \in U$, the forms $\overline{\theta^1_x}$, $\overline{\theta^2_x}$ constitute a dual basis to $f_1(x), f_2(x)$. In fact, for any $Y \in T_x(M)$

$$\begin{aligned}\langle Y, \overline{\theta^1}\rangle &= \langle Y, \psi^*\theta^1\rangle\\ &= \langle \psi_* Y, \theta^1_{x,f_1,f_2}\rangle\\ &= (\pi_* \circ \psi_* Y, f_1)\\ &= (Y, f_1) \qquad \text{since } \pi \circ \psi = \text{id}.\end{aligned}$$

Thus $\langle Y, \overline{\theta^1}\rangle = (Y, f_1)$ and $\langle Y, \overline{\theta^2}\rangle = (Y, f_2)$. Let $\alpha^1 = \pi^*\overline{\theta^1}$ and $\alpha^2 = \pi^*\overline{\theta^2}$. For any $X \in T_{(x,e_1,e_2)}(F(M))$, $\langle X, \alpha^1\rangle = (\pi_* X, f_1)$ and $\langle X, \alpha^2\rangle = (\pi_* X, f_2)$ while $\langle X, \theta^1\rangle = (\pi_* X, e_1)$ and $\langle X, \theta^2\rangle = (\pi_* X, e_2)$. Comparing with (6.1) and (6.2) gives

$$\begin{aligned}\theta^1 &= (\cos\tau)\alpha^1 + (\sin\tau)\alpha^2\\ &\text{and}\\ \theta^2 &= -(\sin\tau)\alpha^1 + (\cos\tau)\alpha^2.\end{aligned} \tag{6.4}$$

If the Riemann metric of M comes from an immersion in Euclidean space, there is a unique form Θ^{12} on $\mathcal{O}(M)$ with

$$\begin{aligned}d\theta^1 &= \Theta^{12} \wedge \theta^2\\ d\theta^2 &= \Theta^{12} \wedge \theta^1.\end{aligned} \tag{6.5}$$

Now we can give an explicit expression for the form Θ^{12} in terms of local coordinates. We will construct locally a form $\Theta^{12'}$ satisfying $d\theta^1 = \Theta^{12'} \wedge \theta^2$ and $d\theta^2 = -\Theta^{12'} \wedge \theta^1$; we then know that $\Theta^{12'} = \Theta^{12'}$ since (6.5) determines Θ^{12} uniquely. This local construction allows us to conclude the existence of Θ^{12} on any two-dimensional Riemann manifold—where the Riemann metric does not necessarily come from an immersion. In fact, if we construct Θ^{12}, locally, on $\pi^{-1}(U)$ the uniqueness guarantees that the Θ^{12} constructed for the various U's fit together to make a global form. We make use of the local expressions (6.3) for θ^1 and θ^2 to construct Θ^{12}. For this purpose we introduce the functions k_1 and k_2 on $\mathcal{O}(M)$ by

$$k_1 = \overline{k_1} \circ \pi \qquad \text{and} \qquad k_2 = \overline{k_2} \circ \pi$$

(6.6) where

$$d\overline{\theta^1} = \overline{k_1}\, \overline{\theta^1} \wedge \overline{\theta^2} \qquad \text{and} \qquad d\overline{\theta^2} = \overline{k_2}\, \overline{\theta^1} \wedge \overline{\theta^2}.$$

Thus

$$d\alpha^1 = k_1\alpha^1 \wedge \alpha^2 \qquad \text{and} \qquad d\alpha^2 = k_2\alpha^1 \wedge \alpha^2.$$

Now

$$d\theta^1 = -\sin\tau\, d\tau \wedge \alpha^1 + \cos\tau\, d\tau \wedge \alpha^2 + (k_1 \cos\tau + k_2 \sin\tau)\alpha^1 \wedge \alpha^2$$
$$d\theta^2 = -\cos\tau\, d\tau \wedge \alpha^1 - \sin\tau\, d\tau \wedge \alpha^2\, (-k_1 \sin\tau + k_2 \cos\tau)\alpha^1 \wedge \alpha^2.$$

Since $\alpha^1 \wedge \alpha^2 = \theta^1 \wedge \theta^2$, a direct verification shows that the form

$$(6.7) \qquad \begin{aligned} \Theta^{12} &= d\tau + k_1\alpha^1 + k_2\,\alpha^2 \\ &= d\tau + (k_1 \cos\tau + k_2 \sin\tau)\theta^1 + (-k_1 \sin\tau + k_2 \cos\tau)\theta^2 \end{aligned}$$

satisfies (6.5).

The form Θ^{12} has a simple geometric interpretation: Suppose we first consider the case where the Riemann metric φ comes from an immersion, φ, of M into Euclidean three space. We then have the vector valued functions Φ, E^1, E^2, E^3 defined on $\mathcal{O}(M)$ by

$$\begin{aligned} \Phi &= \varphi \circ \pi \\ E^1(x, e_1, e_2) &= \varphi_{*x}(e_1) \\ E^2(x, e_1, e_2) &= \varphi_{*x}(e_2) \\ E^3(x, e_1, e_2) &= n(x). \end{aligned}$$

The forms θ^i and Θ^{ij} are then given by

$$\theta^i = (d\Phi, E^i)$$

and

$$\Theta^{ij} = (dE^i, E^j)$$

or

$$\begin{aligned} d\Phi &= \theta^1 E^1 + \theta^2 E^2 \\ dE^i &= \sum \Theta^{ij} E^j. \end{aligned}$$

Let γ be a differentiable curve on M. A differentiable family of unit vectors $e_1(\cdot)$ along γ (where $e_1(s) \in T_{r(s)}(M)$) is the same as a curve C in $\mathcal{O}(M)$ with $\pi \circ C = \gamma$. [Here $C(s) = (r(s), e_1(s), e_2(s))$.] Let us call the family $e_1(s)$ *parallel* if the unit vectors are all changing normal to the surface in three space. In other words, $e_1(s)$ is *parallel* if the vector

$$\frac{d\varphi_{*r(s)}(e_1(s))}{ds}$$

is normal to $\varphi(M)$ for all s. Let us see how to express this condition. Let X_s be the tangent vector to the curve C at $C(s)$. Then, by the definition of dE^1

$$\frac{d}{ds}\varphi_{*r(s)}(e_1(s)) = \langle X_s, dE^1\rangle.$$

Notice $(\langle X_s, dE^1\rangle, E^1(C(s))) = 0$ and $(\langle X_s, dE^1\rangle, E^2(C(s))) = \Theta^{12}$. Now E^1 and E^2 span the tangent space to $\varphi(M)$. Thus for $\langle X_s, dE^1\rangle$ to be orthogonal to $\varphi(M)$ is the same as saying that $\langle X_s, \Theta^{12}\rangle = 0$.

Thus $e_1(s)$ is parallel along γ if and only if $\langle X_s, \Theta^{12}\rangle = 0$.

Similarly on any two-dimensional Riemann manifold we will call a family of unit vectors parallel along a curve γ if $\langle X_s, \Theta^{12}\rangle = 0$. With this definition of parallel translation we can state the following:

THEOREM 6.1. *Let γ be any differentiable curve on M. Given the unit vector $e_1 \in T_{r(0)}(M)$, there is a unique parallel family of unit vectors $e_1(s)$, along γ with $e_1(0) = e_1$. If $e_1'(s)$ is another unit vector of $T_{\gamma(0)}(M)$ differing from e_1 by an angle σ then $e_1'(s)$ differs from $e_1(s)$ by the same angle σ for all s.*

It is clearly sufficient (by breaking γ up into small pieces if necessary) to prove Theorem 6.1 for curves γ lying entirely in a coordinate chart. Then we can use the local expression for Θ^{12}.

Let us rewrite the condition for parallel translation along $\gamma(s)$. In terms of local coordinates, the unit vector $e_1(s)$ is given by a function $\tau(s)$ where

$$e_1(s) = \cos\tau(s)f_1(\gamma(s)) - \sin\tau(s)f_2(\gamma(s)).$$

Then

$$\begin{aligned}\langle X_s, \Theta^{12}\rangle &= \langle X_s, d\tau\rangle + \langle X_s, k_1\alpha^1 + k_2\alpha^2\rangle \\ &= \frac{d\tau(s)}{ds} + \langle X_s, \pi^*(k_1\theta^1 + k_2\theta^2)\rangle \\ &= \frac{d\tau(s)}{ds} - \langle \pi_* X_s, k_1\theta^1 + k_2\theta^2\rangle.\end{aligned}$$

But $\pi_* X_s = Y_s$ is the tangent vector to γ at $\gamma(s)$. Thus

$$\langle X_s, \Theta^{12} \rangle = \frac{d\tau(s)}{ds} - F_\gamma(s)$$

where $F_\gamma(s) = \langle Y_s, \overline{k_1}\overline{\theta^1} + \overline{k_2}\overline{\theta^2} \rangle$ is a function depending only on γ. In particular $e_1(s) \to f_1(s)$ is parallel if and only if

$$\frac{d\tau(s)}{ds} = F_\gamma(s).$$

From this we see that given $e_1(0)$ there is a unique parallel family $e_1(s)$ starting with $e_1(0)$. Furthermore, if $e_1'(0)$ is a second unit vector at $\gamma(0)$, the angle between $e_1(s)$ and $e_1'(s)$ is equal to the angle between $e_1(0)$ and $e_1'(0)$. Thus parallel translation preserves angles, proving Theorem 6.1.

Notice that if M is (locally isometric to) Euclidean space, then we can choose $f_1 = \partial/\partial x^1$ and $f_2 = \partial/\partial x^2$ so that $\overline{\theta^1} = dx^1$ and $\overline{\theta^2} = dx^2$. In this case, $\overline{k_1} = \overline{k_2} = 0$ and τ is just the angle that e_1 makes with $\partial/\partial x^1$, i.e., with the x_1-axis. Thus $\Theta^{12} = -d\tau$, in this case. Then the condition for parallel translation becomes $d\tau/ds = 0$ which coincides with the usual notion of parallelism in Euclidean geometry.

Let M be any two-dimensional Riemann manifold. For any curve γ on M there is an obvious way of choosing unit vectors along γ: Just let $e_1(s)$ be the unit tangent vector to γ at $\gamma(s)$. Thus for every curve γ on M we get a curve, which we shall call $\tilde{\gamma}$ on $\mathcal{O}(M)$. [Here $\tilde{\gamma}(s) = (\gamma(s), e_1(s), e_2(s))$ and $e_1(s)$ is the tangent to $\gamma(s)$.]

We call the form $\tilde{\gamma}^*(\Theta^{12})$ the *geodesic curvature form of* γ. [In the Euclidean case this is just $\tilde{\gamma}^*(d\tau)$ which is the ordinary curvature cf. (4.8).]

Let us consider those curves whose geodesic curvature vanishes; i.e., those curves whose tangent vectors are parallel.

THEOREM 6.2. *A curve γ on M (parametrized by arc length) has the property that its tangent vectors $\gamma'(s)$ are parallel if and only if γ is a geodesic of M.*

Proof: First let us write down the condition for a tangent vector X of $\mathcal{O}(M)$ to be tangent to a curve $\tilde{\gamma}$. Since $\pi \circ \tilde{\gamma} = \gamma(s)$ we have $\pi_* X = \gamma'(s)$. On the other hand, $\tilde{\gamma}(s) = (\tilde{\gamma}(s), \gamma'(s), e_2(s))$ so that $(\pi_* X, \gamma'(s)) = (\pi_* X, e_1(s)) = 1$, $(\pi_* X, e_2(s)) = 0$. Thus $\langle X, \theta^1 \rangle = 1$ and $\langle X, \theta^2 \rangle = 0$. Conversely if $\langle X, \theta^2 \rangle = 0$ it is easy to see that X is tangent to some curve $\tilde{\gamma}(s)$. If in addition γ' is to be parallel we must have $\langle X, \Theta^{12}\rangle = 0$. Thus for X to be tangent to a curve $\tilde{\gamma}$ with γ' parallel we must have

$$\langle X, \theta^1\rangle = 1$$

$$\text{(6.8)} \qquad \text{and} \qquad \langle X, \theta^2\rangle = 0,$$

$$\langle X, \Theta^{12}\rangle = 0.$$

Since θ^1, θ^2 and Θ^{12} are linearly independent, Equations (6.8) determine a vector field X on $\mathcal{O}(M)$. For any curve γ, the family γ' is parallel if and only if $\tilde{\gamma}'$ satisfies (6.8).

Now we can identify $\mathcal{O}(M)$ with the set of unit tangent (or cotangent) vectors since e_2 is determined by e_1. Thus we have a map, ι, of $\mathcal{O}(M) \to T^*(M)$; it is evident from the definitions that $\iota^*\theta = \theta^1$ where θ is the canonical form on $T^*(M)$ cf. Theorem 7.1 of Chapter III. The set of unit vectors is the surface of constant energy $H = 1$. According to the principle of least action, (3.8) of Chapter IV, the geodesic spray is tangent to $\mathcal{O}(M)$ (regarded as a submanifold of $T^*(M)$) and is given by $\langle X, \theta^1\rangle = 1$ and $X \lrcorner\, \iota^* d\theta = X \lrcorner\, d\theta^1 = 0$. But $d\theta^1 = \Theta^{12} \wedge \theta^2$ so $X \lrcorner\, d\theta^1 = 0$ is equivalent to $\langle X, \theta^2\rangle = \langle X, \Theta^{12}\rangle = 0$. Thus conditions (6.8) are the equations for the geodesic spray proving Theorem 6.2.

According to Equation (5.10) $d\Theta^{12} = -(A_{11}A_{12} + A_{12}A_{21})\theta^1 \wedge \theta^2$. Thus by (5.9) we can write

$$\text{(6.9)} \qquad d\Theta^{12} = -\pi^*(K\, dA)$$

where K is the Gaussian curvature.

Let D be a domain with regular boundary on M and let ψ be a map of some neighborhood of $\bar{D} \to F(M)$ satisfying $\pi \circ \psi =$ identity. Then, by Stokes' theorem,

$$\int_{\partial D} \psi^*(\Theta^{12}) = \int_D \psi^*\pi^* K\, dA$$

$$\text{(6.10)} \qquad = -\int_D K\, dA.$$

Let us apply (6.10) to the map ψ constructed as follows: Let Y be a vector field on a compact manifold M which vanishes at only a finite number of points, $p_1, \ldots, p_r$. At any $x \in M$ where $Y_x \neq 0$, put $e_1(x) = Y_x/\|Y_x\|$ so $\psi(x) = (x, e_1(x), e_2(x))$. About each point p_i choose a chart (U_i, h_i) such that U_i contains no other point p_j and $h_i(p_i) = (0, 0)$

in the plane. Let $\gamma_{i,r}$ be $h_i^{-1}(C_r)$ where C_r is the circle of radius r in the plane. [Here r is chosen so small so that C_r lies in $h_i(U_i)$.] Now let $D = M - U_i$ (interiors of $\gamma_{i,r}$). We must compute

$$\int_{\gamma r} \psi^*(\Theta^{12})$$

where γ_r is oriented clockwise rather than anti-clockwise. For this purpose we introduce about p_i the orthonormal frames coming from the coordinates x^1, x^2. Thus we have coordinates x^1, x^2, τ where $\tau(x, e_1, e_2)$ measures the angle that $e_1(x)$ makes with

$$\left.\frac{\partial}{\partial x^1}\right|_x.$$

Thus (taking $\gamma_{i,r,\tau}$ clockwise) $-\int_{\gamma_{i,r}} d\tau(e_1(s)) = 2\pi \cdot$ index of Y at p_i. But $\psi^*\Theta^{12} = \psi^* d\tau + \psi^*(k_1\alpha^1 + k_2\alpha^2)$ so $-\int_{\gamma_{i,r}} \psi^*\Theta^{12} = 2\pi \cdot$index of Y at $p_i + \int_{\gamma_{i,r}} \overline{k_1\theta^1} + \overline{k_2\theta^2}$. Now as $r \to 0$ the second term on the right vanishes and $\int_D \to \int_M$ on the right of (6.10). Thus we have proved the following:

Let Y be a vector field which vanishes at a finite number of points $p_1, \ldots, p_n$. Then

$$\sum_i \text{index}_{\,p_i}(Y) = \frac{1}{2\pi}\int_M K\,dA. \tag{6.11}$$

In particular the sum of the indices of a vector field is independent of the vector field and the total integral of the curvature is independent of the Riemann metric.

Thus, for instance, if M is S^2, the vector field tangent to the meridian circles has zeros only at the north and south poles and the index at each of these points is $+1$. Thus (6.11) says that the sum of indices of *any* vector field on S^2 must equal 2. (In particular it is impossible to construct a vector field on S^2 which doesn't vanish anywhere.) Furthermore, (6.11) says that no matter what Riemann metric we put on S^2 we have

$$\int_{S^2} K\,dA = 4\pi.$$

Similarly, if M is the torus, we can put on it a vector field which doesn't vanish anywhere so the integer in (6.11) is 0.

EXERCISE 6.1. Let M denote the torus with angular coordinates φ_1, φ_2 (defined up to 2π). The map f will denote the immersion of $M \to$ Euclidean three space given by

$$x_1 = (2 + \cos \varphi_1) \cos \varphi_2$$
$$x_2 = (2 + \cos \varphi_1) \sin \varphi_2$$
$$x_3 = \sin \varphi_1.$$

1. What is the Riemann metric on M induced by f? That is, compute $(\partial/\partial\varphi_1, \partial/\partial\varphi_1)_p$, $(\partial/\partial\varphi_1, \partial/\partial\varphi_2)_p$ and $(\partial/\partial\varphi_2, \partial/\partial\varphi_2)_p$ for each $p \in M$. (These can be given as functions of φ_1, φ_2.) Let e_1, e_2 denote the vector fields obtained by orthonormalizing $\partial/\partial\varphi_1, \partial/\partial\varphi_2$. What is the explicit expression for e_1, e_2?

2. What is the total area of the torus relative to this Riemann metric?

3. In terms of the vector fields e_1, e_2 given in 1, we can introduce (angular) coordinates $\varphi_1, \varphi_2, \tau$ on $\mathcal{O}(M)$. In terms of these coordinates, what are the explicit expressions for the vector valued functions Φ, E^1, E^2, E^3 on $\mathcal{O}(M)$? (Each of these vector valued functions should be given in terms of the fixed standard basis of Euclidean three space, i.e., a triplet of functions of $\varphi_1, \varphi_2, \tau$.) Thus

$$\Phi(\varphi_1, \varphi_2, \tau) = ((2 + \cos \varphi_1) \cos \varphi_2, (2 + \cos \varphi_1) \sin \varphi_2, \sin \varphi_1).$$

What are the corresponding expressions for E^1, E^2, and E^3?

4. What are the explicit expressions for the forms θ^1, θ^2, and Θ^{12} on $\mathcal{O}(M)$ given in terms of $d\varphi_1$, $d\varphi_2$, and $d\tau$?

5. What is the Gaussian curvature, K, of M? (Again, K can be given as a function of φ_1, φ_2.) For which points of M is the Gaussian curvature positive and for which points is it negative?

6. On any Riemann manifold there is an obvious way of identifying a linear differential form with a vector field [via the identification of $T_x^*(M)$ with $T_x(M)$ given by the scalar product in $T_x(M)$]. If g is a function, the vector field associated to the form dg is denoted by grad g. Let M be the torus with Riemann as above and let x_1 be the function given by $x_1(\varphi_1, \varphi_2) = (2 + \cos \varphi_1) \cos \varphi_2$ as before. Show that the vector field $X = \operatorname{grad} x_1$ vanishes at exactly four points and compute its index at each of these points.

Now we give another way of computing the integer in (6.11). Let C^1 and C^2 be two curves in $\mathcal{O}(M)$ with $\pi \circ C^1 = \pi \circ C^2$. Thus $C^1(s) = (\gamma(s), e_1(s), e_2(s))$ and $C^2(s) = (\gamma(s), f_1(s), f_2(s))$. Let us suppose that the angle between $e_1(s)$ and $f_1(s)$ is given (locally) by a function $\sigma(s)$. In terms of a local coordinate system on $\mathcal{O}(M)$, we have $\tau(C^1(s)) = \tau(C^2(s)) + \sigma(s)$. If X_s^1 and X_s^2 are the tangent vectors to these two curves we have

$$\langle X_s^1, \Theta^{12}\rangle = \langle X_s^2, \Theta^{12}\rangle + \frac{d\sigma(s)}{ds}. \tag{6.12}$$

Let us apply this remark to the following situation. Let D be a domain lying entirely in a coordinate chart U and whose boundary is a simple closed curve γ. Let ψ be the map of $U \to \mathcal{O}(M)$ obtained by orthonormalization of $\partial/\partial x^1$, $\partial/\partial x^2$. Take

$$C^2(s) = \psi(\gamma(s))$$

and

$$C^1(s) = (\gamma(s), e_1(s), e_2(s))$$

where $e_1(s)$ is the unit tangent vector to $\gamma(s)$. Now the last equation implies that

$$\int_\gamma \tilde{\gamma}^*(\Theta^{12}) = \int_\gamma \psi^*\Theta^{12} + \int_\gamma d\sigma$$

where σ is the angle that $\gamma'(s)$ makes with $\partial/\partial x^1$.

Thus $\int_\gamma d\sigma$ is just the total curvature of γ with respect to the Euclidean metric corresponding to the coordinates x^1, x^2. According to (4.12), this integral is equal to 2π. Thus we have, by (6.10)

$$\int \tilde{\gamma}^*\Theta^{12} = 2\pi - \int_D K\, dA. \tag{6.13}$$

Equation (6.13) generalizes, as (4.13) does, for the case of piecewise differentiable curve $\gamma = C_1, \ldots, C_n$ to

$$\sum \int_{C_i} \tilde{C}_i^*(\Theta^{12}) + \sum \alpha_i = 2\pi - \int_D K\, dA, \tag{6.14}$$

where the α_i are the exterior angles at the corners. If the C_i are all geodesics, the first term on the right of (6.14) vanishes and we get

$$\sum \alpha_i = 2\pi - \int_D K\, dA. \tag{6.15}$$

For instance, if D is a geodesic triangle, and we consider the interior angles $\beta_i = \pi - \alpha_i$, Equation (6.15) becomes

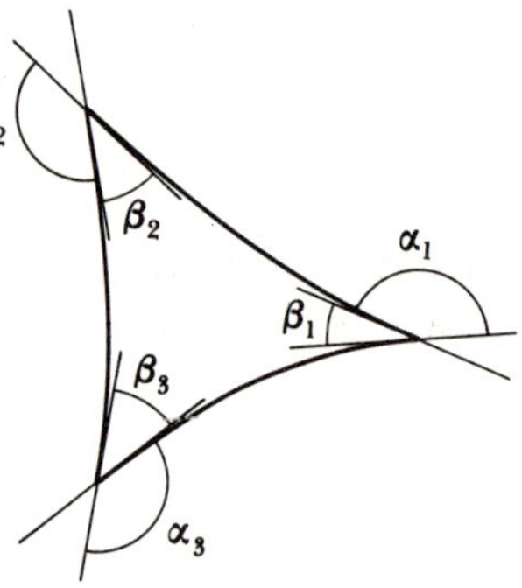

FIGURE 6.9

$$\beta_1 + \beta_2 + \beta_3 = \pi + \int K\, dA. \tag{6.16}$$

(In the case of the sphere, for instance, $K \equiv 1$ and (6.16) is the standard formula in spherical trigonometry.)

Now let us proceed to a computation of the integer in (6.11).

DEFINITION: *A connected closed cellulated surface is a compact differentiable 2-manifold M together with a finite collection of closed subsets (called the cells), $F_1, \ldots, F_{\alpha_2}$, such that*

$$M = \bigcup_{i=1}^{2} F_i$$

satisfying the following:

1. *For each F_i there is a one-to-one bidifferentiable map, f_i of F_i onto a polygon with n_i edges, $(n_i \geqslant 3)$.*

2. *For $i \neq j$, $F_i \cap F_j$ is either empty or $f_i(F_i \cap F_j)$ is a fixed edge or vertex of the corresponding polygon.*

3. *Given F_i, F_j, $i \neq j$ we can find a finite collection of polygons, F_{i_0}, F_{i_s} such that $F_{i_0} = F_i, F_{i_s} = F_j$ and $F_{i_k} \cap F_{i_{k+1}} \neq \varnothing$.*

The distinct points of M which are mapped onto vertices of the finite number of polygons will be labelled in some way as $P_1, \ldots, P_{\alpha_0}$.

The distinct subsets of M which are mapped onto edges of the polygon will be labelled, as $P_i P_j$ (we choose a definite ordering of the vertices), provided that their boundary consists of the points P_i, P_j. The total number of these edges is denoted by α_1, and call these $E_1 \ldots$.

For each F_i having vertices $P_{i_1}, \ldots, P_{i_s}$ choose a definite ordering of the vertices $P_{i_1}, \ldots, P_{i_s}$ so the boundary is transversed cyclicly. Thus we have a definite labelling, once and for all, e.g.,

$$F_i \leftrightarrow P_{i_1}, \ldots, P_{i_s} \qquad i = 1, \ldots, \alpha_2$$
$$E_i \leftrightarrow P_u P_v \qquad i = 1, \ldots, \alpha_1$$
$$P_i \leftrightarrow P_i \qquad i = 1, \ldots, \alpha_0.$$

Now we define some vector spaces as follows:

$$C_0 = \text{real vector space with basis } P_1, \ldots, P_{\alpha_0}$$
$$C_1 = \text{real vector space with basis } E_1, \ldots, E_{\alpha_1}$$
$$C_2 = \text{real vector space with basis } F_1, \ldots, F_{\alpha_2}.$$

Next we define two linear transformations $\partial_2 : C_2 \rightarrow C_1$ by letting

$$\partial_2(P_{i_1}, \ldots, P_{i_s}) = P_{i_1}P_{i_2} + P_{i_2}P_{i_3} + \cdots + P_{i_s}P_{i_1}$$

with the understanding that

$$P_{i_1}P_{i_2} = \begin{cases} E_i \\ -E_i \end{cases} \quad \text{if} \quad E_i = P_{i_2}P_{i_1}.$$

$\partial_1 : C_1 \rightarrow C_0$ by letting

$$\partial_2(P_{i_1}P_{i_2}) = P_{i_2} - P_{i_1}$$

and extend by linearity. By checking on generators of the vector space, it is clear that $\partial_1 \circ \partial_2 = 0$. Thus $\partial_2(C_2) \subset \partial_1^{-1}(0)$. It turns out that

$$H_2(M, R) = \partial_2^{-1}(0)$$
$$H_1(M, R) = \partial_1^{-1}/\partial_2(C_2)$$
$$H_0(M, R) = C_0/\partial_1(C_1).$$

The dimension of these vector spaces are called the Betti numbers of M, denoted by $\beta_2, \beta_1, \beta_0$.

DEFINITION 6.1: $\alpha_0 - \alpha_1 + \alpha_2 = \chi(M)$ *is called the Euler-Poincaré characteristic.*

It is an easy fact (which we shall not prove here)

$$(\chi M) = \beta_0 - \beta_1 + \beta_2.$$

The definition of $\chi(M)$ is highly "non-invariant," i.e., *a priori* it depends strongly on the cellulation. Now we show that it does not. In fact, let us apply (6.14) to the case where D is the face F_i of a cellulation of M. Let $\#F$, $\#E$, and $\#V$ be the number of faces, edges, and vertices. If we add over all faces we get

$$2\pi(\#F) - \int_M K\,dA = \sum \alpha_i \tag{6.15}$$

where the terms involving $\int K_\gamma(s)\,ds$ vanish since each edge is transversed in two directions and where the sum on the right is over all exterior angles of all vertices. Now at each vertex the sum of the exterior angles is clearly equal to $\pi \times$ (the number of edges meeting the vertex) $- 2\pi$. Since each edge meets exactly two vertices, the sum on the right is $2\pi\ \#E - 2\pi(\#V)$. Thus (6.15) becomes the Gauss-Bonnet formula.

$$\begin{aligned} \chi(M) &= \#F - \#E + \#V \\ &= \frac{1}{2\pi}\int_M K\,dA. \end{aligned} \tag{6.16}$$

So far we have seen the Gaussian curvature, K, in two roles: First as the Jacobian of the spherical map and second, [via (6.13)] as a measure of the effect of parallel translating around a small closed curve. Now we will see that it enters also into the second variation of arc length. Let γ be a geodesic of M. Recall that the conjugate point of $\gamma(0)$ along γ was defined as follows. We consider the map $\pi \circ \alpha_t$ of $T_{\gamma(0)}(M) \to M$ where α_t is the geodesic flow. Then the first point where the induced tangential map $\pi_* \circ \alpha_{t*}$ of $T_0(T_{\gamma(0)}(M)) \to T_{\gamma(t)}(M)$ becomes singular is the conjugate point of $\gamma(0)$ along γ. Since the manifold $\iota(\mathcal{O}(M)) \subset T(M)$ is left invariant by the geodesic flow, we can reformulate this definition staying on $\mathcal{O}(M)$: We replace $T_{\gamma(0)}(M)$ by the frames of $T_{\gamma(0)}(M)$, i.e., by the unit vectors in $T(M)$. Let $X_{12}(=\partial/\partial\tau)$ be the unit tangent vector to all the fibers of $\mathcal{O}(M)$, i.e., X_{12} is the vector field given by $\langle X_{12}, \theta^1\rangle = \langle X_{12}, \theta^2\rangle = 0$ and $\langle X_{12}, \Theta^{12}\rangle = 1$. Let us continue to denote by α_t

the restriction of the geodesic flow to $\mathcal{O}(M)$. Then the conjugate point is the first point where

$$\pi_* \alpha_{t*} Y_0 = 0 \tag{6.17}$$

where $Y_0 = X_{12}(\gamma(0), \gamma'(0), e_2(0))$, i.e., Y_0 is the unit tangent vector to the frames at $\gamma(0)$. Now (6.17) is the same as $\langle \alpha_{t*} Y_0, \theta^1 \rangle = \langle \alpha_{t*} Y_0, \theta^2 \rangle = 0$ or

$$\begin{aligned} &\langle Y_0, \alpha_t^* \theta^1 \rangle = 0. \\ (6.18) \qquad \text{and} \qquad & \\ &\langle Y_0, \alpha_t^* \theta^2 \rangle = 0. \end{aligned}$$

Now let X be the infinitesimal generator of α, i.e., let X be the vector field given by (6.8). Then (6.8) implies that

$$\begin{aligned} \mathcal{L}_X \theta^1 &= 0 \\ \mathcal{L}_X \theta^2 &= \Theta^{12} \qquad (6.19) \\ \mathcal{L}_X \Theta^{12} &= -K\theta^2. \end{aligned}$$

Now

$$\frac{d}{dt} \langle Y_0, \alpha_t^* \theta^1 \rangle = \langle Y_0, \alpha_t^* \mathcal{L}_X \theta^1 \rangle = 0$$

and $\langle Y_0, \theta^1 \rangle = 0$ so the first equation of (6.18) holds identically. Let $\xi(t) = \langle Y_0, \alpha_t^* \theta^2 \rangle$. Then

$$\frac{d}{dt} \xi(t) = \langle Y_0, \alpha_t^* \mathcal{L}_X \theta^2 \rangle = \langle Y_0, \alpha_t^* \Theta^{12} \rangle$$

and therefore

$$\begin{aligned} \frac{d^2}{dt^2} \xi(t) &= \langle Y_0, \alpha_t^* \mathcal{L}_X \Theta^{12} \rangle \\ &= \langle Y_0, \alpha_t^*(-K\theta^2) \rangle = -K\xi(t). \end{aligned}$$

Thus the function ξ satisfies (the Jacobi equations)

$$\frac{d^2\xi}{dt^2} + K\xi = 0. \tag{6.20}$$

Thus *the curvature also measures the second variation of arc length.* There are a number of consequences which can be drawn from (6.20). First we state a theorem of Sturm concerning equations of the type (6.20).

LEMMA 6.1. *Let ξ be a solution of* (6.20) *and let η be a non-zero solution of*

$$\frac{d^2\eta}{dt^2} + K'\eta = 0 \tag{6.21}$$

where $K' < K$. Then between every two zeros of η there is at least one zero of ξ.

Proof: We may assume that $\eta(a) = \eta(b) = 0$ and $\eta(t) > 0$ for $a < t < b$. Suppose $\xi \neq 0$ on $a \leqslant t \leqslant b$, so $\xi > 0$, say. Let m = minimum of ξ and multiplying η by a non-zero constant if necessary, we can assume $\max_{a \leq t \leq b}\eta(t) = m$. Thus $\xi - \eta \geqslant 0$ and $\xi(t_0) - \eta(t_0) = 0$ for some point t_0, $a < t_0 < b$. Since $\xi - \eta$ has a minimum at this point we must have $d(\xi - \eta)/dt = 0$ and $d^2(\xi - \eta)/dt^2 \geqslant 0$. But $d^2(\xi - \eta)/dt^2 = -K\xi + K'\eta = (K' - K)\xi(t_0) < 0$ which is a contradiction, proving Lemma 6.1.

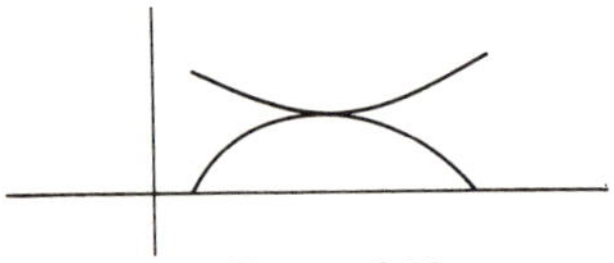

FIGURE 6.10

If we take $K' = C^2$ = const, the solutions of (6.21) are just $\cos Ct$ and $\sin Ct$. Thus if $K > C^2$ every solution of (6.20) must vanish on any interval of length C. This means, that on a surface whose curvature is $> C^2$, the conjugate points along geodesics are at a distance at most πC away. Now goedesics do not minimize distance beyond the conjugate point. Thus no two points can be more than a distance πC apart. Thus we have the

THEOREM 6.3. (Bonnet). *Let M be a connected two-dimensional Riemann manifold whose curvature K satisfies $K \geqslant C^2$. Then any two points are at a distance at most πC apart. In particular, if M is complete, it is compact.*

We leave as an exercise to the reader the proof of

THEOREM 6.4. *Let M be a connected two-dimensional Riemann manifold whose curvature K is always $\leqslant 0$. Then no geodesic of M has any conjugate points. In particular if M is simply connected, it is diffeomorphic to E^2.*

CHAPTER VII

The Geometry of G-Structures

In Chapter II we introduced the bundle of all coframes, $\mathscr{F}^*(M)$ and the bundle of frames, $\mathscr{F}(M)$, of a manifold M. In Chapter VI we studied the bundle of orthonormal frames, $\mathcal{O}(M)$ of a Riemann manifold, M, and the bundle, $\mathcal{O}_\varphi(M)$ arising from an immersion of an n-dimensional manifold, M, into E^{n+k}. All of these bundles have some properties in common—they are all differentiable manifolds with a distinguished map, which we call π, onto M. Furthermore, on each of these manifolds a Lie group acts (to the right). Thus $GL(n)$ acts on $\mathscr{F}(M)$ and $\mathscr{F}^*(M)$ [according to (6.1) et. seq. of Chapter II]; the group $O(n)$ acts in the same way on $O(n)$ and the group $O(n) \times O(k)$ [regarded as a subgroup of $\mathcal{O}(n + k)$] acts in the obvious way on $\mathcal{O}_\varphi(M)$. Finally, in all of these cases, for any $x \in M$, if $y \in \pi^{-1}(x)$ and $z \in \pi^{-1}(x)$ there is a unique A in the Lie group such that $z = yA$. In this chapter we shall study such objects, which are called *principal bundles*, for any Lie group. An important geometrical concept on a principal bundle is the notion of parallel translation: In the last section of Chapter VI we saw that there was a natural way of parallel translating frames of $\mathcal{O}(M)$ along a curve on a two-dimensional Riemann manifold, M. On a general principal bundle there is no distinguished way of parallel translating along a curve. However, a choice of such a method (called a *connection*) is an interesting geometrical object and gives rise to a notion of curvature generalizing the Gaussian curvature of a Riemann 2-manifold. After studying these general objects, we will turn our attention to a more special class of principal bundles which are of particular interest in differential geometry.

1. Principal and associated bundles, connections. Let G be a Lie group and P a differentiable manifold. Recall [cf. Section 6 of Chapter II for $G = GL(n)$] that G is said to act differentiably to the right on P if there is a differentiable map, φ, of $P \times G \to P$ such that for any $a \in G$ the map, R_a, of $P \to P$ given by $p \to \varphi(p, a)$ is a diffeomorphism of P onto itself, and $\varphi(\varphi(p, a), b) = \varphi(p, ab)$ i.e., $R_{ab} = R_b \circ R_a$. The map R_e is the identity if e is the identity of G. For simplicity, we shall usually write $R_a(p) = p \cdot a$ so that we have $p \cdot (ab) = (p \cdot a) \cdot b$.

Definition 1.1. *Let P and M be differentiable manifolds and π a differentiable map of P onto M. Let G be a Lie group acting to the right on P. Then (P, M, π, G) is called a principal G-bundle (or better P is a principal bundle over M with structure group G) if*

1) *G acts freely on P*, i.e., *for any $p \in P$ if $p \cdot a = p$ then $a = e$.*
2) *Let p_1 and p_2 be any two points of P. Then $\pi(p_1) = \pi(p_2)$ if and only if there is an $a \in G$ such that $p_1 \cdot a = p_2$. In other words, M can be regarded (via π) as the "quotient space" of P under the action of G.*
3) *P is locally trivial over M—i.e., any $x \in M$ has a neighborhood U and a diffeomorphism, ψ of $\pi^{-1}(U) \to U \times G$ such that $\psi(p) = (\pi(p), \eta(p))$ where $\eta(p) \in G$ and $\psi(pa) = (\pi(p), \eta(p)a)$.*

If M is just a point then P looks like G itself. For any M, if $x \in M$ then $\pi^{-1}(x)$ is a differentiable submanifold of P, which we call the fiber over x. Thus P looks like a collection of copies of G sitting over M. Thus, in a certain sense, the notion of principal bundle can be regarded as a geometric generalization of the notion of Lie group. We can define what we mean by a *homomorphism* between two principal bundles: Let ρ be a homomorphism of a Lie group G_1 into a Lie group G_2. Let P_1 be a principal G_1 bundle and P_2 a principal G_2 bundle over the same manifold, M. We say that a differentiable map, f, of $P_1 \to P_2$ is a ρ homomorphism if $f(p_1 \cdot a) = f(p_1)\rho(a)$ for all $p_1 \in P_1$ and $a \in G$. A diffeomorphism which is a homomorphism will be called an *isomorphism.*

Consider the case where G_1 is the identity. Then P_1 can be regarded as M itself and f is a map of $M \to P_2$ with $\pi \circ f = \text{id}$. In this case f is called a *cross-section* of the bundle P_2. If a bundle P admits a cross-section, f, then there is an isomorphism φ_f from P to the bundle $M \times G$. In fact for any $p \in P$ with $\pi(p) = x$, there is a unique $a \in G$ with $f(x)a = p$. Just set $\varphi_f(p) = (x, a)$. An easy verification shows that φ_f is differentiable, and, in fact an isomorphism of P with $M \times G$.

What we have shown is that any principal bundle admitting a cross-section is *trivial*, i.e., isomorphic to $M \times G$. Conversely, the bundle $M \times G$ admits the obvious cross-section, f, where $f(x) = (x, g(x))$ and g is any differentiable map of $M \to G$.

We should pause to point out that the question of whether certain bundles are trivial or not is frequently of great geometric interest. As an illustration, let us look at the case of the bundles $\mathcal{O}(M)$, where M is a two-dimensional compact oriented Riemann manifold. To say that $\mathcal{O}(M)$ is trivial, or what is the same, has a cross-section, means that there exists a non-zero vector field on M. This happens only when M is a torus. Furthermore, if M is not a torus, the integer $\chi(M)$ (= sum of indices of any vector field on M) measures, in a sense, how much $\mathcal{O}(M)$ differs from being trivial. The study of the question of measuring "obstruction" to triviality of principal bundle is one of the principal tools used in the theory of fiber bundles. The results thus obtained are crucial to the study of differential topology. For the development of these ideas the reader is referred to Steenrod's book *The Topology of Fiber Bundles* (Princeton University Press, 1951) and to Princeton lecture notes by Milnor— "Lectures on characteristic classes." We will concentrate more on geometric problems associated with principal bundles and not with the topological problems lying outside the scope of this book.

Observe that if P is a principal G bundle over M, and U is some open subset of M then $\pi^{-1}(U)$ is a principal G-bundle over U. Condition 3) of Definition 1.1 asserts that every point $p \in M$ has a neighborhood U such that the bundle $\pi^{-1}(U)$ is trivial.

Let us choose a covering, $\{U_\alpha\}$, of M by such neighborhoods with the corresponding maps, φ_α, of $\pi^{-1}(U_\alpha) \to G$. Then on each intersection, $U_\alpha \cap U_\beta$ the map $p \to \varphi_\beta(p) \cdot (\varphi_\alpha(p))^{-1}$ is constant along fibers so that we get a map $\psi_{\beta\alpha}$ of $U_\alpha \cap U_\beta \to G$. The maps $\psi_{\beta\alpha}$ are called the transition functions of P relative to $\{U_\alpha\}$. They obviously satisfy the rule

$$\psi_{\gamma\beta} \cdot \psi_{\beta\alpha} = \psi_{\gamma\alpha}. \tag{1.1}$$

Conversely, given an open covering $\{U_\alpha\}$ on a manifold M and a set of differentiable maps, $\psi_{\beta\alpha}$, of $U_\alpha \cap U_\beta \to G$, which satisfy (1.1) we can construct a principal bundle, P, having $\psi_{\beta\alpha}$ as transition functions. In fact, the proof, using Theorem 2.1 of Chapter II, proceeds exactly as the argument [for the case of $\mathscr{F}^*(M)$] in Section 6 of Chapter II and will be left to the reader.

Another important case of the notion of homomorphism of $P_1 \to P_2$ is where G_1 is a subgroup of G_2 and ρ the injection homomorphism. Such a homomorphism is called a *reduction* of the G_2 bundle, P_2, to the G_1 bundle, P_1. We say that the G_2 bundle, P_2, is reducible to the subgroup, G_1, if there exists a reduction to a G_1 bundle, P_1. Thus, if we take $G_1 = \{e\}$, a bundle is $\{e\}$ reducible if and only if it is trivial. As additional examples, observe that for any manifold, M, the $GL(n)$ bundle, $\mathscr{F}(M)$ is always reducible to an $O(n)$ bundle. In fact, we can put a Riemann metric on M, and consider the corresponding $O(n)$ bundle, $\mathscr{O}(M)$. It is a submanifold of $\mathscr{F}(M)$ and the identity mapping is obviously an isomorphism of $\mathscr{O}(M)$ into $\mathscr{F}(M)$. Similarly, $\mathscr{F}(M)$ is reducible to a $SL(n)$ bundle if and only if M is orientable.

Let $f : P_1 \to P_2$ be a reduction of P_2 to P_1. Let us choose a covering, U_α, of M and functions $\varphi_\alpha^1 : \pi_1^{-1}(U_\alpha) \to G_1$ corresponding to P_1. Let $\psi_{\beta\alpha}$ be the corresponding transition functions. If π_2 is the projection of $P_2 \to M$, define a map, φ_α^2, of $\pi_2^{-1}(U_\alpha) \to G_2$ as follows: Any $p_2 \in \pi_2^{-1}(U_\alpha)$ can be written as $f(p_1)a$ for some $p_1 \in P_1$ and $a \in G_2$. Put $\varphi_\alpha^2(p_2) = \varphi_\alpha^1(p_1)a$. If we choose some other representation $p_2 = f(p_1')a'$ then $f(p_1')a'a^{-1} = f(p_1)$ so that $p_1'a'a^{-1} = p_1$. Thus $\varphi_\alpha^1(p_1')a' = \varphi_\alpha^1(p_1)a$ so the map φ_α^2 is well defined. We let the reader verify that it defines an isomorphism of $\pi_2^{-1}(U_\alpha)$ with $U_\alpha \times G_2$. Now $\varphi_\beta^2 \cdot (\varphi_\alpha^2(p_2))^{-1} = \varphi_\beta^1 \cdot (\varphi_\alpha^1(p_2))^{-1}$ for $p_2 = f(p_1)$ so that the *transition functions for P_2 are also $\psi_{\beta\alpha}$. Thus the transition functions take values in G_1.* Conversely, assume that there we can choose a covering $\{U_\alpha\}$ of M so that the transition functions of P_2 take values in the subgroup G_1. By the discussion on page 219, the $\psi_{\beta\alpha}$ are differentiable maps into G_1. Since they satisfy (1.1) we can construct the corresponding bundle P_1 and the obvious injection of $P_1 \to P_2$. Thus we have proved

LEMMA 1.1. *Let P_2 be a principal G_2 bundle over M and let G_1 be a subgroup of G_2. Then P_2 is reducible to a G_1 bundle if and only if we can find a covering $\{U_\alpha\}$ of M whose transition functions, $\psi_{\alpha\beta}$ all take their values in G_1.*

Let P be a principal bundle with structure group G. Since G acts on P, every one parameter subgroup of G gives a one parameter group of diffeomorphisms of P. In particular, every left invariant vector field, X, on G gives rise to a vector field, $\tilde{X}$, on P. It is easy to verify that the map $X \to \tilde{X}$ is a homomorphism of the Lie algebra, g, of G into the set of all vector fields of P. It is injective. We can say somewhat more: If $\tilde{X}_p = 0$ for some $p \in P$ then $X = 0$. In fact, if $\tilde{X}_p = 0$ then p is left fixed under the one parameter group generated by $\tilde{X}$ which is just the action of $\exp tX$ on P. By condition 2) of Definition 1.1 we conclude that $\exp tX = e$, thus $X = 0$. Notice that $p \cdot \exp tX$ lies on the same fiber as p so that $\tilde{X}$ is always tangent to the fiber. The map $X \to \tilde{X}_p$ of $g \to T_p(\mathscr{F}(M))$ is an injection into the tangent space to the fiber which has the same dimension as g. Thus given any $Y \in T_p(\mathscr{F}(M))$ tangent to the fiber through p there is a unique $X \in g$ with $\tilde{X}_p = Y$. In other words, there is a canonical way of identifying the tangent space to the fiber through a point of P with the Lie algebra g.

Let a be an element of G, and $X \in g$. Let φ_{1t} be the one parameter group of diffeomorphisms of P generated by $(R_a)^*\tilde{X}$ and φ_{2t} the one parameter group generated by $\tilde{X}$. Now $\tilde{Y} = (R_a)^*\tilde{X}$ and $\tilde{X}$ are R_a-related. [Recall that $R_a^*\tilde{X}_p = (R_a)_*^{-1}\,\tilde{X}_{pa}$ so that $R_{a*}\tilde{Y}_p = \tilde{X}_{pa}$ thus $\tilde{Y}$ and $\tilde{X}$ are R_a-related, cf. Definition 8.3 of Chapter II.] According to Exercise 8.4 of Chapter II we have $R_a \circ \varphi_{1t} = \varphi_{2t} \circ R_a$. Now $\varphi_{2t} = R_{\exp tX}$ so that $\varphi_{1t} = R_a^{-1} R_{\exp tX} R_a = R_{a^{-1}(\exp tX)a} = R_{A_a^{-1}(\exp tX)}$. Now the infinitesimal generator of $A_a^{-1}(\exp tX)$ is just $Ad\,a^{-1}X$. We thus have

$$R_a^*\tilde{X} = (\widetilde{Ad\,a^{-1}X}). \tag{1.2}$$

Let us now discuss the problem of introducing a notion of parallel translation in a principal bundle, P over an n-dimensional manifold M. As we have seen in the case of Riemann two-manifolds, we must choose a way of moving the fiber along any differentiable curve, C, of M. That is, given $p \in P$ with $\pi(p) = C(0)$, we want a unique way of choosing a point p_t with $\pi(p_t) = C(t)$ for all t. Furthermore, we want the rule to be a motion of the fiber, i.e., that $(p \cdot a)_t = p_t \cdot a$ for any $a \in G$. We thus wish to have a rule which assigns, to each differentiable curve C of M and to each $p \in \pi^{-1}(C(0))$, a curve γ_p of P with i) $\pi(\gamma_p(t)) = C(t)$ and ii) $\gamma_{pa}(t) = \gamma_p(t) \cdot a$ for all $a \in G$ and all t in the domain of definition of C. We also make some reasonable assumptions of a differential nature: iii) that the tangent vector $\gamma_p'(0)$ depends only on $C'(0)$—thus we have a map, σ, of $T_x(M) \to T_p(P)$ for each $p \in \pi^{-1}(x)$. It is clear from condition i) that $\pi_{*p} \circ \sigma_p = \text{id}$. We impose the additional condition—iv) that σ_p be linear. Now $\sigma_p(T_x(M))$ is an n-dimensional subspace, H_p, of $T_p(P)$ with the property that $\pi_{*p}H_p = T_x(M)$. Observe that the choice of subspace, H_p, determines the map σ_p since π_{*p} is an isomorphism of H_p onto $T_x(M)$. We can thus give the subspace H_p instead of σ_p, i.e., give an n-dimensional differential system, $\mathscr{H}$, on P and we shall assume—v) $\mathscr{H}$ is differentiable. Condition ii) implies that $R_{a*}H_p = H_{p\cdot a}$. We can thus subsume all of our previous conditions in

DEFINITION 1.2. *Let P be a principal bundle with structure group G over the n-dimensional manifold M. A connection on P is a (differentiable) n-dimensional differential system, $\mathscr{H}$, on P such that*

1) $\pi_* H_p = T_{\pi(x)}(M)$ (*where* $H_p = \mathscr{H}(p)$)

and

2) $R_{a*}H_p = H_{pa}$.

The subspace, H_p, is called the horizontal space of the connection at p.

Notice that at each point, p, the tangent space to the fiber together with the subspace H_p span $T_p(P)$. Thus a choice of H_p gives a projection (along H_p) of $T_p(P)$ onto the tangent space to the fiber. Now we have identified the tangent space to the fiber with the Lie algebra g. We thus get a linear map of $T_p(P) \to g$ which coincides with the previous identification on vectors tangent to the fiber. In short, we get a g-valued differential form, Ω, which is easily seen to be differentiable. It is called the *connection form.*[1] The subspace H_p is just the set of those vectors satisfying $\langle X, \Omega_p\rangle = 0$. Now condition 2) of Definition 1.2 says that if $\langle X, \Omega_p\rangle = 0$ then $\langle R_{a*}X, \Omega_{p\cdot a}\rangle = 0$. Equation (1.2) says that if X is tangent to the fiber then $\langle R_{a*}X, \Omega_{p\cdot a}\rangle = Ad\, a^{-1}\langle X, \Omega_p\rangle$. Thus for any $X \in T_p(P)$ we get $\langle R_{a*}X, \Omega_{p\cdot a}\rangle = Ad\, a^{-1}\langle X, \Omega_p\rangle$. In other words

$$R_a^*\Omega = Ad\, a^{-1}\Omega. \tag{1.3}$$

Conversely, suppose we are given a g-valued form Ω satisfying (1.3) on P and such that $\langle \tilde{X}, \Omega\rangle = X$ for all $X \in g$. Then it is easy to see that the differential system spanned by all vectors annihilated by Ω is a connection.

---

[1] In the case $P = \mathcal{O}(M)$ of a two-dimensional Riemann manifold the form Ω is just the Θ^{12} of Chapter VI. Here the group $O(2)$ is the circle so its Lie algebra can be regarded as R—that is why Θ^{12} is a real-valued form.

It is clear that a connection on P induces a connection on $\pi^{-1}(U)$ for any open subset, U, of M.

Let us now show how to construct a connection on any principal bundle. On the trivial bundle, $U \times G$, there is the obvious connection corresponding to the direct sum decomposition of the tangent space at each point. Let P be a principal bundle over M and $\{U_\alpha\}$ a covering of M such that $\pi^{-1}(U_\alpha)$ is trivial. Let Ω_α be the connection forms on $\pi^{-1}(U_\alpha)$ and let $\{\varphi_\alpha\}$ be a partition of unity subordinate to $\{U_\alpha\}$. Put $\Phi_\alpha = \varphi_\alpha \circ \pi$ and $\Omega = \sum \Phi_\alpha \Omega_\alpha$. Since each of the Ω_α satisfies (1.3) and the Φ_α are constant on fibers, the form Ω satisfies (1.3). Furthermore, for any $X \in g$, we have $\langle \tilde{X}, \Omega \rangle = \sum \Phi_\alpha \langle \tilde{X}, \Omega_\alpha \rangle = \sum \Phi_\alpha X = X$ so that Ω is a connection form.

Let Y be a vector field on M and P a principal bundle over M with a given connection $\mathscr{H}$. Then at each $p \in P$ there is a unique vector $\hat{Y}_p \in H_p$ with $\pi_{*p}\hat{Y}_p = Y_{\pi(p)}$. We thus get a vector field $\hat{Y}$ on P which is π-related to Y. According to Theorem 8.3 of Chapter II, if Y_1 and Y_2 are vector fields on M, $[\hat{Y}_1, \hat{Y}_2] = \widehat{[Y_1, Y_2]}$. Thus the map $Y \to \hat{Y}$ gives an isomorphism of the vector fields on M into the vector fields on P. A vector field on P of the type $\hat{Y}$ will be called *horizontal.*

Using this isomorphism we can define parallel translation along any curve, C, on M. That is, given any $p \in \pi^{-1}(C(0))$ we wish to show that there is a unique horizontal curve γ with $\gamma(0) = p$ and $\pi \circ \gamma = C$. By a horizontal curve we mean a curve whose tangent vector lies in $\mathscr{H}$ for all t. To prove the existence and uniqueness of γ it is sufficient to prove that we can cover the domain of definition of C by a finite number of open intervals and prove existence and uniqueness on each. We can thus assume that C lies entirely in a neighborhood U such that $\pi^{-1}(U)$ can be written as $U \times G$ and such that C, defined for $0 \leqslant t \leqslant 1$ is the integral curve of some vector field, X, on U. Then it is clear that the γ must be the solution curve to the vector field $\hat{X}$ passing through p. This shows uniqueness.

To show existence, we must be sure that the solution to $\hat{X}$ through p is defined for *all* $0 \leqslant t \leqslant 1$.[2] To this effect, we look at the form of $\hat{X}$ in terms of the local product representation. It is given by $X + Y^x$ where Y^x is some vector field on G, for each fixed $x \in U$. In view of (1.3), Y^x is a right invariant vector field on G for each fixed x. Now we

[2] We have to worry about the fact that the standard existence theorem gives only *local* solutions which can not, in general, be extended for all t.

are looking for a solution γ to $\hat{X}$ where $\gamma(t) = (x(t), g(t))$. We must have $x'(t) = X_{x(t)}$ and $g'(t) = Y^{x(t)}_{g(t)}$. Now if we take $x(t) = C(t)$ we get a solution of the first of these equations. We then must solve the system of ordinary equations $g'(t) = Y^{x(t)}_{g(t)}$ on G. If we let $\omega^1, \ldots, \omega^k$ be a basis of right invariant forms on G and put $ds^2 = \omega^{1^2} + \cdots + \omega^{k^2}$. Then $\|Y^{x(t)}_{g(t)}\| < K$ for some sufficiently large K with respect to this Riemann metric. This implies that the solution curve g of $g'(t) = Y^{x(t)}_{g(t)}$ has length $\leqslant Ka$ on any interval $0 \leqslant t \leqslant a$ where it is defined; in particular, as G is complete with respect to this metric, if g is defined for $0 \leqslant t < a$, then $g(t)$ approaches a limit as $t \to a$. Applying the existence and uniqueness theorem to any $a < 1$ we conclude that g is defined for $0 \leqslant t < a + \epsilon$ for some sufficiently small ϵ. In particular, if we take $a = l.u.b.\ \{b \mid g(t)$ is defined for $0 \leqslant t \leqslant b\}$ we conclude that $a = 1$. In other words, $g(t)$, and therefore $\gamma(t)$, is defined for all $0 \leqslant t \leqslant 1$.

It is clear from the construction above that the definition of parallel translation fulfills all the requirements i)–iv) that we made on P. It is also clear that if C_1 and C_2 are two curves which are close in the C^1 topology then if $p_1 = \gamma_1(0)$ and $p_2 = \gamma_2(0)$ are close, the curves γ_1 and γ_2 are close in the C^1 topology.

It is clear how to extend the definition of parallel translation to include the case of piecewise differentiable curves.

Let P be a principal G bundle over M with connection $\mathscr{H}$ and let $p \in P$ with $\pi(p) = x$. Let C be any closed piecewise differentiable curve starting and ending at x. Parallel translation around C is a differentiable diffeomorphism, τ_C, of the fiber $\pi^{-1}(x)$ onto itself and

(1.4) $$\tau_C(q \cdot a) = \tau_C(q)a \qquad \text{for any } q \in \pi^{-1}(x).$$

Let C_1 and C_2 be two curves defined for $0 \leqslant t \leqslant 1$, with $C_1(0) = C_1(1) = C_2(0) = C_2(1) = x$. We can then define the curve $C_3 = C_1 \vee C_2$ by $C_3(t) = C_1(2t)$ for $0 \leqslant t \leqslant \frac{1}{2}$ and $C_3(t) = C_2(2t - 1)$ for $\frac{1}{2} \leqslant t \leqslant 1$. It is clear that $\tau_{C_1 \vee C_2} = \tau_{C_2} \circ \tau_{C_1}$ so that in this way the set of all such diffeomorphisms forms a group called the *holonomy group* at x, denoted by Φ_x. Not only is Φ_x a group, it is a topological group in a natural way: Call an element $\tau \in \Phi_x$ close to the identity if it is of the form τ_C where C is close to the constant curve in the C^1 topology.

Let y be some other point of M and suppose that there is a curve, D, joining x to y. Then any closed curve, C, at x gives rise to a closed curve $D \vee C \vee D^{-1}$ at y and thereby an isomorphism, α_D, between Φ_x

and Φ_y. Thus if M is connected it makes sense to talk of *the* holonomy group, Φ, of the connection, $\mathscr{H}$, on P over M.

Now let us consider the point p in the fiber over x. For any $\tau \in \Phi_x$ we have that $\tau(p) = p \circ a_\tau$ where a_τ is some element of G. Furthermore,

$$\tau_1(\tau_2(p)) = \tau_1(p \cdot a_{\tau_2}) = p \cdot a_{\tau_1} \cdot a_{\tau_2}$$

by (1.4). Thus $a_{\tau_1 \tau_2} = a_{\tau_1} \cdot a_{\tau_2}$. In other words, the map, $\tau \to a_\tau$ is an isomorphism, β_p, of Φ_x into G. It is easy to see that this is a continuous map relative to the topology that we have introduced on Φ_x. Let us choose some other point, q, in the fiber over $y \in M$. If M is connected, we can join x to y by a curve D. Now let p' be the point on $\pi^{-1}(y)$ obtained by parallel translating p along D and let $p' \cdot b = q$. Then $(\alpha_D \tau_C)p'$ is obtained by parallel translating p' along C back to $\pi^{-1}(x)$ to obtain p, then applying τ_C to get $p \cdot a_{\tau_C}$ and then translating back to $\pi^{-1}(y)$ along D to get $p' \cdot a_{\tau_C}$. Thus

$$\beta_{p'} \circ \alpha_D = \beta_p.$$

At the point q we have

$$\alpha_D(\tau)_q = \alpha_D(\tau)p' \cdot b = p' \cdot \beta_p(\tau) \cdot b = p' \cdot b \cdot b^{-1} \beta_p(\tau) b = q \cdot b^{-1} \beta_p(\tau) b.$$

Thus on a connected manifold a point of P gives an isomorphism of the holonomy group into G and any two such isomorphisms differ by an inner automorphism of G.

Since β_p is continuous, we can associate a Lie subalgebra, ϕ_p, to Φ_x according to Lemma 4.4 of Chapter V. If M is connected then ϕ_p and ϕ_q differ from each other by an automorphism $Ad\, b$ and therefore ϕ_p is isomorphic to ϕ_q. We may, in this case, talk of *the* holonomy algebra, ϕ, of the connection, $\mathscr{H}$.

Let us consider the case where M is simply connected, i.e., where every closed curve is homotopic to a constant curve. Then it is easy to see that every piecewise differentiable curve can be deformed in the piecewise C^1 topology to a constant curve. This means that every element of Φ_x can be joined by an arc to the identity, i.e., Φ_x is arc-wise connected. In this case, Exercise 4.2 of Chapter V implies that $\beta_p(\Phi_x) \subset \exp \phi_p$. Thus $\exp \phi_p$ is the smallest Lie subgroup of G containing $\beta_p(\Phi_x)$. Actually, it can be shown that any arc-wise connected subgroup of a Lie group is a Lie group—cf. Yamabe [23]—in particular that $\beta_p(\Phi_x) = \exp \phi_p$ in this case.

More generally, it can be shown that the map, β_p always allows us to identify Φ_x as a Lie subgroup of G, cf. Ambrose-Singer [1]. In the discussion which follows we will assume that the holonomy group is a Lie group. The reader can either take this for granted (or refer to the relevant papers) or always restrict attention to the case *where M is simply connected and replace Φ by the Lie group* $\exp \phi_p$ *in all the following theorems.*

Let P_1 be a principal bundle with structure group G_1 and P_2 a principal bundle with structure group G_2 over a manifold M. Let f be a homomorphism of $P_1 \to P_2$ corresponding to the homomorphism, ρ, of $G_1 \to G_2$. Let $\mathscr{H}_1$ be a connection on P_1. For any $p \in P_1$ define the horizontal space at $f(p)$ by $H_{2f(p)} = f_{*p}H_{1p}$. Notice that if $f(p) = f(p')$ then $p = p' \cdot a$ where $\rho(a) = e$ so that $f_{p*}R_{a*} = f_{p'*}$. Now $H_{1p} = R_{a*}(H_{1p'})$ and $f_{*p}(H_{1p}) = f_{*p}R_{a*}(H_{1p'}) = f_{p'*}(H_{1p'})$. This shows that the definition of $H_{2f(p)}$ does not depend on p. Furthermore, if $p = q \cdot a$, then $f(p) = f(q)\rho(a)$ and $H_{2f(p)} = R_{\rho(a)*}H_{2f(q)}$. In other words, the assignment $f(p) \to H_{2f(p)}$ looks like a connection except that it is only defined on $f(P_1)$. We remedy this as follows: Let $r \in P_2$ with $r = f(p) \cdot a$. Let $H_{2r} = R_{a*}H_{2f(p)}$. Again if $r = f(q) \cdot b$ we get the same value for H_{2r}, showing that the differential system, $\mathscr{H}_2$, sending $r \to H_{2r}$ is well defined. It is easy to check that $\mathscr{H}_2$ is a connection on P_2. The connection $\mathscr{H}_2$ just defined on P_2 is called the image of $\mathscr{H}_1$ under f.

Definition 1.3. *Let P_2 be a G_2 bundle over M and let G_1 be a subgroup of G_2. A connection, $\mathscr{H}_2$, on P_2 is said to be reducible to G_1, if there is an isomorphism, f, of a G_1 bundle, P_1 into P_2 and a connection, $\mathscr{H}_1$ on P_1 so that $\mathscr{H}_2$ is the image of $\mathscr{H}_1$ under f.*

Exercise 1.1. Let f be an isomorphism of P_1 into P_2. Show that a connection $\mathscr{H}_2$ on P_2 is the image under f of a connection on P_1 if and only if $H_{2f(p)} \subset f_*T_p(P_1)$ for all $p \in P_1$.

Theorem 1.1. *Let P be a principal bundle over a connected manifold with connection $\mathscr{H}$ and let Φ be the holonomy group of $\mathscr{H}$. Then $\mathscr{H}$ is reducible to Φ.*

Proof: Choose a point $p \in P$ with $\pi(p) = x$ and consider the group Φ_p which is a Lie subgroup of G. Our first step is to show that P is reducible to a Φ_p bundle. For this purpose we wish to choose a covering, U_α, of

M with transition functions taking values in Φ_p. To this effect, choose an atlas, $\{U_\alpha, h_\alpha\}$, on M so that $h_\alpha(U_\alpha)$ is the unit ball and P is trivial over U_α. Let x_α be the point $h_\alpha^{-1}(0, \ldots, 0) \in U_\alpha$ and for each α let D_α be a curve joining $x = \pi(p)$ to x_α. Let p_α denote the point of $\pi^{-1}(x_\alpha)$ obtained by parallel translating p along D_α. For each $q \in \pi^{-1}(U_\alpha)$ parallel translation along the straight line, $l_{\alpha q}$, joining $\pi(q)$ to x_α gives a point $p_\alpha \cdot a \in \pi^{-1}(x_\alpha)$. We thus get a map from $\pi^{-1}(U_\alpha)$ into G. It is easy to check that this is differentiable and gives us an isomorphism, $q \to (\pi(q), \varphi_\alpha(q))$ of $\pi^{-1}(U_\alpha)$ with $U_\alpha \times G$. To check the transition functions we suppose that $q \in U_\alpha \cap U_\beta$, and examine $\varphi_\beta(q)\varphi_\alpha(q)^{-1}$. Now $\varphi_\beta(q)$ is obtained by parallel translating q along $l_{\beta q}$ to get $p_\beta\varphi_\beta(q)$. Parallel translating q along $l_{\beta q}$ and then D_β to x gives $p\varphi_\beta(q)$ while parallel translating along $l_{\alpha q}$ and D_α to x gives $p\varphi_\alpha(q)$. Thus parallel translating p, around the closed curve $D_\alpha l_{\alpha q}^{-1} l_{\beta q} D_\beta^{-1}$ sends $p \to \varphi_\beta(q)\varphi_\alpha(q)^{-1} \in \Phi_p$. This shows that the transition functions all take their value in Φ_p. Lemma 1.1 says that we can construct a Φ-bundle, P', and an isomorphism, f, of P' into P. In the proof of Lemma 1.1, a point of P' was represented, locally, by the pair (x, a) where $x \in U_\alpha$ and $a \in \Phi$. The map f takes this point into the point, q, with the corresponding representation in $\pi^{-1}(U_\alpha)$. But to say that q has such a representation in $\pi^{-1}(U_\alpha)$ means that q can be parallel translated along *some* curve to p. In fact, by definition, when we parallel translate along l_α to p_α we get the element, a, of $\Phi_{p\alpha} = \Phi_p$. This means that parallel translating along some closed curve at x_α takes p into $p \cdot a$; then translating along D_α takes us to $p_\alpha \cdot a$, and finally along l_α takes us into q. Thus $f(P')$ consists of the set of all points of P which can be joined to p by horizontal curves. This implies that any horizontal curve starting at a point of $f(P')$ is completely contained in $f(P')$. In particular, every horizontal

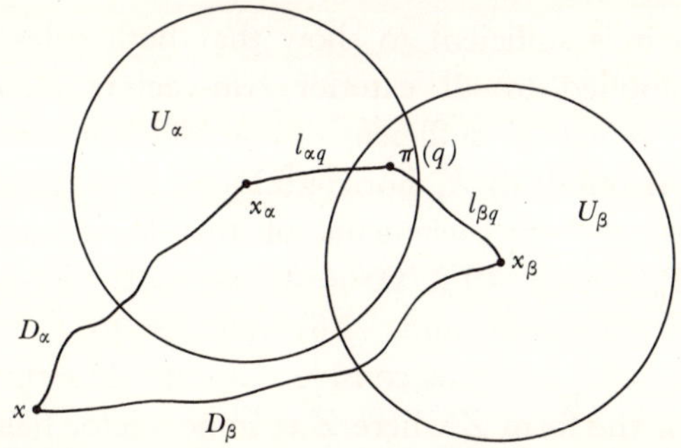

tangent vector at a point $f(q)$ for $q \in P'$ lies in $f_{*q}(T_q(P'))$. By Exercise 1.1 this implies that $\mathscr{H}$ is the image of a connection on P', proving Theorem 1.1.

Let us now discuss the infinitesimal version of parallel translation around closed curves. Thus we wish to consider what happens when we traverse the "infinitesimal parallelogram" spanned by two horizontal tangent vectors, X and Y at a point, p, of P. In view of the heuristic discussion on page 103, we should expect that the effect of parallel translating around this infinitesimal parallelogram should be given by $\langle X \wedge Y, d\Omega\rangle$ where Ω is the connection form. Comparing this with Equation (6.9) of Chapter VI we call this the curvature of $\mathscr{H}$ at $X \wedge Y$. More precisely,

DEFINITION 1.4. *Let $\mathscr{H}$ be a connection on a principal bundle P, whose connection form is Ω. Let $\hbar$ denote the projection, at each point, of $T_p(P)$ onto H_p. (Thus $\hbar(X) = 0$ for X tangent to the fiber and $\hbar(X) = X$ for horizontal X.) For any differential k-form, θ, defined on M, the form $D\theta$ given by*

$$\langle X_1 \wedge \cdots \wedge X_{k+1} \mid D\theta\rangle = \langle \hbar X_1 \wedge \cdots \wedge \hbar X_{k+1} \mid d\theta\rangle \tag{1.5}$$

is called the covariant differential of θ. In particular, the form $D\Omega$ is called the curvature of the connection $\mathscr{H}$.

Let us now prove the fundamental structure equation

$$d\Omega = \frac{-1}{2}[\Omega \wedge \Omega] + D\Omega. \tag{1.6}$$

To verify (1.6) it is sufficient to show that both sides give the same answer when applied to all exterior two-vectors $X_p \wedge Y_p$ at every $p \in M$. For this, it will be sufficient to consider the various cases i) X_p and Y_p both horizontal, ii) X_p horizontal and Y_p vertical, and iii) both X_p and Y_p vertical [since elements of this form span $\wedge^2(T_p(M)]$. In case i) $\langle X_p, \Omega\rangle = \langle Y_p, \Omega\rangle = 0$ so that $\langle X_p \wedge Y_p|\Omega \wedge \Omega\rangle = [\langle X_p, \Omega\rangle, \langle Y_p, \Omega\rangle] = 0$ and (1.6) reduces to (1.5) since $\hbar(X_p) = X_p$ and $\hbar(Y_p) = Y_p$. Let us consider case ii). For this extend X_p to a vector field of the form $\hat{Z}$ where Z is some vector field on M. [This can always be done—just extend $\pi_*(X)$ to a vector field, Z, on M.] Also extend Y_p to a vector field $\tilde{A}$ where $A \in g$. By Equation (1.10) of Chapter III (applied to vector-valued forms) we have

$$\langle \tilde{A} \wedge \hat{Z} | d\Omega \rangle = \tilde{A}\langle \hat{Z}, \Omega \rangle - \hat{Z}\langle \tilde{A}, \Omega \rangle - \langle [\hat{Z}, \tilde{A}], \Omega \rangle.$$

Now $\langle \hat{Z}, \Omega \rangle \equiv 0$ since $\hat{Z}$ is horizontal and $\hat{Z}\langle \tilde{A}, \Omega \rangle = 0$ since $\langle \tilde{A}, \Omega \rangle = A = \text{const}$. Thus the first two terms on the right vanish. The last term on the right also vanishes. In fact

$$0 = \mathscr{L}_{\tilde{A}}\langle \hat{Z}, \Omega \rangle = \langle \mathscr{L}_{\tilde{A}}\hat{Z}, \Omega \rangle + \langle \hat{Z}, \mathscr{L}_{\tilde{A}}\Omega \rangle.$$

But $\mathscr{L}_{\tilde{A}}\Omega = -[A, \Omega]$ by (the infinitesimal version of) (1.2). Thus $\langle \mathscr{L}_{\tilde{A}}\hat{Z}, \Omega \rangle = 0 = \langle [\hat{Z}, \tilde{A}], \Omega \rangle$. Thus in case ii) the left-hand side of (1.6) vanishes. But so does the right since $\langle \hat{Z}, \Omega \rangle = 0$ and $h(\tilde{A}) = 0$. We are thus left with case iii). We then extend X_p and Y_p to vector fields $\tilde{A}$ and $\tilde{B}$. Then $\langle \tilde{A} \wedge \tilde{B} | D\Omega \rangle = 0$ and (1.6) reduces to

$$\langle \tilde{A} \wedge \tilde{B} | d\Omega \rangle = -\tfrac{1}{2}([A, B] - [B, A]) = -[A, B].$$

On the other hand, Equation (1.10) of Chapter III asserts that

$$\langle \tilde{A} \wedge \tilde{B} | d\Omega \rangle = \tilde{A}\langle \tilde{B}, \Omega \rangle - \tilde{B}\langle \tilde{A}, \Omega \rangle - \langle [\tilde{A}, \tilde{B}], \Omega \rangle.$$

But $\langle \tilde{B}, \Omega \rangle = B = \text{const}$ so $\tilde{A}\langle B, \Omega \rangle = 0$ and similarly $\tilde{B}\langle \tilde{A}, \Omega \rangle = 0$. But $\widetilde{[A, B]} = [\tilde{A}, \tilde{B}]$ so that $\langle [\tilde{A}, \tilde{B}], \Omega \rangle = [A, B]$. This proves (1.5) in case iii). It should be observed that (1.5) in case iii) is nothing other than the equation of structure of the Lie group G [Equation (6.6) of Chapter V] carried over to the fiber of P.

It is an instructive exercise for the reader to compare Equation (1.6) with Equations (2.10) and (3.11) of Chapter VI. The forms Ω_M^{ij} introduced there on the bundle $\mathcal{O}(M)$ do indeed define a connection and the forms $d\Omega_M^{ij} - \sum_{r=1}^{k} \Omega_M^{ir} \wedge \Omega^{rj}$ are the curvature forms for the connection. Equation (2.10) is just the equation of structure for this particular connection lifted to $\mathcal{O}_\varphi(M)$. We can thus reformulate condition (3.11) as asserting that the curvature vanishes.

We now show the relation between the curvature and the holonomy algebra of a connection.

THEOREM 1.2. (Ambrose-Singer). *Let Ω be the connection form of a connection, $\mathscr{H}$ on a principal bundle P. The holonomy algebra, ϕ_p, at a point $p \in P$ is spanned by the set of all elements of g of the form $\langle X \wedge Y, D\Omega \rangle_q$ where q is any point that can be obtained from p by parallel translation along any curve of M and where X and Y are horizontal tangent vectors at q.*

Proof: We first observe that it is sufficient to prove Theorem 1.2 for case that the holonomy group is all of G. In fact, by Theorem 1.2 we can find a reduction of the bundle to P^1 whose structure group is the holonomy group and so that the connection comes from P^1. If the theorem is true for the reduced bundle it is clearly true for P.

If the holonomy group is all of G then every point of P can be obtained from p by parallel translation, and the holonomy algebra is all of g. Any horizontal vector at q can be written in the form $\hat{W}_q$ where W is a vector field on M. For two such vector fields,

$$\langle \hat{W} \wedge \hat{Z} | D\Omega \rangle = \langle \hat{W} \wedge \hat{Z} | d\Omega \rangle = \hat{W}\langle \hat{Z}, \Omega \rangle - \hat{Z}\langle \hat{W}, \Omega \rangle$$
$$- \langle [\hat{W}, \hat{Z}], \Omega \rangle = -\langle [\hat{W}, \hat{Z}], \Omega \rangle.$$

We thus wish to show that g is spanned by all vectors of the form $\langle [\hat{W}, \hat{Z}], \Omega \rangle_q$ where q is an arbitrary point of p and W and Z are arbitrary vector fields of M. Let h be the subalgebra of g generated by the elements of this form. Consider the corresponding subalgebra, $\tilde{h}$ of $\tilde{g}$ consisting of vertical vector fields on P. It will be sufficient to show that $\tilde{g} = \tilde{h}$. To this effect, consider the differential system, $\mathscr{D}$, on P spanned by all vector fields of the form $\hat{W} + \tilde{A}$ where W is a vector field on M and $A \in h$. *It is completely integrable.* In fact, the vertical component of $[\hat{W}, \hat{Z}]$ lies in $\tilde{A}$ by definition and $[\tilde{A}, \hat{W}]$ is horizontal. Furthermore $[\tilde{A}, \tilde{B}] = \widetilde{[A, B]}$ which lies in $\tilde{h}$ since h is a subalgebra. Now if h were a proper subalgebra of g, $\mathscr{D}$ would be a nontrivial distribution. Since any horizontal curve is an integral curve of $\mathscr{D}$, the maximal integral manifold passing through a point $p \in P$ would contain all points which could be joined to p by parallel translation. This contradicts the assumption that every point of p can be so joined to P, proving the theorem.

It should be remarked that in practice the main use of Theorem 1.2 is to reduce the bundle. That is, we use Theorem 1.2 to compute the holonomy algebra and then apply Theorem 1.1. As the extreme case we have:

EXERCISE 1.2. Let P be a principal bundle over a simply connected manifold M. Suppose $\mathscr{H}$ is a connection on P with zero curvature. Then P is equivalent to the trivial bundle $B \times G$ and $\mathscr{H}$ is equivalent to the trivial connection on $B \times G$.

For any principal bundle we can define the analogue of the "quantities" of Chapter I and of Section 6 of Chapter II. We must merely replace the general linear group by the structure group of the bundle; more precisely,

DEFINITION 1.5. *Let P be a principal G-bundle over a manifold M and let Q be a differentiable manifold on which G acts differentiably to the left. On $P \times Q$ the group G acts differentiably to the right by the rule $R_a(p, q) = (pa, a^{-1}q)$. The quotient space of $P \times Q$ under the action of G (we identify two points if one can be carried into the other by an element of G) is called the Q associated bundle of P, and is denoted by $Q(P)$.*

We let the reader verify that $Q(P)$ is a differentiable manifold with a natural projection, π_Q, onto M. We also leave to the reader the statements and proofs of the obvious generalizations of the propositions of Section 6 of Chapter II to the present situation of arbitrary principal bundles. In particular, each $p \in P$ with $\pi(p) = x$ gives a diffeomorphism denoted by p, of Q onto $\pi_Q^{-1}(x)$ and

$$p \cdot a\,(q) = p(a \cdot q). \tag{1.7}$$

We single out some cases of associated bundles for special mention: If Q is a vector space on which G acts linearly then $Q(P)$ is called a vector bundle with structure group G.

If $Q = G$, and G acts on itself by left multiplication, then $Q(P)$ can be identified with P. In fact, in each equivalence class of $P \times Q = P \times G$ there is exactly one element of the form (p, e). Notice that G acts to the right on $P \times G$—just define $(p, g_1)g_2 = (p, g_1g_2)$. This action commutes with the equivalence (1.7) relation:

$$(R_a(p, q_1)q_2) = (pa, a^{-1}g_1)g_2 = (pa, a^{-1}g_1g_2) = R_a[(p, g_1)g_2].$$

Thus G acts to the right on $Q(P) = P$. What is this action? It is nothing other than the standard action of G on P, once we have identified $Q(P)$ with P. (The reader should verify this as an exercise.) All of this sounds completely redundant, which it is. However, it can be generalized slightly to yield an important (but also slightly redundant) observation. Suppose that G_1 is a (Lie) subgroup of G_2 and P_1 is a principal G_1 bundle. Then G_1 acts by left multiplication on G_2 and G_2 acts by right multiplication on $P_1 \times G_2$. This action commutes with the equivalence relation on $P_1 \times G_2$ [just equation (1.7) again but this time $g_1 \in G_1$ and $g_2 \in G_2$]. Thus G_2 acts on the quotient space $G_2(P_1)$. The reader can verify that this makes $G_2(P_1)$ a principal G_2-bundle. *In other words, given a G_1-bundle we can always enlarge it in a canonical way to a G_2-bundle.*

Let us now suppose that we have some connection, $\mathscr{H}$ on the principal bundle P, and let $Q(P)$ be some associated bundle. Let y be some point of $Q(P)$ with $\pi_Q(y) = x$ and let $p_0 \in \pi^{-1}(x) \subset P$. Since p_0 determines a diffeomorphism of Q onto $\pi_Q^{-1}(x)$, there is a unique $q \in Q$ with $p_0(q) = y$. The point q determines a map of P into $Q(P)$: Simply map each $p \in P$ into $p(q)$. Let us denote this map by σ_{p_0}. Let $\mathscr{D}_y = \sigma_{p_0}^*(\mathscr{H}_{p_0})$. Let us show that $\mathscr{D}_y$ does not depend on the choice of p_0. If we replace p_0 by p_0a then (1.6) says that we must replace q by $a^{-1}q$. Thus $\sigma_{p_0a}(p) = p(a^{-1}q) = (R_{a^{-1}}p)q$ by (1.1). Thus

$$\sigma_{p_0a} = \sigma_{p_0} \circ R_{a^{-1}}.$$

Thus

$$\sigma_{p_0a*}(\mathscr{H}_{p_0a}) = \sigma_{p_0}^* \circ R_{a^{-1}*}\mathscr{H}_{p_0a} = \sigma_{p_0}^*\mathscr{H}_{p_0}$$

so $\mathscr{D}_y$ does not depend on p_0. It is easy to check that the $\mathscr{D}$ so defined is a differential system on $Q(P)$ and that $\mathscr{D}_y$ is a complement to tangent space to the fiber at y. Furthermore, let C be a curve on M with $C(0) = x_0$ and $C(1) = x_1$ and let $y_0 \in Q(P)$ be a point of $\pi_Q^{-1}(x_0)$. Choose a $p_0 \in \pi^{-1}(x_0)$ as before and let γ be the horizontal curve in P lying over C with $\gamma(0) = P_0$. Then the curve $\sigma_{p_0} \circ \gamma$ is clearly a solution curve of the differential system $\mathscr{D}$. We have thus shown that

Given a connection, $\mathscr{H}$, on P there is defined a differential system, $\mathscr{D}$, on $Q(P)$ such that

1) *$\mathscr{D}_y$ is a complement to the vertical at each y*

and

2) *for any curve C on M with $C(0) = x_0$ and $C(1) = x_1$ and to any $y_0 \in \pi_Q^{-1}(x_0)$. There is a unique solution curve γ to $\mathscr{D}$ with $\gamma(0) = y_0$ and $\pi_Q \circ \gamma = C$.*

[Condition 2) thus says that we can parallel translate the fibers of $Q(P)$ along any curve of M.]

We shall call a differential system, $\mathscr{D}$, on $Q(P)$ satisfying 1) and 2) a connection on $Q(P)$. Thus *every connection on P determines a connection on*

$Q(P)$. Since the action of G on Q can be quite trivial, it is clear that the converse is not true. However,

EXERCISE 1.3. Show that if G acts freely (effectively) on Q then every connection on $Q(P)$ comes from a unique connection on P.

2. G-structures. Among the class of all fiber bundles there are certain ones (essentially reductions of the bundle of frames) which play a central role in differential geometry. The rest of this book will be devoted to the study of these objects. Before making the definitions we first introduce some notation. Let M be a differentiable n-dimensional manifold and let $\mathscr{F}(M)$ be its bundle of frames. Recall that $\mathscr{F}(M)$ is a principal $GL(n)$ bundle over M. A point, p, of $\mathscr{F}(M)$ with $\pi(p) = x$ is a basis of the tangent space $T_x(M)$. It is now somewhat convenient to change our viewpoint slightly. Let V be a standard (i.e., fixed) n-dimensional vector space with a distinguished basis. (Thus we may take $V = R^n$ with its standard basis.) Then a basis, p, of $T_x(M)$ defines an isomorphism of V onto $T_x(M)$; we shall denote this isomorphism by p. We can also regard $GL(n)$ as the group, $GL(V)$, of all linear transformations of V. In this way, conforming to the language of Section 1, the bundle $\mathscr{F}(M)$ is a principal $GL(V)$ bundle and the tangent bundle, $T(M)$ is the associated bundle $V(\mathscr{F}(M))$. In particular, according to Equation (1.7), if $p \in \mathscr{F}(M)$, $v \in V$ and $a \in GL(V)$, we have

$$pa(v) = p(av). \tag{2.1}$$

On the bundle of frames there sits a canonically defined V-valued linear differential form, ω. It is defined as follows: Let p be a point of $\mathscr{F}(M)$ with $\pi(p) = x$ and let $X \in T_p(\mathscr{F}(M))$. Then $\pi_* X \in T_x(M)$ and we put

$$\langle X, \omega \rangle = p^{-1}(\pi_*(X)). \tag{2.2}$$

The reader can verify that ω, as given by (2.2), is a differentiable linear differential form on $\mathscr{F}(M)$. Let us now see how ω behaves under the action of $GL(V)$. Let $X \in T_p(\mathscr{F}(M))$ and $a \in GL(V)$. Then $\pi_*(X) = \pi_*(R_{a*}X)$ since π commutes with R_a. Thus

$$\begin{aligned} \langle R_{a*}X, \omega \rangle = (pa)^{-1}(\pi_* X) &= a^{-1}(\pi_* X) \qquad \text{(by (2.1))} \\ &= a^{-1}\langle X, \omega \rangle. \end{aligned}$$

We thus have the formula

$$R_a^* \omega = a^{-1} \omega. \tag{2.3}$$

The term $a^{-1}\omega$ is just the result of applying the linear transformation a to the V-valued linear differential form, ω. Now the Lie algebra, $gl(V)$ acts on V and so we can write down the infinitesimal version of (2.3): For any $A \in gl(V)$ we have

$$\mathscr{L}_{\tilde{A}}\omega = -A\omega. \tag{2.4}$$

Now $\pi_* \tilde{A} = 0$ so that $\langle \tilde{A}, \omega \rangle = 0$. By Equation (1.9) of Chapter III we can rewrite (2.4) as

$$\tilde{A} \lrcorner\, d\omega = -A\omega. \tag{2.5}$$

We now turn to the definition of G-structures.

Definition 2.1. *Let G be a subgroup of $GL(V)$. A G-structure on an n-dimensional manifold, M, is a reduction of $\mathscr{F}(M)$ to the group G. Thus a G-structure, B_G, on a manifold, M, is a submanifold of $\mathscr{F}(M)$ with the property that for any $p \in B_G$ and any $a \in GL(V)$ the point $p \cdot a$ belongs to B_G if and only if $a \in G$.*

Examples of G-structures

1) $G = \{e\}$. In this case a G-structure on M is a choice of frame at each point of M. With such a structure we have an identification of the tangent space at each point with V. We thus can "parallel translate" tangent vectors from one point to another on M. [That is, $X \in T_x(M)$ is identified with $p^{-1}(X) \in V$ which in turn is identified with $q(p^{-1}(X) \in T_y(M))$ if p is the frame at x and q is the frame at y.] For this reason an $\{e\}$-structure is sometimes called a complete parallelism on M. The most important special case of a complete parallelism is when M is a Lie group. As we have seen in Chapter V, if M is a Lie group then the left invariant vector fields give an identification of the tangent space at each point of M with the Lie algebra of M.

2) Let V_1 be a k-dimensional subspace of V and let G be the group of all linear transformations leaving V_1 invariant. Let us consider a G-structure, B_G, with this group G. Let $p \in B_G$ and $x = \pi(p)$. Define the subspace $\mathscr{D}(x) \subset T_x(M)$ by $\mathscr{D}(x) = p(V_1)$. For any $a \in G$, we have $pa(V_1) = p(aV_1) = p(V_1)$, so that $\mathscr{D}(x)$ does not depend on the choice of $p \in \pi^{-1}(x)$ in B_G. Thus a B_G gives a k-dimensional differential system on M. Conversely, let $\mathscr{D}$ be a differential system on M. Let B_G be the

submanifold of $\mathscr{F}(M)$ consisting of those frames p with $p^{-1}(\mathscr{D}(x)) = V_1$ (where $x = \pi(p)$). Then B_G is easily seen to be a G-structure. Thus in this case a G-structure is the same as a k-dimensional differential system.

3) Let (,) be a positive definite scalar product on V and G the set of isometries of V under this scalar product. (Thus $G = O(n)$.) In this case, a B_G is just a Riemann metric on M. In fact, a Riemann metric picks out the bundle of orthogonal frames, $\mathscr{O}(M)$, which is exactly an $O(n)$ structure. Conversely, given a family of frames at the tangent space at each point each one differing from another by an orthogonal transformation defines a scalar product on each tangent space—i.e., a Riemann metric. Thus giving an $O(n)$ structure is the same as giving a Riemann metric.

4) Let (,) be a scalar product on V and let G be the conformal group of (,). That is, a $T \in GL(V)$ belongs to V if and only if $(Tu, Tv) = \lambda_T(u, v)$ for all $u, v \in V$ where λ_T depends only on T. A G-structure corresponding to this group is called a *conformal structure.* If M possesses a conformal structure then it makes sense to speak of the (cosine of the) angle between two tangent vectors at the same point of M.

5) Let (,) be a non-degenerate *antisymmetric* bilinear form on V. (Thus V is even dimensional, say of, dim $2n$.) Let $G = Sp(n)$ be the symplectic group, i.e., the group of all automorphisms of (,). Thus a $T \in GL(V)$ belongs to $Sp(n)$ if and only if $(Tu, Tv) = (u, v)$ for all u and $v \in V$. Given a $Sp(n)$ structure, $B_{Sp(n)}$, on a manifold M we can define a non-degenerate antisymmetric bilinear form on the tangent space at each point of M. In fact, for each $x \in M$, let p be a point of $B_{Sp(n)}$ with $\pi(p) = x$. For any two tangent vectors, X and Y, at x put $(X, Y) = (p^{-1}(X), p^{-1}(Y))$. For any other point $pT \in \pi^{-1}(x)$ in $B_{Sp(n)}$ we have, by (2.1), $((pT)^{-1}(X), (pT)^{-1}(Y)) = (X, Y)$ so that this definition is independent of the choice of p. It is easily seen that, conversely, a non-degenerate antisymmetric bilinear form on the tangent space at each point (varying differentiably from point to point) i.e., *an exterior two form of maximal rank* gives a $Sp(n)$ structure. Thus giving an $Sp(n)$ structure on a manifold M is the same as giving an exterior two from, Ω, on M. For this reason a $B_{Sp(n)}$ structure is sometimes called an *almost Hamiltonian structure.* A Hamiltonian structure satisfies the additional condition $d\Omega = 0$ (cf. Definition 7.1 of Chapter III).

6) Let V be a $2n$-dimensional vector space and let J be a linear transformation on V with $J^2 = -I$. Let $GL(n, C)$ denote the subgroup

of $GL(V)$ consisting of those linear transformations, $T \in GL(V)$ satisfying $TJ = JT$. The meaning of this group is the following: The vector space V is a vector space over the real numbers. The linear transformation, J, allows us to make V into a vector space over the complex numbers. In fact, we must define multiplication by i. We do this by setting $iv = Jv$ for all $v \in V$. Since $J^2 = -I$, the reader can easily check that this defines V as a complex vector space. Now not every linear transformation of V as a real vector space is a complex linear transformation; it must commute with multiplication by i. The condition is thus $TJ = JT$, i.e., that $T \in GL(n, C)$. A $GL(n, C)$ structure on a manifold M is called an *almost complex* structure. An almost complex structure on M is the same as giving a linear transformation J_x on each tangent space $T_x(M)$ (varying differentiably with x) and such that $J_x^2 = -I$. In fact, given $p \in B_{GL(n,C)}$ with $\pi(p) = x$ define $J_x = pJp^{-1}$. By (2.1) we have $(pT)J(pT)^{-1} = p(TJT^{-1})p^{-1} = pJp^{-1}$ for $T \in GK(n, C)$ so that J_x is well defined. Conversely, such a J_x at each x defines a $GL(n, C)$ structure as the reader can easily verify. Thus if a manifold, M, possesses an almost complex structure each tangent space becomes a vector space over the complex numbers. The reason for the word "almost" is that the most important case is that of a *complex structure.* It is defined as follows: A manifold M is said to possess a complex structure if we are given an atlas $(\{U_\alpha, h_\alpha)\}$ with $h_\alpha = (x_\alpha^1, \ldots, x_\alpha^n, y_\alpha^1, \ldots, y_\alpha^n)$ so that if we write $z_\alpha^1 = x_\alpha^1 + iy_\alpha^1, \ldots, z_\alpha^n = x_\alpha^n + iy_\alpha^n$ and write the transition functions

$$x_\alpha^i = x_\alpha^i(x_\beta^1, \ldots, x_\beta^n, y_\beta^1, \ldots, y_\beta^n),$$
$$y_\alpha^i = y_\alpha^i(x_\beta^1, \ldots, x_\beta^n, y_\beta^1, \ldots, y_\beta^n)$$

as $z_\alpha^i = z_\alpha^i(x_\beta^1, \ldots, x_\beta^n, y_\alpha^1, \ldots, y_\beta^n)$ then the z_α^i are actually complex analytic functions $z_\alpha^i = z_\alpha^i(z_\beta^i, \ldots, z_\beta^n)$ of the z_β^j. In short, a complex structure on a manifold, or (more succinctly) a complex manifold, is one on which we can introduce complex coordinates and such that the transition functions are complex analytic. The reader can easily verify that a complex structure defines an almost complex structure. On the other hand, not every almost complex structure comes from a complex structure.

7) Let V be an (arbitrary) n-dimensional vector space and $SL(V)$ the set of all linear transformations of determinant 1. (That is, the set of all T with $\wedge^n(T) = \text{id.}$) The reader should verify that giving an $SL(V)$ structure on a manifold is the same as giving a (nowhere zero) exterior differential form of degree n on M.

8) Let V be a vector space with a choice of orientation. Let $GL(V)^+$ be subgroup of orientation preserving transformations. The reader should verify that a $GL(V)^+$ structure is the same as an orientation of M.

REMARKS:

a) It is clear from the above examples that just about any "structure" on a manifold that depends only on "first-order" properties, e.g., is described by a field of quantities as in Section 6 of Chapter II, corresponds to some G-structure. Much of the techniques we will develop can be easily extended to cover the case of "higher-order" structures. We shall restrict ourselves to first-order structures for the sake of simplicity.

b) Given a manifold, M, and a subgroup, G, of $GL(V)$ there does not necessarily exist a G-structure on M. Thus, for instance, if M is not orientable, it cannot carry a structure of Example 8 or 7 (or for that matter 1, 5, or 6). As we saw in the last section of Chapter VI, the sphere, S^2, does not carry a non-vanishing vector field so *a fortori* does not possess a structure of Example 1. On the other hand, we have proved that every manifold can be given a Riemann structure and therefore also a conformal structure. In general, the problem of when a manifold can be given a G-structure is a delicate topological question and is outside the scope of this book.

c) If B_G is a G-structure over M and U is an open subset of M then $\pi^{-1}(U)$ is a G-structure over U (where π is the projection of B_G onto M). We shall, for the most part, be concerned with local properties of G-structures and hence frequently restrict our G-structures to smaller and smaller neighborhoods.

Let φ be a diffeomorphism of a manifold M_1 onto a manifold M_2. Then, just as in Section 6 of Chapter II, φ induces a diffeomorphism, φ_*, of $\mathscr{F}(M_1)$ onto $\mathscr{F}(M_2)$. If B_G^1 and B_G^2 are G-structures over M_1 and M_2 we can ask whether or not $\varphi_*(B_G^1) = B_G^2$.

DEFINITION 2.2. *Let B_G^1 and B_G^2 be G-structures on M_1 and M_2. A diffeomorphism, φ, of $M_1 \to M_2$ is called an isomorphism of B_G^1 onto B_G^2 if $\varphi_*(B_G^1) = B_G^2$. We say that B_G^1 and B_G^2 are isomorphic if there exists an isomorphism of B_G^1 onto B_G^2. An isomorphism of B_G^1 with itself is called an automorphism.*

Thus an isomorphism of two Riemann structures is an isometry, and isomorphism of two almost Hamiltonian structures on M_1 and M_2 is a diffeomorphism, φ, such that $\varphi^*\Omega_2 = \Omega_1$ (where Ω_i gives the almost Hamiltonian structure on M_i).

DEFINITION 2.3. *Let B_G^1 and B_G^2 be G-structures on M_1 and M_2. Let $x \in M_1$ and $y \in M_2$. We say that B_G^1 and B_G^2 are locally equivalent at (x, q) if there exist open neighborhoods U of x and V of y and an isomorphism, φ, of $B_{G|U}^1$ onto $B_{G|V}^2$ such that $\varphi(x) = y$. (Here $B_{G|U}^1$ denotes the restriction of B_G^1 to the open set U and $B_{G|V}^2$ the restriction of B_G^2 to V.)*

The general equivalence problem is to find criteria (for a group G) to decide whether or not two G-structures are locally equivalent.

We should remark that many of the theorems of this book are solutions of the equivalence problem for special cases. For instance, Frobenius' theorem (Theorem 5.1 of Chapter III) says that a differential system (i.e., a G-structure as in Example 2 above) is locally equivalent to a particularly simple kind of differential system (given by the tangent to $x^{n+1} = \text{const}$, and $x^n = \text{const}$) provided that a certain integrability condition [Equation (5.2) of Chapter III] is satisfied. Darboux' theorem (Theorem 6.1 of Chapter III) says that any two Hamiltonian structures [i.e., $Sp(n)$ structures as in Example 5 satisfying the additional condition that $d\Omega = 0$] are locally equivalent. Proposition 3.1 of Chapter VI says that any Riemann manifold whose bundle of orthogonal frames has form Ω_M^{ij} satisfying (3.7) and (3.11) is locally isomorphic to E^n. Theorem 2.4 of Chapter V says that any two Lie groups (i.e., $\{e\}$-structures of a certain type) are locally equivalent if they have the same constants of structure.[3]

[3] In many cases the problem of interest is a little different from the equivalence problem, namely what we might call the *conjugate* equivalence problem. Let G_1 and G_2 be conjugate subgroups of $GL(V)$. We say that $B_{G_1}(M)$ is conjugate to $B_{G_2}(M)$ if there exists an $a \in GL(V)$ such that $R_a(B_{G_1}(M)) = B_{G_2}(M)$. (Thus in particular, $a^{-1}G_1a = G_2$.) The conjugate equivalence problem is to decide when one G-structure is equivalent to a conjugate of another. In the case of the $\{e\}$-structures of Lie groups, for instance, one is interested in local isomorphism (which is such a conjugate equivalence) and not equivalence which involves a choice of basis for the Lie algebra. For a discussion of this point cf. Guillemin and Sternberg's "An algebraic model of transitive differential geometry," Bull. Amer. Math. Soc., vol. 70 (1964) pp. 16–47.

There is a difference between the first three theorems cited and the last. In the first three theorems we obtained necessary and sufficient conditions for the G-structures to be locally equivalent to a G-structure of a particularly simple type. Let us make the general definition. Notice that for any G the vector space V considered as a differentiable manifold, carries a natural G-structure. In fact, the vector space structure gives an identification of $T_v(V)$ with V for each $v \in V$; i.e., a complete parallelism or $\{e\}$-structure on V. If we now consider, at each v, the set of all frames differing from the standard frame by the action of an element of G, we get a G-structure on V. This G-structure is called the *(standard) flat G-structure.* Thus, for example, the flat $O(n)$ structure is Euclidean space.

DEFINITION 2.4. *A G-structure is called locally flat if it is locally equivalent to the flat G-structure.*

Thus the first three theorems cited above give necessary and sufficient conditions for the corresponding G-structures to be locally flat. Theorem 2.4 of Chapter V is not of this nature since the flat $\{e\}$-structure is V (a commutative Lie group) and the general (non-commutative) Lie group is not locally flat.

EXERCISE 2.1. Let B_G be a G-structure over M. Show that B_G is locally flat if and only if there is an atlas $\{(U_\alpha, h_\alpha)\}$ on M with $h_\alpha = x_\alpha^1, \ldots, x_\alpha^n$ and the frames

$$\left(\frac{\partial}{\partial x_\alpha^1}(x), \ldots, \frac{\partial}{\partial x_\alpha^n}(x)\right)$$

lie in B_G for each $x \in U_\alpha$.

We now proceed to study methods that are useful in the handling of the general equivalence problem. We first observe that since B_G is a submanifold of $\mathscr{F}(M)$ we can restrict the form ω so as to obtain a V-valued form on B_G. When there is no danger of confusion we will continue to call this restricted form ω. (This is consistent with the notation in Chapter V for the $\{e\}$-structure of a Lie group. In Chapter VI we denoted the corresponding form for a Riemann structure on M by ω_M.) Equations (2.1) to (2.5) still hold for the case of any G-structure but for $a \in G$ in (2.1) and (2.3) and $A \in g$ in (2.4) and (2.5) where g is the Lie algebra of G (and thus a subalgebra $gl(V)$). We use these equations to construct a (vector valued) function which is intrinsically defined on B_G.

Let B_G be a G-structure on a manifold M and let $p \in B_G$. Let $H_1 \subset T_p(B_G)$ be a complement to the vertical at p. Thus H_1 is an n-dimensional subspace of the tangent space which, together with the tangent to the fiber, span $T_p(B_G)$. (For instance, if $\mathscr{H}$ were a connection on B_G the space $\mathscr{H}(p)$ would be such an H_1.) The linear form ω maps $T_p(B_G)$ onto V and the kernel of this map is the subspace consisting of those vectors X satisfying $\pi_* X = 0$, i.e., the tangent space to the fiber. Thus ω gives an isomorphism of H_1 onto V. Now let H_2 be a second complement to the vertical at p. Let v be an element of V and X_1 be the element of H_1 with $\langle X_1, \omega \rangle = v$ and $X_2 \in H_2$ with $\langle X_2, \omega \rangle = v$. Now by the definition, (2.2), of ω

$$p(v) = \pi_*(X_1) = \pi_*(X_2).$$

Thus $\pi_*(X_2 - X_1) = 0$ or $X_2 - X_1$ is vertical. Since we have a natural identification of the tangent space to the fiber with the Lie algebra, g, we can regard $X_2 - X_1$ as an element of g. Thus,

Given the pair H_1 and H_2 we have defined a map, S_{H_1, H_2} of $V \to g$.

It is easy to check that this map is linear. Furthermore

for fixed H_1, the subspace H_2 is uniquely determined by S_{H_1, H_2},

as can be readily verified. Finally, it is easy to see that

given H_1 and $S \in \operatorname{Hom}(V, g)$ there exists an H_2 such that $S = S_{H_1, H_2}$.

In other words, having fixed an H_1, the set of all complements to the vertical can be parametrized by $\operatorname{Hom}(V, g)$.

Let us now consider the exterior differential form, $d\omega$. For each $p \in B_G$, the form $d\omega$ gives a map of $\wedge^2(T_p(B_G)) \to V$. Let H be some complement to the vertical at p. Then $\wedge^2(H)$ is a subspace of $\wedge^2(T_p(B_G))$ and so the restriction of $d\omega$ to $\wedge^2(H)$ gives a map of $\wedge^2(H)$ into V. But we have identified $\wedge^2(H)$ with $\wedge^2(V)$. We thus get a map of $\wedge^2(V) \to V$ which we shall denote by c_H. By definition, if u and v are vectors in V then

$$c_H(u \wedge v) = \langle X \wedge Y | d\omega \rangle \tag{2.6}$$

where X and Y are elements of H such that $\langle X, \omega \rangle = u$ and $\langle Y, \omega \rangle = v$.

We have thus assigned a $c_H \in \mathrm{Hom}(V \wedge V, V)$ to each complement to the vertical at p. Let us now see how c_H depends on H. For this purpose, let H_1 and H_2 be two complements to the vertical at p. Then

$$c_{H_2}(u \wedge v) - c_{H_1}(u \wedge v) = \langle X_2 \wedge Y_2 | d\omega\rangle - \langle X_1 \wedge Y_1 | d\omega\rangle$$

where X_i and Y_i lie in H_i with $\langle X_i, \omega\rangle = u$ and $\langle Y_i, \omega\rangle = v$. We can rewrite this as

$$\begin{aligned} c_{H_2}(u \wedge v) &- c_{H_1}(u \wedge v) \\ &= \langle (X_2 - X_1) \wedge Y_2 | d\omega\rangle + \langle X_1 \wedge (Y_2 - Y_1) | d\omega\rangle \\ &= \langle Y_2, (X_2 - X_1) \lrcorner\, d\omega\rangle - \langle X_1, (Y_2 - Y_2) \lrcorner\, d\omega\rangle. \end{aligned}$$

Now $X_2 - X_1$ is a vertical vector and is, by our definition, given by $S_{H_1, H_2}(u)_p$. (Recall that $S_{H_1,H_2}(u)$ is an element of g and $S_{H_1, H_2}(u)_p$ is the corresponding vertical vector at p.) Now by Equation (2.5), $S_{H_1, H_2}(u) \lrcorner\, d\omega = -S_{H_1,H_2}(u)\omega$. Applying the same argument to $Y_1 - Y_2$, we get, since $\langle Y_2, \omega\rangle = v$ and $\langle X_1, \omega\rangle = u$

(2.7) $$c_{H_2}(u \wedge v) - c_{H_1}(u \wedge v) = S_{H_1,H_2}(v)\cdot u - S_{H_1,H_2}(u)\cdot v.$$

[Recall that $S_{H_1, H_2}(v)$ is an element of g and g acts on V so that $S_{H_1, H_2}(v)\cdot u$ is an element of V.] We can rewrite Equation (2.7) in a slightly different form. Let us consider the space $\mathrm{Hom}(V, g)$. On the space there is an antisymmetrization operator, $\mathscr{A}$, defined as follows: for each $S \in \mathrm{Hom}(V, g)$ put

$$\mathscr{A}(S)(u)v = S(u)v - S(v)u.$$

(Since g is a subspace of $\mathrm{Hom}(V, V)$ to give $\mathscr{A}(S)(u)$ it is sufficient to give $\mathscr{A}(S)(u)v$ for all $v \in V$.) It is clear that $\mathscr{A}(S)(u)v = -\mathscr{A}(S)(v)u$ so that $\mathscr{A}(S)$ gives an antisymmetric bilinear map of V into V or, in short, $\mathscr{A}(S)$ can be regarded as an element of $\mathrm{Hom}(V \wedge V, V)$. In other words, $\mathscr{A}$ is a map from $\mathrm{Hom}(V, g)$ to $\mathrm{Hom}(V \wedge V, V)$. We can thus write (2.7) as

(2.8) $$c_{H_2} - c_{H_1} = -\mathscr{A}(S_{H_1, H_2}).$$

Now $\mathscr{A}$ $(\mathrm{Hom}(V, g))$ is a subspace of $\mathrm{Hom}(V \wedge V, V)$. Let ρ denote the projection of $\mathrm{Hom}(V \wedge V, V)$ onto the quotient space $\mathrm{Hom}(V \wedge V, V)/\mathscr{A}(\mathrm{Hom}(V, g))$. Equation (2.8) says that $c_{H_1} - c_{H_2} \in \mathscr{A}(\mathrm{Hom}(V, g))$ i.e., that $\rho(c_{H_1}) = \rho(c_{H_2})$. *Thus* $\rho(c_H)$ *is independent of the choice of* H *at* p. We thus have a well defined function, c, on B_G

$$c : B_G \to \mathrm{Hom}(V \wedge V, V)/\mathscr{A}(\mathrm{Hom}(V, g))$$

given by

$$c(p) = \rho(c_H) \tag{2.9}$$

for any H complement to the vertical at p. *This function is called the (first-order) structure function of* B_G. The reason for this nomenclature is that c generalizes the constants of structure of a Lie group. In fact, if $G = \{e\}$ then $g = 0$ and c maps $B_{\{e\}}$ into $\operatorname{Hom}(V \wedge V, V)$. [This can be seen directly since there is only one H at each $p \in B_{\{e\}}$ so that $c(p) = c_H$.] In the case that $B_{\{e\}}$ structure is a Lie group this function is a constant. Since, in this case, V is the Lie algebra of the Lie group, it should come as no great surprise that the map of $V \wedge V \to V$ that we get is just the Lie bracket operation on V (regarded as an antisymmetric bilinear map). The reader should verify these facts as a very instructive exercise.

Let us see how the structure function varies along the fiber. For this purpose let us choose an H at p and consider the space $R_{a*}H$ which is a complement to the vertical at $p \cdot a$. For $u, v \in V$ choose X and $Y \in H$ such that $\langle X, \omega\rangle = u$ and $\langle Y, \omega\rangle = v$. According to (2.3), for any $a \in G$ we have

$$\langle R_{a*}X, \omega\rangle = a^{-1}u \quad \text{and} \quad \langle R_{a*}Y, \omega\rangle = a^{-1}v.$$

On the other hand,

$$\langle R_{a*}X \wedge R_{a*}Y | d\omega\rangle = \langle X \wedge Y | R_{a*}d\omega\rangle = a^{-1}\langle X \wedge Y | d\omega\rangle.$$

Thus

$$c_{R_{a*}H}(a^{-1}u \wedge a^{-1}v) = a^{-1}c_H(u \wedge v)$$

or

$$c_{R_{a*}H}(u \wedge v) = a^{-1}c_H(au \wedge av). \tag{2.10}$$

Now the group G acts on V and therefore on $V \wedge V$. Therefore G acts on $\operatorname{Hom}(V \wedge V, V)$ as follows: For each $a \in G$ and $L \in \operatorname{Hom}(V \wedge V, V)$ let $\sigma(a)L$ be given by

$$\sigma(a)L(u \wedge v) = aL(a^{-1}u \wedge a^{-1}v).$$

This is indeed a representation of G on $\operatorname{Hom}(V \wedge V, V)$ as the reader can readily check. Equation (2.10) then can be written as

$$c_{R_{a*}H} = \sigma(a^{-1})c_H.$$

The reader can also easily check that the space $\mathscr{A}(\operatorname{Hom}(V, g))$ is an invariant subspace of $\operatorname{Hom}(V \wedge V, V)$ under the action of G. We therefore get a representation of G on the quotient space which we will also denote by σ. By (2.9) and (2.10) we then obtain

$$c(p \cdot a) = \sigma(a)^{-1}c(p). \tag{2.11}$$

The structure function was intrinsically defined. Therefore (the reader should give a direct proof using the definitions) we have

THEOREM 2.1. *Let B_G^1 and B_G^2 be G-structures over M_1 and M_2. If φ is an isomorphism of M_1 onto M_2 then*

$$c^2 \circ \varphi_* = c^1$$

where c^1 is the structure of B_G^1 and c^2 the structure function of B_G^2.

An important consequence of Theorem 2.1 is

THEOREM 2.2. *If a G-structure is locally flat then its structure function vanishes identically.*

To prove Theorem 2.2 we must simply show that the structure function of the standard flat G-structure vanishes identically. Now (2.11) implies that if c vanishes at any point of the fiber it vanishes identically on the fiber. By (2.9), if $c_H = 0$ for some H at p then $c(p) = 0$. Now B_G, by construction, contains the standard $\{e\}$-structure over V as a submanifold. We will choose our p so as to belong to this submanifold and our H to be tangent to it. It therefore suffices to show that $d\omega$ vanishes identically on the canonical complete parallelism of V. Since the group is $\{e\}$, the projection π is a diffeomorphism of this $B_{\{e\}}$ with V. Then $(\pi^{-1})^*\omega$ is a form on V. By definition, $(\pi^{-1})^*\omega = d(\text{id})$ where id is the identity map of V onto itself (regarded as a vector valued form). Therefore $d\omega = 0$ proving Theorem 2.2.

The converse of Theorem 2.2 is not true in general. In fact, as we shall see at the end of this section, for many important groups, G, the first-order structure function vanishes identically for all B_G. If this happens, the first-order structure function gives no information and we must then look at higher-order invariants.

The converse of Theorem 2.2 is true for some groups however, and we now give some examples. Suppose the group G leaves invariant a subspace $V_1 \subset V$. There is an obvious projection τ of $\text{Hom}(V \wedge V, V)$

onto $\mathrm{Hom}(V_1 \wedge V_1, V/V_1)$ namely, for $T \in \mathrm{Hom}(V \wedge V, V)$ and $v_1, v_2 \in V_1$ we put $\tau(T)(v_1 \wedge v_2)$ = image of $T(v_1 \wedge v_2)$ in V/V_1. The kernel of τ consists of all those T such that $T(v_1 \wedge v_2) \in V_1$ for all $v_1, v_2 \in V_1$. Now if g leaves V_1 invariant, then $Av_1 \in V_1$ for all $v_1 \in V_1$. Then for any $S \in \mathrm{Hom}(V, g)$,

$$\mathscr{A}(S)(v_1 \wedge v_2) = S(v_1)v_2 - S(v_2)v_1 \in V_1$$

for $v_1, v_2 \in V_1$. Thus,

$$\mathscr{A}(S) \subset \mathrm{Ker}\,\tau.$$

We therefore have a projection, η,

$$\frac{\mathrm{Hom}(V \wedge V, V)}{\mathscr{A}(\mathrm{Hom}(V, g))} \xrightarrow{\eta} \mathrm{Hom}(V_1 \wedge V_1, V/V_1) \tag{2.12}$$

and so $\eta(c)$ is a function on B_G with values in $\mathrm{Hom}(V_1 \wedge V_1, V/V_1)$.

Exercise 2.2. Show that if G is the group of all transformations leaving V_1 invariant (i.e., B_G is a structure as in Example 2) then $\mathscr{A}(\mathrm{Hom}(V, g)) = \mathrm{Ker}\,\tau$; thus η is an isomorphism. Show that in this case Theorem 2.2 and its converse is precisely Frobenius' theorem.

Exercise 2.3. Let B_G be an almost Hamiltonian structure given by a form Ω on M. Show that the structure function on B_G vanishes if and only if $d\Omega = 0$. Thus in this case Theorem 2.2 and its converse is Darboux' theorem (Theorem 6.1 of Chapter III).

Another important case where the converse of Theorem 2.2 is true is for almost complex structures. The assertion is that an almost complex structure whose structure function vanishes is complex. This is a theorem of Newlander-Nirenberg [15] and depends on some deep techniques in the theory of partial differential equations.

Let us now return to Theorem 2.1. It asserts that if we are to find an isomorphism between two G-structures B_G^1 and B_G^2 then we must be able to find a map, φ_*, of B_G^1 onto B_G^2 such that $c^2 \circ \varphi_* = c^1$. In general, this kind of problem is very difficult. That is, given two manifolds and a function on each, to decide whether there is a map carrying one function into the other is very hard. The only really easy (and trivial) case is where one of the functions is a constant. Then the other function must also be a constant; and, in fact, the same constant. We can thus state

Corollary 2.1. *Let B_G be a G-structure with a constant structure function. Any G-structure locally equivalent to B_G must have the same constant as its structure function.*

Exercise 2.4. Call a G-structure, B_G, locally frame transitive if it has the following property: For any pair of points p and $q \in B_G$ with $\pi(p) = x$ and $\pi(q) = y$ there is local (x, y) automorphism, φ, of B_G such that $\varphi_*(p) = q$. Show that every frame transitive G-structure has constant structure function. (Here also, the converse is false in general.)

Even if the structure function is not constant we have some tools at our disposal to implement the necessary condition of Theorem 2.1. The main idea is to use Equation (2.11) carefully. Instead of trying to formulate a broad theorem covering all situations, we describe a general procedure for using (2.11) and Theorem 2.1 to obtain necessary conditions for equivalence and illustrate this procedure by a special case.

Since the group G acts (via the representation σ) on $\mathrm{Hom}(V \wedge V, V)/\mathscr{A}(\mathrm{Hom}(V, g))$, this space decomposes into orbits under the action of G. That is, let us introduce the equivalence relation $w_1 \sim w_2$ if there is an $a \in G$ such that $\sigma(a)w_1 = w_2$. Then each equivalence class turns out to be a differentiable manifold (called an orbit of G in $\mathrm{Hom}(V \wedge V, V)/\mathscr{A}(\mathrm{Hom}(V, g))$. These orbits will be of varying dimension. For instance, since the representation is linear, the origin is left fixed by all elements of G and is thus a zero-dimensional orbit. What usually happens is the following: If we consider the orbits of maximal dimension either they are discrete or they form (locally) a continuous family. Let us first consider the case where the orbits are discrete (and finite in number). The orbits of maximal dimension are then open sets of $\mathrm{Hom}(V \wedge V, V)/\mathscr{A}(\mathrm{Hom}(V, g))$. Now the structure function, c, of a G-structure, B_G, is continuous. Therefore, if $c(p)$ lies in an orbit of maximal dimension, then $c(q)$ will lie in the same orbit for all $q \in \pi^{-1}(U)$ where U is some neighborhood of $\pi(p)$. Let us call a point, x, of M *generic* if $c(p)$ lies in an orbit of maximal dimension. Suppose we restrict attention to the local equivalence problem for G-structures at generic points, $x \in M_1$ and $y \in M_2$. Then Theorem 2.1 implies that c^1 *and* c^2 *take values in the same orbit of* G *in* $\mathrm{Hom}(V \wedge V, V)/\mathscr{A}(\mathrm{Hom}(V, g))$. Let w be some fixed point of this orbit and let H be the isotropy subgroup of w, i.e., the subgroup of G, consisting of those $a \in G$ such that $\sigma(a)w = w$. Let B_H^1 be the submanifold of B_G^1 consisting of those p such that $c(p) = w$. (Observe that since G acts transitively on the orbit, Equation (2.11) implies that there is some p on each fiber with $c(p) = w$. The discussion on page 219 can be used to conclude that B_H^1 is a submanifold of B_G^1.) The submanifold, B_H^1, is actually an H-structure. Similarly, we construct the H-structure, B_H^2. Now $c^2 \cdot \varphi_* = c^1$ implies that $\varphi_*[(c^1)^{-1}(w)] = (c^2)^{-1}(w)$ or $\varphi_*(B_H^1) = B_H^2$. Thus in order for φ to be a local isomorphism of the G-structure B_G^1 onto B_G^2 it must also be a local isomorphism of the H-structure B_H^1 onto B_H^2. The converse is obvious: If φ is an H-structure isomorphism then by enlarging the group we get a G-structure isomorphism. We have thus reduced the problem of equivalence to the smaller group, H.

In the case of continuous orbits[4] a similar analysis can be made. In this case the orbits of maximal dimension are grouped into a finite number of families. The union of the members of each family is an open set in $\mathrm{Hom}(V \wedge V, V)/\mathscr{A}(\mathrm{Hom}(V, g))$. Each of these open sets can be regarded as a "fiber space" whose fiber is an orbit. Then instead of picking a fixed element in a finite number of maximal orbits as in the discrete case, we must choose a "cross-section" for this fiber space of appropriate type. That is, we pick a submanifold, W, of $\mathrm{Hom}(V \wedge V, V)/\mathscr{A}(\mathrm{Hom}(V, g))$ with the property that it meets each maximal orbit, $\mathscr{O}$, at one point. The subgroup leaving that point fixed will depend on the point, in general. Then we proceed as before, reducing the equivalence problem to the submanifold, $c^{-1}(W)$. Here $c^{-1}(W)$ is not a G-structure in the sense we have defined since the group will now vary from point to point.[5] In order to study the equivalence problem in full generality one must therefore consider these more general objects. For the sake of simplicity we will not consider them here. We should remark that for many "nice" Lie groups, standard theorems in the theory of Lie groups imply that we can choose W so that $c^{-1}(W)$ is a G-structure.

In either case, we have reduced the local equivalence problem (in the generic case) to the subgroup H. We can now construct the structure function for the H-structures in question and apply Theorem 2.1 again. We can continue to apply this procedure until no further reduction of the group is possible. There are two advantages to reducing the group.

---

[4] We will not prove any of these facts relating to the continuous case. We are just describing the general procedure. In any specific case, all of the objects mentioned in this discussion must be constructed explicitly.

[5] The appropriate geometrical setting is the notion of differentiable groupoid of Ehresmann.

First, we get more and more refined necessary conditions from the new structure functions introduced at each stage. These might allow us to distinguish between non-isomorphic G-structures. Secondly, if the group is "small enough" (in a sense to be made precise in the next section) we can actually get necessary and sufficient conditions for local isomorphisms.

Let us illustrate[6] this reduction procedure by applying it to the case of differential systems on manifolds of dimension 5. (We are, in effect, translating a short passage from Cartan's paper "Sur les systemes de Pfaff à cinq variables" [6].) As we have already seen, a differential system is a G-structure where G is the group leaving invariant a subspace, V_1, of V. In this case the space $\mathrm{Hom}(V \wedge V, V)/\mathscr{A}(\mathrm{Hom}(V, g))$ can be identified with the space $\mathrm{Hom}(V_1 \wedge V_1, V/V_1)$. The group G acts by restriction on V_1 and by projection on V/V_1 and therefore also on $\mathrm{Hom}(V_1 \wedge V_1, V/V_1)$. This action (as the reader should check) is just the action, σ, of G on $\mathrm{Hom}(V \wedge V, V)/\mathscr{A}(\mathrm{Hom}(V, g))$ under the identification of this space with $\mathrm{Hom}(V_1 \wedge V_1, V/V_1)$. Let us first look at the case where V_1 is of codimension one in V (and dim V is arbitrary). In this case, V/V_1 is the real line and G acts by real multiplication. Thus $\mathrm{Hom}(V_1 \wedge V_1, V/V_1)$ is just the space of antisymmetric bilinear forms on V_1. Two such forms $(\, , \,)_1$ and $(\, , \,)_2$ lie in the same orbit under G if there is a scalar, λ, and a non-singular linear transformation, T, on V_1 such that $(u, v)_1 = \lambda(Tu, Tv)_2$ for all $u, v \in V_1$. Now Theorem 5.1 of Chapter I implies that $(\, , \,)_1$ and $(\, , \,)_2$ lie in the same orbit if and only if they have the same rank. Thus there are a finite number of orbits. The orbit of maximal dimension corresponds to maximal rank, the orbit of zero dimension to the zero element. Now a slight extension of Theorem 6.1 of Chapter III (Darboux' theorem) says that the rank is the only invariant of a differential system; i.e., two differential systems of the same rank are locally equivalent. Notice that in this theorem we *do* allow the structure function, c, to take values in a singular orbit (i.e., an orbit which is not of maximal dimension). What we do insist on (to apply Darboux' theorem) is that if c has values in a singular orbit that it have values in that orbit on a whole open set. Thus, for instance, in the case $c = 0$ the system is completely integrable and Darboux' theorem reduces to Frobenius' theorem. If c is allowed to take on singular values at isolated points the situation becomes complicated and very little is known. Now let us return to the case of V_1 of arbitrary codimension. It turns out that the first situation that cannot be completely analyzed on the basis of Frobenius' and Darboux' theorems occurs when V is 5-dimensional. We therefore study this case. If V_1 is of dimension 1 there is nothing to say (all are equivalent)

---

[6] The reader who does not wish to go through this rather lengthy discussion of a special example can skip directly to the general discussion on page 327.

and if V_1 is dimension 4 the situation is covered by Darboux' theorem. The interesting case is then when dim $V_1 = 2$ or 3. Let us first consider the case dim $V_1 = 2$. The group, G, is the group of all non-singular linear transformations leaving the subspace V_2 invariant. If we choose a basis, e_1, e_2, e_3, e_4, e_5 of V such that e_4, e_5 span V then G can be identified with the group of all non-singular matrices of the form

$$\left(\begin{array}{ccc|cc} \sim & \sim & \sim & 0 & 0 \\ \sim & \sim & \sim & 0 & 0 \\ \sim & \sim & \sim & 0 & 0 \\ \hline \sim & \sim & \sim & \sim & \sim \\ \sim & \sim & \sim & \sim & \sim \end{array}\right)$$

i.e., with zeros in the upper right-hand block and arbitrary entries elsewhere. The group G is 19-dimensional. Since we will be repeatedly reducing the group (and G-structure) let us label this group and the succeeding ones by their dimension. Thus we call our group G_{19}. The structure function of G_{19} takes values in $\mathrm{Hom}(V_1 \wedge V_1, V/V_1)$. Since V_1 is two-dimensional, $V_1 \wedge V_1$ is one-dimensional and V/V_1 is three-dimensional. Therefore $\mathrm{Hom}(V_1 \wedge V_1, V/V_1)$ is three-dimensional. The group G_{19} acts transitively on the non-zero vectors of this space, as can easily be checked. In other words, there are exactly two orbits: The zero-dimensional manifold consisting of the zero vector and the three-dimensional manifold consisting of all non-zero vectors. If the structure function vanishes identically we are in the completely integrable situation and all such structures are locally equivalent. If the structure function vanishes at a point but not identically in a whole neighborhood, our procedure breaks down. We therefore consider the case where the structure function does not vanish. Let us pick a definite element of $\mathrm{Hom}(V_1 \wedge V_1, V/V_1)$: In terms of the basis $e_1, \ldots, e_5$ let us pick that element, w, of $\mathrm{Hom}(V_1 \wedge V_1, V/V_1)$ which sends $e_4 \wedge e_5 \to \{e_3\}$ (where $\{e_3\}$ is the equivalence class of e_3 mod V_1). Our procedure then yields a reduction of $B_{G_{19}}$ to $B_{G_{16}}$. Here G_{16} is that subgroup of G_{19} consisting of all $a \in G_{19}$ with

$$w(au \wedge av) = aw(u \wedge v)$$

for all $u, v \in V_1$ (and where the action of a on the right-hand side of the equation is the induced action on the quotient space). In terms of the particular w that we have chosen, and the basis $e_1, \ldots e_5$, the group G_{16} can be regarded as the group of all non-singular matrices of the form

$$(G_{16}): \left(\begin{array}{ccc|cc} \sim & \sim & 0 & 0 & 0 \\ \sim & \sim & 0 & 0 & 0 \\ \sim & \sim & \Delta & 0 & 0 \\ \hline \sim & \sim & \sim & \alpha & \beta \\ \sim & \sim & \sim & \gamma & \delta \end{array}\right) \qquad \text{where} \quad \Delta = \alpha\delta - \beta\gamma$$

The Lie algebra, g_{16}, of G_{16} is the set of all matrices of the form

$$g_{16}: \begin{pmatrix} - & - & 0 & 0 & 0 \\ - & - & 0 & 0 & 0 \\ - & - & a_1 + a_4 & 0 & 0 \\ - & - & - & a_1 & a_2 \\ - & - & - & a_3 & a_4 \end{pmatrix}$$

(where the dashes denote arbitrary elements). We must now examine the range of the structure function, G_{16}, for $B_{G_{16}}$. This time the structure function takes values in

$$\operatorname{Hom}(V \wedge V, V/V_1)/\mathscr{A}(\operatorname{Hom}(V, g_{16})).$$

According to our general procedure we should study the action of G on this space. As this involves a rather messy computation we proceed a little differently. Observe that G_{16} leaves invariant the three-dimensional subspace spanned by e_3, e_4 and e_5. (It leaves w invariant and therefore the one-dimensional subspace, $w(V_1)$, of V/V_1. The space spanned by e_3, e_4 and e_5 is just the inverse image of $w(V_1)$ in V, and therefore invariant.) Let us call this space W. Since G_{16} leaves W invariant there is a natural projection, χ, of $\operatorname{Hom}(V \wedge V, V)/\mathscr{A}(\operatorname{Hom}(V, g_{16}))$ onto the space $\operatorname{Hom}(W \wedge W, V/W)$. As G_{16} is not the group of all linear transformations leaving W invariant, this projection is not an isomorphism. Nevertheless, we can consider the function $\chi \circ c_{16}$ to reduce the group. (Thus we will use only a "piece" of the structure function to reduce the group. This trick is convenient in many cases.) Since W is three-dimensional, so is $W \wedge W$, and V/W is two-dimensional. An element of $\operatorname{Hom}(W \wedge W, V/W)$ is therefore of rank 0, rank 1 or rank 2. [If $\chi \circ c_{16}$ vanishes identically then the differential system corresponding to W is completely integrable. It then is an elementary consequence of Darboux' and Frobenius' theorems that all such structures are locally equivalent. If $\chi \circ c_{16}$ is of rank 1 identically a case analysis again shows (after a further group reduction) that the situation can be completely handled by Frobenius and Darboux.] The generic case is where $\chi \circ c_{16}$ is of rank 2 everywhere. Notice that the element $\chi \circ c_{16}$ is not arbitrary. By the very construction of $B_{G_{16}}$, the value of $\chi \circ c_{16}$ (which is an element of $\operatorname{Hom}(W \wedge W, V/W)$) must send $e_4 \wedge e_5$ into zero. The reader can verify that G_{16} acts transitively on the (four-dimensional) manifold of all elements of rank 2 of $\operatorname{Hom}(W \wedge W, V/W)$ which vanish on $e_4 \wedge e_5$. Let us pick a fixed element, z, of this manifold: In terms of our basis let $z(e_3 \wedge e_4) = \{e_1\}$ and $z(e_3 \wedge e_5) = \{e_2\}$ where $\{\ \ \}$ this time means coset mod W. The subgroup of G_{16} leaving z fixed is then a 12-dimensional group G_{12}. The reader can easily verify that the matrix form of G_{12} is

$$G_{12}: \begin{pmatrix} \Delta\alpha & \Delta\beta & 0 & 0 & 0 \\ \Delta\gamma & \Delta\delta & 0 & 0 & 0 \\ - & - & \Delta & 0 & 0 \\ - & - & - & \alpha & \beta \\ - & - & - & \gamma & \delta \end{pmatrix} \qquad \text{where} \quad \Delta = \alpha\delta - \beta\gamma$$

and the dashes denote arbitrary elements. The Lie algebra, g_{12}, of G_{12} is the set of all matrices of the form

$$g_{12}: \left(\begin{array}{cc|c|cc} 2a_1 + a_4 & e_2 & 0 & 0 & 0 \\ a_3 & a_1 + 2a_4 & 0 & 0 & 0 \\ \hline - & - & a_1 + a_4 & 0 & 0 \\ \hline - & - & - & a_1 & a_2 \\ - & - & - & a_3 & a_4 \end{array}\right).$$

Before proceeding with the application of our general method to the case of two-dimensional differential systems, let us look at three-dimensional systems. In this case we are given a three-dimensional subspace, W_1, of V and G is the group of all non-singular linear transformations of V leaving W_1 invariant. This group is also 19-dimensional. If we choose a basis $e_1, \ldots, e_5$ of V such that e_3, e_4, and e_5 lie in W_1 then this group corresponds to the group of all non-singular matrices of the form

$$\left(\begin{array}{c|ccc} \sim\sim & 0 & 0 & 0 \\ \sim\sim & 0 & 0 & 0 \\ \hline \sim\sim & & \sim\sim & \\ \sim\sim & & \sim\sim & \\ \sim\sim & & \sim\sim & \end{array}\right).$$

This time, $W_1 \wedge W_1$ is three-dimensional and $\mathrm{Hom}(W_1 \wedge W_1, V/W_1)$ is six-dimensional. Here the group acts transitively on the elements of rank 2 (the single orbit of maximal dimension). Let us choose a fixed element of this space. Now a mapping of rank two on a three-dimensional space vanishes on a line. A line of $W_1 \wedge W_1$ corresponds to a two-dimensional subspace of W_1 (since W_1 is three-dimensional). Then we have picked out a two-dimensional subspace of W_1 invariant under the (13-dimensional) subgroup of G leaving z invariant. In particular, we have associated a two-dimensional differential system to every generic three-dimensional system. If we carry out a similar reduction on this two-dimensional invariant subspace we get G_{12} again. Thus *the problem of classifying generic three-dimensional and two-dimensional differential systems on a five manifold are completely equivalent.*

The reductions from G_{12} down get more complicated. By straightforward but lengthy and messy calculations one can carry out this reduction procedure twice more to get a seven-dimensional group, G_7, whose Lie algebra, g_7, is given by

$$g_7 : \left(\begin{array}{ccc|cc} 2a_1 + a_4 & a_2 & 0 & 0 & 0 \\ a_3 & a_1 + 2a_4 & 0 & 0 & 0 \\ a_5 & a_6 & a_1 + a_4 & 0 & 0 \\ \hline a_7 & 0 & 4/3a_6 & a_1 & a_2 \\ 0 & a_7 & -4/3a_5 & a_3 & a_4 \end{array}\right).$$

No further reductions are possible via the method we have outlined above. To study this problem further one must look at the second-order invariants as described in the next section.

Let us return to the general case. The structure functions, c, for a G-structure, B_G, takes values in

$$\mathrm{Hom}(V \wedge V, V)/\mathscr{A}(\mathrm{Hom}(V, g)).$$

Let us choose a complement, C, to $\mathscr{A}(\mathrm{Hom}(V, g))$ in $\mathrm{Hom}(V \wedge V, V)$. Thus $\mathrm{Hom}(V \wedge V, V) = C + \mathscr{A}(\mathrm{Hom}(V, g))$. If we identify C with the quotient space $\mathrm{Hom}(V \wedge V, V)/\mathscr{A}(\mathrm{Hom}(V, g))$ then c can be regarded as taking values in C.

In this way we obtain a function on B_G with values in $\mathrm{Hom}(V \wedge V, V)$. Let us denote this function by $\bar{c}$. (We repeat that the function $\bar{c}$ depends on the choice of the complement C. We will frequently make such a choice of complement for the various groups, G, that arise.) Equation (2.11), interpreted in terms of $\bar{c}$, becomes

$$\bar{c}(pa) = \sigma(a)\bar{c}(p) + \mathscr{A}(t) \tag{2.13}$$

where t is some element of $\mathrm{Hom}(V, g)$ (depending on p and a).

We should also observe that the choice of the complement, C, picks out a distinguished class of horizontal subspaces at each point, p, of B_G. In fact, by the definition of C and $\bar{c}$ there will exist horizontal subspaces at p such that

$$c_H = \bar{c}(p). \tag{2.14}$$

In general there will be many such horizontal subspaces, H. In fact, if H_1 satisfies (2.14) then so will H_2 whenever $\mathscr{A}(S_{H_1, H_2}) = 0$. This is an immediate consequence of (2.8). We will return to this point in the next section.

Notice that if H satisfies (2.14) at p then $R_{a*}H$ will satisfy (2.14) at pa if and only if $\sigma(a)\bar{c}(p) \in C$. This follows immediately from (2.13). In particular,

PROPOSITION 2.1. *If C is invariant under $\sigma(G)$ (i.e., $\sigma(a)C = C$ for all $a \in G$), then if H satisfies (2.14) so will $R_{a*}H$ for all $a \in G$.*

In order to facilitate the reading of Cartan's works on the equivalence problem it might be convenient to translate some of the main formulae of this section into notation somewhat closer to his. Let us first choose a basis $e_1, \ldots, e_n$ of V. In terms of this basis, the V valued form ω can be regarded as an n-tuplet of real valued linear forms: $\omega = (\omega^1, \ldots, \omega^n)$. Let us choose a local product representation for B_G; i.e., we restrict to a small neighborhood, U, of M and a local identification of $\pi^{-1}(B_G)$ with $U \times G$. Now G, as a Lie group, carries a canonical left invariant g-valued form. Under the projection of $U \times G$ onto G this form pulls back to a form defined (locally) on B_G. Let us denote this form by Π. Let $A_1, \ldots, A_r$ be a basis of g. Then we can write $\Pi = (\Pi^1, \ldots, \Pi^r)$. The forms $\omega^1, \ldots, \omega^n$, $\Pi^1, \ldots, \Pi^r$ then span the cotangent space at each point of B_G. Notice that whereas the forms $\omega^1, \ldots, \omega^n$ are intrinsically defined on B_G, the forms, $\Pi^1, \ldots, \Pi^r$ are not—they depend on the choice of the local product representation. They are not completely arbitrary, however; they are determined on the fiber. In fact, for any $A \in g$ we have $\langle \tilde{A}, \Pi \rangle = A$ so that

$$\langle \tilde{A}_\alpha, \Pi^\beta \rangle = \delta^\beta_\alpha \,. \tag{2.15}$$

In what follows the only property of the Π's that we will use is that they satisfy (2.15). In fact, we shall have occasion to modify the Π's several times during the discussion. Notice that since the ω^i vanish on the fibers, if the Π^β satisfy (2.15) so will $\Pi^\beta + \sum d^\beta_i \omega^i$ for any functions d^β_i.

We should also observe that for a fixed choice of Π's, a horizontal subspace, H is given by $r \times n$ numbers $d^\beta_i(H)$. That is, a tangent vector X at p lies in H if and only if

$$\langle X, \Pi^\beta \rangle = \sum d^\beta_i \langle X, \omega^i \rangle$$

where $d^\beta_i = d^\beta_i(H)$. In this way we have a parametrization of all horizontal subspaces at each point.

Let us define the constants $a^i_{j\alpha}$ by

$$A_\alpha e^i = \sum a^i_{j\alpha} e^j \tag{2.16}$$

(Recall that $e^1, \ldots, e^n$ is a basis of V and that $A_\alpha \in g$ acts on V so (2.16) makes sense.) An immediate transcription of (2.5) yields

$$d\omega^i = \sum c^i_{jk} \omega^1 \wedge \omega^k + \sum a^i_{j\alpha} \omega^j \wedge \Pi^\alpha \tag{2.17}$$

where the c^i_{jk} are some functions on M. To verify (2.17) we observe that since $d\omega$ vanishes on the fiber there can be no terms involving $\Pi^\alpha \wedge \Pi^\beta$. The only thing to check in (2.17) is that $a^i_{j\alpha}$ is the coefficient of $\omega^i \wedge \Pi^\alpha$. But this is exactly what (2.5) says (using (2.15) and (2.16)).

In Equation (2.17) the functions c^i_{jk} depend on the choice of the Π^α. In fact, the values $c^i_{jk}(p)$ of these functions at a point, p, correspond to the c_H introduced earlier. Here the horizontal subspace, H, is the one given by $\langle X, \Pi^\beta \rangle = 0 \quad (\beta = 1, \ldots, r)$ and $c_H = \sum c^i_{jk}(p) e_i \otimes e^{j*} \wedge e^{k*}$ [where we have identified $\text{Hom}(V \wedge V, V)$ with $V \otimes V^* \wedge V^*$] or $c_H(e_j \wedge e_k) = \sum c^i_{jk} e_i$. Replacing H by a second horizontal subspace is the same as replacing Π^β by $\Pi^\beta + \sum d^\beta_i \omega^i$, and this, of course, modifies the c^i_{jk} according to the rule given in (2.7). (The reader can write down the indicial form of this rule if he so desires.)

Now we have already seen that choosing a complement, C, to $\mathscr{A}(\text{Hom}(V, g))$ picks out a distinguished class of subspaces at each point and fixed the function $\bar{c}$. How do we express this in indicial language? Well, choosing the subspace C amounts to imposing a number of linear relations on the c^i_{jk}. (These are just the conditions that $\sum c^i_{jk} e_i \otimes e^{j*} \wedge e^{k*}$ lie in C. Frequently, it is convenient to get these conditions (i.e., choose C) so that they are of the form $c^{i'}_{j'k'} \equiv 0$ for a suitable collection of indices c', j', k'. This isn't always the most convenient choice of C. Frequently the heart of many of Cartan's computations lies in making a clever choice of C.) At any rate, once we have chosen C, the functions c^i_{jk} become well determined. The forms Π^β are not completely determined. They suffer the same ambiguity that the horizontal subspaces satisfying (2.14) do. We can, however, make some choice of Π^β so that (2.15) and (2.17) hold and so that the c^i_{jk} satisfy other linear conditions. The functions c^i_{jk} may be regarded as an indicial form of our structure function, c.

Equation (2.17) shows that the structure function is not arbitrary. That is, not every $\text{Hom}(V \wedge V, V)/\mathscr{A}(\text{Hom}(V, g))$ valued function on a principal G-bundle satisfying (2.11) is the tensor of struction of a G-structure. In fact, if we take the exterior derivative of (2.17) we get an expression involving (sums of exterior products of) the forms dc^i_{jk}, $d\omega^i$, ω^j, Π^α and $d\Pi^\alpha$ which must vanish. We can replace the $d\omega^i$ by their values given by (2.17). We can't say too much about the $d\Pi^\alpha$ (except that (2.15) implies that we know the terms of the form $\Pi^\beta \wedge \Pi^\gamma$ from the structure constant of G.) However the $d\Pi^\alpha$ are always prefaced by a suitable coefficient $a^i_{j\alpha}$. Let us write

$$(2.18) \quad d\Pi^\alpha = \sum_{r,s=1}^{n} A^\alpha_{r s}\omega^r \wedge \omega^s + \sum B^\alpha_{r\beta}\omega^r \wedge \Pi^\beta + \sum \gamma^\alpha_{\beta\gamma}\Pi^\beta \wedge \Pi^\gamma,$$

for suitable functions A^α_{rs} and $B^\alpha_{\beta\gamma}$ and let us proceed to the computation. Taking the exterior derivative of (2.17) gives[7]

$$0 = \sum dc^i_{jk}\omega^j \wedge \omega^k + \sum c^i_{jk}\, d\omega^j \wedge \omega^k - \sum c^i_{jk}\omega^j \wedge d\omega^k$$
$$+ \sum a^i_{j\alpha}\, d\omega^j \wedge \Pi^\alpha - \sum a^i_{j\alpha}\omega^j \wedge d\Pi^\alpha.$$

---

[7] To simplify notation we adopt the convention that all repeated indices under a summation sign are summed. The Latin indices range from 1 to n and the Greek from 1 to r.

We now substitute $d\omega^j$ and $d\omega^k$ from (2.17) and $d\Pi^\alpha$ from (2.18). Let us define the functions $c^i_{jk,l}$ and $c^i_{jk,\beta}$ by

$$dc^i_{jk} = \sum c^i_{jk,l}\omega^l + \sum c^i_{jk,\beta}\Pi^\beta.$$

If we then collect the coefficients of $\omega^j \wedge \omega^k \wedge \omega^l$ $(j < k < l)$ and set this term equal to zero we get

$$\text{(2.19)} \quad \begin{aligned} (c^i_{jk,l} + c^i_{kl,j} + c^i_{lj,k}) + 2(c^i_{rl}c^r_{jk} + c^i_{rk}c^r_{lj} + c^i_{rj}c^r_{kl}) \\ - (a^i_{j\alpha}A^\alpha_{kl} + a^i_{k\alpha}A^\alpha_{lj} + a^i_{l\alpha}A^\alpha_{jk}) = 0. \end{aligned}$$

If we collect the coefficients of $\omega^j \wedge \omega^k \wedge \Pi^\alpha$ $(j < k)$ and set this term equal to zero we obtain

$$\text{(2.20)} \qquad c^i_{jk,\alpha} + 2(a^r_{j\alpha}c^i_{rk} + a^r_{k\alpha}c^i_{rs} - a^i_{r\alpha}c^r_{jk}) - a^i_{j\beta}B^\beta_{k\alpha} + a^i_{j\beta}B^\beta_{k\alpha} = 0.$$

Comparing the coefficients of $\omega^j \wedge \Pi^\alpha \wedge \Pi^\beta$ just gives the equations of structure of g back again. The astute reader should observe (exercise!) that (2.20) does not really give us any new information. It is essentially a consequence of (2.11) and (2.8). Equation (2.19) does give us information that we hadn't obtained before. Equation (2.19) should probably be called the *Bianchi-Cartan identities*.

Let us see what the Bianchi-Cartan identities imply for the case of a constant structure function. In this case the c^i_{jk} are constant so that the terms $c^i_{jk,l}$ vanish and (2.19) becomes

$$\text{(2.21)} \qquad 2(c^i_{rl}c^r_{jk} + c^i_{rk}c^r_{lj} + c^i_{rj}c^r_{kl}) = a^i_{j\alpha}A^\alpha_{kl} + a^i_{k\alpha}A^k_{lj} + a^i_{l\alpha}A^\alpha_{jk}.$$

We thus obtain a set of quadratic necessary conditions on the c^i_{jk} [namely the existence of suitable A^α_{kl} so that (2.21) holds]. Let us formulate this condition in index-free language in terms of the function $\bar{c}$ (which is assumed to be a constant).

We first observe that there is a natural bilinear map, λ, of $\text{Hom}(V \wedge V, V) \times \text{Hom}(V \wedge V, V)$ into $\text{Hom}(V \wedge V \wedge V, V)$. It is defined as follows: For $C_1, C_2 \in \text{Hom}(V \wedge V, V)$

$$\begin{aligned} \lambda(C_1, C_2)(v_1 \wedge v_2 \wedge v_3) &= C_1(C_2(v_1 \wedge v_2) \wedge v_3) \\ &\quad + C_1(C_2(v_2 \wedge v_3) \wedge v_1) + C_1(C_2(v_3 \wedge v_1) \vee v_2). \end{aligned}$$

The reader should verify that the left-hand side of (2.21) is just the representation in terms of the basis $e_1, \ldots, e_n$ of the element $\lambda(\bar{c}, \bar{c})$. On the other hand, let us consider the space $\text{Hom}(V \wedge V, g)$. There is a natural linear mapping, μ, of this space into $\text{Hom}(V \wedge V \wedge V, V)$ defined by

$$\mu(\tau)(v_1 \wedge v_2 \wedge v_3) = \tau(v_1 \wedge v_2)v_3 + \tau(v_2 \wedge v_3)v_1 + \tau(v_3 \wedge v_1)v_2$$

for any $\tau \in \text{Hom}(V \wedge V, g)$ and $v_1, v_2, v_3 \in V$. In terms of the bases of V and g that we have chosen, an element of $\text{Hom}(V \wedge V, g)$ is given by its coefficients A^α_{kl}. Then to say that there exist A^α_{kl} so that (2.21) is the same as saying that there is a $\tau \in \mathrm{Hom}(V \wedge V, g)$ such that $\lambda(\bar{c}, \bar{c}) = \mu(\tau)$. Thus (2.21) becomes

$$\lambda(\bar{c}, \bar{c}) \in \mu(\mathrm{Hom}(V \wedge V, g)). \tag{2.22}$$

Let us see what all of this reduces to in the case of the simplest G-structure—the $\{e\}$-structure of a Lie group. Then V is the Lie algebra of the Lie group and the $c = \bar{c} \in \mathrm{Hom}(V \wedge V, V)$ is just the bracket operation

$$c(v_1 \wedge v_2) = [v_1, v_2]$$

on this algebra. In this case $g = 0$ so that condition (2.21) says $\lambda(\bar{c}, \bar{c}) = 0$. This just says $[[v_1, v_2]v_3] + [[v_2, v_3]v_1] + [[v_3, v_1]v_2] = 0$. Thus in the case of the $\{e\}$-structure of a Lie group, the Bianchi-Cartan identities reduce to the Jacobi identities.

3. Prolongations. In the previous section we studied the equivalence problem for G-structures via the first-order structure function, c. We now wish to point out that for certain groups, G, the first-order structure function gives no information whatsoever; in fact, the structure function vanishes identically for all G-structures. Since c takes values in

$$\mathrm{Hom}(V \wedge V, V)/\mathscr{A}(\mathrm{Hom}(V, g))$$

it will suffice to exhibit algebras for which $\mathscr{A}(\mathrm{Hom}(V, g)) = \mathrm{Hom}(V \wedge V, V)$.[8] This will happen when the dimension of $\mathrm{Hom}(V, g)$ is large and the dimension of the kernel of $\mathscr{A}$ is small. We are thus led to consider the kernel of $\mathscr{A}$ which we shall denote by $g^{(1)}$. Since this space will be of fundamental importance in what follows we give a slightly more general definition:

DEFINITION 3.1. *Let V and W be vector spaces and let g be a subspace of* $\mathrm{Hom}(V, W)$. *Let $g^{(1)}$ be the subspace of* $\mathrm{Hom}(V, g)$ *consisting of those T such that*

$$T(u)v = T(v)u \tag{3.1}$$

for all $u, v \in V$. (Recall that $T(u) \in g$ so that $T(u)v \in V$.) The space $g^{(1)}$ is called the first prolongation of g. Set $g^{(2)} = g^{(1)(1)}$ and, in general, $g^{(i+1)} = g^{(i)(1)}$. The space $g^{(i)}$ is called the i'th prolongation of g.

[8] A trivial example where this happens is when $g = gl(V)$ (the algebra of all linear transformations of V). This isn't too interesting since all $GL(V)$ structures are locally equivalent. We wish to exhibit cases where the first-order structure function cannot distinguish between two non-equivalent structures.

In the case at hand, $V = W$ and g is actually a subalgebra, not merely a subspace, of $\mathrm{Hom}(V, V)$. The definition of $\mathscr{A}$ shows that $g^{(1)} = \mathrm{Ker}\,\mathscr{A}$.

Exercise 3.1. If we identify $\mathrm{Hom}(V, W)$ with $W \otimes V^*$ then g is a subspace of $W \otimes V^*$ and $\mathrm{Hom}(V, g)$ can be regarded as a subspace of $W \otimes V^* \otimes V^*$. Show that then

$$g^{(1)} = \mathrm{Hom}(V, g) \cap W \otimes S^2(V).$$

Exercise 3.2. Under the same kind of identification show that

$$g^{(k)} = g \otimes \underbrace{V^* \otimes \cdots \otimes V^*}_{k\text{-times}} \cap W \otimes S^{k+1}(V^*).$$

Exercise 3.3. On a vector space V, any vector field can be regarded as a V-valued function. We can therefore talk of the Jacobian of a vector field (i.e., the Jacobian of the corresponding map of $V \to V$). Under the identification of the tangent space at every point of V with V itself, we can regard this Jacobian as a function on V with values in $\mathrm{Hom}(V, V)$. We have thus associated a $\mathrm{Hom}(V, V)$-valued function, which we shall denote by $\mathrm{Jac}(X)$ to each vector field, X, on V. (In terms of a basis, if $X = (X^1, \ldots, X^n)$ then the matrix of $\mathrm{Jac}(X)$ is $(\partial X^i/\partial x^j)$.) Show that the vector field X is an infinitesimal automorphism of the standard flat G-structure on V if and only if $\mathrm{Jac}(X)(v) \in g$ for all $v \in V$.

Exercise 3.4. Given any vector field, X, on V regarded as a V-valued function, we can compute its Taylor expansion about any point. The terms of degree $k + 1$ in the Taylor expansion can be regarded as an element of $V \otimes S^{k+1}(V^*)$. Show that if X is an infinitesimal automorphism of the flat g-structure then the term of degree $k + 1$ in its Taylor expansion about any point) lie in $g^{(k)}$. Conclude that if $g^{(k)} = 0$ for some k then the algebra of all infinitesimal automorphisms of the flat G-structure is finite-dimensional.

Definition 3.2. *A subspace, g, of* $\mathrm{Hom}(V, W)$ *such that $g^{(k)} = 0$ for some (and hence all larger) k is called of finite type. If $g^{(k)} \neq 0$ for all k then g is said to be of infinite type.*

Let us illustrate the concept of prolongations by examining several algebras. Let us first consider the case of the orthogonal algebra, $o(n)$.

In this case we have a symmetric, non-degenerate[9] bilinear form, $(\, , \,)$, on V. The algebra $o(n)$ consists of those $A \in gl(V)$ such that

$$(Au, v) + (u, Av) = 0 \qquad \text{for all } u, v \in V. \tag{3.2}$$

PROPOSITION 3.1. *For the algebra* $o(n)$ *we have*

$$o(n)^{(1)} = 0,$$

i.e., its first prolongation vanishes.

If we introduce a basis, the proof of Proposition 3.1 reduces to Lemma 1.2 of Chapter VI. For convenience, we reproduce the proof here. Let u, v, and w be any three vectors in V, and let $T \in o(n)^{(1)}$. Then

$$\begin{aligned}
(T(u)v, w) &= (T(v)u, w) && \text{(by (3.1))}\\
&= -(u, T(v)w) && \text{(by (3.2))}\\
&= -(u, T(w)v) && \text{(by (3.1))}\\
&= (T(w)u, v) && \text{(by (3.2))}\\
&= (T(u)w, v) &&\\
&= (v, T(u)w) && \text{(by symmetry of } (\, , \,))\\
&= -(T(u)v, w). &&
\end{aligned}$$

Thus $(T(u)v, w) = -(T(u)v, w) = 0$. Since w is arbitrary and $(\, , \,)$ is non-singular, $T(u)v = 0$. Thus $T(u) = 0$ for all u, i.e., $T = 0$.

Let us derive some important consequences from Proposition 3.1. Observe that $\dim \operatorname{Hom}(V \wedge V, V) = n \times (n(n-1)/2)$. On the other hand, $\dim o(n) = n(n-1)/2$.[10] Therefore $\dim \operatorname{Hom}(V, o(n)) = n^2(n-1)/2 = \dim \operatorname{Hom}(V \wedge V, V)$. Since $\ker \mathscr{A} = o(n)^{(1)} = 0$ we can conclude that

$$\mathscr{A}(\operatorname{Hom}(V, o(n))) = \operatorname{Hom}(V \wedge V, V). \tag{3.3}$$

In particular, the first-order structure function gives absolutely no information in the case of Riemannian geometry.

We can also derive an important positive consequence from (3.3). In fact, in this case there is no problem with the choice of a complement

[9] The algebra $o(n)$ comes from a positive definite bilinear form. All our present discussion, however, is valid for any non-degenerate symmetric form, not necessarily positive definite.

[10] *Proof*: The form $(\, , \,)$ allows an identification of V with V^* and those of $gl(V) = V \otimes V^*$ with $V^* \otimes V^*$: The element $A \in gl(V)$ goes into the bilinear form $\hat{A}$ where $\hat{A}(u, v) = (Au, v)$. Then (3.2) shows that $o(n)$ goes over to $V^* \wedge V^*$ thus $\dim o(n) = n(n-1)/2$.

C, to $\mathscr{A}(\mathrm{Hom}(V, o(n)))$, it is the zero subspace. By the discussion at the end of Section 2, at each point of $B_{o(n)}$ ($= \mathcal{O}(M)$) there is a horizontal subspace, H, such that $c_H = 0$. By (2.11) and Proposition 3.1 this horizontal subspace is unique. According to (2.11) this choice of horizontal subspace at each point is a connection. We have thus proved,

THEOREM 3.1. (Levi-Civita). *Given any Riemann manifold, M, there is a canonical connection on $\mathcal{O}(M)$. The horizontal subspace, H, at each point is characterized by $c_H = 0$.*

To give a connection on $\mathcal{O}(M)$ is the same as giving an $o(n)$-valued form. In terms of the standard basis of $o(n)$, this amounts to giving the forms Ω_M^{ij}. This completes the discussion on page 250. The reader can verify that Equation (2.17) reduces, in this case, to Equation (3.8) of Chapter VI.

EXERCISE 3.5. Compare Equation (5.10) of Chapter VI with Equation (1.6) (the structure equation of a connection). Conclude that the form $\sum_{k,l} A_{ik}\mathcal{O}^k \wedge A_{jl}\mathcal{O}^l$ is $\rho^*(D\Omega)$ in terms of a basis. (The map ρ is defined on page 249.) Thus Levi-Civita's theorems gives a new proof and an interpretation of Gauss' *theorem egregium.*

The following exercises illustrate the notion of prolongation with various other algebras.

EXERCISE 3.6. Let us consider the Lie algebra, $co(n)$, of the conformal group. It consists of all $A \in gl(V)$ satisfying

$$(Au, v) + (u, Av) = \lambda_A(u, v)$$

for all $u, v \in V$ where (,) is a non-degenerate symmetric bilinear form. Map $co^{(1)} \to V^*$ by sending each $T \in co(n)^{(1)}$ into the linear form $u \to \lambda_{T(u)}$. Show that this is an isomorphism, i.e., we can identify $co(n)^{(1)}$ with V^*.

EXERCISE 3.7. Now let us consider $co(n)^{(2)}$. For each $S \in g^{(2)}$ and $u, v \in V$, the linear transformation $S(u, v)$ lies in $co(n)$. Therefore for any pair of vectors $x, y \in V$

$$(S(u, v)x, y) + (x, S(u, v)y) = \lambda_{uv}(x, y)$$

where λ_{uv} is a symmetric bilinear form depending on S. If λ_{uv} vanishes for all u and v then $S \in o(n)^{(2)}$ and therefore $S = 0$. Show that $\lambda(u, v) = 0$ if $\dim V \geqq 3$. [*Hint*: since $\lambda(u, v) = \lambda(v, u)$ it suffices to show that $\lambda(u, u) = 0$. Let u and x be a pair of orthogonal vectors $-(u, x) = 0$. Then

$$\begin{aligned}\lambda(u, u)(x, x) &= 2(S(u, u)x, x) = 2(S(u, x)u, x)\\ &= -2(S(u, x)x, u)\\ &= -2(S(x, x)u, u)\\ &= -\lambda(x, x)(u, u).\end{aligned}$$

In particular, if u and x are orthogonal unit vectors $((u, u) = (x, x) = 1)$ we have $\lambda(u, u) = -\lambda(x, x)$. If there were three or more pairwise orthogonal unit vectors this would imply that $\lambda(u, u) = 0$. Thus $co(n)$ is of finite type if $n \geqq 3$.] The algebra $co(2)$ is of finite type.

EXERCISE 3.8. Let (,) be a non-degenerate antisymmetric bilinear form on V and $sp(n)$ the algebra of all matrices satisfying

$$(Au, v) + (u, Av) = 0.$$

The bilinear form (,) induces an identification of V with V^* and therefore of $V \otimes V^*$ with $V^* \otimes V^*$. Show that under this identification, the algebra $sp(n) \subset V \otimes V^*$ $(= \mathrm{Hom}(V, V))$ goes over into the space of symmetric tensors, $S^2(V^*)$. More generally, we have an identification of $V \otimes S^{k+1}(V^*)$ with $V^* \otimes S^{k+1}(V^*)$. Show that under this identification $sp(n)^{(k)}$ goes over into $S^{k+2}(V^*)$—thus $sp(n)$ is of infinite type.

EXERCISE 3.9. Let (,) and V be as in Exercise 3.8. Let g be the "conformal symplectic algebra" consisting of all $A \in gl(V)$ satisfying

$$(Ax, y) + (x, Ay) = \lambda_A(x, y).$$

Show that $g^{(1)} = sp(n)^{(1)}$.

We now return to general considerations. We study a G-structure and assume that we have chosen a complement, C, once and for all for G. Then at each point of B_G we have a distinguished class of horizontal subspaces: The H satisfying $c_H \in C$. Now any horizontal subspace, H, at a point $p \in B_G$ induces an isomorphism of $T_p(B_G)$ with $V + g$. In fact, we already have an isomorphism of the tangent space to the fiber with g. Since ω induces an identification of H with V we get an isomorphism of $T_p(B_G)$ with $V + g$. To say that we have a family of horizontal subspaces at p means that we have a family of frames of B_G at p. How do two such frames differ from each other? According to (2.8), if H_1 and H_2 both satisfy $c_{H_i} \in C$ then $S_{H_1, H_2} \in g^{(1)}$. Let z_1 and z_2 be the frames corresponding to H_1 and H_2. By definition, for any $v \in V$, we have

$$z_1(v) = z_2(v) + S_{H_1, H_2}(v).$$

On the other hand, $z_1(A) = z_2(A)$ for any $A \in g$. We therefore make the following definition:

DEFINITION 3.3. *We call $G^{(1)}$ the group of all linear transformations of $V + g$ of the form a_T (for $T \in g^{(1)}$) where*

$$a_T(A) = A \qquad \textit{for } A \in g$$

(3.4) *and*

$$a_T(v) = v + Tv \qquad \textit{for } v \in V.$$

In terms of a "matrix" representation, $G^{(1)}$ consists of all linear transformations of the form

$$\begin{array}{cc} & \begin{array}{cc} V & g \end{array} \\ \begin{array}{c} V \\ g \end{array} & \begin{pmatrix} I & T \\ 0 & I \end{pmatrix}. \end{array}$$

We can now assert

THEOREM 3.2. *A choice of complement, C for the algebra g picks out a $G^{(1)}$ structure on each G-structure B_G called the first prolongation of B_G. In order for two G-structures, B_G^1 and B_G^2 to be locally equivalent it is necessary and sufficient that the $G^{(1)}$ structures over B_G^1 and B_G^2 be locally equivalent (as $G^{(1)}$ structures).*

We have already established the first assertion. Let B_G^1 and B_G^2 be two G-structures over M_1 and M_2 and let φ be an isomorphism of B_G^1 onto B_G^2. If H_1 is a horizontal subspace at $p_1 \in B_G^1$ and $H_2 = (\varphi_*)_*(H_1)$ is the corresponding subspace at $p_2 \in B_G^2$ then $c_{H_2} = c_{H_1}$. In particular, $c_{H_2} \in C$ if and only if $c_{H_1} \in C$. Thus $(\varphi_*)_*$ maps the frames of the $G^{(1)}$ structure on B_G^1 onto those of the $G^{(1)}$ structure over B_G^2.

Conversely, let ψ be a diffeomorphism of B_G^1 onto B_G^2 which is a $G^{(1)}$ isomorphism. Then $\psi_*(\tilde{A}^1) = \tilde{A}^2$ where $\tilde{A}^1$ and $\tilde{A}^2$ are the vector fields on B_G^1 and B_G^2 corresponding to an $A \in g$. In particular, φ commutes with the action of G on B_G. It therefore defines a differentiable map, φ, of M_1 onto M_2, such that

$$\varphi \circ \pi_1 = \pi_2 \circ \psi \tag{3.5}$$

where π_1 and π_2 are the projections of B_G^1 and B_G^2 onto M_1 and M_2. Let p_1 be a point of B_G^1 (i.e., a frame of M) and H_1 be a horizontal subspace at p_1. To say that ψ is a $G^{(1)}$ isomorphism implies that the isomorphism of V with H_1 carries over to the isomorphism of V with $H_2 = \psi_*(H_1)$.

Now for $v \in V$ the frame corresponding to H_1 of the $G^{(1)}$ structure assigns to v a vector $X \in H_1$ with $\pi_{1*}(X) = p_1(v)$ [here p_1 is regarded as an isomorphism of V with $T_{\pi_1(p)}(M)$]. Similarly the frame corresponding to H_2 assigns to v a vector $Y \in H_2$ such that $\pi_{2*}(Y) = p_2(v)$. As ψ is to be an isomorphism of the $G^{(1)}$ structures we must have $\psi_*(X) = Y$. Now (3.5) implies $\varphi_* \circ \pi_{1*} = \pi_{2*} \circ \psi_2$ so that $\varphi_* p_1(v) = p_2(v)$. But this just says that $\varphi_*(p_1) = p_2$ or in other words $\varphi_* = \psi$, proving Theorem 3.2.

We can now apply all the methods of Section 2 to determine whether the two $G^{(1)}$ structures are equivalent, i.e., construct its structure function, reduce the groups, etc.

Before discussing this procedure we remark that a slight adaptation of our method allows us to consider "higher order" structures. Instead of considering the full $G^{(1)}$ structure associated to B_G we might study a substructure. This amounts to imposing a "second order" condition (corresponding to a subgroup of $G^{(1)}$) in addition to the "linear" conditions imposed by the G-structure B_G. A typical illustration would be to pick out an $\{e\}$-structure as a reduction of the $G^{(1)}$ structure. This amounts to picking one horizontal subspace at each point of B_G out of the distinguished collection. This might be done for instance by picking a connection on B_G. Thus the theory of G-connections can be subsumed under our method. Another case of second-order structure is the so-called projective differential geometry. For a treatment of these subjects from the present point of view see Guillemin [8].

Let us see what the structure function of the first prolongation looks like. For this we must examine the Lie algebra of $G^{(1)}$. From Definition 3.3 it is clear that it consists of all linear transformations of the form A_T (for $T \in g^{(1)}$) where

$$A_T(A) = 0 \qquad \text{for } A \in g$$

and

$$A_T(v) = T(v) \qquad \text{for } v \in V.$$

In short, the Lie algebra of $G^{(1)}$ can be regarded as $g^{(1)}$ where a $T \in g^{(1)}$ acts on $V + g$ by acting trivially on g, i.e., $T(A) = 0$ for $A \in g$. In other words, we have a specific representation of $g^{(1)}$ on $V + g$. The new structure function lies in a quotient space of

$$\mathrm{Hom}((V + g) \wedge (V + g), V + g).$$

Again, as on page 327, if we choose a complement we can regard the structure function as lying in a subspace of

$$\mathrm{Hom}((V + g) \wedge (V + g), V + g).$$

Now an element of this space gives, via the direct sum decomposition of $V + g$, a number of pieces. First of all, we get an element of $\mathrm{Hom}(V \wedge V, V)$. The reader should verify as an instructive exercise that this is nothing other than the old structure function of B_G lifted up to the $G^{(1)}$ structure (which is after all a bundle with a projection onto B_G). We also get an element of $\mathrm{Hom}(g \wedge g, g)$. The reader can verify that this is just the Lie bracket operation on g. What is "new" are two bilinear maps—an element of $\mathrm{Hom}(V \otimes g, g)$ and an element of $\mathrm{Hom}(V \wedge V, g)$.

EXERCISE 3.10. Show that if $g^{(1)} = 0$ and the $G^{(1)}$ structure is a connection on B_G (as is in the case of Riemannian geometry) the "new" element lying in $\mathrm{Hom}(V \wedge V, g)$ can be regarded as the curvature of this connection.

We can continue the procedure. We thus obtain $G^{(2)}, G^{(3)}, \ldots, G^{(k)} \ldots$ structures which are canonically associated to the original G-structures. They are called the second, third...kth prolongation of B_G. In this way we obtain a sequence of refined necessary conditions for the equivalence problem. The Lie algebra of $G^{(k)}$ is a representation of $g^{(k)}$. In particular, if $g^{(k+1)} = 0$ then $G^{(k+1)} = \mathrm{id}$. The $G^{(k+1)}$ structure is on the kth prolongation if B_G is an $\{e\}$-structure, i.e., a complete parallelism.

Thus the equivalence problem for structures of finite type can be reduced to the equivalence problem for $\{e\}$-structures by passing to a suitable prolongation.

As an application of the preceding, let us consider the problem of flatness. As is easily seen, the structure function of all the prolongations of the flat G-structures are constant. [The crucial pieces of these functions vanish. For instance, on the first prolongation the structure function (which is constant) is an element of

$$\mathrm{Hom}((V + g) \wedge (V + g), V + g).$$

The parts lying in $\mathrm{Hom}(V \wedge V, V)$ and $\mathrm{Hom}(V \otimes g, g)$ vanish. The term in $\mathrm{Hom}(g \wedge g, g)$ is the structure constant of g and will not vanish. However, this piece is the same for all G-structures and is thus uninteresting.] We now have

THEOREM 3.3. *Let G be a group of finite type. A necessary and sufficient condition for a G-structure to be locally flat is that the structure function of all prolongations of G be constant and equal to the corresponding structure constants of the flat G-structure.*

(The equality in Theorem 3.3 need only be verified up to that k for which $g^{(k)} = 0$.)

The necessity is an immediate consequence of Theorem 3.2 and Corollary 2.1. The proof of the sufficiency goes as follows: At the kth stage we get two complete parallelisms with the same constant structure function. We now simply apply Theorem 2.4 of Chapter V. Here the group is the group of all automorphisms of the flat G-structure and the manifold is the kth prolongation of B_G.

It should be remarked that Theorem 3.3 is valid under far more general hypotheses than the assumption that G is of finite type. For instance, it has also been proved (in the paper by Singer and Sternberg mentioned in the Preface) if G is completely reducible or for any G if the G-structure B_G is assumed to be analytic. It is to be *conjectured* that the theorem is true for all groups G and all G-structures B_G but I know of no proof as of yet.

In the next section we shall study complete parallel structures and therefore structures of finite type in some detail. We close this section with the remark that in dealing with the equivalence problem a fruitful method is to try to reduce the group to one of finite type by the methods of Section 2 and then to apply the procedure of this and the next section.

Thus the group G_7 obtained in Section 2 from the study of distributions on 5 space is of finite type. In fact, $g_7^{(2)} = 0$. On the other hand the initial group was of infinite type.

4. Structures of finite type. We have seen in the last section that the equivalence problem for structures of finite type can be reduced to the local equivalence problem for $\{e\}$-structures, i.e., complete parallelisms. We are therefore led to the study of the equivalence problem for complete parallelisms (or in Cartan's terminology the *restricted* equivalence problem). We will solve this problem (under fairly general assumptions) by using the methods of Section 2, mainly Theorem 2.1. In our present situation $g = 0$ so that the structure function, c, takes values in $\mathrm{Hom}(V \wedge V, V)$. Also we can identify $B_{\{e\}}$ with the base manifold, M, so that *c can be regarded as a map of M into* $\mathrm{Hom}(V \wedge V, V)$. Let us examine the two extreme cases: 1) c is a constant or 2) c is an immersion of M into $\mathrm{Hom}(V \wedge V, V)$. In the first case we have already handled the local equivalence problem: *If M_1 and M_2 are two manifolds with complete parallelisms and the structure function of M_1 is constant then they are locally equivalent if and only if the structure function of M_2 is constant and equal to that of M_1.* The necessity is just Theorem 2.1 and the sufficiency follows from Theorem 2.7 of Chapter V. (The proof there just uses the constancy of the c^i_{jk}.) Let us now examine the other extreme (and generic) case, let us assume that c is an immersion of M into $\mathrm{Hom}(V \wedge V, V)$. In the neighborhood of a point, p, of M, let us choose an n-dimensional subspace, W, of $\mathrm{Hom}(V \wedge V, V)$ so that a projection, ρ, of $c(M)$ onto W is non-singular in a neighborhood, U, of $c(p)$.

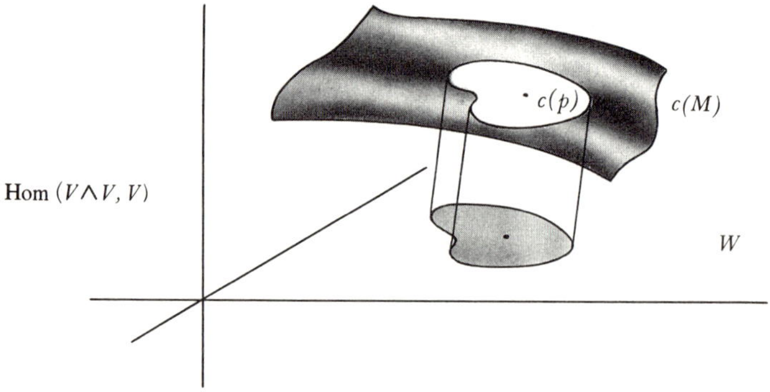

In this way we get a distinguished chart $(U, \rho \circ c)$ about p. Now let M' be a second manifold with a parallel structure and assume that φ is a local isomorphism of M with M' defined on U with $\varphi(p) = q$. By Theorem 2.1 we must have $c' \circ \varphi = c$, where c' is the structure function of M'. Thus $c'(\varphi(U)) = c(U)$ and, in particular c' is an immersion at q and ρ is non-singular about q. We can thus write $(\rho \circ c') \circ \varphi = \rho \circ c$. Since $\rho \circ c$ and $\rho \circ c'$ are diffeomorphisms, we can write $\varphi = (\rho \circ c')^{-1} \circ (\rho \circ c)$. In particular φ is uniquely determined. We can thus formulate the conditions for local equivalence in the generic case as follows:

Let M have a complete parallelism whose structure function, c, is an immersion about a point p. Choose an n-dimensional subspace W with a projection, ρ, of $\mathrm{Hom}(V \wedge V, V)$ *onto W which is non-singular on $c(M)$ near $c(p)$. Let M' be a second parallel structure and $q \in M'$. In order for there to exist a local isomorphism, φ, of $M_1 \to M_2$ with $\varphi(p) = q$ the structure function, c', of M' must be an immersion at q. Furthermore, ρ must be non-singular on $c'(M')$ near q. If these conditions are satisfied, the only possibility for φ is $\varphi = (\rho \circ c')^{-1} \circ (\rho \circ c)$. In order for this map φ to be a local isomorphism we must have $c' \circ \varphi = c$.*

That these conditions are sufficient will be seen later in a more general setting.

We should remark that for the study of G-structures of finite type, we cannot be satisfied with the last result despite the fact that it applies to the generic parallel structure. In fact, in the $\{e\}$-structure that arises by successive prolongations of a finite G-structure, the structure function will have certain pieces (namely the structure constants of the Lie algebras involved) which are constant. Thus the structure function will not be an immersion.

In the "intermediate" case, where c is neither an immersion nor a constant we must examine the "higher order" structure functions as in Section 3. In order to present the results in a form most useful for computation we will use indicial notation and introduce a few definitions. Let us choose a basis for V so that the V-valued form ω giving the complete parallelism on M can be written as $\omega = (\omega^1, \ldots, \omega^n)$. For any function, f, on M, we can write

$$df = \sum f_i \omega^i. \tag{4.1}$$

The function f_i is called the ith *covariant derivative of f*. If we have a function depending on some indices we will preface the index indicating covariant differentiation by a comma. We will thus write, for instance,

$$dc^i_{jk} = \sum_l c^i_{jk,l} \omega^l. \tag{4.2}$$

Let $\mathscr{F}$ be any collection of functions on M. By the *rank* of $\mathscr{F}$ at $p \in M$, denoted by $r_p(\mathscr{F})$, is meant the dimension of the subspace of $T^*_p(M)$ spanned by df for $f \in \mathscr{F}$. In classical language, $r_p(\mathscr{F})$ is the degree of functional independence of $\mathscr{F}$ at p. The family $\mathscr{F}$ is called *regular* at p if $r_q(\mathscr{F}) = r_p(\mathscr{F})$ for all q in a neighborhood of p. Suppose that $\mathscr{F}$ is regular at p with $r_p(\mathscr{F}) = k$. Then we can choose $f_1, \ldots, f_k \in \mathscr{F}$ so that $df_1, \ldots, df_k$ are independent in a neighborhood of p. Extend the f_i to a coordinate system about p. That is, choose a coordinate system $x^1, \ldots, x^n$ about p such that $x^1 = f_1, \ldots, x^k = f_k$. For any function $f \in \mathscr{F}$, df must be a linear combination of $dx^1, \ldots, dx^k$, by the regularity of $\mathscr{F}$. Thus every $f \in \mathscr{F}$ can be written as $f = f(x^1, \ldots, x^k)$. To resume, we have established

LEMMA 4.1. *Let $\mathscr{F}$ be a regular family of functions of rank k at p. We can then choose coordinates $x^1, \ldots, x^n$ about p so that every $f \in \mathscr{F}$ is of the form $f = f(x^1, \ldots, x^k)$. Such a coordinate system is called a coordinate system adapted to $\mathscr{F}$.*

Let us now return to the study of complete parallelisms. In terms of our basis of V, the structure function c takes the form c^i_{jk} where

$$d\omega^i = \sum c^i_{jk}\omega^j \wedge \omega^k.$$

We thus get a family of $n(n(n-1))/2$ functions on M. We denote this family of functions by $\mathscr{F}_0$. Let $\mathscr{F}_1$ denote the family of $(n(n-1)/2)(n + n^2)$ functions given by

$$\mathscr{F}_1 = \{c^i_{jk}, c^i_{jk,l}\}$$

where the $c^i_{jk,l}$ are given by (4.2). More generally, let $\mathscr{F}_s$ be the family of $(n(n-1)/2)(n + n^2 + \cdots + n^s)$ functions,

$$\mathscr{F}_s = \{c^i_{jk}; c^i_{jk,i_1}; c^i_{jk,i_1,i_2}; \ldots c^i_{jk,i_1,\ldots,i_s}\}.$$

Let us set

$$k_s(p) = r_p(\mathscr{F}_s). \tag{4.3}$$

Since $\mathscr{F}_s \subset \mathscr{F}_{s+1}$ we can write

$$0 \leqslant k_0(p) \leqslant k_1(p) \leqslant \cdots \leqslant k_s(p) \leqslant n. \tag{4.4}$$

LEMMA 4.2. *Suppose that $\mathscr{F}_{s+1}$ is regular at p. If $k_s(p) = k_{s+1}(p)$ then $k_s(p) = k_u(p)$ for all $u > s$.*

Proof: Let $k = k_s(p)$ and let $x^1, \ldots, x^n$ be an adapted coordinate system at p. Since $k_s(p) = k_{s+1}(p) = k$ we may assume further that

$x^1, \ldots, x^k$ are elements of $\mathscr{F}_s$ (not merely elements of $\mathscr{F}_{s+1}$). Define the functions $x^i_{,j}$ by

$$dx^i = \sum x^i_{,j}\omega^j.$$

By the definition of $\mathscr{F}_{s+1}$, the functions $x^i_{,j}$ lie in $\mathscr{F}_{s+1}$. We can therefore write

$$x^i_{,j} = x^i_{,j}(x^1, \ldots, x^k).$$

Now for any $f \in F_{s+1}$ we have $f = f(x^1, \ldots, x^k)$ so that

$$df = \sum_1^k \frac{\partial f}{\partial x^i}\, dx^i = \sum_{j=1}^n \left(\sum_{i=1}^k \frac{\partial f}{\partial x^i}\, x^i_{,j}\right)\omega^j$$

or

$$f_{,j} = \sum_{i=1}^k \frac{\partial f}{\partial x^i}\, x^i_{,j}.$$

All the functions on the right-hand side of this equation are functions of $x^1, \ldots, x^k$. Thus for any $f \in F_{s+1}$, all the covariant derivatives of f are functions of $x^1, \ldots, x^k$. Thus the rank of F_{s+2} is k. Proceeding in this way by induction yields Lemma 4.2.

Definition 4.1. *Let $\omega^1, \ldots, \omega^n$ define a complete parallelism on M. Let p be a point of M. If, for some s, $\mathscr{F}_{s+1}$ is regular at p and $k_s(p) = k_{s+1}(p)$ then the parallelism is said to be regular at p. The smallest such s is called the order, $r(p)$, of the parallelism at p and the integer $k(p) = k_s(p)$ is called the rank of the parallelism at p. A chart adapted to $\mathscr{F}_s$ at p will simply be called an adapted chart at p.*

It should be remarked that if p is a "generic" point of M then $\mathscr{F}_i$ will be regular at p for all i. In this case, the only way that i can be less than the order of the parallelism at p is for $k_i(p)$ to be less than $k_{i+1}(p)$. Thus for such a point p we can write

$$r(p) \leqslant k(p) \leqslant n.$$

This means that in the study of "generic" G-structures of finite type only a finite number of differentiations are necessary for the equivalence problem and this integer is determined by G.

Let $\omega^1, \ldots, \omega^n$ be a regular complete parallelism on M, of order r and rank k. Each of the functions of $\mathscr{F}_r$ has a determined set of indices.

In particular, if $x^1, \ldots, x^n$ is an adapted chart, one can consider the indices of each of the functions $x^1, \ldots, x^k$ of $\mathscr{F}_s$. Each of the other functions of $\mathscr{F}_s$ is a function of $x^1, \ldots, x^k$. We are now in a position to state necessary and sufficient conditions for equivalence.

THEOREM 4.1. *Let M be a manifold with a complete parallelism given by $\omega^1, \ldots, \omega^n$ which is regular of order r and rank k at $p \in M$. Let $\bar{M}$ be a second manifold with a complete parallelism given by $\bar{\omega}^1, \ldots, \bar{\omega}^n$ and let $\bar{p}$ be a point of $\bar{M}$. In order for there to be a local isomorphism, φ, of $M \to \bar{M}$ carrying p into $\bar{p}$ it is necessary and sufficient that all of the following conditions be satisfied:*

i) *the complete parallelism on $\bar{M}$ is of order r and rank k at $\bar{p}$.*

ii) *Let $x^1, \ldots, x^n$ be an adapted chart at p. Let $\bar{x}^1, \ldots, \bar{x}^k$ be the functions of $\bar{\mathscr{F}}_r$ with the same index as $x^1, \ldots, x^k$. Then $\bar{x}^1, \ldots, \bar{x}^k$ are independent, i.e., can be extended to an adapted chart on $\bar{M}$ at $\bar{p}$, and $\bar{x}^i(\bar{p}) = x^i(p)$, $i = 1, \ldots, k$.*

iii) *Let f be a function in $\mathscr{F}_{r+1}$ and $\bar{f}$ the function with the same indices in $\bar{\mathscr{F}}_{r+1}$. The function f can be expressed as a function of $x^1, \ldots, x^k$, say $f = g(x^1, \ldots, x^k)$. Then $\bar{f}$ is the same function of $\bar{x}^1, \ldots, \bar{x}^k$, i.e., $\bar{f} = g(\bar{x}^1, \ldots, \bar{x}^k)$.*

If φ is a local isomorphism then

$$\bar{f} \circ \varphi = f \tag{4.5}$$

for all f and $\bar{f}$ of the same index in $\mathscr{F}_{r+1}$ and $\bar{\mathscr{F}}_{r+1}$.

Remark: If $r = 0$ and $k = n$ then Theorem 4.1 is just a restatement of the condition on page 341. If $r = 0$ and $k = 0$ Theorem 4.1 reduces to Theorem 2.4 of Chapter V. Our present proof will proceed more or less as the proof of Theorem 2.4 of Chapter V.

We just observe that to say that φ is a local isomorphism is the same as saying

$$\varphi^* \bar{\omega}^i = \omega^i, \qquad i = 1, \ldots, n. \tag{4.6}$$

In particular, if φ is a local isomorphism then

$$d\omega^i = \varphi^* \, d\bar{\omega}^i = \sum (\bar{c}^i_{jk} \circ \varphi) \varphi^* \omega^j \wedge \varphi^* \, \omega^k = \sum c^i_{jk} \omega^j \wedge \omega^k$$

so that

$$\bar{c}^i_{jk} \circ \varphi = c^i_{jk}.$$

This implies

$$d(\bar{c}^i_{jk} \circ \varphi) = \sum (\bar{c}^i_{jk,l} \circ \varphi)\varphi^*(\bar{\omega}^l)$$
$$= \sum c^i_{jk,l}\omega^l$$

so that

$$\bar{c}^i_{jk,l} \circ \varphi = c^i_{jk,l}.$$

Proceeding in this way we verify (4.5). Since φ is a diffeomorphism, Equation (4.5) clearly implies conditions i), ii) and iii), proving the necessity part of the theorem.

Now let us prove the sufficiency. Let us choose an adapted chart $x^1, \ldots, x^n$ at p and an adapted chart, $\bar{x}^1, \ldots, \bar{x}^n$, of the same indices at $\bar{p}$. According to (4.5) we must have $\bar{x}^i \circ \varphi = x^i$ for $i = 1, \ldots, k$. Now instead of looking directly for the map φ let us look for its graph which will be a submanifold of $M \times \bar{M}$.

The manifold $M \times \bar{M}$ has projections, π_1 and π_2, onto the first and second factors. Let us put

$$\theta^i = \pi_1^*\omega^i \qquad \text{and} \qquad \bar{\theta}^i = \pi_2^*\bar{\omega}^i.$$

Similarly, in a neighborhood of $(p, \bar{p})$ define the functions $y^i, \bar{y}^i$ by

$$y^i = x^i \circ \pi_1 \qquad \text{and} \qquad \bar{y}^i = \bar{x}^i \circ \pi_2.$$

The forms $\theta^1, \ldots, \theta^n, \bar{\theta}^1, \ldots, \bar{\theta}^n$ are clearly independent on $M \times \bar{M}$ and $y^1, \ldots, y^n, \bar{y}^1, \ldots, \bar{y}^n$ is a coordinate system about $(p, \bar{p})$. If φ is a local isomorphism then the graph, N, of φ [the submanifold of $M \times \bar{M}$ consisting of all points $(x, \varphi(x))$] is an n-dimensional submanifold of $M \times \bar{M}$. If ι is the injection of $N \to M \times \bar{M}$ then (4.6) and (4.5) imply that

$$\iota^*(\theta^i - \bar{\theta}^i) = 0, \qquad i = 1, \ldots, n \tag{4.7}$$

and

$$\iota^*(y^i - \bar{y}^i) = 0, \qquad i = 1, \ldots, n. \tag{4.8}$$

Conversely, let N be an n-dimensional submanifold of $M \times \bar{M}$ passing through $(p, \bar{p})$ satisfying (4.7). Since N is n-dimensional and $\theta^1, \ldots, \theta^n$, $\theta^1 - \bar{\theta}^1, \ldots, \theta^n - \bar{\theta}^n$ are independent on $M \times \bar{M}$, we can conclude that $\iota^*(\theta^1), \ldots, \iota^*(\theta^n)$ are independent on N. Similarly, $\iota^*(\bar{\theta}^1), \ldots, \iota^*(\bar{\theta}^n)$ are independent on N. This implies that both π_1 and π_2 are non-singular when restricted to N. Then (4.7) implies that $\varphi = \pi_{2|N} \circ (\pi_{1|N})^{-1}$ satisfies (4.6).

Now the equations $y^i - \bar{y}^i = 0$ $(i = 1, \ldots, k)$ define a $2n - k$ dimensional submanifold, K, of $M \times \bar{M}$. We shall show that the Equations (4.7) restricted to K, define a completely integrable differential system of dimension n. The integral manifold passing through $(p, \bar{p})$ will then be the graph of the desired map φ.

Let us solve the equations

$$dx^i = \sum_{j=1}^{n} x^i_{,j}\, \omega^j, \qquad i = 1, \ldots, k$$

for k of the ω's in terms of the dx^i's and the remaining ω's. By renumbering the ω's, if necessary, we can write

$$\omega^i = \sum_{j=1}^{k} a^i_j\, dx^j + \sum_{j=k+1}^{n} b^i_j\, \omega^j, \qquad i = 1, \ldots, k. \tag{4.9}$$

The functions a^i_j and b^i_j occurring in (4.9) are rational functions of the $x^i_{,j}$ and are therefore functions of $x^1, \ldots, x^k$. In view of conditions ii) and iii) we can solve for the same $\bar{\omega}^i$ in terms of the $d\bar{x}^i$ and obtain (4.9) with bars over everything. Lifting to $M \times \bar{M}$ we obtain

$$\theta^i - \bar{\theta}^i = \sum_{j=1}^{k} (a^i_j\, dy^j - \bar{a}^i_j\, d\bar{y}^j) + \sum_{j=k+1}^{n} b^i_j\, \theta^j - \bar{b}^i_j\, \bar{\theta}^j$$

where the a^i_j, b^i_j are functions of $y^1, \ldots, y^k$ and $\bar{a}^i_j$, $\bar{b}^i_j$ are the *same* functions of $\bar{y}^1, \ldots, \bar{y}^k$. In particular, on the manifold K the first expression on the right vanishes and $b^i_j = \bar{b}^i_j$ so that on K we can write

$$\theta^i - \bar{\theta}^i = \sum_{j=k+1}^{n} b^i_j(\theta^j - \bar{\theta}^j), \qquad i = 1, \ldots, k.$$

Thus at most $(n - k)$ of the forms $\theta^l - \bar{\theta}^l$ are linearly independent on K at each point of K. On the other hand, since all the $\theta^l - \bar{\theta}^l$ are independent on $M \times \bar{M}$ and K is of codimension k, at least $(n - k)$ of these forms must be independent on K. Thus the Equations (4.7), when restricted to K define a differential system of dimension $(2n - k) - (n - k) = n$.

Let us show that this system is completely integrable. Observe that

$$d\omega^i = \sum c^i_{jk}\, \omega^j \wedge \omega^k \quad \text{and} \quad d\bar{\omega}^i = \sum \bar{c}^i_{jk}\, \bar{\omega}^j \wedge \bar{\omega}^k.$$

Therefore

$$d(\theta^i - \bar{\theta}^i) = \sum c^i_{jk}\, \theta^j \wedge \theta^k + \sum \bar{c}^i_{jk}\, \bar{\theta}^j \wedge \bar{\theta}^k,$$

where c^i_{jk} are functions of $y^1, \ldots, y^k$ and $\bar{c}^i_{jk}$ are the *same* functions of $\bar{y}^1, \ldots, \bar{y}^k$. In particular, on K we can write $c^i_{jk} = \bar{c}^i_{jk}$ and so

$$d(\theta^i - \bar{\theta}^i) = \sum c^i_{jk}(\theta^j - \bar{\theta}^j) \wedge \theta^k + \sum c^i_{jk}\bar{\theta}^j \wedge (\bar{\theta}^k - \theta^k)$$

proving the complete integrability, and thus proving Theorem 4.1.

Let us now consider local automorphisms of a G-structure. Let us call a point of M *completely regular* if it is regular and has the same order (as well as rank) as all nearby points. Then Theorem 4.2 has as an immediate

COROLLARY 4.1. *Let p be a completely regular point of M of rank k, and let $x^1, \ldots, x^n$ be an adapted chart at p. Then in a neighborhood of p, any point on the manifold $x^1 = \text{const}, \ldots, x^k = \text{const}$ can be carried into any other such point by a local automorphism of M.*

So far, all our considerations have been purely local. Let us now study the group of all automorphisms of a manifold M with a complete parallelism. A moment's reflection shows that the structure of this group will depend heavily on global properties of M. [For instance, removing a single point from M might have the effect of drastically reducing the group. Thus completeness of M (with respect to the metric $(\omega^1)^2 + \cdots + (\omega^n)^2$ say) plays a crucial role.]

Let M be a manifold with a complete parallelism and G its group of automorphisms.

Let us assume, for simplicity, that M is connected (although this doesn't really matter in the proof). Then the forms $\omega^1, \ldots, \omega^n$ define a Riemann metric on M and thus makes M into a metric space. This, in turn, puts a topology on G : $a \in G$ and $b \in G$ are close if $d(x, (ab^{-1})x) < \epsilon$ for all $x \in M$ where $d(\ ,\ )$ is a distance function on M.

THEOREM 4.2. (Kobayashi). *Let M be a connected manifold with a complete parallelism and let G be its group of automorphisms. For any point $p \in M$ the map, φ_p, of $G \to M$ given by*

$$\varphi_p(a) = a(p)$$

for all $a \in G$ is one-to-one and bicontinuous. The set $\varphi_p(G)$ is a closed submanifold of M and the differential structure induced by φ_p on G makes G into a Lie group. In short, the group of automorphisms of a complete parallelism is a Lie group.

If we apply Theorem 3.2 we obtain, as an immediate consequence,

COROLLARY 4.2. *The group of automorphisms of a H-structure of finite type is a Lie group.*

In the special case of Riemann structures, this result was first proved by Myers and Steenrod.

For the proof of Theorem 4.2 let us introduce some notation. The complete parallelism gives an identification of V with the tangent space at each point of M. Thus a vector, $v \in V$, gives use to a vector field, which we shall call $\hat{v}$ on M. The vector field $\hat{v}$ defines a local one parameter group which we shall call $\exp tv$. Thus $\exp tv$ is defined at $p \in M$ for sufficiently small t. Or, what is the same, at a point $p \in M$, the local diffeomorphism, $\exp v$ ($= \exp 1 \cdot v$), are defined for all v near 0 in V. The map $p \to (\exp v)(p)$ is a diffeomorphism of a neighborhood of 0 in V onto a neighborhood of p in M.

If a is an element of G then $a^*\hat{v} = \hat{v}$ for all $v \in V$. Thus we can write

$$a \circ \exp v = (\exp v) \circ a \tag{4.10}$$

wherever the local diffeomorphism, $\exp v$, is defined.

LEMMA 4.3. *The group G acts without fixed points on M. In other words, if for any $a \in G$ we have $a(p) = p$ for any point $p \in M$ then $a = identity$.*

In fact, the set of fixed points of a is closed. On the other hand, (4.10) implies that this set is open. Since M is connected this shows that the set of fixed points is either empty or is all of M.

LEMMA 4.4. *Let a_k be a sequence of elements of G such that for some point, $p \in M$, $a_k(p) \to a(p)$ for some $a \in G$. Then $a_k \to a$ in the topology of G.*

Again we use a connectedness argument. The set of points q where $a_k(q) \to a(q)$ converges is closed:

if $q_r \to q$ and $a_r(q_k) \to a(q_k)$ then first choose k such that $d(q_k, q) < \epsilon$.

Next choose r so large that

$$d(a_r(q_k), a(q_k)) < \epsilon.$$

Then

$$\begin{aligned} d(a_r(q), a(q)) &\leqslant d(a_r(q_k), a_r(q)) + d(a(q_k), a(q)) \\ &\qquad + d(a_r(q_k), a(q_k)) \\ &\leqslant 3\epsilon. \end{aligned}$$

On the other hand, (4.10) shows that the set is open. Thus $a_k(p) \to a(p)$ for all $p \in M$. Now choose an everywhere dense sequence of points, $\{p_k\}$ on M. For any $\epsilon > 0$ choose a sufficiently large k so that every point of M is within ϵ distance of some p_i, $(i \leqslant k)$. Then for r sufficiently large, $d(a_r(p_i), a(p_i)) < \epsilon$ for all $i \leqslant k$. Then $d(a_r(p), a(p)) < 3\epsilon$ for all $p \in M$, i.e., $a_r \to a$ in the topology of G.

LEMMA 4.5. *Let $\{a_k\}$ be a sequence of elements of G such that $a_k(p) \to q$ for some p and $q \in M$. Then there exists an $a \in G$ such that $a(p) = q$.*

We wish to define an automorphism, a, of M such that $a(p) = q$. To this effect, first define the diffeomorphism, a, in a neighborhood, U, of p by setting $a((\exp v)(p)) = \exp v(q)$. At all points of U the map a is a pointwise limit of the a_i. If we extend the definition of a in this way along any curve, the map so obtained is still the pointwise limit of the a_i and thus does not depend on the choice of curve. Thus the connectedness of M allows us to define the automorphism a.

So far we have proved that the map, φ_p, of $G \to M$ is one to one and bicontinuous and that the image of G is closed. We now must show that $\varphi_p(a)$ is a submanifold of M. For this purpose we first construct the tangent space to this submanifold.

A sequence $a_k \to e$ of elements of G will be said to converge tangentially to the identity at p if the sequences

$$\frac{x^i(a_k(p)) - x^i(p)}{[\sum (x^j(a_k(p)) - x^j(p))^2]^{1/2}}$$

converges in terms of some (and hence every) coordinate chart. This sequence then defines a tangent vector at p. By using the same arguments as in Section 4 of Chapter V, we conclude that the set of these vectors form a completely integrable differential system, $\mathscr{D}$, on M.

Let N be the maximal integral manifold of this differential system passing through p. We now show that $N \subset \varphi_p(G)$. We will show that $\varphi_p(G) \cap N$ is an open subset of N. Since $\varphi_p(G)$ is closed this will imply that $N \subset \varphi_p(G)$.

Let X_q be a tangent vector belonging to $\mathscr{D}$ at q. Then by (4.10) the sequence, $\{a_k\}$, defining X_q defines the vector $(\exp v)_* X_q$ at $\exp v(q)$ for small $v \in V$. We thus get a (locally defined) vector field X such that

(4.11) $$X_{\exp v(q)} = (\exp v)_* X_q.$$

The set of all such vector fields clearly span $\mathscr{D}$ in a neighborhood of q. If q lies on N, every point near q on N is of the form $\exp X(q)$.[11] We shall show that all such points are of the form $b(q)$ for some $b \in G$. In particular, if $q \in \varphi_p(G) \cap N$, i.e., if $q = a(p)$ then all points near q will be of the form $ba(p)$, proving that $\varphi_p(G) \cap N$ is open in N. For this purpose, let $v_1 \in V$ such that $\hat{v}_{1q} = X_q$. Extend v_1 to a basis $v_1, \ldots, v_n$ of V. Then the map $v \to \exp v(q)$ introduces a coordinate system, $x^1, \ldots, x^n$, about q—the point $\exp(x^1 v_1 + \cdots + x^n v_n)q$ having coordinates $x^1, \ldots, x^n$. The coordinates of q are $(0, \ldots, 0)$. Let us compute the coordinates of $\exp X(q)$. Now

$$\begin{aligned} X_{\exp tv_1(q)} &= (\exp tv_1)_* X_q \\ &= (\exp tv_1)_* \hat{v}_{1q} \\ &= \hat{v}_{1\,\exp(tv_1)(q)}. \end{aligned}$$

In other words, the vector fields X and $\hat{v}_1$ agree *along the curve* $\exp tv_1(q)$. Thus we have

$$\exp X(q) = \exp \hat{v}_1(q)$$

and the coordinates of $\exp X(q)$ are $(1, 0, \ldots, 0)$. Now let $\{a_k\}$ be a sequence of elements of G such that $a_k \to e$ and such that

(4.12) $$\left| \frac{x^i(a_k(q))}{[\sum x^j(a_k(q))^2]^{1/2}} - \delta_1^i \right| < \epsilon$$

for large k. Such a sequence exists since the coordinates of q are $(0, 0, \ldots, 0)$ and the coordinates of X are $(1, 0, \ldots, 0)$.

The expression $\sum x^j(a_k(q))^2$ tends to zero. Thus (4.12) implies that for large enough k, we can find an integer N such that

(4.13) $$| N x^i(a_k(p)) - \delta_1^i | < \epsilon.$$

Now (4.10) implies that if

$$(\exp v)q = aq$$

---

[11] That is, can be reached from q by integrating a suitable X. This exp shouldn't be confused with the map $V \to \exp v(q)$, mapping $V \to M$.

then

$$\begin{aligned}(\exp v)q = (\exp v)^2 q &= (\exp v)aq \\ &= a(\exp \hat{v})q \\ &= a^2 q.\end{aligned}$$

Or, more generally $(\exp Nv)q = a^N q$. In view of the definition of the coordinates $x^1, \ldots, x^n$ we can conclude that $Nx^1(a_k(q))$ are the coordinates of $a_k^N(q)$. Thus (4.13) can be written as

$$| x^i(a_k^N(q)) - x^i(\exp X(q)) | < \epsilon.$$

In other words, we can find elements of G which map q arbitrarily close to $\exp X(q)$. By Lemma 4.5 this implies the existence of an $a \in G$ with $a(q) = \exp Xq$.

We have thus proved that $N \subset \varphi_p(G)$. Let G_0 be the component of the identity in G. We shall show that $N = \varphi_p(G_0)$. Since $\varphi_p(G_0)$ contains N it will be sufficient to show that $\varphi_p(G_0)$ cannot contain any other integral manifold of $\mathscr{D}$. [This would be sufficient since $\varphi_p(G_0)$, just as it contains N, would have to contain the integral manifold of $\mathscr{D}$ passing through any point lying in $\varphi_p(G_0)$.] By connectedness, if $\varphi_p(G_0)$ contains any other integral manifold, we could find a point $q \in N$ which is the limit of a sequence of points, q_k^1 not lying on N. By replacing each q_k^1 by a q_k lying on the integral manifold of $\mathscr{D}$ passing through q_k^1, we may assume that the sequence q_k converges to q in a direction which is not tangential to N.

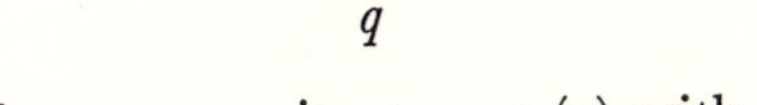

On the other hand, we can write $q_k = a_k(q)$ with $a_k \in G_0$. But by the very definition of $\mathscr{D}$ we can choose a subsequence of the a_k which converge tangentially to N. This contradiction shows that $N = \varphi_p(G_0)$. This in turn implies that $\varphi_p(G)$ is a submanifold of M. The rest of the assertions of Kobayashi's theorem now follow directly.

5. Connections on G-structures. In this section we shall study the geometry of connections on G-structures. On the one hand, as a

G-structure is a principal bundle, the theory can be regarded as a particular case of the theory of connections on principal bundles as developed in Section 1. On the other hand, as was pointed out in Section 3, a connection on a G-structure over a manifold, M, is a special kind of "second-order" structure on M. The interplay of these two theories yields the interesting geometrical properties of a connection on a G-structure. A connection on $\mathscr{F}(M)$ (= the $GL(n)$-structure) is called a *linear connection* on M. By abuse of language we shall call a connection on a G-structure over M a G-connection on M. Thus giving a G-connection on M amounts to giving two objects: First a specific G-structure, B_G, on M and then a connection on B_G. If $G_1 \subset G_2$ then the procedure of enlarging the group (cf. Section 1) shows how to pass from a G_1-connection to a G_2-connection. In particular, any G-connection gives rise to a linear connection.

A G-connection allows us to consider parallel translation of the fiber of B_G along any curve in the base. It also introduces a notion of parallel translation in any associated bundle of B_G. In particular, we have a notion of parallel translation on $T(M)$: If $C : [0, 1] \to M$ is any curve on M and $X_0 \in T_{C(0)}(M)$ there is a unique $X_t \in T_{C(t)}(M)$ for all $t \in [0, 1]$. The map $X_0 \to X_t$ is linear and carries frames of B_G into frames of B_G. Thus all the "additional structure" imposed by the group G on the tangent space is also preserved by parallel translation. For instance, if $G = O(n)$ so that M is a Riemann manifold, we can write $(X_t, Y_t) = (X, Y)$ for all t.

The notion of parallel translation of tangent vectors allows us to introduce the concept of covariant derivatives of a vector field along a curve. Let $C : [a, b] \to M$ be a smooth curve on M and let $\{Y_t \in T_{C(t)}(M)\}$ be a family of tangent vectors along this curve. If we parallel translate Y_t back to $C(0)$ we obtain a tangent vector, $\bar{Y}_t \in T_{C(0)}$. We can then consider the limit

$$\left.\frac{DY_t}{dt}\right|_{t=0} = \lim \frac{\bar{Y}_t - Y_{C(0)}}{t} \tag{5.1}$$

which will exist whenever Y_t is differentiable [as a curve in $T(M)$].

If X and Y are two vector fields on M we can introduce the covariant derivative of Y with respect to X, denoted by $\nabla_X Y$. It is defined as follows: At any point $x \in M$ let C be the integral curve of X passing through p and let $Y_t = Y_{C(t)}$. Then

$$(\nabla_X Y)_x = \left.\frac{DY_t}{dt}\right|_{t=0}. \tag{5.2}$$

It is convenient to have several other expressions for $\nabla_X Y$. We first observe that there is a one to one correspondence $X \leftrightarrow f_X$ between vector fields, X, on M and V-valued functions, f, on $\mathscr{F}(M)$ satisfying $f(p \cdot a) = a^{-1}f(p)$. In fact, such a function, f clearly gives a vector field. If we are given a vector field X on M, let $\hat{X}$ be the lifted vector field on B_G (with respect to the connection). Then $f_X = \langle \hat{X}, \omega \rangle$ is the desired function on B_G.

Let X and Y be vector fields on M and $p \in B_G$ with $\pi(p) = x$. Then

(5.3) $$(\nabla_X Y)_x = p \cdot (\hat{X} f_Y)(p).$$

(In this formula $\hat{X} f_Y$ is a V-valued function on B_G so that $\hat{X} f_Y(p) \in V$ and thus $p(\hat{X} f_Y(p)) \in T_x(M)$). To verify (5.3) let C be the solution curve of X through x and let $\bar{C}$ be the horizontal curve passing through p and lying over C. Then by the definition of connection on associated bundles (cf. the end of Section 1) Equations (5.1) and (5.2) yield

$$\begin{aligned}(\nabla_X Y)_x &= \lim_{t \to 0} \frac{1}{t} [\bar{C}(0) \cdot \bar{C}(t)^{-1} \cdot Y_t - Y_0] \\ &= \bar{C}(0) \lim_{t \to 0} \frac{1}{t} [\bar{C}(t)^{-1} Y_t - \bar{C}(0)^{-1} Y_0] \\ &= \bar{C}(0) \lim_{t \to 0} \frac{1}{t} [f_Y(\bar{C}(t)) - f_Y(\bar{C}(0))] \\ &= p \hat{X}_p f_Y(p)\end{aligned}$$

since $\bar{C}(0) = p$ and X_p is tangent to $\bar{C}$ at p. This proves (5.3).

Now the map $X \to \hat{X}$ is linear and $\widehat{gX} = (\pi^* g)\hat{X}$. Also the map $Y \to f_Y$ is linear and satisfies $f_{gY} = \pi^*(g) f_Y$. This fact, together with (5.3) implies

(5.4) $$\nabla_{X_1 + X_2} Y = \nabla_{X_1} Y + \nabla_{X_2} Y,$$

(5.5) $$\nabla_{gX} Y = g \nabla_X Y,$$

(5.6) $$\nabla_X (Y_1 + Y_2) = \nabla_X Y_1 + \nabla_X Y_2,$$

and

(5.7) $$\nabla_X g Y = g \nabla_X Y + (Xg) Y.$$

Let us now give an expression for the covariant derivative in terms of local coordinates. Let $x^1, \ldots, x^n$ be a system of coordinates in a neighborhood, U, on M. We can then write

(5.8) $$\nabla_{\partial/\partial x^i} \left(\frac{\partial}{\partial x^j} \right) = \sum_k \Gamma^k_{ij} \frac{\partial}{\partial x^k},$$

where the Γ_{ij}^k are suitable functions on U. Using Equations (5.4)–(5.7) we can use the Γ_{ij}^k to compute the covariant derivative of any vector field with respect to another in U. Let $y^1, \ldots, y^n$ be a second coordinate system with functions $\Gamma_{ij}'^k$ on V. Then if $U \cap V$ we can write

$$\frac{\partial}{\partial y^i} = \sum_{l=1}^{n} \frac{\partial x_l}{\partial y^i} \frac{\partial}{\partial x_l}.$$

Substituting this into

$$\nabla_{\partial/\partial y^i}\left(\frac{\partial}{\partial y^j}\right) = \sum \Gamma_{ij}'^k \frac{\partial}{\partial y^k}$$

yields

$$(5.9) \qquad \Gamma_{ij}'^k = \sum_{l=1}^{n} \frac{\partial^2 x^l}{\partial y^i \partial y^j} \frac{\partial y^k}{\partial x^l} + \sum_{r,s,t=1}^{n} \Gamma_{rs}^t \frac{\partial x^r}{\partial y^i} \frac{\partial x^s}{\partial y^j} \frac{\partial y^k}{\partial x^t}.$$

Suppose then we are given a collection of functions Γ_{ij}^k for each chart of an atlas of M so that in the overlap of two charts the Γ's are related as in (5.9). It is easy to check that this defines a covariant derivative $\nabla_X Y$ for any X and Y. In fact, locally we use (5.4) to (5.8) to compute $\nabla_X Y$ in any chart. Equation (5.9) guarantees that the expressions obtained agree on the overlap of two charts and are independent of the local coordinate system. Let us now show that such Γ_{ij}^k actually define a linear connection on M. For this purpose let us recall the local coordinates we have introduced on $\mathscr{F}(M)$: If $x^1, \ldots, x^n$ is a coordinate system on $U \subset M$ then a point $p \in \mathscr{F}(M)$ with $\pi(p) = x \in U$ is a frame $(X_1, \ldots, X_n)$ where

$$X_i = \sum_{k=1}^{n} X_i^k \left(\frac{\partial}{\partial x^k}\right)$$

where the matrix (X_i^k) is non-singular. Let (Y_j^i) denote the inverse matrix so that (X_i^k) and (Y_j^i) are matrix-valued functions on $\pi^{-1}(U)$. It is then easy to check that

$$(5.10) \qquad \omega^i = \sum Y_j^i \pi^* dx^j$$

on $\pi^{-1}(U)$ [where $\omega = (\omega^1, \ldots, \omega^n)$ is the fundamental V-valued form on $\mathscr{F}(M)$]. Now set

$$(5.11) \qquad \Omega_i^k = \sum_{j=1}^{n} Y_j^k \left(dX_i^j + \sum_{k,l=1}^{n} \Gamma_{kl}^j X_i^l \pi^* dx^k\right).$$

The reader can verify, as an exercise, that the forms Ω_i^k are globally defined and are independent of the local coordinate system. Let $v^1, \ldots, v^n$ be the basis of V and $A^i_j \in gl(V)$ be given by $A^i_j v^k = \delta^i_k v^j$. Then, as can be easily verified, the form

$$\Omega = \sum \Omega^j_i A^i_j$$

defines a connection on $\mathscr{F}(M)$ and the Γ^k_{ij} of this connection are precisely the ones we started on. To resume, we have

THEOREM 5.1. *Let M be C^∞ (or analytic) manifold. Then the following three objects on M are equivalent*

i) *a linear connection on M*

ii) *a map $X, Y \to \nabla_X Y$ for any pair of C^∞ (analytic) vector fields on M satisfying* (5.4)–(5.7)

or

iii) *a collection of functions Γ^k_{ij} for each chart of an atlas of M transforming as in* (5.9).

So far, we have considered $T(M)$ as an associated bundle of B_G and studied the parallel translation on $T(M)$ given by a G-connection. Now any tensor bundle is an associated bundle and we can consider the parallel translation on it. We can thus consider covariant differentiation of a tensor field along a curve or with respect to a vector field. Thus ∇_X becomes a linear operator mapping the space of tensor fields of a given type onto itself. The appropriate generalization of (5.7) is that ∇_X is a derivation of the algebra of all tensor fields. An important associated notion is that of *covariant differential* of a tensor field. Let t be a tensor field of contravariant degree r and covariant degree s [of degree (r, s) for short]. Then for any vector field X, the tensor field $\nabla_X t$ is also of degree (r, s). We thus have a rule which associates a tensor field of degree (r, s) to each vector field—thus a tensor field of degree $(r, s + 1)$. We denote this tensor field by ∇t. The case where t is contravariant of degree one will come up several times later on. Suppose that t is of degree $(1, s)$. If $X_1, \ldots, X_s$ are vector fields then $t(X_1, \ldots, X_s)$ is a vector field. The tensor field, ∇t, of type $(1, s + 1)$ is then given by

$$(5.12) \qquad \begin{aligned} (\nabla t)(X_1, \ldots, X_s, X) &= \nabla_X(t(X_1, \ldots, X_s)) \\ &\quad - \sum_{i=1}^{s} t(X_1, \ldots, \nabla_X X_i, \ldots, X_s). \end{aligned}$$

The choice of a connection on B_G induces a complete parallelism on B_G—an identification of the tangent space at each point with $V + g$. In fact, the definition of a G-structure allows an identification of the tangent to the fiber with g and a choice of horizontal subspace, H, induces an identification of H with V. Every $A \in g$ and every $v \in V$ gives rise to a vector field on B_G. In Section 1 we decided to denote the vector field corresponding to A by $\tilde{A}$. In accordance with this (and in slight contrast to the notation of Section 4) we will denote the vector field corresponding to v by $\tilde{v}$. By their very definition we have

$$\langle \tilde{A}, \omega \rangle = 0 \quad \text{and} \quad \langle \tilde{A}, \Omega \rangle = A \tag{5.13}$$

and

$$\langle \tilde{v}, \omega \rangle = v \quad \text{and} \quad \langle \tilde{v}, \Omega \rangle = 0, \tag{5.14}$$

where ω is the fundamental V-valued form on B_G and Ω is the (g-valued) connection form. We now wish to compute the Lie brackets of the various vector fields $\tilde{A}$ and $\tilde{v}$. If $A \in g$ and $B \in g$ we have

$$[\tilde{A}, \tilde{B}] = \widetilde{[A, B]}.$$

Let us verify that

$$[\tilde{A}, \tilde{v}] = -\widetilde{A \cdot v} \tag{5.15}$$

for $A \in g$ and $v \in V$. To see this let a_t be the one parameter subgroup of G generated by A. Then if $p \in B_G$ we have $a_{t*}\tilde{v}_p = \widetilde{(a_t^{-1}v)}_{a_t p}$ according to (2.3.) Passing to the infinitesimal yields (5.15). We now wish to compute $[\tilde{v}_1, \tilde{v}_2]$ for $v_1, v_2 \in V$. For this purpose we introduce

DEFINITION 5.1. *The form $D\omega$ is called the torsion of the connection.*

Recall that according to the definition of D given in Section 1 we have

$$\langle X \wedge Y | D\omega \rangle = \langle hX \wedge hY | d\omega \rangle$$

for any vector fields, X and Y on B_G. In view of (2.3) we can write

$$R_a^*(D\omega) = a^{-1} D\omega. \tag{5.16}$$

In analogy with Equation (1.6) we can write

$$d\omega = -\Omega \wedge \omega + D\omega. \tag{5.17}$$

The verification of (5.17) proceeds as did the verification of (1.6). One uses Equation (1.10) of Chapter III applied to suitable choices of vector fields $\tilde{A}$ and $\tilde{v}$. The details will be left as an exercise to the reader. If we apply (5.17) to $\tilde{v}_1 \wedge \tilde{v}_2$ we obtain

$$\begin{aligned}\langle \tilde{v}_1 \wedge \tilde{v}_2 | D\omega \rangle &= \langle \tilde{v}_1 \wedge \tilde{v}_2 | d\omega \rangle \\ &= \tilde{v}_1 \langle \tilde{v}_2, \omega \rangle - \tilde{v}_2 \langle \tilde{v}_1, \omega \rangle - \langle [\tilde{v}_1, \tilde{v}_2], \omega \rangle\end{aligned}$$

or, since $\langle \tilde{v}_2, \omega \rangle = v_2 = \text{const}$ and $\langle \tilde{v}_1, \omega \rangle = v_1 = \text{const}$, we have

$$\langle [\tilde{v}_1, \tilde{v}_2], \omega \rangle = \langle \tilde{v}_1 \wedge \tilde{v}_2 | D\omega \rangle.$$

Thus the horizontal component of $[\tilde{v}_1, \tilde{v}_2]$ is $\langle v_1 \wedge v_2, D\omega \rangle$. Similarly, we have

$$\langle [\tilde{v}_1, \tilde{v}_2], \Omega \rangle = -\langle \tilde{v}_1 \wedge \tilde{v}_2, D\Omega \rangle.$$

We can thus write

$$(5.18) \qquad [\tilde{v}_1, \tilde{v}_2] = \widetilde{\langle \tilde{v}_1 \wedge \tilde{v}_2 | D\omega \rangle} + \widetilde{\langle \tilde{v}_1 \wedge \tilde{v}_2 | D\Omega \rangle}.$$

To resume the table of Lie brackets, we have

$$[\tilde{A}, \tilde{B}] = \widetilde{[A, B]},$$

$$(5.15) \qquad [\tilde{A}, \tilde{v}] = -\widetilde{A \cdot v},$$

and

$$(5.18) \qquad [\tilde{v}_1, \tilde{v}_2] = \widetilde{\langle \tilde{v}_1 \wedge \tilde{v}_2 | D\omega \rangle} + \widetilde{\langle \tilde{v}_1 \wedge \tilde{v}_2 | D\Omega \rangle}.$$

The torsion and curvature introduced above are differential forms on B_G. They also can be regarded as tensor fields on M. To see this let us notice that there is a correspondence between certain kinds of forms on B_G and tensor fields on M. We have already seen that correspondence between certain V-valued functions on B_G and vector fields on M. Let us call a differential form, θ, on B_G horizontal if $\langle X_1 \wedge \cdots \wedge X_k | \theta \rangle = 0$ if one of the X_i is vertical.

Proposition 5.1. *There is a one-to-one correspondence between tensor fields, t, of type $(1, s)$ on M which are antisymmetric in the covariant variables and the set of horizontal V-valued s-forms, θ, on B_G satisfying $R_a^*\theta = a^{-1}\theta$. If $\pi(p) = x$ and $X_1, \ldots, X_s \in T_p(B_G)$ then the correspondence is given by*

$$t_x(\pi_*(X_1), \ldots, \pi_*(X_s)) = p\langle X_1 \wedge \cdots \wedge X_s | \theta \rangle.$$

The proof of the proposition is obvious and will be left as an exercise to the reader. As a particular case, the torsion form, $D\omega$, gives rise to a tensor field, T, of type (1, 2) on M; it is called the torsion *tensor* field on M. It assigns a vector field, $T(X, Y)$, to every pair of vector fields, X and Y on M, and

$$T(X, Y) = -T(Y, X).$$

Let us consider, for the moment, a tensor field, t, of type $(1, s)$ on M. If we apply it, by contraction on the last $s - 1$ indices, to vector fields $X_1, \ldots, X_{s-1}$ we obtain a tensor field, $t(X_1, \ldots, X_{s-1})$ of type (1, 1) on M. Now a tensor field of type (1, 1) gives a linear transformation of $T_x(M)$ for each $x \in M$. Via the G-structure, B_G, the group G can be regarded as a group of linear transformations of $T_x(M)$; and thus the algebra g can be regarded as an algebra of linear transformations of $T_x(M)$ for each $x \in M$. We shall say that a tensor field of type (1, 1) is *g-valued* if the linear transformation it gives on $T_x(M)$ lies in g for all $x \in M$. We shall say that a tensor field, t, of type $(1, s)$ is g-valued if $t(X_1, \ldots, X_{s-1})$ is a g-valued tensor field of type (1, 1) for all vector fields $X_1, \ldots, X_{s-1}$ on M. In analogy to Proposition 5.1 we can now state

PROPOSITION 5.2. *There is a one-to-one correspondence between the g-valued tensor fields, t, of type $(1, s)$ which are antisymmetric in their last $s - 1$ indices and the g-valued $(s - 1)$ forms, θ, on B_G satisfying*

$$R_a^*\theta = ad(a^{-1})\theta.$$

If $\pi(p) = x$ and $X_1, \ldots, X_{s-1} \in \wedge^{s-1}(T_p(B_G))$ then the correspondence is given by

$$t_x(T_*(X_1), \ldots, \pi_*(X_{s-1}))Z = p(\langle X_1 \wedge \cdots \wedge X_{s-1}|\theta\rangle(pZ))$$

for any tangent vector $Z \in T_x(M)$.

(To give t_x it is sufficient to give the linear transformation $t_x(Y_1, \ldots, Y_{s-1})$ for $Y_1, \ldots, Y_{s-1} \in T_x(M)$. To give this linear transformation it is sufficient to know its value on any tangent vector $Z \in T_x(M)$.) The proof of this proposition will also be left as an exercise to the reader. In particular the form $D\Omega$ gives rise to a tensor field, R, of type (1, 3) on M called the *curvature tensor* field. It assigns to every pair of vector fields, X and Y on M a tensor field $R(X, Y)$ belonging to g at each point. We have

$$R(X, Y) = -R(Y, X).$$

We can give a simple expression for the torsion and curvature tensors in terms of the operator ∇:

$$T(X, Y) = \nabla_X \cdot Y - \nabla_Y \cdot X - [X, Y], \tag{5.19}$$

and

$$R(X, Y) = [\nabla_X, \nabla_Y] - \nabla_{[X,Y]}. \tag{5.20}$$

In (5.20) the expression $[\nabla_X, \nabla_Y]$ means the bracket of ∇_X and ∇_Y as operators on the space of all vector fields. That is, in terms of the action of $R(X, Y)$ on a vector field, Z, we have

$$R(X, Y)Z = \nabla_X(\nabla_Y \cdot Z) - \nabla_Y(\nabla_X \cdot Z) - \nabla_{[X,Y]}Z.$$

Let us verify (5.19). By Proposition 5.1,

$$\begin{aligned} T(X, Y)_x &= p \cdot (\langle \hat{X} \wedge \hat{Y} | D\omega \rangle_p) \\ &= p(\langle \hat{X} \wedge \hat{Y} | d\omega \rangle_p) \\ &= p(\hat{X}_p \langle \hat{Y}, \omega \rangle - \hat{Y}_p \langle \hat{X}, \omega \rangle - \langle [\hat{X}, \hat{Y}]_p, \omega \rangle). \end{aligned}$$

Now $\pi_*[\hat{X}, \hat{Y}] = [X, Y]$ so that we can substitute $\widehat{[X, Y]}$ for $[\hat{X}, \hat{Y}]$ in the last term on the right. According to (5.3), if $\pi(p) = x$, we can write

$$(\nabla_X \cdot Y)_x = p(\hat{X}_p f_Y) = p \cdot (X_p \langle \hat{Y}, \omega \rangle).$$

Making the same remark for $\nabla_Y X$ proves (5.19).

Let us now prove (5.20). Let Z be a vector field on M and $f = \langle Z, \omega \rangle$ the corresponding function on B_G. Then if $\pi(p) = x$, we have

$$([\nabla_X, \nabla_Y]Z - \nabla_{[X,Y]} \cdot Z)_x = p(\hat{X}_p(\hat{Y}f) - \hat{Y}_p(\hat{X}f) - H([X, Y])_p f).$$

But $\hat{X}_p(\hat{Y}f) - \hat{Y}_p(\hat{X}f) = [\hat{X}, \hat{Y}]_p f$ so the expression in the parenthesis is $([\hat{X}, \hat{Y}] - h([\hat{X}, \hat{Y}]))f$; in other words, the vertical component of $[\hat{X}, \hat{Y}]$ applied to f. But $\langle \hat{X} \wedge \hat{Y} | D\Omega \rangle - \hat{X}\langle \hat{Y}, \Omega \rangle - \hat{Y}\langle \hat{X}, \Omega \rangle - \langle [\hat{X}, \hat{Y}], \Omega) = -\langle [\hat{X}, \hat{Y}], \Omega \rangle$ so that the vertical component of $[\hat{X}, \hat{Y}]_p$ is just $\hat{A}$ where $A = -\langle \hat{X} \wedge \hat{Y} | D\Omega \rangle$. On the other hand, in view of the transformation law $R_a^* f = a^{-1} f$, we can write $\hat{A}f = -Af$. Thus, we can write

$$\begin{aligned} [\nabla_X, \nabla_Y]Z - \nabla_{[X,Y]}Z &= p(\langle \hat{X} \wedge \hat{Y} | D\Omega \rangle f(p)) \\ &= p(\langle \hat{X} \wedge \hat{Y} | D\Omega \rangle p^{-1} Z) \end{aligned}$$

since $f(p) = p^{-1}Z$. Comparing this with Proposition 5.2 proves (5.20). Equation (5.20) says that the curvature measures how much the map $X \to \nabla_X$ fails to be a Lie algebra homomorphism.

EXERCISE 5.1. Using (5.19) show that if we write, in terms of local coordinates, $T(\partial/\partial x^i, \partial/\partial x^j) = \sum T^k_{ij}(\partial/\partial x^k)$ then

$$T^k_{ij} = \Gamma^k_{ij} - \Gamma^k_{ji}.$$

EXERCISE 5.2. Using (5.20) show that if we write

$$R\left(\frac{\partial}{\partial x^j}, \frac{\partial}{\partial x^k}\right)\cdot\frac{\partial}{\partial x^l} = \sum R^i_{ljk}\frac{\partial}{\partial x^i}$$

then

$$R^i_{ljk} = \frac{\partial\Gamma^i_{kl}}{\partial x^j} - \frac{\partial\Gamma^i_{jl}}{\partial x^k} + \sum_{r=1}^{n}(\Gamma^i_{jr}\Gamma^r_{kl} - \Gamma^i_{kr}\Gamma^r_{jl}).$$

We now derive some classical identities involving the torsion and curvature tensors and their covariant derivatives. For any function of three variables, t, we shall denote by $\mathfrak{S}t$ the cyclic sum of t. Thus if t is a tensor of type $(1, 3)$ the expression $\mathfrak{S}t(X, Y, Z) = t(X, Y, Z) + t(Y, Z, X) + t(Z, X, Y)$.

Since ∇_Z is a derivation of the tensor algebra we can write

$$(5.21) \qquad (\nabla_Z\cdot T)(X, Y) = \nabla_Z\cdot T(X, Y) - T(\nabla_Z X, Y) - T(X, \nabla_Z Y).$$

Using (5.19) and the antisymmetry of T we can write

$$T(T(X, Y), Z) = T(\nabla_X\cdot Y, Z) + T(Z, \nabla_Y\cdot X) - T([X, Y], Z).$$

Summing cyclically and using (5.21) we obtain

$$\mathfrak{S}T(T(X, Y), Z) = \mathfrak{S}\{-(\nabla_Z\cdot T)(X, Y) + \nabla_Z T(X, Y) - T([X, Y], Z)\}.$$

Now

$$\begin{aligned}\mathfrak{S}(\nabla_Z T(X, Y) - T([X, Y], X)) &= \mathfrak{S}(\nabla_Z\nabla_X\cdot Y - \nabla_Z\cdot\nabla_Y\cdot X - \nabla_Z[X, Y] - \nabla_{[X,Y]}Z \\ &\qquad + \nabla_Z[X, Y] + [[X, Y], Z]) \\ &= \mathfrak{S}(\nabla_X\cdot\nabla_Y Z - \nabla_Y\nabla_X Z - \nabla_{[X,Y]}Z)\end{aligned}$$

since $\mathfrak{S}[[X, Y], Z] = 0$ (Jacobi's identity). Using (5.20) we obtain

$$(5.22) \qquad \mathfrak{S}(R(X, Y)\cdot Z) = \mathfrak{S}\{T(T(X, Y), Z) + (\nabla_X T)(Y, Z)\}.$$

In particular, if $T = 0$ we have

$$(5.23) \qquad \mathfrak{S}(R(X, Y)\cdot Z) = 0.$$

By similar arguments, which we shall leave as an exercise to the reader, we can obtain Bianchi's identity,

$$(5.24) \qquad \mathfrak{S}\{(\nabla_Z\cdot R)(X, Y)\} + \mathfrak{S}\{R(T(X, Y), Z)\} = 0.$$

In particular, if $T = 0$,

$$\mathfrak{S}\{(\nabla_Z \cdot R)(X, Y)\} = 0. \tag{5.25}$$

6. The spray of a linear connection. We first observe that a linear connection on a manifold gives rise to a spray on M: At any point $z \in T(M)$ there is a unique horizontal vector, W_z, which satisfies $\pi_*(W_z) = z$. We thus have a vector field, W, on $T(M)$ which is a second order differential equation on M. To show that it is a spray we must verify that Equation (6.2) of Chapter IV holds. This can be verified directly if the reader so desires. We shall derive it as a consequence of a remark which is fundamental for what follows. Instead of considering the full tangent bundle it will be convenient to restrict attention to the bundle $T^+(M)$ of all non-zero tangent vectors. Each non-zero $v \in V$ induces a map, φ_v, of $\mathscr{F}(M)$ onto $T^+(M)$ given by

$$\varphi_v(p) = p(v). \tag{6.1}$$

Let us fix v once and for all and write φ for φ_v. Then

$$\varphi_*(\hat{v}_p) = W_{\varphi(p)}. \tag{6.2}$$

In fact, $\varphi_*(\hat{v}_p)$ is horizontal and $\pi_*\varphi_*(\hat{v}_p) = \pi_*(\hat{v}_p) = p \cdot v$. In short, the vector fields $\hat{v}$ and W are φ-related. Let R_t denote right multiplication on $\mathscr{F}(M)$ by the scalar linear transformation tI, i.e., $R_t = R_{tI}$. Then

$$\varphi \circ R_t(v) = p(tv) = \mu_t(p(v))$$

where μ_t is multiplication by t on $T^+(M)$ (cf. the beginning of Section 6 of Chapter IV). According to (2.3)

$$R_t^* \hat{v} = \frac{1}{t} \hat{v}.$$

Using (6.2) we obtain that $\frac{1}{t} \hat{v}$ and $\mu_t^* W$ are φ-related or

$$\mu_t^* W = \frac{1}{t} W.$$

Thus W is a spray.

Since W is a spray, we can transcribe all of the results of Chapter IV on sprays to the case of linear connections. We thus have an exponential map, normal coordinates geodesics, etc. By the definition of W we see that a curve, C, of M *is a geodesic if and only if its tangent vectors are parallel along* C.

It is natural to expect that the torsion and the curvature should give information about the spray. We shall see that they give the equations of variation of W, cf. Section 5 of Chapter IV. The point is that Equations (5.15) and (5.16) give the equations of variation of the flow $\hat{v}$ on $\mathscr{F}(M)$. Using (6.2), we get the equation of variation of W. First some preliminary remarks.

Let α_t denote the (local) flow generated by W on $T(M)$. Choosing a fixed geodesic, C, on M, we are interested in describing the flow α_{t*} which maps $T_{C'(0)}(T(M))$ onto $T_{C'(t)}(T(M))$. For this purpose we must have a way of parametrizing the vector space $T_{C'(t)}(T(M))$. This can be done in terms of the parallel translation given by the connection. In fact, the space $T_{C'(t)}(T(M))$ has two distinguished n-dimensional subspaces. The vertical n-dimensional subspace tangent to the fiber and the horizontal subspace is given by the connection. The projection, π_*, allows us to identify the horizontal subspace with $T_{C(t)}(M)$. This space, in turn, can be identified with $T_{C(0)}(M)$ via parallel translation along C. The vertical component is the tangent space of $T_{C(t)}(M)$ at the point $C'(t) \in T_{C(t)}(M)$. Via the standard identifications of the tangent space of a vector space with the vector space, and of $T_{C(t)}(M)$ with $T_{C(0)}(M)$ we get an identification of $T_{C'(t)}(T(M))$ with $T_{C(0)}(T(M))$ which is in turn identified with $T_{C(0)}(M) + T_{C(0)}(M)$. In short, we have identified the restriction of $T(T(M))$ along C' with $(T_{C(0)}(M)(T_{C(0)}(M)) \times R$. If we are given any curve, X_t, of tangent vectors along C in M, we obtain a curve X_t, DX_t/dt in $T_{C'(0)}(T(M))$. (The reader should go back to Section 5 of Chapter IV and observe that precisely this identification was what was needed for the argument there with the second variation.)

If we choose a $p_0 \in \mathscr{F}(M)$ with $\pi(p_0) = C(0)$ then p gives an identification of $T_{C(0)}(M)$ with V. We then have an identification of the restriction of $T(T(M))$ to $C'(t)$ with $(V \times V \times R)$. Since the reader is probably confused by all of these identifications we restate them in the form that we shall use them. A vector $\xi \in T_{C'(t)}(T(M))$ corresponds to the point $(a, b; t)$ where a is the vector in V obtained by parallel translating $\pi_*\xi$ back to $T_{C(0)}(M)$ and then mapping $T_{C(0)}(M)$ onto V by p_0. If p_t is the horizontal curve lying over $C(t)$ in $\mathscr{F}(M)$ and passing through p_0, then this is the same as writing

$$a = p_t^{-1}(\pi_*\xi).$$

The vector b is given by taking the vertical component, vert ξ, of ξ, identifying it with a vector in $T_{C(t)}(M)$ via $l_{C'(t)}$ (the standard identification of the tangent space of a vector space with a vector space) and then applying p_t. Thus

$$b = p_t^{-1} \circ l_{C'(t)}(\text{vert } \xi).$$

Let us now consider the map φ_* of $T_{p_t}(\mathscr{F}(M))$ onto $T_{C'(t)}(T(M))$. By the definitions of the vector fields $\tilde{a}$ and $\tilde{A}$ and via our identification, we have

$$\varphi_*(\tilde{a}_{p(t)}) = (a, 0; t) \tag{6.3}$$

and

$$\varphi_*(\tilde{A}_{p(t)}) = (0, A \cdot v; t). \tag{6.4}$$

We can now apply (5.15) and (5.18) to compute the equations of variation. In fact, if we write $(a_t, b_t; t) = \alpha_{t*}(a_0, b_0; 0)$ we get

$$\frac{da_t}{dt} = b - \langle \tilde{a}_{tp_t} \wedge \tilde{v}_{p_t} | D\omega \rangle$$

(6.5) and

$$\frac{db_t}{dt} = -\langle \tilde{a}_{tp_t} \wedge \tilde{v}_{p_t} | D\Omega \rangle \cdot v.$$

We can rewrite this in more instructive forms. Let $X_t = C'(t)$ be the tangent vector to C and let $Y_t = p_t(a_t)$ be the projection onto $T_{C(t)}(M)$ of a tangent vector moving according to the flow of α_{t*}. Let us assume for simplicity that the torsion vanishes. Then using Proposition 5.1 and 5.2 we can write

$$\frac{D^2 Y_t}{dt^2} + R(Y_t, X_t) \cdot X_t = 0. \tag{6.6}$$

The most important case where (6.6) is used is in Riemannian geometry. As we have seen, there is a unique connection on $\mathcal{O}(M)$ and thus on $\mathscr{F}(M)$ with torsion zero. This defines a spray on $T(M)$. *This spray coincides with the geodesic spray introduced in Chapter IV.* The proof is the same as that of Theorem 6.2 of Chapter VI. In fact, since parallel translation preserves length, both the geodesic spray and the spray of the connection leave invariant the submanifold, $S(M)$, of $T(M)$, consisting of those vectors of unit length. It will be sufficient to prove that the two vector fields coincide when restricted to $S(M)$. According to (3.8) of Chapter IV, the geodesic spray restricted to $S(M)$ is the vector field satisfying

$$X \lrcorner\ d\theta = 0 \qquad \langle X, \theta \rangle = 1 \tag{6.7}$$

where θ is the fundamental form of $T^*(M)$ brought back to $T(M)$ and restricted to $S(M)$. That is, for any $z \in S(M)$ and $Y \in T_z(S(M))$,

$$\langle Y_t, \theta \rangle = (\pi_* Y, z). \tag{6.8}$$

Now let us choose v of length one in V so that the map $\varphi = \varphi_v$ of $\mathscr{F}(M)$ onto $T^+(M)$ maps $\mathscr{O}(M)$ onto $S(M)$. Let ψ denote the restriction of φ to $\mathscr{O}(M)$. Let us evaluate the form $\psi^*(\theta)$ on $\mathscr{O}(M)$. For any tangent vector, $X \in T_p(\mathscr{O}(M))$ we have

$$\begin{aligned}
\langle X, \psi^*\theta \rangle &= \langle \psi^* X, \theta \rangle \\
&= (\pi_* \psi_* X, \pi(\psi(p))) \\
&= (\pi_* X, p(v)) \\
&= (p^{-1}(\pi_* X), v) \qquad \text{(scalar product in } V) \\
&= (\langle X, \omega \rangle, v).
\end{aligned}$$

Thus

$$\psi^*\theta = (\omega, v).$$

Now $\tilde{v}$ satisfies $\langle \tilde{v}, \psi^*\theta \rangle = (\langle \tilde{v}, \omega \rangle, v) = (v, v) = 1$. Also

$$\begin{aligned}
\psi^*\, d\theta &= (d\omega, v) \\
&= (\Omega \wedge \omega, v).
\end{aligned}$$

Since $\tilde{v}$ is horizontal, $\langle \tilde{v}, \Omega \rangle = 0$. Thus for any $X \in T(\mathscr{O}(M))$ we have

$$\langle v \wedge X | \Omega \wedge \omega \rangle = -\langle X, \Omega \rangle \cdot v.$$

Since $\langle X, \Omega \rangle$ is an element of the orthogonal algebra, $(\langle X, \Omega \rangle \cdot v, v) = 0$. Thus

$$\tilde{v} \lrcorner\ \psi^*\, d\theta = 0.$$

Using (6.2), we conclude that W satisfies (6.7), proving our assertion.

Equation (6.6) together with the results of Chapter IV show that global properties of a Riemann manifold can be deduced from information about its curvature tensor. The general idea is, that one develops comparison theorems analogous to Lemma 6.1. One then uses these comparison theorems to deduce estimates for the distance between two conjugate points. To give a simple prototype of these theorems we first examine the expression $(R(Y, X)\cdot X, Y)$ for a pair of orthogonal unit vectors, X and Y. This expression, as can easily be checked, depends only on the plane spanned by X and Y. It is called the *sectional curvature* of the plane. To see its meaning, let us consider immersion, f, of a two-dimensional surface, S, into M. This immersion picks out a manifold $\mathcal{O}_f(S)$ consisting of pairs $(x; e_1, \ldots e_n)$ where $x \in S$ and $(e_1, \ldots, e_n)$ is a frame at $f(x)$ whose first two vectors are tangent to $f(S)$ and whose last $n - 2$ vectors are orthogonal to $\varphi(S)$. [Compare with the construction of $\mathcal{O}_\varphi(M)$ in Section 2 of Chapter VI.] There is an obvious immersion, φ of $\mathcal{O}_f(S)$ into $\mathcal{O}(M)$ and an obvious projection, π_f of $\mathcal{O}_f(S)$ onto S such that $f \circ \pi_\varphi = \pi \circ \hat{f}$. Also, the first two components e_1, e_2 of $e_1, \ldots, e_n$ give an orthogonal frame on S relative to the induced Riemann metric. We thus get a projection, φ, of $\mathcal{O}_f(S)$ onto $\mathcal{O}(S)$. In short, we get a diagram just as in Section 3 of Chapter VI:

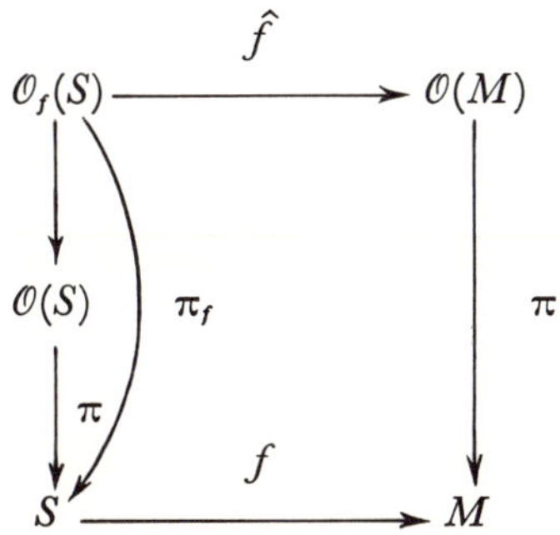

Let us put $\theta^i = \hat{f}^*(\omega^i)$ and $\hat{\Theta}^{ij} = \hat{f}^*(\Omega^{ij})$, where $\omega = (\omega^1, \ldots, \omega^n)$ is the fundamental form on $\mathcal{O}(M)$ and $\Omega = (\Omega^{ij})$ is the connection form of the Levi-Civita connection on M. Then it is easy to check (just as in Section 3 of Chapter VI) that

$$\theta^1 = \rho^*(\omega_S^1), \quad \theta^2 = \rho^*(\omega_S^2) \quad \text{and} \quad \theta^i = 0 \qquad i > 2 \tag{6.9}$$

where (ω_S^1, ω_S^2) is the fundamental form on $\mathcal{O}(S)$.

Now $d\omega^i = \sum \Omega^{ij} \wedge \omega^j$ so that $d\theta^i = \sum \hat{\Theta}^{ij} \wedge \theta^j$. Substituting into (6.9) for $i > 2$ gives

$$0 = \hat{\Theta}^{i1} \wedge \theta^1 + \hat{\Theta}^{i2} \wedge \theta^2.$$

By Cartan's lemma, this implies that, for $i > 2$,

$$\hat{\Theta}^{i1} = A^{i\,11}\theta^1 + A^{i\,12}\theta^2$$

(6.10) and

$$\hat{\Theta}^{i2} = A^{i\,21}\theta^1 + A^{i\,22}\theta^2$$

with $A^{i\,12} = A^{i\,21}$. If we substitute $d\theta^i = \sum \hat{\Theta}^{ij} \wedge \theta^j$ into (6.9) for $i = 1, 2$, we conclude that $\Theta^{12} = \rho^*(\Theta^{12})$ [where Θ^{12} is the connection form on $\mathcal{O}(S)$]. Now according to Equation (6.9) of Chapter VI $d\Theta^{12} = -K\omega_S^1 \wedge \omega_S^2$ where K is the Gaussian curvature of S. On the other hand,

$$d\hat{\Theta}^{12} = \sum \hat{\Theta}^{ij} \wedge \hat{\Theta}^{j2} + \sum R^{12}_{kl}\theta^k \wedge \theta^l$$

where $D\Omega^{ij} = \sum R^{ij}_{kl}\omega^j \wedge \omega^l$. Using (6.10) we obtain

$$K = -R^{12}{}_{12} + \sum_{i=3}^{n} (A^{i\,11}A^{i\,22} - (A^{i\,12})^2). \tag{6.11}$$

The expression $-R^{12}{}_{12}$ is clearly the sectional curvature of the tangent plane to $\varphi(S)$. The second expression on the right of (6.11) plays a role similar to that of the second fundamental form. It measures how much the surface is "bent" relative to M. We shall see, if all the geodesics of S issuing from a point x are also geodesics of M then this term vanishes. In fact, let v be some linear combination of e_1 and e_2, say $v = e_1$. Then by the definition of $\mathcal{O}_f$, we see that for $p_1, p_2 \in \mathcal{O}_f(S)$

$$\psi\hat{f}(p_1) = p_1(v) = p_2(v) = \psi \circ \hat{f}(p_2)$$

whenever $\rho(p_1) = \rho(p_2)$. The geodesic flow on S is the vector field on $\mathcal{O}(S)$ given by

$$\langle X, \omega_S^1\rangle = 1, \quad \langle X, \omega_S^2\rangle = \langle X, \Theta^{12}\rangle = 0.$$

Let $\hat{X}_p$ be a tangent vector of $\mathcal{O}_f(S)$ with $\rho_\alpha(\hat{X}_p) = X_{(p)}$. Then $\psi_*\hat{f}_*(\hat{X})_p$ is tangent to a geodesic of M if and only if

$$\langle \psi_*\hat{f}_*(\hat{X}_p), \omega^1\rangle = 1, \quad \langle \psi_*\hat{f}_*, \Omega^{ij}\rangle = 0$$

or

$$\langle \hat{X}_p, \Theta^{ij}\rangle = 0, \quad \langle \hat{X}_p, \theta^1\rangle = 1.$$

Comparing the two equations we see that they are the same if and only if the second term on the right side of (6.11) vanishes. Thus for any plane, $P \subset T_x(M)$, *the sectional curvature of P is the Gaussian curvature of the surface* $\exp(P)$ *at the origin.*

If we examine the above discussion we see that if a surface M contains just *two* geodesics passing through a point x, then its Gaussian curvature at x equals the sectional curvature of its tangent plane at x. Let us apply this remark to the following situation. Let $C(t)$ be a geodesic of M, and let X_t be a parallel family of tangent vectors along C. This defines a two-dimensional surface in the normal bundle of C. Applying the exponential map we get a two-dimensional surface containing C which at each point of C contains two geodesics. Thus the Gaussian curvature of this surface equals the sectional curvature along C. In particular, if the sectional curvature of M is bounded from below, say $\geq K$, then C does not minimize distance on the surface beyond a length of π/K according to Theorem 6.2 of Chapter VI. *A fortiori*, C does not minimize distance in M. Thus

If M is complete and its sectional curvature is bounded from below by a positive constant then M is compact.

The problem of deducing further conclusions from positive sectional curvature remains open. A powerful and striking result along this line is a theorem of Rauch which asserts that if the sectional curvature M lies between $\frac{1}{4}$ and 1 then M is homeomorphic to a sphere. (Rauch [18] had intrically proved the theorem for a constant $h > \frac{1}{4}$. Berger [2] and Klingenberg [11] have improved the result to $\frac{1}{4}$ and shown that this is best possible.)

APPENDIX I

Two Existence Theorems

In this appendix we state the implicit function theorem and the existence theorem for ordinary differential equations, and outline their proofs. Both of these are standard and the proofs that we outline aren't the best as we shall use more stringent conditions than are necessary. For proofs under more general hypothesis the reader is referred to Carathéodory [4]. For the details of proofs similar to the ones we outline here, couched in more geometrical language, the reader is referred to Lang [12]. Let us start with

THE IMPLICIT FUNCTION THEOREM. *Let $F^1, \ldots, F^k$ be C^k $(k \geqslant 1)$ (or analytic) functions of $y^1, \ldots, y^k, x^1, \ldots, x^n$ defined in some neighborhood of the origin in $E^k \times E^n$. Let $F^i(0, \ldots, 0; 0, \ldots, 0) = 0$. Suppose further that*

$$\det\left(\frac{\partial F^i}{\partial y^j}\right)(0, \ldots, 0; 0, \ldots, 0) \neq 0. \tag{1}$$

Then in a sufficiently small neighborhood of the origin in E^n there are functions $\varphi^1, \ldots, \varphi^k$ which are uniquely determined by the equations

$$F^i(\varphi^1(x^1, \ldots, x^k), \ldots, \varphi^k(x^1, \ldots, x^n); x^1, \ldots, x^k) \equiv 0, \qquad i = 1, \ldots, k \tag{2}$$

and $\varphi^i(0,\ldots,0) = 0$. *The functions* φ^i *are as differentiable as the* F*'s.*

Let us make some simplifying remarks. Let (β^i_j) be the inverse matrix to $(\partial F^i/\partial y^j)|_{(0,\ldots,0;\,0,\ldots,0)}$. Replacing the functions F^i by $G^i = \sum \beta^i_j F^j$ does not change the meaning of (2) but does allow us to assume that the matrix $(\partial G^i/\partial y^j)$ is the identity matrix at the origin. Thus we can write $G^i = y^i - f^i$ where $\partial f^i/\partial y^j = 0$ at the origin for all i and j. Equation (2) now becomes

$$(3) \qquad \varphi^i(x^1,\ldots,x^n) = f^i(\varphi^1(x^1,\ldots,x^n),\ldots \varphi^k(x^1,\ldots,x^n);\, x^1,\ldots,x^n).$$

Let us assume for a moment that we have proved the existence, uniqueness, and continuity of functions φ satisfying (3). Then the φ all have derivatives at the origin. In fact, by the mean value theorem and since the φ are continuous

$$\begin{aligned}\varphi^i&(0,\ldots,x^r,\ldots,0)\\ &= f^i(\varphi^1(0,\ldots,x^r,\ldots,0)\ \ldots\ \varphi^k(0,\ldots,x^r,\ldots,0);\,0,\ldots,x^r,\ldots,0),\\ &= \sum \frac{\partial f^i}{\partial g^j}\,\varphi^j(0,\ldots,x^r,0,\ldots,0) + \frac{\partial f^i}{\partial x^j}\,x^r\end{aligned}$$

where $\partial f^i/\partial x^r$ each $\partial f^i/\partial y^j$ is evaluated at a point tending to zero with x^r. Regarding this as a system of linear equations for the $\varphi^i(0,\ldots,x^r,\ldots,0)$, solving and dividing by x^r we obtain

$$\frac{\varphi^i(0,\ldots,x^r,\ldots,0)}{x^r} = \frac{\partial f^i}{\partial x^r} + \sum a^i_j \frac{\partial f^j}{\partial x^r}$$

where the a^i_j are functions tending to zero with x^r. Letting $x^r \to 0$ we see that $(\partial\varphi^i/\partial x^r)(0;\ldots,0)$ exists and is equal to $(\partial f^i/\partial x^r)(0,\ldots,0;\,0,\ldots,0)$. Since the origin is no better than any other nearby point and since we have uniqueness we conclude that the partial derivatives of the φ^j exist at all nearby points. Furthermore, these partial derivatives are the (unique) solutions of the linear equations

$$\sum \frac{\partial F^i}{\partial y^j}\frac{\partial \varphi^j}{\partial x^r} + \frac{\partial F^i}{\partial x^r} = 0.$$

They are therefore continuous. In addition they are as smooth as the partial derivatives of F.

In short, to prove the implicit function theorem it is enough to show that there are unique continuous φ satisfying (3). For any neighborhood, U, of the origin in E^n, let $\Phi(U)$ be the space of all k-tuplets of functions $\varphi = (\varphi^1, \ldots, \varphi^k)$ defined on U. Let us set

$$\|\varphi - \psi\| = \sum_{i=1}^{k} \sup_{x \in D} |\varphi^i(x^1, \ldots, x^n) - \psi^i(x^1, \ldots, x^n)|.$$

Let T denote the mapping of a subset of $\Phi(U)$ into itself given by $T_\varphi = \psi = (\psi^1, \ldots, \psi^k)$ where

$$\psi^i(x^1, \ldots, x^n) = f^i(\varphi^1(x^1, \ldots, x^n), \ldots, \varphi^k(x^1, \ldots, x^k); x^1, \ldots, x^n).$$

It is defined for φ with $\|\varphi\|$ sufficiently small.

Let us estimate $\|T\varphi\|$ in terms of $\|\varphi\|$, for small $\|\varphi\|$ and small U. We first observe that since $f^i(0, \ldots, 0; 0, \ldots, 0, 0) = 0$ we can assume that $|f^i(y^1, \ldots, y^k; x^1, \ldots, x^n)| < \epsilon$ for $|x^i|$ and $|y^i| < \delta$. Also, since $\partial f^i/\partial y^j = 0$ at the origin we can assume that $|\partial f^i/\partial y^j| < \epsilon$ for $|x^i|$ any $|y^i| < \delta$ and that $|\partial f^i/\partial x^r| < K$. Then the mean value theorem implies that

$$|f^i(\varphi^1(x^1, \ldots, x^n), \ldots, \varphi^k(x^1, \ldots, x^n); x^1, \ldots, x^n)| \leqslant \sum \frac{\partial f^i}{\partial y^j} \varphi^j + \sum \frac{\partial f^i}{\partial x^r} x^n.$$

Thus $\|T\varphi\| < k\epsilon\|\varphi\| + knK\delta < \epsilon\|\varphi\| + nK\delta$. Suppose T is defined on the set of those $\varphi \in \Phi(U)$ with $\|\varphi\| \leqslant \eta$. Then choose U so small that δ and ϵ satisfy

$$k\epsilon\eta + knK\delta < \eta.$$

For instance, arrange that $\epsilon < 1/2k$ and $\delta < \eta/2knK$. Then T maps the ball of radius η in $\Phi(U)$ (i.e., the set of all φ with $\|\varphi\| \leqslant \eta$) into itself. Furthermore, the mean value theorem also implies that

$$\|T\varphi - T\psi\| < k\epsilon\, \|\varphi - \psi\|.$$

We now apply

Proposition 1. *Let T be a map of a complete metric space, S, into itself satisfying*

$$d(T\varphi, T\psi) < \kappa d(\varphi, \psi), \qquad \kappa < 1, \tag{4}$$

where d is the distance function. Then there is a unique fixed point of T. This fixed point is $\lim_{n\to\infty} T^n\varphi$ for any $\varphi \in S$.

The proof of Proposition 1 is immediate: Uniqueness—if there were two fixed points, φ and ψ we would have that $d(T\varphi, T\varphi) = d(\varphi, \psi)$ contradicting (4). Existence—$d(T^{n+1}\varphi, T^n\varphi) < \kappa^n d(T\varphi, \varphi)$, so that $\lim T^n\varphi$ exists.

Applying the proposition to the map T we get a unique solution of Equation (3). Starting with a continuous φ we obtain the fixed point as a uniform limit of continuous functions and is therefore continuous. This proves the implicit function theorem.

Existence Theorem for Ordinary Differential Equations. *Let $f^1(x^1, \ldots, x^n, t), \ldots, f^n(x^1, \ldots, x^n, t)$ be C^k (analytic) functions defined in some neighborhood of the origin in $E^n \times R$. Then there exist neighborhoods, U, of the origin in E^n and I of 0 in R such that for any $(x_0^1, \ldots, x_0^n) \in U$ and all $t \in I$ there are unique functions $\varphi^1(t; x_0^1, \ldots, x_0^n), \ldots, \varphi^n(t; x_0^1, \ldots, x_0^n)$ such that*

$$\frac{d\varphi^i}{dt} = f^i(\varphi^1, \ldots, \varphi^n; t)$$

(5) *and*

$$\varphi^i(0, x_0^1, \ldots, x_0^n) = x_0^i.$$

The functions φ^i are of class C^{k+1} (analytic) in t and of class C^k (analytic) in x.

One first shows the existence of solutions of (5). For this one writes it in "solved" form so as to apply Proposition 1; Equation (5) is equivalent to

$$\varphi^i(t; x_0^1, \ldots, x_0^n) = \int_0^t f^i(\varphi^1, \ldots, \varphi^n; t) + x_0^i. \tag{6}$$

Let us assume that the f^i are defined for $|x^i| < \eta$ and $|t| < a$, and that $\sup_{i,x,t} |f^i| < M$ on this region. Let $C(a; b)$ be the set of n-tuples of functions $\varphi = (\varphi^1, \ldots, \varphi^n)$ defined for $|t| \leqslant a$ and satisfying $|\varphi^i| \leqslant b$. If $|x_0^i| < \eta/2$ $(i = 1, \ldots, \eta)$ and $|t| < 1/2M$ then for one $\varphi \in C(a; b)$ the element $T\varphi = \psi = (\psi^1, \ldots, \psi^n)$ where

$$\psi^i(t; x_0^1, \ldots, x_0^n) = \int_0^t f^i(\varphi^1, \ldots, \varphi^n, t) + x_0^i$$

is well defined and lies in $C(a; b)$. Setting $\|\varphi\| = \sum_{i=1}^{n} \sup_{t,t} |\psi^i|$ we can see that

$$\|T\psi - T\varphi\| < naK\|\varphi - \psi\|$$

where $K = \sup_{i,j,x} |\partial f^i/\partial x^j|$ as follows immediately from the mean value theorem. Thus if a is chosen $< 1/nK$ we can apply Proposition 1 to get the existence and uniqueness of the solution of (6)—and therefore (5).

We must then prove the smoothness of these functions. Continuity and differentiability in t follows immediately from (6). The first step is to prove continuity in the x_0^i. To this effect we observe that the mean value theorem yields

$$\begin{aligned} &|\varphi^i(t; x_0^1, \ldots, x_0^n) - \varphi^i(t; y_0^1, \ldots, y_0^n)| \\ &\qquad < K \int_0^t \sum |\varphi^j(s; x_0^1, \ldots, x_0^n) - \varphi^j(s; y_0^1; y_0^k)|\, ds. \end{aligned} \tag{7}$$

This easily implies that

$$|\varphi^i(t; x_0^1, \ldots, x_0^n) - \varphi^i(t; y_0^1, \ldots, y_0^n)| < e^{nK\Sigma_j |x_0^j - y_0^j|}$$

establishing continuity.

To establish differentiability, one first observes that if the φ^i were differentiable in the x^i, then (5) implies that the partial derivatives $\partial\varphi^i/\partial x^j$ must satisfy the linear differential equations (the equations of variation—cf. Chapter IV)

$$\frac{d}{dt}\left(\frac{\partial\varphi^i}{\partial x_0^j}\right) = \sum_{i=1}^{n} \frac{\partial f^i}{\partial x^k}\frac{\partial\varphi^k}{\partial x_0^j}. \tag{8}$$

The $\partial f^i/\partial x^k$ occurring in (8) are functions of t for fixed $x_0^1, \ldots, x_0^n$. To show that the φ are actually differentiable with respect to the x^i, we compare the difference quotient

$$\frac{\varphi^i(t, x_0^1, \ldots, x_0^j + h, \ldots, x^n) - \varphi^i(t, x_0^1, \ldots, x_0^n)}{h} \tag{9}$$

with the solutions of the (7). Again one uses the mean value theorem to obtain an inequality similar to (7) for the difference of (9) and the solutions of (8). One then concludes that the φ^i are differentiable. Further smoothness then follows without difficulty. For details see [12].

APPENDIX II

Outline of Theory of Integration on E^n

In this appendix we present, in outline, the theory of integration on Euclidean space. In the standard calculus texts, the theory, especially the law for change of variables is only developed for two or three dimensions. In our treatment we follow the exposition given by J. Schwartz in the American Mathematical Monthly, vol. 61, 1954, pp. 81–85.

By a rectangle $\square_a^b$ where $a = (a^1, \ldots, a^n)$ and $b = (b^1, \ldots, b^n)$ is meant the set of all $x = (x^i, \ldots, x^n)$ with $a^i \leqslant x^i \leqslant b^i$. (We will also allow "partly open" rectangles where some of the inequalities are strict and allow some of the a^i or b^i to be ∞ or $-\infty$.)

A theory of integration on E^n consists of giving a collection of bounded sets $\mathfrak{D}$, a collection of bounded functions $\mathscr{F}$, and a rule $\int$, which assigns to each $D \in \mathfrak{D}$ and $f \in \mathscr{F}$ a number $\int_D f(x)\, dx$ subject to the following axioms:

1. Axioms on $\mathfrak{D}$:

   a) If $D_1 \in \mathfrak{D}$ and $D_2 \in \mathfrak{D}$ then $D_1 \cup D_2 \in \mathfrak{D}$,

$$D_1 \cap D_2 \in \mathfrak{D} \quad \text{and} \quad D_1 - D_2 \in \mathfrak{D}.$$

b) $\mathfrak{D}$ contains all bounded rectangles.
c) If $D \in \mathfrak{D}$ and T is a Euclidean motion then $TD \in \mathfrak{D}$.

or

c′) If $D \in \mathfrak{D}$ and T is a translation then $TD \in \mathfrak{D}$.

2. Axioms on $\mathscr{F}$.
a) $\mathscr{F}$ is a vector space.
b) $\mathscr{F}$ contains all constants.
c) If $f \in \mathscr{F}$, then for any Euclidean motion, T, the function $f_T \in \mathscr{F}$ where $f_T(x) = f(Tx)$.

or

c′) If $f \in \mathscr{F}$, for any translation, T, the function $f_T \in \mathscr{F}$.

3. Axioms on $\int$.
a) For each fixed D, the integral $\int_D f(x)\,dx$ is linear in f.
b) If $D_1 \cap D_2 = \varnothing$ then $\int_{D_1 \cup D_2} f(x)\,dx = \int_{D_1} f(x)\,dx + \int_{D_2} f(x)\,dx$ for any f.
c) $\int_{TD} f(T^{-1}x)\,dx = \int_D f(x)\,dx$ for every Euclidean motion T.

or

c′) $\int_{TD} f(T^{-1}x)\,dx = \int_D f(x)\,dx$ for every translation T.
d) For any $D \in \mathfrak{D}$, if $f(x) \leqslant g(x)$ for all $x \in D$ then $\int_D f(x)\,dx \leqslant \int_D g(x)\,dx$.
e) $\int_{\square_{(0,\ldots,0)}^{(1,\ldots,1)}} 1\,dx = 1.$

For any D, define $\mu(D) = \int_D 1\,dx$.

A set S will be called *paved* if it is the union of a finite number of rectangles.

EXERCISE 1. Show that if D is the family of all bounded paved sets then D satisfies conditions a), b), and c′).

A finite collection of rectangles $\{\square_i\}$ such that $\square_i \cap \square_j = \varnothing$ if $i \neq j$ and $E^n = \bigcup \square_i$ is called a *paving* of E^n. For example:

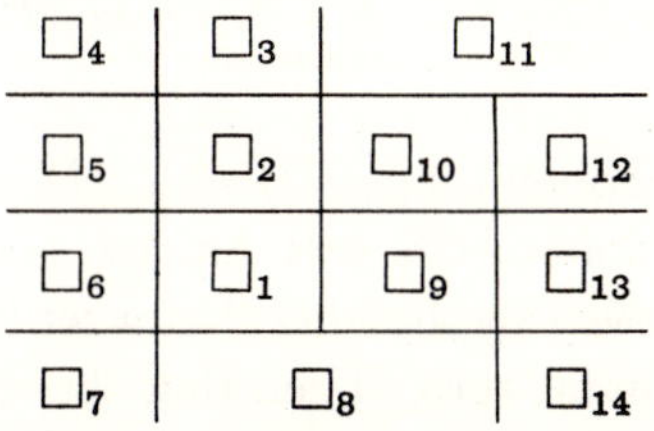

A function f is said to be paved if there exists a paving $\{\square_i\}$ of E^n such that f is constant on each $\square_i$.

EXERCISE 2. Show that if F is the set of all paved functions then F satisfies Axiom 2.

EXERCISE 3. Show that there is one and only one rule, $\int$, which, on the above F and D, satisfies Axioms 3a), b), c′), d) and e). The formula for this $\int$ is as follows: Let f be a paved function with paving $\{\square_i\}$ $i = 1, \ldots, k$ such that $f(x) = c_i$ for $x \in \square_i$. Let D be a paved set $D = \bigcup \square_j'$.

Then

$$(1) \qquad \int_D f(x)\, dx = \sum_{i=1,\, j=1}^{i=k,\, j=1} C_i \mu(\square_i \cap \square_j')$$

where

$$(2) \qquad \mu(\square_x^y) = |y_1 - x_1|\, |y_2 - x_2| \cdots |y_n - x_n|$$

for any rectangle $\square_x^y$.

Hint: First try to verify (2). To do this, verify (2) for the rectangle

$$\square_{(0,\ldots,0)}^{(\frac{1}{2^r},\ldots,\frac{1}{2^r})},$$

using 3b), 3c′) and 3e). Then for any rectangle $\square_x^y$ where $y - x = (\frac{1}{2^r}, \ldots, \frac{1}{2^r})$ by c′). Then for any rectangle whatsoever using 3b), 3c′) and 3d). Then prove (1).

For any bounded set D and any paving $P = \{\square_i\}$ we put

$$I_P(D) = \bigcup_{\substack{\square_i \in P \\ \square_i \subset D}} \square_i \quad \text{and} \quad O_P(D) = \bigcup_{\substack{\square_i \in P \\ \square_i \cap D \neq \varnothing}} \square_i$$

We define

$$\mu_-(D) = \underset{\substack{\text{all pavings} \\ P}}{\text{l. u. b.}} \ \mu(I_P(D))$$

$$\mu_+(D) = \underset{\substack{\text{all pavings} \\ P}}{\text{g. l. b.}} \ \mu(O_P(D)).$$

A set D is said to have content if $\mu_-(D) = \mu_+(D)$. Its content is denoted by $\mu(D)$. The collection of all sets having content satisfies axioms 1a), b), c′). This is proved by showing that a set has content if and only if its boundary has content zero. Then that the sets of content zero are invariant under Euclidean motion and that boundary $(D_1 \cup D_2) \subset$ boundary $D_1 \cup$ boundary D_2 etc.

Exercise 4. Show that a set D has content if and only if its boundary has content zero.

Exercise 5. Let S be a bounded subset of E^n which lies entirely in a subspace of dimension $n - 1$. Show that S has content zero.

Exercise 6. Let D be a set with content zero in E^n and let φ be a differentiable map of $E^n \to E^n$. Show that $\varphi(D)$ has content zero.

Exercise 7. Let D be a set whose boundary is the union of a finite number of sets D_i where each $D_i = \varphi_i(\square_i)$, where $\square_i$ is a rectangle in some E^m $(m < n)$ and φ a differentiable map of $E^m \to E^n$. Show that D has content.

For any bounded function f and any set D with content we define

$$\underline{\int_D} f(x)\, dx = \text{l. u. b.} \int_{I_P(D)} s(x)\, dx$$

where the l. u. b. is taken over all paved functions s (where P is the paving of s) such that $s(x) \leqslant f(x)$ for all x, and

$$\overline{\int_D} f(x)\, dx = \text{g. l. b.} \int_{O_P(D)} t(x)\, dx$$

over all paved functions, t, with $t(x) \geqslant f(x)$ for all x. The function f is called integrable on D if $\underline{\int_D} f(x)\, dx = \overline{\int_D} f(x)\, dx$. Its integral is denoted by $\int_D f(x)\, dx$. A function is called integrable if $\underline{\int_D} f(x)\, dx = \overline{\int_D} f(x)\, dx$ for all sets with content. The collection of all integrable functions satisfies Axioms 2a), b), c′) (Proof?).

Exercise 8. If f is almost continuous (i.e., continuous except at a set of content zero) then f is integrable.

It is then easy to prove.

Theorem 1. *Let $\mathfrak{D}$ be a collection of domains satisfying Axioms* 1a), b), c′), *containing all paved sets and contained in the collection of all sets with content. Let $\mathscr{F}$ be a collection of functions satisfying Axioms* 2a), b), c′), *containing all paved functions and contained in the collection of all integrable functions. Then there is a unique* $\int$ *satisfying Axioms* 3a), b), c′), d), e) *on* $\mathfrak{D}$ *and* $\mathcal{F}$.

Changing the value of a function on a set of content zero does not affect the value of its integral. In particular, a function can be left undefined on a set of content zero and it still makes sense to talk of its integral.

For any subset S of E^n, the characteristic function of S, denoted by φ_S is defined by

$$\begin{aligned} \varphi_S(x) &= 1 \qquad \text{if } x \in S \\ \varphi_S(x) &= 0 \qquad \text{if } x \notin S. \end{aligned}$$

EXERCISE 9. Show that $\varphi_{S_1 \cap S_2} = \varphi_{S_1} \cdot \varphi_{S_2}$ for any two sets S_1 and S_2.

EXERCISE 10. Show that $\varphi_{S'} = 1 - \varphi_S$ for any S.

EXERCISE 11. Express $\varphi_{S_1 \cup S_2}$ and $\varphi_{S_1 - S_2}$ in terms of φ_{S_1} and φ_{S_2}.

EXERCISE 12. Show that D has content if and only if φ_D is continuous outside of a set of content zero.

EXERCISE 13. Let $D_1 \subset D_2$ be two domains having content and let f be an integrable function. Show that

$$\int_{D_1} f(x)\, dx = \int_{D_2} f(x)\varphi_{D_1}(x)\, dx.$$

A function has *compact support* if $f(x) = 0$ for x outside some bounded set. If f has compact support we define

$$\int_{E^n} f(x)\, dx = \int_D f(x)\, dx$$

where D is a rectangle containing all points where $f(x) \neq 0$. In particular, since we can write $\int_D f(x)\, dx = \int_{E^n} \varphi_D f(x)\, dx$ we need consider only integrals over E^n when this is more convenient.

Let $E^n = E^k \times E^l$ so every $x \in E^n$ can be written as $x = (y, z)$ with $y \in E^k$, $z \in E^l$. For any set $D \subset E^n$ define the set $D_y \subset E^l$ as $D_y =$ the set of all z such that $(y, z) \in D$.

Suppose $\mathfrak{D}$ is a collection of domains satisfying 1a), b), c′) and $\mathscr{F}$ a collection of domains satisfying 2a), b), c′) such that

A) Let D be any domain in $\mathfrak{D}$. Then for almost all y (i.e., outside of a set of content zero in E^k) the set D_y has content (as a subset of E^l).

B) For any $f \in \mathscr{F}$ the function $f(y, \cdot)$ (a function of $z \in E^l$) is integrable (as a function on E^l) for almost all y,

and

C) For any $f \in \mathscr{F}$ and any $D \in \mathfrak{D}$ the function (of y) given by $\int_{D_y} f(y, z)\, dz$ is integrable. (Notice that since D is bounded it automatically has compact support.)

For such collections $\mathfrak{D}$ and $\mathscr{F}$ the rule which assigns to each $f \in \mathscr{F}$ and each $D \in \mathfrak{D}$ the number

$$\int_{E^k}\left(\int_{D_y} f(y, z)\, dz\right) dy$$

satisfies Axioms 3a), b), c′), d), e). Thus by the uniqueness part of Theorem 1 we have

$$\int_D f(x)\, dx = \int_{E^k}\left(\int_{D_y} f(y, z)\, dz\right) dy.$$

Here are two $\mathfrak{D}$'s and $\mathscr{F}$'s which obviously satisfy A), B) and C).

1) $\mathfrak{D}_1$ = all paved sets and $\mathscr{F}_1$ = all paved functions.

2) $\mathscr{F}_2$ = set of all functions f of compact support such that there is a set $C_f \subset E^k$ of content zero (in E^k) so that if $y \notin C_f$ there is a function $\delta_y(\epsilon)$ and a set $S_y(\epsilon) \subset E^l$ with $\mu_{E^l}(S_y(\epsilon)) < \epsilon$ (here μ_{E^l} means the content as a set in E^l) such that $|f(y, z) - (y', z)| < \epsilon$ whenever $\|y - y'\| < \delta(\epsilon)$ and $z \notin S_y(\epsilon)$.

$$\mathfrak{D}_2 = \text{the set of all } D \text{ such that } \varphi_D \in \mathscr{F}_2.$$

Lemma. *Any non-singular linear transformation can be written as the product of linear transformations of the form*

$$T_\lambda(x_1, \ldots, x_i, x_{i+1}, \ldots, x_n) = (x_1, \ldots, \lambda x_i, x_{i+1}, \ldots, x_n)$$

and

$$S_{ij}(x_1, \ldots, x_n) = (x_1, \ldots, x_i + x_j, x_{i+1}, \ldots, x_n).$$

Outline of proof: First the transformation

$$(x_1, \ldots, x_n) \to (x_2, x_1, x_3, \ldots, x_n)$$

can be obtained this way. In fact (ignoring the remaining variables which are left fixed)

$$\begin{aligned}(x_1, x_2) &\xrightarrow{S_{12}} (x_1 + x_2, x_2) \xrightarrow{T_{-1}} (-x_1 - x_2, x_2)\\ &\xrightarrow{S_{21}} (-x_1 - x_2, -x_1) \xrightarrow{T_{-1}} (-x_1 - x_2, x_1)\\ &\xrightarrow{S_{12}} (-x_2, x_1) \xrightarrow{T_{-1}} (x_2, x_1).\end{aligned}$$

The same argument shows that we can interchange any pair of variables this way. Thus any linear transformation which consists of a permutation of the variables can be obtained this way. If (a_{ij}) is a non-singular matrix then at least one $(n - 1) \times (n - 1)$ submatrix must be non-singular. By applying a suitable permutation of the variables, we can arrange that the upper left-hand $(n - 1) \times (n - 1)$ submatrix is non-singular. Iterating this procedure implies that up to multiplication by a permutation transformation, every linear transformation T has a matrix

$$\begin{pmatrix} a_{11} & \cdots & a_{1n} \\ \vdots & & \vdots \\ a_{n1} & \cdots & a_{nn} \end{pmatrix}$$

such that each of the submatrices

$$\begin{pmatrix} a_{11} & \cdots & a_{1k} \\ \vdots & & \vdots \\ a_{k1} & \cdots & a_{kk} \end{pmatrix} \qquad 1 \leqslant k \leqslant n$$

is non-singular.

We must thus show that such a linear transformation can be written as a product of the T_λ's and S_{ij}'s. This goes as follows:

$$\begin{aligned}(x_1, \ldots, x_n) &\to (a_{11}x_1, x_2, \ldots, x_n)\\ &\left.\begin{aligned}&\to (a_{11}x_1, a_{12}x_2, x_2, \ldots, x_n)\\ &\to (a_{11}x_1 + a_{12}x_2, a_{12}, \ldots, x_n)\\ &\to (a_{11}x_1 + a_{12}x_2, x_2, \ldots, x_2)\end{aligned}\right\} \quad \text{if } a_{12} \neq 0\end{aligned}$$

until $(a_{11}x_1 + \cdots + a_{1n}x_n, x_2, \ldots, x_n)$. Then since $a_{11} \neq 0$, we can get $(a_{11}x_1 + \cdots + a_{1n}x_n, a_{12}x_2 + a_{22}x_2, \ldots)$ etc. In general this will work since

$$\begin{pmatrix} a_{11} & \ldots & a_{1k} \\ \vdots & & \vdots \\ a_{k1} & \ldots & a_{kk} \end{pmatrix}$$

is non-singular.

We now apply the lemma to conclude that

$$\int_{TD} f(Tx)|\det T|\, dx = \int_D f(x)\, dx$$

for any linear transformation T. In fact, first check the formula for the T_λ's and S_{ij}'s by using multiple integration and verifying the appropriate one-dimensional assertions. Then the general case follows from the lemma and the multiplication rule for determinants.

COROLLARY 1. $\mu(TD) = |\det T|\mu(D)$.

COROLLARY 2. *The* $\int$ *on the set of all domains with content and all integrable functions satisfies Axiom* 3c.

Notation: If ψ is a differentiable map of $E^n \to E^n$ we denote its Jacobian linear transformation by J_ψ. The matrix of $J_\psi(x)$ is $(j_{ik}(x))$ where $j_{ik} = \partial\psi^i/\partial x^k$.

THEOREM 2. *Let* ψ *be a* 1–1 *map of* $D_1 \to D_2$ *such that* ψ *and* ψ^{-1} *are* C^1. *If* f *is an integrable function defined on* D_2 *then* $f \circ \psi$ *(where* $f \circ \psi(x) = f(\psi(x))$*) is integrable in* D_1 *and*

$$\int_{D_2} f(y)\, dy = \int_{D_1} f(\psi(x))|\det J_\psi(x)|\, dx.$$

Proof: First some notation

$$\text{if} \quad x = (x_1, \ldots, x_n) \quad \text{put} \quad |x| = \max_{i=1,\ldots,n} |x_i|$$

$$\text{if} \quad A = (a_{ij}) \quad \text{put}$$

$$|A| = \max_{i=1,\ldots,n} \sum_j |a_{ij}|$$

so

$$|Ax| \leqslant |A|\ |x|.$$

Now if $x = (x_1, \ldots, x_n)$ and $p = (p_1, \ldots, p_n)$ then

$$\psi_i(x) - \psi_i(p) = \sum_{k=1}^{n} j_{ik}[p + \theta_i(x - p)](x_k - p_k)$$

where $\theta_i \leqslant 1$.

Thus if C is the cube given by $|x - p| \leqslant s$

$$|\psi(x) - \psi(p)| \leqslant s \max_{z \in C} |J_\psi(y)| \qquad \text{for } x \in C.$$

Thus $\psi(C)$ lies in the cube

$$|y - \psi(p)| \leqslant s \max_{z \in C} |J_\psi(y)|.$$

Thus

$$\mu(\psi(C)) \leqslant \Big(\max_{y \in C} |J_\psi(y)| \Big)^n \mu(C). \tag{1}$$

For any linear transformation, A, and any set T

$$\mu(A^{-1}S) = |\det A^{-1}| \mu(S).$$

Thus

$$|\det A|^{-1} \mu(\psi(C)) = \mu(A^{-1}\psi(C)).$$

Substituting this into Equation (1) applied to the map $A^{-1} \circ \psi$ (instead of ψ) we get

$$|\det A|^{-1} \mu(\psi(C)) \leqslant \Big(\max_{y \in C} |J_{A^{-1} \circ \psi}(y)| \Big)^n \mu(C).$$

Now $J_{A^{-1} \circ \psi}(y) = A^{-1} J_\psi(y)$ by the composition law for Jacobians. The last equation becomes

$$\mu(\psi(C)) \leqslant |\det A| \Big(\max_{y \in C} |A^{-1} J_\psi(y)| \Big)^n \mu(C). \tag{2}$$

Let S be any subset of D having content and choose a paving of S by cubes of edge size δ. Let $I_P(S) = \bigcup C_i$ and x_i be a point in C_i. Apply (2) to each of the cubes, C_i, taking $A = J_\psi(x_i)$. Summing, we obtain

$$\mu(\psi(I_P(S))) \leqslant \sum_{C_i \in I_P(S)} |\det J(x_i)| \Big(\max_{z \in C_i} |J_\psi(x_i)^{-1} J_\psi(z)| \Big)^n \mu(C_i).$$

Now $J_\psi(x)$ is a continuous matrix valued function of x. Thus $J_\psi(x_i)^{-1} J_\psi(z) \to I$ as $z \to x_i$ where I is the identity matrix. Thus $\max_{z \in C_i} |J(x_i)^{-1} J(z)| \to 1$ (and in fact uniformly outside a set of arbitrarily small content). Thus the right-hand side of the last inequality tends to $\int_S |\det J_\psi(x)|\, dx$ so we have

$$\mu(\psi(S)) \leqslant \int_S |\det J_\psi(x)|\, dx. \tag{3}$$

Assume now that $f \geqslant 0$.

Let $Q = \{\square_i\}$ be a paving of E^n and $t \leqslant f$ a non-negative paved function defined on D_2. Then apply (3) to each of the $S = \psi^{-1}(\square_i)$, sum and pass to the limit yields

$$\int_{D_2} f(y)\, dy \leqslant \int_{D_1} f(\psi(z))|\det J_\psi(x)|\, dx.$$

Interchanging D_2 and D_1 and replace $f(y)$ by $g(x) = f(\psi(x))|\det J_\psi(x)|$ and use the composition law for Jacobians and the multiplication rule for determinants to conclude that

$$\int_{D_2} f(y)\, dy = \int_{D_1} f(\psi(x))|\det J_\psi(x)|\, dx$$

for all non-negative functions. But this implies the equality for all functions proving the theorem.

References

1. Ambrose, W. and Singer, I. M., "A theorem on holonomy," Trans. Amer. Math. Soc., vol. 75 (1953) pp. 428–443.

2. Berger, M., "Sur quelques variétés Remanniennes suffisamment pincées," Bull. de la Soc. Math. de France, vol. 88 (1960) pp. 57–71.

3. Bernard, D., "Sur la geometrie differentielle des G-structures," Ann. de l'Institute Fourier, vol. 10 (1960) pp. 151–270.

4. Carathéodory, C., *Variationsrechnung und partielle Differentialgluchungen erster Ordenung*, Leipzig, 1956.

5. Cartan, E., *Leçons sur les invariants intégraux*, Paris, Hermann, 1922.

6. ———, *Oeuvres completes*, Partie II, vol. 2, Paris, Gautier-Villars, 1953.

7. Chern, S. S. and Lashof, R. K., "The total curvature of immersed manifolds," Amer. J. Math., vol. 79 (1957) pp. 308–318.

8. Guillemin, V., *Theory of finite G-structures*, Ph.D. Thesis, Harvard, 1962.

9. Haefliger, A., *Plongements de variétés dans le domaine stable*, Seminaire Bourbaki (1962).

10. Hartman, P. and Nirenberg, L., "On spherical image maps whose Jacobians do not change sign," Amer. J. Math., vol. 81 (1959) pp. 901–920.

11. Klingenberg, W., "Contributions to Riemannian geometry," Ann. of Math. (2) vol. 69 (1959) pp. 654–666.

12. Lang, S., *Introduction to differential manifolds*, New York, Interscience, 1962.

13. Morrey, C. B., "The analytic embedding of abstract real-analytic manifolds," Ann of Math. (2) vol. 68 (1958) pp. 159–201.

14. Morse, A. P., "The behaviour of a function on its critical set," Ann. of Math. (2) vol. 40 (1939) pp. 62–70.

15. Newlander, A. and Nirenberg, L., "Complex analytic coordinates in almost complex manifolds," Ann. of Math. (2) vol. 65 (1957) pp. 391–404.

16. Nijenhuis, A., *Theory of the geometric object*, Ph.D. Thesis, University of Amsterdam, 1952.

17. Nomizu, K., *Lie groups and differential geometry*, Math. Soc. Japan, 1956.

18. Rauch, H. E., *Geodesics and curvature in differential geometry in the large*, New York, 1959.

19. deRham, G., *Variétés differentiables*, Hermann, Paris, 1955.

20. Sard, A., "The measure of the critical values of differentiable maps," Bull. Amer. Math. Soc., vol. 48 (1942) pp. 883–890.

21. Whitney, H., "A function not constant on a connected set of critical points," Duke Math. J., vol. 1 (1935) pp. 514–517.

22. ———, *Geometric Integration Theory*, Princeton, N.J., Princeton University Press, 1957.

23. Yamabe, H., "On an arcwise connected subgroup of a Lie group," Osaka Math. J. (Japan), vol. 2 (1950) pp. 13–14.

Index

D

E

F

G

R. Narasimhan *Analysis on real and Complex Manifolds* Adv. Stud. in Pure Math Vol I

North Holland 1968

48